TROPICAL FORAGE PLANTS:

Development and Use

TROPICAL FORAGE PLANTS:
Development and Use

Edited by
Antonio Sotomayor-Ríos, Ph.D.
W.D. Pitman, Ph.D.

CRC Press
Boca Raton London New York Washington, D.C.

Library of Congress Cataloging-in-Publication Data

Tropical forage plants : development and use / edited by Antonio Sotomayor-Rios and W.D. Pitman.

p. cm.

Includes bibliographical references and index.

ISBN 0-8493-2318-5

1. Forage plants—Tropics. 2. Pastures—Tropics. 3. Tropical crops. I. Sotomayor-Rios, Antonio. II. Pitman, William Don, 1949–

SD193.3.T76 T73 2000

633.2′00913—dc21 00-058574

CIP

Direct all inquiries to CRC Press LLC, 2000 N.W. Corporate Blvd., Boca Raton, Florida 33431.

International Standard Book Number 0-8493-2318-5
Library of Congress Card Number 00-058574
Printed in the United States of America 1 2 3 4 5 6 7 8 9 0
Printed on acid-free paper

Preface

This book, *Tropical Forage Plants: Development and Use*, covers forage plant research and resulting pasture development, which have undergone dramatic changes during the past few decades. This multiauthored publication provides a global perspective, but attempts to complement rather than duplicate the coverage previously given to tropical forages from some geographical regions, particularly Australia. Thus the dominance of this book by American authors reflects both the limited previous emphasis on compiling the specifically tropical contributions of forage researchers in the Western Hemisphere and the drastic recent reductions in numbers of forage scientists available for such tasks in some tropical regions.

While attempts have been made to provide a consistent presentation format, similar efforts have been made to retain the individual style and perspective of authors of the various chapters. For example, from reading an individual chapter, the conclusion might be drawn that either nitrogen fertilization of grass pastures or grass-legume mixtures without additional fertilizer nitrogen are superior in the tropics. While many factors would affect such a comparison, the reader should not interpret the enthusiasm of presentation of individual authors as a dismissal of other options, especially under different circumstances of available resources and economic opportunities. Other examples include authors' choices of species for illustrations and enthusiasm for technologies successfully experienced.

There is some inconsistency among chapters in scientific names used and particularly in common names. The chapters "Germplasm Resources of Tropical Forage Grasses" and "Germplasm Resources of Tropical Forage Legumes" provide complete listings of the various names and authorities for all scientific names used. Other chapters have generally used both local common names and provided reference to the locally used and recognized scientific name. Recent taxonomic revisions, and even some not so recent, contribute to the less than universal agreement on taxonomic nomenclature of a few species.

Some duplication among chapters occurs. This was necessary to allow thoroughness and consistent treatment of topics within individual chapters. Also, the perspective of the individual authors and context of the information allow different concepts to be developed from the same information in some cases.

The apparent inconsistencies in nomenclature, recent availability of much of the technology, and even questions of acceptability of tropical pasture development as a viable land use option illustrate the current dynamic circumstances engulfing tropical forage science. The technology and the global context affecting use of this tropical forage and pasture technology are addressed in this book.

Antonio Sotomayor-Ríos
W.D. Pitman

Editors

Antonio Sotomayor-Ríos, Ph.D., is Dean and Director of the College of Agricultural Sciences at the University of Puerto Rico, Mayagüez Campus. He earned his B.S. degree from the University of Puerto Rico and his M.S. and Ph.D. degrees from Purdue University.

Dr. Sotomayor-Ríos' early research on tropical forage included introduction, field evaluation, cytology, and taxonomy of tropical grasses. His research and teaching endeavors have included assignments at the University of Puerto Rico Rio Piedras Agricultural Experiment Station, Corozal Substation, and Mayagüez Campus, and the Universidad Nacional Autonoma de Honduras.

Before his return to the University of Puerto Rico, Dr. Sotomayor-Ríos was a research geneticist for more than 20 years with the USDA-ARS Tropical Agriculture Research Station in Mayagüez. His research during those years included improvement of maize, sorghum, millet, and forage grasses and legumes for the tropics. He also served as research leader for the USDA-ARS Germplasm Introduction and Research Unit at Kingshill, St. Croix, U.S. Virgin Islands, and directed a training program in agricultural technology for college graduates in the Caribbean Basin for several years.

Dr. Sotomayor-Ríos' numerous professional activities include membership in the Sorghum Crop Advisory Committee, National Plant Genetic Resources Board, Crop Science Society of America, and American Society of Agronomy (ASA), serving as chairman of the ASA Membership Committee from 1978 to 1987. He is past president of the Caribbean Food Crops Society and was editor and coordinator of the *Revista del Colegio de Agronomos de Puerto Rico* from 1983 to 1988. Additionally, he has authored and coauthored more than 120 papers, abstracts, and technical publications.

W.D. (Buddy) Pitman, Ph.D., is Associate Professor of the Louisiana State University Agricultural Center at the Rosepine Research Station. Dr. Pitman earned his B.S. degree from East Texas State University, his M.S. degree from Texas Tech University, and his Ph.D. from Texas A&M University.

From 1980 to 1993, Dr. Pitman conducted research at the University of Florida Agricultural Research Center at Ona, Florida. His research focused on tropical forage legume germplasm evaluation and grazing management of tropical grass and grass-legume pastures.

At his current position since 1994, Dr. Pitman researches subtropical pasture species and integration of temperate forage species into forage systems for cool-season pasture. Results of these programs have been presented at international, national, and local meetings of forage specialists and livestock producers. More than 240 publications have resulted from this research, including more than 50 scientific journal articles and a similar number of publications in proceedings of scientific meetings, book chapters, and experiment station bulletins and circulars.

Dr. Pitman's research has also included his participation in cooperative projects evaluating forage germplasm in several locations in the American tropics and participation in trips to collect tropical forage legume germplasm.

Contributors

Martin B. Adjei, Ph.D.
University of Florida
Agricultural Research and Education Center
Ona, Florida

Friedrich H. Beinroth, Ph.D.
Professor of Soil Science
Department of Agronomy and Soils
University of Puerto Rico
Mayagüez, Puerto Rico

Byron L. Burson, Ph.D.
USDA, ARS
Department of Soil and Crop Sciences
Texas A&M University
College Station, Texas

Raymond N. Gallaher, Ph.D.
Department of Agronomy
University of Florida
Gainesville, Florida

Wayne W. Hanna, Ph.D.
Research Geneticist
USDA, ARS, South Atlantic Area
Coastal Plain Experiment Station
Tifton, Georgia

Raymond J. Jones, Ph.D.
CSIRO
Division of Tropical Crops and Pastures
Townsville, Queensland, Australia

Albert E. Kretschmer, Jr., Ph.D.
Professor of Agronomy
University of Florida
Indian River Research and Education Center
Fort Pierce, Florida

Stephen L. Lapointe, Ph.D.
USDA-ARS
U.S. Horticultural Research Lab
2001 South Rock Road
Fort Pierce, Florida

Carlos E. Lascano, Ph.D.
Ruminant Nutritionist
Tropical Grasses and Legume Project
CIAT
Cali, Colombia

James P. Muir, Ph.D.
Texas A&M University
Agricultural Research and Extension Center
Stephenville, Texas

W.D. Pitman, Ph.D.
Associate Professor
Louisiana State University Agricultural Center
Rosepine Research Station
Rosepine, Louisiana

Kenneth H. Quesenberry, Ph.D.
Agronomy Department
University of Florida
Gainesville, Florida

José San José, Ph.D.
Centro de Ecologia y Ciencias Ambientales
Instituto Venezolano de Investigaciones Cientificas
Caracas, Venezuela

Stan C. Schank, Ph.D.
(deceased)
University of Florida
Gainesville, Florida

Eduardo C. Schröder, Ph.D.
Department of Agronomy and Soils
University of Puerto Rico
Mayagüez, Puerto Rico

Rex L. Smith, Ph.D.
Plant Science Laboratory
Agronomy Department
University of Florida
Gainesville, Florida

Ronald M. Sonoda, Ph.D.
University of Florida
Indian River Research and Education Center
Fort Pierce, Florida

Antonio Sotomayor-Ríos, Ph.D.
Dean and Director
College of Agricultural Sciences
University of Puerto Rico, Mayagüez Campus
Mayagüez, Puerto Rico

Salvio Torres-Cardona
Administrative Officer
Tropical Agricultural Research Station
USDA, ARS
Mayagüez, Puerto Rico

José Vicente-Chandler, Ph.D.
Department of Agronomy and Soils
University of Puerto Rico
Mayagüez, Puerto Rico

Sherlie H. West, Ph.D.
USDA, ARS
Agronomy Seed Laboratory
University of Florida
Gainesville, Florida

David S. Wofford, Ph.D.
Agronomy Department
University of Florida
Gainesville, Florida

Bruce A. Young, Ph.D.
Grassland, Soil, and Water Research Laboratory
Temple, Texas

Contents

Section I

Introduction to the Tropical Environment

1 Land Resources for Forage Production in the Tropics

Friedrich H. Beinroth

CONTENTS

I. INTRODUCTION

Land integrates the biophysical environment at the earth's surface and comprises climate, vegetation, geomorphology, soils, and hydrology. It is thus a broader concept than either soil or terrain; but purely economic and social characteristics are not included as these form part of the socioeconomic context. For the purpose of this discussion, tropical land is considered as the land that lies in the tropics, i.e., in a belt between latitudes 23° 27′ N and S.

It is not the intent of this chapter to present an exhaustive account of the land resources of the tropics. Rather, this is an attempt to convey, within the imposed limits of space, the essence of the

0-8493-2318-5/01/$0.00+$.50

nature and properties of tropical land resources in general, and how they affect forage production in particular.

This brief treatise inevitably lacks specificity and comprehensiveness. Moreover, in the process of selecting some key features for discussion, other important properties and issues may have been omitted. The chapter may nevertheless create awareness of the enormous soil and land diversity in the tropics and the covariant variability in productivity. As holistic, systems-based approaches to the assessment and management of forage production systems are more widely implemented, knowledge of land resources becomes more and more important.

II. LANDSCAPE AND SOIL DIVERSITY IN THE TROPICS

A. Physiography

Physical geographers divide continents into two broad structural regions, continental platforms and orogenic belts, each having a distinct geomorphic history of mountain building, erosion, and sedimentation.[1]

Continental platforms are the most extensive region in the tropics. They form massive and stable blocks of the earth's crust that have not suffered diastrophisms since Cambrian times and consist of shields and associated basins. Examples are the Australian Shield, the Deccan Shield in India, the African Shield, and the Brazilian and Guiana Shields in South America. In several instances, tectonic movements caused fracturing of the shields resulting in the formation of rift valleys, composed of grabens and horsts, that are often accompanied by active volcanic systems. The Rift Valley in East Africa is a prominent example. Orogenic belts, on the other hand, consist of mountain ranges and the subsiding zones that usually border them. The mountain ranges belong to the Alpine Fold Belts that originated mainly in the Cretaceous and Tertiary periods and comprise the Andes, the mountain ranges in Central America and the Antilles, the Himalayas, and most of southeast Asia.[1] The Amazon Basin east of the Andes and the Ganges Valley south of the Himalayas typify subsiding zones.

The attendant geologic and geomorphic processes of volcanism, erosion, and the sedimentation of unweathered debris in the subsiding zones and the accumulation of preweathered sediments in the continental basins all have a profound effect on the nature and quality of the soil and land resources in the tropics. (For a more complete discussion of these phenomena, reference is made to the excellent and recent text by A. Van Wambeke.[1])

It should thus be obvious that there is much scope for landform variability in the tropics. There are the wide and open landscapes of low relief on the uplands of the pre-Cambrian shields, the extensive plains in the continental basins, the graben/horst landscapes of rift valleys, some of the world's largest drainage systems, the landforms created by recent volcanisms, dissected uplands, peneplains, tropical karst, and lacustrine and littoral landscapes. Notably absent from this geomorphic panacea are the landforms associated with recent glaciation, such as U-shaped valleys and loess deposits, as Pleistocene glaciation in the tropics affected but the highest peaks and thus is of very limited areal extent.

B. Soil Diversity

Rationalizing soil variability in terms of the environmental factors first postulated by Dokuchaev a century ago, continues to be a unifying philosophy in pedology. Viewed from this perspective, the great diversity of soils in the tropics is an inevitable consequence of the enormous diversity of ecosystems found in the intertropical areas. Both the driest and wettest areas on earth are in the tropics, namely the Atacama Desert, where only sporadic traces of rainfall occur, and Mt. Waialeala in Hawaii, where more than 11,700 mm have been recorded. Mean annual temperatures vary from more than 30°C at the low elevations to below 0°C on the snowcapped mountains of South America and East Africa. Covariant with this climatic variability, a multitude of ecosystems, ranging from deserts to rainforests, occur in the tropical belt. With the exception of glacial formations, all rock types are found in the tropics resulting in a wide range of soil parent materials. And the oldest geomorphic surfaces are found in the intertropical areas, some of them as old as 25 million years, dating back to the mid-Tertiary.

Guy D. Smith (personal communication) distinguished four broad geologic-pedologic provinces in the tropics:

1. Africa and India: generally acid parent rocks, but with the superimposed influence of calcium-rich dust from bordering deserts;
2. South America and Southeast Asia: generally acid parent rocks;
3. Oceania: largely basic volcanic rocks and limestones;
4. Areas of recent volcanism: young volcanic material.

In view of the immense environmental diversity encountered in the tropics, often within short distances, the complexity and variability of the resulting soil and landscape patterns is not surprising. The small island of Puerto Rico may serve as an example: in an area of < 9,000 km^2, soils representing 10 of the 11 orders currently recognized in *Soil Taxonomy*[2] have been identified. A recent article by Eswaran et al.[3] provides more complete information on soil diversity in the tropics.

C. Temperate vs. Tropical Soils

The term *laterite* readily comes to mind when the subject of tropical soils is raised. Since the soils that have been referred to as laterite are confined to the tropics, the underlying implication is that soil formation in the tropics is somehow different from that in the higher latitudes. This concept is largely erroneous, however, as the basic processes and reactions of pedogenesis are the same everywhere. Processes like lessivage, pedoturbation, eluviation and illuviation, and decalcification are universal in nature. Reactions such as hydrolysis, oxidation, and reduction have no geographic boundaries.

Figure 1.1 provides a comparison of two important soil parameters, organic carbon content and cation exchange capacity, in tropical and temperate pedons stratified by soil orders. Viewed from this perspective, the differences are not striking; however, the frequency distribution of the individual values is quite dissimilar in the two regions.

Yet, while there is no real difference in the kind of soil-forming processes operating in the tropics, there may be significant differences in degree. In parts of the tropics, the combination of continually high temperatures, copious amounts of rainfall, and geomorphic stability over millions of years have allowed the pedogenic processes to produce extreme manifestations of soil formation, namely the Oxisols. The conditions conducive to their formation are not ubiquitous, however, and Oxisols therefore account for only about 23% of the intertropical land area. Conversely, 77% of the soils of the tropics have counterparts in the temperate region. Oxisols are nevertheless the single most extensive soil order in the tropics and, because they are unique to the tropics, this is the only region in the world where all 11 orders of *Soil Taxonomy*[2] occur.

III. GEOGRAPHY AND PROPERTIES OF TROPICAL SOILS

This section provides general information on the geography and properties of the dominant kind of soils of the tropics with an emphasis on those soils that represent major land resources. An in-depth discussion is beyond the scope of this chapter; the reader interested in more specific information may consult the books by Wilding et al.,[4] Sanchez,[5] and Van Wambeke,[1] which provide comprehensive and current coverage of the subject of formation, classification, and management of the soils of the tropics.

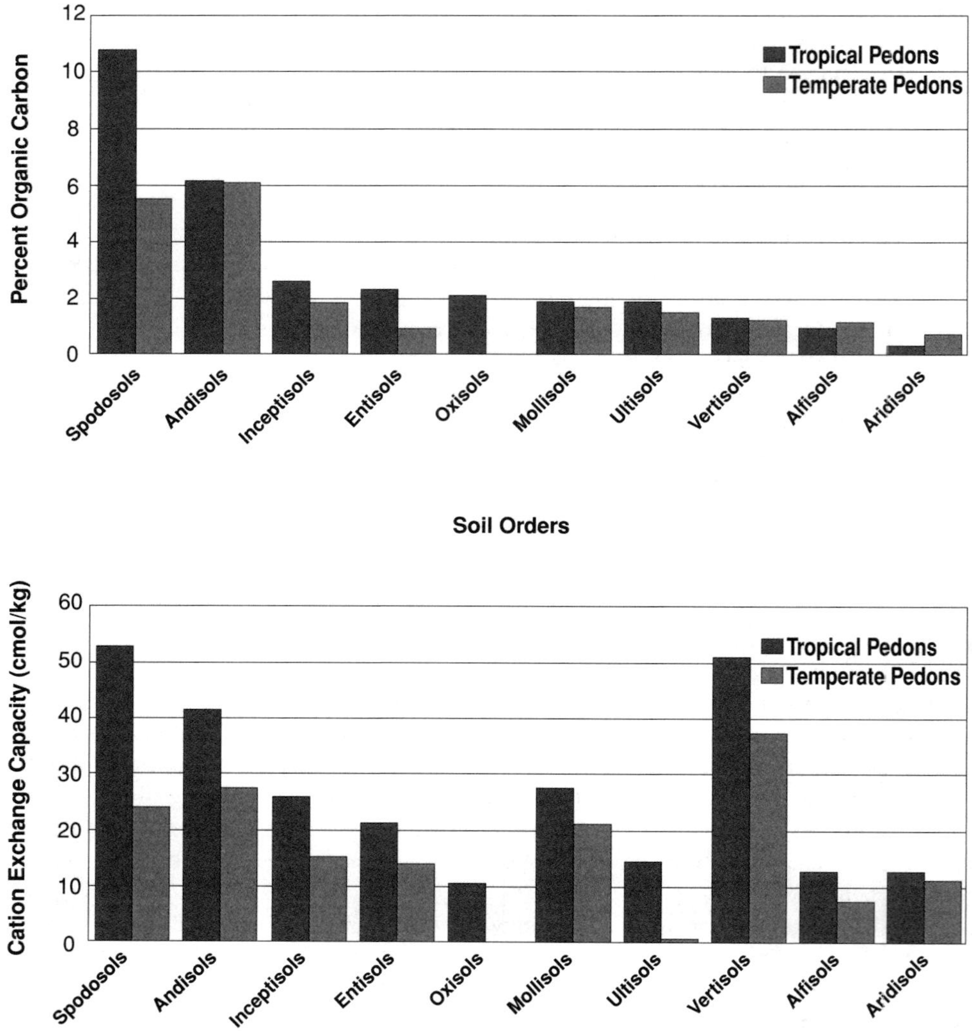

FIGURE 1.1 Organic carbon content and cation exchange capacity in the top 30 cm of tropical vs. temperate pedons, by orders of *Soil Taxonomy*. (Courtesy of National Soil Survey Laboratory, Soil Conservation Service, U.S. Department of Agriculture.)

Four soil orders, namely the Oxisols, Ultisols, Aridisols, and Alfisols, account for nearly three quarters of the soils of the tropics. Of the remaining seven orders, the Histosols, Mollisols, and Spodosols are of negligible extent.

Table 1.1 shows the area of the soils of the tropics, at the order and suborder level of *Soil Taxonomy*, listed in order of decreasing area. As these estimates are based on areas delineated on small-scale soil maps, it is inevitable that the indicated areas include soils other than the ones belonging to the named order or suborder. The areas indicated in Table 1.1 should, therefore, be considered as areas where the respective order or suborder dominates over other kinds of soil that may be found there.

A. Oxisols

Although the Oxisols cover less than a quarter of the tropical land area, they are still the most extensive soils of the tropics. They occur mostly in South America and Africa and, on both

TABLE 1.1
Extent and Distribution of Soils in the Tropics

Order/Suborder	Area 000 km²	Percent of Tropical Area
Oxisols	11,512	23.2
Aquox	54	0.1
Ustox	4,153	8.4
Perox	948	1.9
Udox	6,357	12.8
Aridisols	9,117	18.4
Salids	134	0.3
Gypsids	386	0.8
Calcids	3,564	7.2
Argids	1,145	2.3
Cambids	3,888	7.8
Ultisols	9,018	18.2
Aquults	501	1.0
Humults	72	0.2
Udults	4,021	8.1
Ustults	4,421	8.9
Xerults	3	< 0.1
Alfisols	6,411	12.9
Aqualfs	69	0.1
Ustalfs	5,545	11.2
Xeralfs	349	0.7
Udalfs	448	0.9
Inceptisols	4,565	9.2
Aquepts	2,050	4.1
Udepts and Ustepts	2,515	5.1
Entisols	3,256	6.6
Aquents	30	0.1
Psamments	2,795	5.6
Fluvents	121	0.2
Orthents	310	0.6
Vertisols	2,189	6.6
Aquerts	84	0.2
Torrerts	736	1.5
Usterts	1,343	2.7
Uderts	26	0.1
Andisols	1,683	3.3
Cryands	158	0.3
Torrands	220	0.5
Xerands	11	< 0.1
Ustands	555	1.1
Udands	739	1.5
Histosols	286	0.6
Hemists	92	0.2
Saprists	194	0.4
Mollisols	234	0.5
Ustolls	164	0.3
Udolls	70	0.1

TABLE 1.1 *(continued)*
Extent and Distribution of Soils in the Tropics

Order/Suborder	Area 000 km²	Percent of Tropical Area
Spodosols	40	0.1
Humods	25	< 0.1
Orthods	15	< 0.1
Shifting Sands	1,172	2.4
Water	186	0.4
Total	49,669	100.0

Source: World Soil Resources, Natural Resources Conservation Service, U.S. Department of Agriculture.

continents, predominantly in the southern hemisphere. Oxisols typically developed *in situ* on the old geomorphic surfaces of the pre-Cambrian shields, but are also found in continental depressions or subsidence zones in preweathered sediments derived from old regoliths, or on ultrabasic rocks that weather rapidly. Their natural ecosystems range from rainforest to savannah; where they are found in aridic climates, they are considered relics.

Oxisols represent advanced stages of rock weathering and soil formation under conditions of free drainage in the humid tropics. The specific pedogenetic process, known as laterization, results in soil material that is mainly composed of oxides or hydroxides of iron and aluminum, 1:1 clay minerals such as kaolinite, and other nonweatherable minerals like quartz. This assemblage causes some of the unique physical and chemical characteristics of the Oxisols. Prominent among these are high structural stability, a low water-holding capacity, a high hydraulic conductivity, a low cation exchange capacity (by definition, less than 16 cmol(+)kg^{-1} per 100 g of clay), an anion exchange capacity that may exceed the cation exchange capacity, and a pH-dependent charge.

B. Ultisols

Ultisols account for about 18% of the soils of the tropics and occur extensively in south-central South America, east-central Africa, northeast India, southwest China, southeast Asia, and northeastern Australia. They normally occur in regions of high precipitation, some of which may have seasonally concentrated rainfall.

A dominant soil forming process in Ultisols is leaching. As these soils occupy geomorphologically stable surfaces, leaching occurs over long periods of time resulting in usually deep and acid soils, with low Ca, Mg, K, and P contents, low base saturation, and low pH values. At low pH, the Al released through hydrolysis of weatherable minerals diffuses in the soil solution, accounting for the normally high Al saturation of these soils.

The physical attributes of Ultisols vary widely with clay activity and particle size distribution. Those with kandic horizons approach the Oxisols in their rheologic behavior, and light-textured surface soils have a tendency to seal and form crusts when they become dry.

C. Aridisols

Aridisols cover about the same area in the tropics as the Ultisols and occur mainly in the sub-Saharan region and east Africa, along the Caribbean and Pacific coasts of South America, and in Pakistan and Australia.

The properties of Aridisols are controlled, in large measure, by the aridic soil moisture regime. Soil-forming processes are sporadic, but intense at times and lead, over time, to the accumulation in the subsoil of clay, carbonates, gypsum, and soluble salts. Lack or scarcity of vegetation exposes the surface soil, which renders it susceptible to erosion by wind.

D. Alfisols

The Alfisols account for about 13% of the tropical land area and are most extensive in a sub-Saharan belt around 10° N latitude, and in east and southeast Africa, Brazil, and India. The majority of them are in savanna regions with climates characterized by seasonal dryness.

The tropical Alfisols share many properties with the Ultisols, but differ from them in that they have a high base saturation. In the instance of tropical Alfisols, this differentiae is of limited significance, however. Because the majority of tropical Alfisols generally have low activity clays having a CEC of less than 24 $cmol(+)kg^{-1}$ per 100 g of clay, their volumetric content of bases may be quite low and thus rapidly depleted under cultivation. Nevertheless, as leaching in the area of tropical Alfisols is less intense, the levels of Ca, Mg, and K are usually higher in the Alfisols than in the Ultisols.

E. Inceptisols

About one tenth of the tropics is dominated by Inceptisols. They occur in a variety of climates mainly on the recent geomorphic surfaces that may be either erosional or aggradational and are associated with mountain ranges and subsiding zones of the orogenic belts. The Andes, Central America, the Caribbean, southeast Asia, and the Orinoco, Amazon, and Ganges valleys typify these structural regions.[1]

Inceptisols show incipient stages of soil formation on the evolutionary path to soils carrying the marks of distinct pedogenetic processes. Because there are many pathways in pedogenesis, there is a wide variety of Inceptisols with an equally wide variability in physical and chemical properties. They may be deep or shallow, clayey or sandy, moist or wet, and acid or alkaline.

F. Entisols

The Entisols dominate about 6.5% of the tropics and most of them occur in Australia, southwest Africa, and in smaller areas in South America. Many kinds of Entisols are found in the tropics, but the most extensive ones belong to the suborder Psamments, which consist mostly of wind-transported sand and are located mainly in arid and semiarid regions on or near continental shields.

The chemical properties of sandy Entisols are controlled by their mineralogical composition. For example, Quartzipsamments consist almost exclusively of quartz sand; Ustipsamments may have feldspars and smectite; while Tropopsamments generally lack weatherable minerals. The physical properties of Psamments are controlled by their high sand content and typified by a high hydraulic conductivity and a very low water-holding capacity.

G. Vertisols

Vertisols occupy about 4.5% of the tropical land area and are extensive in Australia, India, and the Sudan. They typically are found in savannah ecosystems and invariably have subhumid or semiarid climates with pronounced dry seasons.

The most characteristic property of Vertisols is their high content of clay minerals of the smectite group that swell and shrink with seasonal changes in soil moisture. The stresses associated with volume change produce a typical soil structure, shear planes known as slickensides that form when one soil mass slides past another, and crack patterns resulting in gilgai microrelief, which consists of a sequence of microbasins and microknolls, or microvalleys and microridges depending on slope. When wet, Vertisols are nearly impermeable, sticky, and plastic, but they become very hard and massive when dry. The Vertisols have high cation exchange capacities and almost always have a high base saturation with calcium and magnesium dominating the exchange complex.

H. Andisols

The geographic distribution of Andisols in the tropics is closely linked to areas of active volcanism in the circum-Pacific ring that encompasses the western coasts of the Americas, the Philippines, Indonesia, and the Pacific islands, the Lesser Antilles in the Caribbean, and the rift valleys in Africa. The Andisols account for about 3% of the tropical land area and occur mainly in humid and perhumid climates.

The unique properties of Andisols are their low bulk density of 0.90 g cm^{-3} or less and their andic soil properties. They reflect either the presence of volcanic ejecta (ash, pumice, cinders, lava) in the soil or indicate the presence of amorphous clays (allophane, imogolite) that results from rapid weathering of volcanic material in humid climates.

Owing to the dominance of amorphous clays, Andisols are "variable charge" soils in which the surface charges are pH-dependent. Their capacity to retain phosphate is high. Within these general conditions, there is still much scope for variability in chemical and physical properties resulting from differences in the composition of the volcanic material, and the soil climate that controls weathering intensity and rates of leaching and desilication.[1]

I. Histosols, Mollisols, and Spodosols

Soils of these orders cover a combined total of 1.1% of the tropical land area. Although it is recognized that they may nonetheless be of significant local importance, they are excluded from this discussion.

IV. LAND-RELATED CONSTRAINTS TO FORAGE PRODUCTION

A. Kind of Constraints

The Food and Agriculture Organization (FAO)[6] defines land "as an area of the earth's surface, the characteristics of which embrace all reasonably stable, or predictably cyclic attributes of the biosphere above and below this area including those of the atmosphere, the soil and underlying geology, the hydrology, the plant and animal populations, and the results of past and present human activity, to the extent that these attributes exert a significant influence on present and future uses of the land by man." FAO[6] also advanced the concept of land qualities. These are attributes that influence the suitability of land for a specific kind of use and that integrate and can be assessed by measurable soil and site characteristics. Land-related stresses occur when the requirements of a specific land use are not matched by land qualities.

Forage production in the lower latitudes is constrained by a multitude of adverse land conditions. There are, in fact, virtually no land resources in the tropics that have no constraints and that are equally well suited for all forage species. An assessment of stress factors is further complicated by the fact that most lands have multiple rather than single stress conditions. Moreover, stress is not neutral but use-specific and what may be harmful for one plant species may be beneficial for another. Further complication is caused by the time dependence of some constraints. A seasonally high water table, for instance, may result in oxygen deficiency for some months, but pose no problems for the rest of the year.

Beinroth et al.[7] grouped land-related stresses in two basic categories: intrinsic and induced factors. Intrinsic factors are chemical, physical, biological, and holistic conditions that are inherent properties of the land, such as nutrient deficiencies, shallow sola, and low organic carbon content. Induced factors, on the other hand, are caused by external phenomena, mainly climate and human

TABLE 1.2
Biophysical Stress Factors Constraining Rainfed Forage Production

Category	Examples of Stress Factors
	Intrinsic Stresses
Chemical conditions	Nutrient deficiencies
	Excess of soluble salts
	Low base saturation
	Aluminum toxicity
Physical conditions	High susceptibility to erosion
	Steep slopes, shallow soils
	Surface crusting
Biological conditions	Low organic matter content
	High termite population
Holistic conditions	Low soil resilience
	Natural soil degradation
	Induced Stresses
Chemical conditions	Oxygen deficiency
	Acidification
	Contamination with toxicants
Physical conditions	Accelerated soil erosion
	Soil compaction
Climate-controlled conditions	Soil moisture deficiency
	Unsuitable temperature regime
	Insufficient length of growing season
Biological conditions	High incidence of pests and diseases
	Allelopathy
	Elimination of predators
Catastrophic events	Floods and droughts
	Earthquake-induced landslides
	Volcanic activity
Holistic conditions	Poor soil heath
	Lack of aesthetic value of agricultural landscape

activity, and include moisture stress, soil compaction, soil erosion, pests, and plant diseases. Table 1.2 lists intrinsic and induced stresses and provides examples.

B. Nature and Extent of Constraints

A detailed discussion of the land-related constraints to forage production in the tropics would be beyond the scope of this chapter, but the major stress categories that are briefly described here may suffice to convey a general idea of the limitations of tropical land resources. A more detailed account of the management of tropical soils for forage production is presented in this book in Chapter 10 authored by José Vicente-Chandler.

1. Moisture Stress

Insufficient soil moisture is the most widespread constraint to forage production in the tropics and affects about 53% of the land. Drought is a permanent feature of the Aridisols and seasonal moisture stress of various degrees of severity occurs in all of the suborders listed in Table 1.1 that contain the syllables "ust," "torr," or "xer." In the Inceptisols and Entisols, the soil moisture regime is diagnostic below the suborder level, and the dry soils of these orders are, therefore, not included in the computation. Moreover, soil moisture regimes derived from atmospheric data tend to overestimate available soil moisture in soils of low water retention capacity, such as sandy soils, and in soils where the difference in the amount of water held at field capacity and at the permanent wilting point is small, as in most Oxisols. The total area of moisture-deficient land, therefore, exceeds 53% and likely amounts to about 60%.

In about 6.5% of the tropics, forage production is impeded by reduced availability of oxygen to roots caused by a seasonally or permanently high water table. The soils affected by excess water have an aquic soil moisture regime. They are normally differentiated at the suborder level and identified by the prefix "aqu" in the suborder name (see Table 1.1).

2. Nutrient Deficiencies

Low levels of plant nutrients constrain plant growth in more than half (60%) of the tropics. Nutrient stresses may have several causes. One reason is the low cation exchange capacity, which results in a reduced capacity to retain cations and, conversely, in a small nutrient reserve and thus a low nutrient supplying power. This property is diagnostic for the Oxisols, and the "Kandi" taxa of Alfisols and Ultisols, and is typical for the quartzitic Psamments. Further, in the humid and perhumid tropics, abundant amounts of rainfall cause strong leaching conducive to a paucity of bases in the Udox, Perox, Humults, Udults, Umbrepts, and some Tropepts. Still another reason is the presence of a positive charge in the exchange complex of some Oxisols and Andisols. This leads to the strong fixation of anions, notably phosphates and sulfates, rendering these nutrients unavailable to plants.

3. Soil Acidity and Toxicities

The soils of about one third of the land area in the tropics are generally acid in their natural state with reactions of < pH 5. At these low pH values, exchangeable aluminum, if present in the soil, diffuses into the soil solution where it may reach toxic levels and restrict root growth of many plant species. This is a typical feature of the Ultisols and, to a lesser degree, Oxisols and some Tropepts. Aluminum toxicity affects about one third of all tropical land and may become a factor limiting plant growth when aluminum saturation exceeds about 50%. Critical levels, however, differ with the tolerance of the species. *Stylosanthes guianensis* and *Desmodium uncinatum*, for example, tolerate high levels of aluminum quite well, whereas *Medicago sativa* does not.[5]

Similarly, manganese is very soluble at pH values of less than 5.5. If appreciable amounts of this element are present, manganese toxicity can occur and plant growth may be adversely affected. Also, at pH 5 or below, the concentration of H ions could suppress the uptake of both Ca and Mg and result in deficiencies of these elements, particularly in soil with very low cation exchange capacity.[8]

The reader seeking more complete and specific information about the fertility and management of tropical soils is referred to the definitive books by Adams,[9] Sanchez,[5] and Van Wambeke,[1] and a perceptive article by Sanchez and Logan.[10]

C. State of Tropical Land Resources

Grassland is one of the five major types of global ecosystems that also include forests, freshwater systems, coastal/marine habitats, and agricultural land. Grassland is particularly extensive in the tropics where half of the world's permanent pastures are located.[5] Almost invariably, the land resources that support this ecosystem show signs of various degrees of degradation. The Journal of Land Degradation and Rehabilitation defines land degradation "as the loss of utility or potential utility through reduction of, or damage to physical, social or economic features and/or reduction of ecosystem diversity." Land degradation is a major threat to development and the Brundtland

Commission noted that "there is a growing realization in national and multinational institutions that not only many forms of economic development erode the environmental resources upon which they are based, but at the same time environmental degradation can undermine economic development."[11]

In an effort to establish baseline data about the state of land degradation, the United Nations Environment Program (UNEP) commissioned the International Soil Reference and Information Centre (ISREC) to initiate a project entitled Global Assessment of Human-induced Soil Degradation (GLASOD). The project recently published a *World Map of the Status of Human-induced Soil Degradation*. According to the explanatory note that accompanies the maps,[12] the percentage of land affected by soil degradation is as follows:

Africa	17%
Asia	18%
South America	14%
Central America	21%
Australasia	12%
WORLD	15%

The GLASOD project arrived at these estimates by evaluating land degradation resulting from water erosion, wind erosion, and chemical and physical deterioration. The study identified the causative factors of human-induced degradation as deforestation, overgrazing, agricultural mismanagement, overexploitation, and bioindustrial activities.

In a recent and more ambitious effort to assess the state of the world's ecosystems, the United Nations, the World Bank, and the World Resources Institute sponsored a $4 million study called a Pilot Analysis of Global Ecosystems (PAGE). (The findings of the project will be presented at a special millennial session of the UN in September 2000 and published as a World Resources Report.) The PAGE study, which was previewed in a special issue of TIME magazine, estimates that 80% of the world's grasslands are affected by deteriorating soil conditions. This estimate may be correct for all stages of degradation, but the percentage of severely degraded grassland is considerably lower. The PAGE report nevertheless draws attention to an alarming situation and is particularly pertinent for the tropics. For example, Costa and Rehman[13] estimate that in Central Brazil about half the area under sown pasture is degraded.

A distinction should be made between pasture degradation and land degradation. The former process is defined as "... the loss of vigor, productivity, and natural capacity for recovery, in order to sustain production and quality of grass required by animals, and to overcome the detrimental effects of insects, diseases and weeds ...."[13] The primary cause of pasture degradation in the tropics is year-round overgrazing at stocking rates that exceed carrying capacity. The biological deterioration, however, can also impact the soil as it is conducive to wind and water erosion, and nutrient depletion. Ironically, and as pointed out by Vera and Rivas,[14] pasture degradation may be an ambiguous concept. The invasion of woody species that reduce grass areas is viewed by the farmer as detrimental, whereas the ecologist considers it a successional process leading to the reestablishment of the original habitat. A further problem is that the objectives of cattle producers do not necessarily conform to long-term environmental concerns. Costa and Rehman[13] found that in Central Brazil, one of the most important objectives of cattle producers was to maximize the number of cattle, which leads, of course, to pasture and land degradation. They also found that farmers have only fragmental knowledge of the complex livestock systems. In particular, farmers have not been aware of overgrazing and its consequences, and their perception has been marred by a general lack of knowledge on this issue.

Although environmental damage caused by forage-consuming animals is certainly a major factor in land degradation, properly managed pasture and range definitely have a positive effect on landscape stability and carbon sequestration. The latter tends to reduce carbon dioxide levels in the atmosphere and thus helps mitigate the global warming attributed to the "greenhouse effect."

V. LAND EVALUATION FOR FORAGE PRODUCTION

FAO must be credited with spearheading modern land evaluation. The publication of the *Framework for Land Evaluation*[6] and, subsequently, the *Guidelines: Land Evaluation for Rainfed Agriculture*[15] mark methodological breakthroughs. Central to the FAO approach is the comparison of land qualities with the requirements of a specific landuse. If there is a mismatch, and the land qualities are not in balance with the requirements, the production system is stressed and its performance suboptimal.

FAO recently published a case study of land evaluation for livestock productivity in Kenya that exemplifies and explains this approach. The interested reader is urged to consult this excellent study by Kassam et al.[16]

The successful implementation of this methodology depends upon two conditions: (1) the availability of adequate knowledge of the factors that govern growth and development of the species under consideration, and (2) site-specific environmental characterization data. Regarding the former, owing to the large number of forage species and their numerous cultivars and accessions, cultivar-specific, quantitative knowledge of the environmental requirements is still empirical and incomplete in many instances. Frequently, however, land evaluation is hampered by inadequate environmental information, especially soil and weather data. Reliable area- or site-specific primary data obviously provide the best database, but in their absence, default procedures may produce satisfactory surrogate data. Where soil surveys employing *Soil Taxonomy* classes are available, reasonable inferences about soil properties can be made from the name of the taxon that identifies a map unit.[7]

The advent of digital information technologies such as relational databases, simulation models, expert systems, and decision support systems have all contributed to transforming land resource evaluation from an art to a science. The greatest impact, however, has come through advances in two areas: (1) the scientific and commercial development of geographical information systems (GIS) and companion digital terrain models (DTM), and (2) the development of deterministic crop and environmental simulation models, and related rule-based systems. Linking GIS and the attendant databases to models that simulate plant growth allows rapid assessments of land suitability for specific forage species over long time horizons. Until globally applicable models that incorporate the genetic coefficients of major forage species become available, rule-based systems can be employed to generate the parameters required by the models or other pertinent information. The systems approach is philosophically similar to the FAO methodology, but differs from it in that it employs land and soil characteristics rather than land qualities, and computer models rather than intuitive expert knowledge, to produce performance estimates.

For a comprehensive discussion of the principles and methods of land evaluation, the reader is referred to the classic framework,[6] the subsequent guidelines,[15] and a recent book by Davidson[17] that is noteworthy for its breadth and depth.

VI. CONCLUSION

Competition for land in the tropics now used for grazing is expected to increase markedly in the near future, as most of the escalating global population growth is predicted to occur in tropical countries. Another factor is that as some of the so-called third world countries become more affluent, their diets will become less grain-based and the wide gap in beef consumption, 16 kg/capita in tropical America vs. 2 kg/capita in the Far East in 1965,[14] will narrow. A likely consequence of this is that forage production, often already relegated to less productive soils, will be shifted further onto more marginal and fragile lands. As these areas are also more susceptible to degradation, careful management becomes imperative if sustainable forage production is to be achieved.

It is encouraging, therefore, that convincing arguments have been advanced by Brazilian researchers in favor of intensification of beef and milk production in tropical forest areas.[14] The combination of policies and agrosilvopastoral systems could thus control the expansion of the agricultural frontier at the expense of tropical forests and achieve environmental protection without sacrificing agricultural production.[14]

As with other agricultural uses, the judicious husbandry of land resources for forage production requires the generation, mobilization, and integration of environmental data and process knowledge for scientific land use planning that takes full advantage of advances in information science and digital information technology. It also requires visionary scientific and political leadership that precipitates public policy and action programs that ensure a healthy future for the land.

REFERENCES

1. Van Wambeke, A., *Soils of the Tropics — Properties and Appraisal*, McGraw-Hill, New York, 1991.
2. Soil Survey Staff, *Soil Taxonomy — A Basic System of Soil Classification for Making and Interpreting Soil Surveys*, U.S. Dept. of Agriculture, Agriculture Handbook 436, U.S. Government Printing Office, Washington, D.C., 1975.
3. Eswaran, H., Beinroth, F. H., Kimble, J., and Cook, T., Soil diversity in the tropics — Implications for development, in *Myth and Science of Soils of the Tropics*, Lal, R. and Sanchez, P. A., Eds., special publ. no 29, Soil Science Society of America, Madison, WI, 1992, chap. 1.
4. Wilding, L. P., Smeck, N. E., and Hall, G. F., Eds., *Pedogenesis and Soil Taxonomy*, Vols. I and II, Elsevier, New York, 1983.
5. Sanchez, P. A., *Properties and Management of Soils in the Tropics*, John Wiley & Sons, New York, 1976.
6. FAO, *A Framework for Land Evaluation*, FAO Soils Bulletin no. 32, FAO, Rome, 1976.
7. Beinroth, F. H., Eswaran, H., Reich, P. F., and Van den Berg, E., Land related stresses in agroecosystems, Proc. Symp. Agroclimatology and Sustainable Agriculture in Stressed Environments, ICRISAT, Hyderabad, 1993.
8. Kamprath, E. J., Crop response to lime on soils in the tropics, in *Soil Acidity and Liming*, Agronomy Monograph no. 12, 2nd ed., Adams, F., Ed., ASA-CSSA-SSSA, Madison, WI, 1984, 349.
9. Adams, F., Ed., *Soil Acidity and Liming*, Agronomy Monograph no. 12, ASA-CSSA-SSSA, Madison, WI, 1984.
10. Sanchez, P. A. and Logan, J. P., Myths and science about the chemistry and fertility of soils in the tropics, in *Myth and Science of Soils of the Tropics*, Lal, R. and Sanchez, P. A., Eds., special publ. no 29, Madison, WI, 1992, chap. 3.
11. World Commission on Environment and Development, *Our Common Future*, Oxford University Press, Oxford, U.K., 1987.
12. Oldeman, L. R., Hakkeling, R. T. A., and Sombroek, W. G., *World Map of the Status of Human-Induced Soil Degradation: An Explanatory Note*, International Soil Reference Centre, Wageningen, 1990.
13. Costa, F. P. and Rehman, T., Exploring the link between farmers' objectives and the phenomenon of pasture degradation in the beef production systems of Central Brazil, *Agricultural Systems*, 61, 135, 1999.
14. Vera, R. R. and Rivas, L., Grasslands, cattle and land use in the Neotropics and subtropics, in *Proc. XVIII Int. Grassland Congress*, vol. 3, Winnipeg, Manitoba and Saskatoon, Saskatchewan, Canada, 1997, 545.
15. FAO, *Guidelines: Land Evaluation for Rainfed Agriculture*, FAO Soils Bulletin no. 52, FAO, Rome, 1984.
16. Kassam, A. H., van Velthuizen, H. T., Sloane, P. H., Fischer, G. W., and Shah, M. M., *Agroecological Land Resources Assessment for Agricultural Development Planning — A Case Study of Kenya, Resources Data Base and Land Productivity, Technical Annex 5, Livestock Productivity*, FAO and International Institute for Applied Systems Analysis, Rome, 1991.
17. Davidson, D. A., *The Evaluation of Land Resources*, John Wiley & Sons, New York, 1992.

2 Environmental Constraints to Tropical Forage Plant Adaptation and Productivity

W.D. Pitman

CONTENTS

I. INTRODUCTION

Plant environments are primarily defined by the climatic and edaphic factors inherent within particular geographic locations. Biological factors further define the plant environment at a given location. Within the naturally occurring range of climatic conditions, limited practical opportunity exists for management to enhance survival or production of tropical forages. Irrigation is the primary means of altering climatic constraints in agriculture, but physical and economic considerations often limit the use of irrigation on tropical forages.

Several approaches are available to modify the edaphic conditions limiting productivity of tropical forages. While such management inputs can also affect survival of particular forage plants in some situations, survival of tropical forages based on management inputs is typically temporary and does not often contribute to development of sustainable forage systems. Soils can be managed to enhance fertility and other chemical characteristics, alter structure, modify drainage, and increase rate of water infiltration. Such management of soils for tropical forage production may not generally be economical; however, insufficient information is available to characterize the potentially useful management options or their economic feasibility in many cases.

The limited availability of options for practical and economic modification of the climatic and edaphic factors restricting tropical forage adaptation and production contrasts distinctly with the availability of biological options. The biological options include a tremendous array of plant germplasm, the natural and manipulated constitution of this germplasm into diverse plant communities, and imposition of forage management practices to influence production and harvest of the resulting plant growth. This chapter will provide an introduction to the climatic and edaphic limitations typically encountered with tropical forage production. The currently prevailing option for addressing most environmental limitations to tropical forage production is the selection of

0-8493-2318-5/01/$0.00+$.50

appropriate plant genotypes combined with manipulation of grazing livestock to optimize the resulting environmental conditions for plant growth.

II. CLIMATE

Climate is considered to be the long-term manifestation of weather conditions.[1] Weather is the set of atmospheric conditions on a day-to-day basis.[1] Climate can determine the general adaptation and production levels of plants at particular locations. Weather affects factors such as the rate of growth and even survival of individual plants and plant stands on a daily basis.

Agriculture in many tropical regions has not developed to the extent typical of temperate regions. Food crops and systems of agriculture in the tropics based on these crops and use of the natural vegetation by livestock have often not been supplemented with the extensive use of introduced pasture plants. Thus, climatic adaptation ranges of tropical forage plants are generally not as well defined as those of temperate species. Classification of climates can provide an initial indication of potential adaptation ranges of particular plant species based on their areas of current and historic occurrence.

The geographical tropics are the region between the Tropic of Cancer (23° 27′ N) and the Tropic of Capricorn (23° 27′ S). Oceans moderate climate, and maritime influences extend tropical climatic conditions to latitudes beyond the geographical tropics. The simple classification of geographical regions into tropical, temperate, and frigid zones has been subdivided by various classification systems to attain greater usefulness. Tropical climates have been divided into tropical yearlong rainy climates, tropical humid summer climates with either humid winters or comparatively dry winters, alternately wet and dry tropical climates, tropical dry climates with either humid winters or dry winters, tropical semi-desert climates, and tropical desert climates.[2] Application of various systems of climate classification to tropical forages and pastures has been presented previously.[3,4] The 1933 classification by Thornthwaite[5] was an early effort to describe subclimate regions in terms of predominant vegetational differences due primarily to moisture availability. This classification system has been widely used in assessing plant adaptation.

A somewhat more recent classification system by Holdridge[6] distinguishes different environments as life zones. An initial six temperature categories of tropical, subtropical, warm temperate, cool temperate, boreal, and subpolar were further divided by rainfall and vegetation type. The life zones in the tropics consist of tropical desert scrub, tropical thorn woodland, tropical very dry forest, tropical dry forest, tropical moist forest, tropical wet forest, and tropical rain forest. The lack of a grassland designation in the tropics and subtropics by Holdridge is perhaps of particular significance to tropical forage and pasture management. In contrast to temperate regions where grasslands are suggested to represent the ecological climax community or the climatic potential of major land areas, the climate provides potential for woody plant growth throughout the tropics and subtropics. The existence of tropical grasslands is not contested, however, constraints to woody plant growth, which result in maintenance of these grasslands are suggested to be other than climatic. These constraints include edaphic factors such as infertility, soil acidity, nutrient toxicity, and flooding along with additional factors such as fire (natural and anthropogenic) and other anthropogenic suppression of woody plants. Thus, within the designated categories or life zones, considerable variation in plant adaptation exists. Much of this variation is due to soil type. Even though soil type is affected by climate, geological processes produce variations in soils within climatic zones. These zones actually merge together with no discreet boundaries, further masking responses of plants to defined climatic conditions.

Despite the lack of a distinct, readily defined relationship between either plant adaptation or potential plant growth and some quantifiable measure of the overall climate, several readily quantifiable climatic variables directly affect plant growth and survival. Temperature, precipitation, day length, humidity, solar radiation, air movement and their variations and interactions greatly affect plant responses.

Precipitation effects depend on amount, distribution, runoff, infiltration, evaporation, and transpiration. The actual annual average amount of rainfall in tropical regions is often less predictive of plant responses than is characterization of the typical patterns of moist and dry periods. Length, severity, and frequency of dry seasons are often associated with adaptation and productivity of particular tropical forage species or varieties. The concept of wetness of months, as described by Mohr and Van Baren[7] and Ochse et al.[8] can contribute to an understanding of tropical forage plant responses to moisture. Months when more than 100 mm of rain are typically received are designated wet months. Moist months receive 60 to 100 mm of rain, and those receiving less than 60 mm are designated dry months. Such variables as moisture storage in the soil and rates of infiltration, evaporation, and transpiration alter the actual availability of moisture. Plants themselves influence moisture availability and effectiveness of its use. Despite these associated variables, typical length of the principal dry season has been used to classify tropical climates as humid (dry season less than 2.5 months), intermediate humid (2.5 to 5 month dry season), intermediate dry (5 to 7.5 month dry season), semiarid (7.5 to 10 month dry season), and arid (10 to 12 month dry season).[2]

Variations in temperature and solar radiation are normally less dramatic in their effects on plants in the tropics than are variations in moisture, however, such effects can be meaningful. At higher latitudes and at high elevations within the tropics, seasonally low minimum temperatures and associated frosts determine plant adaptation and season of production of adapted species. In addition to temperature, photoperiod can be a major determinant of plant adaptation and growth, especially for temperate species at high elevations in the tropics. Plant responses themselves combine with the climatic variables to form a highly complex interactive system designated as the microclimate.[9]

While delineation of climatic zones can provide guidelines for tropical forage research and development activities, tropical forage research is based on the biological sciences rather than geographic limitations. Thus, the usefulness of adaptation and plant growth assessments of tropical forages extend from the geographic tropics through the rather ambiguously defined subtropics into warm temperate regions in some instances. Tropical pastures, then, become defined by the plant species rather than the geographic location. The plant species include grasses utilizing the C_4 photosynthetic pathway with adaptations to high temperatures and limited yearly variations in day length. These grasses typically lack tolerance for freezing temperatures. Although the tropical legumes do not distinctly differ from temperate legumes in photosynthetic pathway, they express distinct adaptations to high temperatures, day length conditions of the low latitudes, and lack of cold tolerance. As in temperate geographic regions, the temperate forage species prove superior in cooler climates of high elevations even in the geographic tropics. Exceptions to the adaptation of temperate species to cool climates at high elevations in the tropics include those due to inadequate variations in day length required for physiological photoperiod responses.

III. SOILS

Chapter 1, in this text, by Friedrich H. Beinroth describes basic aspects of the development of soils and land resources in the tropics. This chapter presents aspects of the application of soil variations and their interactions with climate to the adaptation and production of tropical forage plants.

From a global perspective, there are two systems of soil classification that may be widely encountered in the process of evaluating tropical forage plants. The UNESCO, FAO approach currently is primarily used as a reference among soil scientists.[10] Buol[10] suggested that this system does not provide the needed detail for work at individual sites. The second broad system is the soil classification system of the United States Department of Agriculture. This hierarchical system is divided into six categories. The highest category is currently represented by eleven soil orders.[10] A taxonomic key is used to classify a particular soil into the appropriate order by sequential assessment. The soil orders in the appropriate sequence are: Histolsols, organic soils; Andisols, other soils of volcanic ash; Spodosols, other soils with humus/amorphous subsoils; Oxisols, other soils with oxide rich subsoils; Vertisols, other soils with extreme shrink–swell properties; Aridisols, other soils with less than 90 days of moisture; Ultisols, other soils with acid subsoils; Mollisols, other soils with thick, dark colored surfaces; Alfisols, other soils with slightly acid subsoils; Inceptisols, other soils with weak subsoil development; and Entisols, other soils. Some forage germplasm development programs in the tropics have been based on the broad classification of soil orders in the target area as illustrated by Toledo.[11] Categories below the soil order are suborder, great group, subgroup, family, and series. Many countries have developed unique soil classification systems for internal use.[10]

Large areas of soils in the tropics are more extensively weathered than is typical of temperate soils. Rather than the soil-building process of podzolization, which predominates in temperate regions, laterization of soils has been extensive in the tropics. This process involves long-term leaching by rainfall, with soluble bases gradually removed. Oxides and hydroxides of iron and aluminum accumulate. Such laterized soils are typically acidic to very acidic with low fertility. Extensive areas of such soils, illustrated by the Oxisols and Ultisols of South America, present a considerable challenge to the development of productive, sustainable tropical pastures.

Classification systems contribute substantially to international communication, large-scale determination of potential plant adaptation, and enhanced documentation of results from tropical forage germplasm evaluations. They can be, however, quite cumbersome and intimidating in the process of applying the technology in the field. From this perspective, Fisher[12] has described a group of soil properties that are readily observable and closely associated with plant responses.

Soil color is associated with organic matter, drainage, and aeration. Up to about 8% organic matter, soil becomes darker with increasing organic matter. A black soil is typically 8% or higher in organic matter and may be poorly drained. A brown soil will generally have slightly less than 8% organic matter and be well drained. Red and yellow soils are low in organic matter, well drained, and contain varying amounts of oxidized iron, from which they obtain their color. Gray soils are low in organic matter and poorly drained, with the color due to excess water and poor aeration. Organic matter of aerated soils in the tropics decomposes readily due to high microbial activity and oxidation, thus the extent of dark colored soils is limited. Organic material in tropical soils is highly dynamic and levels are maintained only by continual replacement by plant material from each year's growth. Such dynamics have not allowed organic matter to greatly affect soil formation in the tropics as it has in temperate regions. Organic matter can, however, be readily depleted from a somewhat stable level in tropical soils by excessive harvest of plant growth.

Soil texture is determined by the percentages of sand, silt, and clay in the topsoil. Sands allow rapid infiltration of water but have limited ability to retain water or nutrients. Clay soils retain water effectively, but the soil surface may seal over and produce excessive runoff. Clay soils such as many included in the Vertisol order often contract when dry and expand when wet. This produces a difficult soil–water environment for plants and can physically damage plant roots. The formation of very stable peds by aggregation of clay particles in highly weathered Oxisols increases pore space of these clay soils. Loamy soils, which are combinations of the soil particle sizes, typically provide superior soils for plant growth.

Soil structure refers to the nature of soil layers below the topsoil. Subsoils range in structure from individual grains of sand, as occur in sand dunes, to massive structures formed by clay particles compacted tightly together with little pore space or other material. Sandy soils with some organic matter and/or clay included adhere together in small clumps to form a granular soil structure. The massive soil structure of clays becomes blocky as organic matter and larger sized soil particles are included. These blocky soils have increased pore space, which improves water and air movement, as well as plant root penetration, compared to a massive structure. The concepts of topsoil and subsoil are not highly useful with many tropical soils, which are not in the distinct horizontal zones or layers typical of soil profiles of temperate regions. Where pasture development follows or is rotated with crops, such distinctions can be useful. The plowed layer may be considered as a topsoil. Where extremely high sand or clay proportions in this layer present excessive risk of establishment failure, residue from crops can be incorporated into the upper layer or topsoil to modify the inherent structure. At greater soil depths, little can be done to economically alter soil structure for tropical forage production, although soil modifications such as deep tillage can be useful for higher value crops.

Soil depth is an additional readily distinguishable characteristic of soils which distinctly affects plant adaptation and productivity. Moisture and nutrient supplying potential of soils can be greatly affected by soil depth or effective rooting depth of a particular plant in a soil. Waterlogging and associated aeration can be determined by soil depth in some situations.

Although slope was not listed by Fisher, erosion hazard and moisture conditions can be greatly affected by slope. Wet, waterlogged, or periodically flooded conditions can result from lack of slope. Limited infiltration and increased erosion are considerations with sloping sites.

In addition to the visibly discernable characteristics of soil color, soil texture, soil structure, and soil depth, Fisher[12] listed the chemical response of soil pH as a readily measurable soil property of particular consequence in plant adaptation. Soil pH affects solubility and availability of plant nutrients and potential toxicity of aluminum and manganese ions in soil solution. Although any soil pH below the neutral measure of 7.0 is considered acidic, many tropical soils are highly acidic with pH ranges of 4.0 and even lower, particularly Oxisols and Ultisols of the hot humid tropics. At the other extreme, Vertisols in semi-arid regions may be near neutral to slightly alkaline, while soils in arid regions, typical of some Aridisols, may have even more alkaline soil reactions. Arid soils, and those previously irrigated, may also have soil salinity of sufficient levels to affect plant adaptation and productivity.

Inherent fertility of soils often has greater significance with tropical forages than with other crops. Economics, infrastructure, and local political circumstances often preclude fertilization as an option in tropical forage production. Thus, selection of forage plants adapted to sustainable nutrient levels in the soil can be critical to success. Careful selection of fertile sites for growth of particularly productive and nutritive forages may also be considered for intensive forage-based enterprises such as milk production in some tropical regions.

IV. PLANT ADAPTATION

General patterns of climate and large regions of similar soils provide a basis for characterizing environments in the tropics. These classifications can be effectively used to indicate broad, general ranges of adaptation of tropical forage grasses. Such an approach allows the general association of the genus *Panicum* with fertile soils over a broad range of rainfall levels, while the genus *Pennisetum* has typically been associated with fertile soils in moist climates. The genus *Cenchrus* has primarily been used in dry, fertile environments. The recent taxonomic reclassification of *Cenchrus ciliaris* to *Pennisetum ciliare* may be of value to taxonomists and geneticists, but it appears to complicate rather than clarify plant adaptation considerations. The genus *Cynodon* has been particularly useful in fertile, humid environments with adaptations of some species in the genus to cooler climates and alkaline and even saline soils. Recent developments with *Andropogon* and *Brachiaria* emphasize their adaptations to acid, infertile soils. These recent characterizations add to the past recognition of *Brachiaria* as adapted to moist sites and *Andropogon* as containing species suited to dry and even warm temperate climates. *Paspalum* species typically have been useful in moist environments of subtropical and warm temperate regions. Such predictable relationships of tropical grasses and environmental conditions have allowed the widespread use and even naturalization of introduced tropical grasses.

Some specific adaptations and limitations of particular tropical legumes have been determined, but reliable prediction of potential adaptation has not resulted. Tropical legume species and genotypes within species typically have very specific requirements for adaptation. Some of this specificity is associated with rhizobial associations for effective nitrogen fixation. Some tropical legumes also are more distinctly affected by photoperiod, especially in relation to flowering. The legumes are also affected by general soil fertility, soil pH, and moisture conditions, as are the tropical grasses. In addition, however, survival of tropical legumes can be determined by very subtle differences such as visually indiscernible differences in elevation over short distances. Such differences may be associated with soil moisture or with nutrient availability due to erosion and/or deposition of soil and plant materials. Slight differences in duration of flooding or waterlogging can determine legume survival. At high elevations and subtropical latitudes, only a degree or two of difference in temperature, often associated with frost, can determine the fate of tropical legumes. Combining these very specific environmental requirements for survival with competition from aggressive grasses and defoliation by grazing livestock has contributed to the lack of widespread success of tropical pasture legume technology.

V. SUMMARY AND CONCLUSIONS

Limited variation in day length among seasons of the year and restriction of low temperatures to high elevations gives the superficial impression of similar climate throughout most of the tropics. Tremendous ranges in rainfall and associated cloud cover and humidity plus differences in air movement, solar radiation, maximum temperature, and evaporation rates provide a great variety of largely moisture-driven climates throughout the tropics. Landscape and soil variations both result from and further enhance the differences in tropical environments associated with moisture conditions. Not only do the obvious differences in rainfall among regions affect plant growth, but many rather inconsequential appearing environmental variations can determine adaptation and productivity of various tropical forage species.

Within the broad climatic regions of the tropics, it is often soil conditions that define environments with greatest effect on forage plants. Particular groups of plants are typically associated with extremely acid and infertile soils, heavy clay soils, coarse sands, seasonally flooded soils, waterlogged soils, and even fertile soils. A striking aspect of tropical forage plant geography is the rather extensive adaptation of the superior tropical grasses and the contrasting highly specific and localized adaptation of most tropical legumes. In addition to greater dependence on a complex biological rhizosphere community, tropical legumes appear to be highly adapted to specific microenvironments defined by climatic, edaphic, and biological conditions.

Research and forage development programs with goals and objectives based on forage improvement across large regions may make greater progress for the overall region with emphasis on the appropriate forage grasses. Efforts targeting tropical forage legumes may need to be rather narrowly focused on specific environments where available germplasm demonstrates particular promise. The probability of developing new commercially successful tropical forage cultivars from superior germplasm is greater with grasses than with legumes due to their differing adaptabilities to variations in environmental conditions.

ACKNOWLEDGMENT

Appreciation is expressed to Dr. Franklin W. Martin of the University of Puerto Rico, Mayagüez, who provided the initial format for this chapter prior to his retirement.

REFERENCES

1. Shaw, R. H., Meterology, in *Encyclopedia of Agricultural Science*, Vol. 3, Arntzen, C. J., Ed., Academic Press, San Diego, California, 1994, 39.

2. Sprague, H. B., *Management of Rangelands and Other Grazing Lands of the Tropics and Subtropics for Support of Livestock Production*, Technical Series Bulletin No. 23, U.S. Agency for International Development, Washington, D.C., 1979.
3. Whiteman, P. C., *Tropical Pasture Science*, Oxford University Press, Oxford, U.K., 1980.
4. Ripley, E. A., Grassland climate, in *Natural Grasslands: Introduction and Western Hemisphere*, Ecosystems of the World 8A, Coupland, R. T., Ed., Elsevier, Amsterdam, 1992, 7.
5. Thornwaite, C. W., The climates of the world, *Geographic Review*, 23, 433, 1933.
6. Holdridge, L. R., The life zone system, *Adansonia*, 6, 199, 1966.
7. Mohr, E. C. J. and Van Baren, F. A., *Tropical Soils*, Interscience, New York, 1954.
8. Ochse, J. J., Soule, M. J., Jr., Dijkman, M. J., and Wehlburg, C., *Tropical and Subtropical Agriculture*, Vol. 1, Macmillan Co., New York, 1961.
9. Kurata, K., Microclimate, in *Encyclopedia of Agricultural Science*, Vol. 3, Arntzen, C. J., Ed., Academic Press, San Diego, California, 1994, 51.
10. Buol, S. W., Soil genesis, morphology, and classification, in *Encyclopedia of Agricultural Science*, Vol. 4, Arntzen, C. J., Ed., Academic Press, San Diego, California, 1994, 101.
11. Toledo, J. M., Pasture development for cattle production in the major ecosystems of the tropical American lowlands, in *Proc. XV International Grassland Congress*, The Japanese Society of Grassland Sciences, Nishi-nasuno, Tochigi-ken, Japan, 1985, 74.
12. Fisher, R. F., Forest ecology, in *Encyclopedia of Agricultural Science*, Vol. 2, Arntzen, C. J., Ed., Academic Press, San Diego, California, 1994, 421.

Section II

The Tropical Forage Plants and Their Development

3 Germplasm Resources of Tropical Forage Grasses

Albert E. Kretschmer, Jr. and W.D. Pitman

CONTENTS

I. INTRODUCTION

The era of collecting exotic tropical forages for evaluation at research centers began in earnest about 50 years ago. The initial forage plant germplasm collection trips were made by CSIRO-ATFGRC (Commonwealth Scientific and Industrial Research Organization–Australian Tropical Forages Genetic Resources Center, Brisbane, Queensland) and other Australian researchers in Africa and the American tropics. This initial germplasm exploration was followed by others, including collection trips by the USA–ARS–SRPIS (United States Department of Agriculture–Agriculture Research Service–Southern Regional Plant Introduction Station), Griffin, GA, CIAT (Centro Internacional de Agricultura Tropical), Cali, Colombia, EMBRAPA's CENARGEN (Central Nacional de Recursos Geneticos e Biotecnologia) Brasilia, Brazil, and ILCA (International Livestock Center for Africa), Addis Ababa, Ethiopia. What began as a "collect whatever is available" project has been largely replaced by specific site collections as information on adaptability of species has become known.[1] This increasing focus on tropical grass germplasm exploration is illustrated with the specific collection of *Panicum* and *Andropogon* species in Africa.[2,3]

During this development period, forage germplasm banks, mostly of rudimentary facilities, were maintained primarily for local storage. In 1970, it was estimated that there were less than 10 tropical forage gene banks in the world, increasing in number to more than 100 presently.[4] More sophisticated methods of storage, annotation of passport and evaluation data, distribution of genetic resources, and evaluation procedures also developed.

More recently, with a system of national and international research facilities emphasizing the need to preserve forage germplasm, local networks and international cooperation among forage germplasm collectors and evaluators spontaneously and formally developed.[1,4-17] At present, the stage of development has been reached where the availability of tropical forage germplasm resources worldwide is substantially greater than the extent of evaluation and use of the germplasm. Furthermore, there is now an expanded knowledge of the tropical grass genera and species that may have potential for commercial use. An evaluation or reevaluation of genetic resources should be a priority, with less emphasis on collection, except for specific purposes or certain genera.

0-8493-2318-5/01/$0.00+$.50

The extensive collections of tropical forage species and tremendous diversity of these species suggest the need for a more efficient approach to evaluation for use as forage than that typically followed for temperate species. This need is associated with the extensive knowledge base and history of use of most temperate forage species compared with very limited information concerning area of adaptation, appropriate management, and genotypic variability of most tropical species. There are many untested species and ecotypes of tropical forages. Breeding is generally unnecessary at this time, except in special instances. Precision of evaluation may not be critical, especially as broad ranges of germplasm are evaluated. Because of the high biomass production of many of the tropical species, stem-to-leaf ratios are high resulting in a disproportionate amount of stemmy, less nutritious, and inedible material left on plants after grazing. Measuring the yield of these tall stemmy types of grasses such as napier grass (*Pennisetum purpureum* Schumach) and star grass (*Cynodon nlemfuensis* Vanderyst) does not represent the edible yield. This has already been recognized with woody legume species such as leucaena [*Leucaena leucocephala* (Lam) de Wit], where forage production is estimated by harvesting what more nearly approaches the edible plant portions. Aspects of evaluation techniques have been thoroughly reported and will not be discussed in this report.[1,12,18-20]

The objective of evaluation of potential pasture plants is to develop the potential persistent cultivar as quickly and efficiently as possible. As long as the species is neither toxic nor a potential weed pest, even having a somewhat higher frequency of failure of preliminary selections in subsequent steps of the evaluation process may not be as detrimental as spending substantially more resources and time in thorough initial screening of a very limited portion of the potentially useful germplasm. For tropical grasses, the primary aspects of persistence involve ability to tolerate grazing and compete with weeds. Thus, initial identification of competitive, grazing-tolerant species could effectively precede more detailed assessments of forage yield and quality in many instances.

Prior to the disciplined approach to grass cultivar development, carpetgrass (*Axonopus* species), naturalized guinea grass (*Panicum maximum* Jacq.), and other natural or native grasses (*Acroceras* species, *Echinochloa* species, and *Brachiaria mutica*) were used for grazing. Later, selections of the more vigorous genotypes were made. Many American and other tropical countries began their pasture improvement or diversification programs using vegetatively planted species, i.e., *Digitaria* (digitgrass) and *Cynodon* (Bermuda grasses), because of low-cost labor and the higher quality and productivity of these grasses than of species such as carpetgrass, common Bermuda grass [*Cynodon dactylon* (L.) Pers.], and jaraguágrass [*Hyparrhenia rufa* (Nees.) Stapf]. Although guinea grass is still used extensively in Brazil and elsewhere, recently *Brachiaria* species and gambagrass (*Andropogon gayanus* Kunth) were introduced and are often the preferred seeded grasses. This change from vegetatively propagated to seeded grasses was aided by and contributed to the rapid expansion of plantings in large areas in Latin America and Australia. Vegetative plantings require more time and are more expensive than seeded plantings. In Florida and the Caribbean, where improved pasture development is restricted in area, vegetatively planted grasses are still being used to a large extent for the "higher quality" species. Even so, Bahia grass pastures presently comprise an estimated 85% of all improved pastures in Florida.[21] The primary attributes favoring Bahia grass are persistence and low maintenance, yet it has not been used as extensively in tropical American, Australian, or Asian countries as in the southern U.S.

II. COLLECTION AND STORAGE

Collection strategies for temperate and tropical forages are similar. Better mapping, climatic adaptation, and species distribution and collection information is presently available compared with that of several decades ago, when tropical forage germplasm collection began. This information should be used to plan and conduct future collection trips. Selected references dealing with collection are: Blumenstock,[22] tropical climate characteristics and distribution; McWilliam,[23] response to various temperatures; Mott,[19] general; Reid,[3] use of climatic data for collecting; Clements and Cameron[2]; and Reid and Strickland,[24] collecting review.

The genetic resource, gene, or germplasm bank has the responsibility to maintain vegetative or seed-propagated genotypes in a viable condition. In the past, large numbers of collections have been lost through improper storage. Additionally, precise collection data (passport data) was missing or incomplete.

Several major germplasm banks have large collections of tropical grasses and legumes. The U.S. national germplasm system and the disposition of major collections have been described.[7,14] Contributions of the introduced grass germplasm to development of cultivars has been presented by Asay.[8] The scientific management of germplasm and seed enhancement have been described by Stalker and Chapman.[15] Cohen et al.[11] summarized the present status of international genetic resource banks and the types of crops stored. In addition to CIAT and ICARDA (International Center for Agricultural Research in Dry Areas), which have a large collection of legumes for forage and human consumption, there are several other centers. The USDA-SRPIS; University of Florida, IFAS, Indian River Research and Education Center, Fort Pierce, Florida, U.S.; CSIRO-ATFGRC; ILCA; and EMBRAPA-CENARGEN have moderate to large and well-managed forage germplasm banks. Additionally, a germplasm bank sponsored by CGIAR (Consultative Group for International Agricultural Research) is being built to serve 10 countries in southern Africa, with no charge to the participants. Gibbens[4] reported that more than 100 gene banks exist worldwide, from sophisticated to rudimentary. Belgium has an IBPGR (International Board for Plant Genetic Resources, Rome) bank described by Vanderborght.[25] This is an example of a smaller, more specific germplasm bank containing mostly tropical legumes.

III. PRESENT AVAILABILITY OF GERMPLASM

Of approximately 10,000 species of *Gramineae*, only a few have been used in the tropics and subtropics. Grasses for tropical environments are primarily from the warm-season tribes *Andropogoneae, Paniceae, Chlorideae,* and *Eragrosteae*. Most of the tropical grass species originated in Africa, while two important genera, *Paspalum* and *Axonopus*, are native to tropical and subtropical America. Further information on history, descriptions, and use of tropical grasses is available in the following publications: Anonymous,[26] arid and semi-arid regions of Africa; Loch and Ferguson,[27] worldwide use and seed production; Bogdan,[28] grass descriptions and use; Clayton,[10] history and development; Cooper,[29] physiology and energy conversion; Eyles et al., [30] northern Australian history, review, and descriptions; Gonzalez,[18] adaptation to drought/rainfall parameters; Humphreys and Riveros,[31] seed production; Kretschmer and Pitman,[32] brief description and use; Loch,[33] history and present seed production in Australia and in the American and Asian tropics; 't Mannetje and Jones,[34] thorough description and use; Minson,[35,36] nutritive differences between tropical and temperate forages and general nutrition; O'Reilly,[37] description of commercial species; Serrao and Neto,[20] adaptation to humid tropics; Skerman and Riveros,[38] descriptions and use; Watson and Dallwitz,[39] species description; and Zeven,[40] legume and grass centers of origin.

The number of grass species and total accessions (about 17,000 as shown in Table 3.1) of selected genera of germplasm held in the major resource banks of the world show that most accessions are in the *Brachiaria, Cenchrus, Digitaria, Panicum, Paspalum,* and *Pennisetum* genera. Many duplicate genotypes among banks exist because of the interchange of germplasm. Most are not native to tropical America, even though many have been more extensively planted there than elsewhere. Of naturalized species, *Panicum* genotypes have been very popular in Brazil and in other Latin American countries. There is little present interest in further evaluations of *Axonopus, Eragrostis, Hyparrhenia,* and *Melinis*. Generally, the most water-tolerant species are found in *Brachiaria, Echinochloa, Hyparrhenia, Paspalum,* and *Setaria*, while most drought-tolerant species are included in *Andropogon, Bothriochloa, Cenchrus, Panicum,* and *Pennisetum*. A few important species are not included

TABLE 3.1
List of Selected Tropical Grasses in Germplasm Banks

	Germplasm Bank								
	SRPIS		CSIRO		CIAT		ILCA		CENARGEN
Genus	Spp.[a,b]	Acc.	Spp.	Acc.	Spp.	Acc.	Spp.	Acc.	Acc.
Andropogon	12	41	24	106	4	115	6	45	94
Axonopus	4	24	4	5	3	4	—	—	88
Bothriochloa	22	696	13	214	2	20	—	—	—
Brachiaria	14	82	23	169	28	1035	27	658	417
Cenchrus	7	826	11	536	2	73	4	114	217
Chloris	25	247	23	193	6	60	7	104	24
Cynodon	12	499	7	70	4	41	3	107	38
Digitaria	41	661	46	425	9	30	13	53	20
Echinochloa	—	—	10	63	4	10	5	54	—
Eragrostis	—	—	17	140	7	56	13	59	6
Hemarthria	3	64	2	7	1	10	—	—	52
Hyparrhenia	—	—	9	59	13	58	8	38	—
Melinis	1	4	5	24	1	17	—	15	26
Panicum	51	756	58	632[b]	9	536	16	197	603
Paspalum	67	1488	50	339	9	85	9	63	1500
Pennisetum	24	658	20	334	8	66	18	210	82
Setaria	—	—	35	268[b]	7	53	8	65	20
Urochloa	6	45	7	217	5	24	7	33	—
Total	289	6091	364	3801	122	2293	144	1815	3187

[a] Spp. = number of species; Acc. = number of accessions. Total numbers of species and accessions may be somewhat misleading because of considerable duplication among the germplasm banks listed. Some collections have also increased to some extent since information was made available.
[b] Forage species only.

Source: Appreciation is extended to M. Spinks, USDA, SRPIS, Griffin, GA; B. C. Pengelly, CSIRO-ATFGRC, Brisbane, Australia; J. Hansen, ILCA, Addis Ababa, Ethiopia; and R. Schultze-Kraft, Universitat Hohenheim, Stuttgart, Germany for assisting with the compilations.

in Table 3.1. *Hymenachne amplexicaulis* (Rudge) Nees, was found to be a highly water-tolerant, seed-producing ecotype, adapted to ponded pasture management systems in Australia.[41] An example of a weedy grass that is dominant in many large areas of the tropics, and where different grazing management practices have been tried, is *Imperata cylindrica* (L.) Raeuschel. Another, less noxious, is torpedograss (*Panicum repens* L.) in Florida and elsewhere, where it has become naturalized. Torpedograss is well accepted by cattle during certain periods of the year.

Additional grass species, most likely in the genera listed in Table 3.1, will be available in the near future. For example, *Paspalum atratum* Swallen (atra paspalum), was released as cultivar 'Suerte' by the University of Florida in 1997.[42,43] It has entirely different growth and quality characteristics than does Bahia grass (*Paspalum notatum* Fluegge). There has been moderate success in developing seed-producing grasses having the characteristics of the better quality vegetatively planted cultivars.[44] Examples are *Digitaria milanjiana* (Endle) Stapf cultivars 'Jarra'[45] and 'Arnhem'[46] which were recently released in Australia. They have many of the same growth characteristics as Pangola digitgrass (*Digitaria eriantha* Steud.), but produce viable seed. Grasses in the *Bothriochloa* genus, although not highly productive, can survive overstocking and extreme sustained drought without noticeable decrease in plant populations. This provides sufficient cover to reduce soil erosion caused by loss of native vegetation, which has resulted from overstocking.

IV. IMPORTANT GRASS SPECIES

The following are synopses of some of the more important grass species presently being used in the tropics and subtropics. For a more detailed description, the reader is referred to Chippendahl,[47] 't Mannetje and Jones, [34] Kretschmer and Pitman,[32] and to the books previously cited. In many instances, a single genotype has been developed and given a cultivar name by the developing country. Thereafter, the same genotype has sometimes been given new cultivar names by other countries when the grass or legume becomes successful.

Andropogon gayanus Kunth (gambagrass) is a tropical (about 15° N to 25° S), tufted, rhizomatous African species with a short day flowering response. It can reach a height of about 3 m. The most important of four recognized botanical varieties is *bisquamulatus*. The cultivar 'Planaltina' from Brazil and cultivar 'Carimagua' from Colombia are different names for the same genotype simultaneously released in 1980. This is the most important genotype of the species because of its widespread use (with other synonymous cultivar names in other countries) in most Latin American countries.[48-50] It associates well with some twining or erect/branching tropical legumes. It is best adapted to warm areas with 400 to 1500 mm rainfall and can withstand long dry seasons, while tolerating flooding, fires, and acid and low fertility soils. It does not grow well on heavy clay soils. It is better adapted to subhumid climates. Thus, the niche for gambagrass is probably limited to the subhumid areas with low-input systems. There is at least one large collection of gambagrass accessions that may contain a majority of its diversity. Because leaf-to-stem ratios decrease during the growing season, crude protein (CP) and digestibility decline substantially. This results in ranges of about 2 to 13% CP and 40 to 55% digestibility. In the areas where grown, gambagrass P and Ca concentrations have been 0.08 to 0.14 and 0.27 to 0.39%, respectively.

The genus *Brachiaria* is one of the most important seeded grass genera for primarily tropical subhumid and humid regions. It includes nearly 100 species. Species of this genus have been widely used in high and seasonal rainfall areas, with some species still favored by growers. There are large numbers of accessions held in germplasm banks at CIAT, CSIRO, ILCA, and EMBRAPA. Evaluation of the extensive collections of various species of this genus is still in progress.[51-53] Except for *B. ruzizienses* Germain and Evrard, which is cross-pollinated, all other named species are apomictic.

Brachiaria brizantha (A. Rich.) Stapf (palisadegrass) is a tufted, prostrate, semierect, short-rhizomed perennial. It can reach a height of 30 to 200 cm. Sometimes *B. brizantha* is difficult to distinguish from *B. decumbens* Stapf, which has a similar area of adaptation of humid to subhumid climate, acid to slightly acid soils, and dry seasons up to five months. It does not tolerate poorly drained soils and requires more fertile soils than does *B. decumbens*. Palisadegrass can be severely damaged by spittlebugs (*Aeneolamia, Deois,* and *Zulia* species), however, cultivar 'Marandu' from Brazil is reported to be resistant. Cultivar 'La Libertad' is a subsequent release from Colombia. Palisadegrass is palatable to cattle but not to horses. The CP and digestibility vary from about 4 to 19 and 50 to 75%, respectively, for 12 and 2 weeks of regrowth, respectively. Annual liveweight gain of steers has been 400 to 500 kg ha^{-1} at stocking rates of 1.5 and 2.5 steers ha^{-1} in dry and rainy seasons, respectively. It has higher forage quality than does *B. decumbens* or *B. humidicola* (Rendle) Schwiek. Presently, palisadegrass is preferred over *B. decumbens* and *B. humidicola* by Brazilian ranchers.

Brachiaria decumbens Stapf (signalgrass) is a low to moderately high growing perennial, rhizomatous/stoloniferous grass that has the same general area of adaptation as *B. brizantha*. *B. decumbens* has lower forage quality than does *B. brizantha*. Signalgrass was the first of this genus to be widely used in Brazil, but its susceptibility to spittlebug attack caused growers to change to *B. humidicola*. Photosensitization of skin and hepatic disorders can be problems for young cattle, goats, and sheep when signalgrass is grazed as a pure stand. Signalgrass thrives best with annual rainfall above about 1250 mm and a maximum dry season of 5 months. It can grow well on acid, highly Al-saturated soils or on fertile soils. It tolerates high stocking rates and is competitive with weeds and associated legumes. Little variation exists among accessions.[27,51-53] 'Basilisk,' released in 1967 (Australia), 'IPEAN,' released in the 1960s (Brazil), and cultivars from several other countries released from 1987 to 1991 probably come from one genotype (this is only one of many examples of several cultivars representing one genotype).

Brachiaria dictyoneura (Fig. & De Not.) Stapf is distributed in southern and eastern Africa. It is densely tufted and sparingly stoloniferous with short rhizomes and has many of the characteristics of *B. humidicola*. Slightly higher in nutritive value than *B. humidicola*, it is less stoloniferous and less competitive. Both are adapted to the humid and subhumid tropics and to a wide range of soils, including acid/Al-saturated soils. Association with legumes is reported to be better with *B. dictyoneura* than with *B. humidicola*. Also, *B. dictyoneura* recovers from spittlebug damage faster than *B. decumbens*. The genotype released as cultivar 'Llanero' from Colombia and under other names in other countries represents the limited variability among the few available accessions. It has recently been suggested that the released variety should be classified as *B. humidicola*.[54]

Brachiaria humidicola (Rendle) Schweik. (creeping signalgrass, koroniviagrass) is a highly stoloniferous and prostrate grass. The cultivar 'INIAP-NAPO 701' released in 1983 (Ecuador) is probably the same genotype as CIAT 679, Tully (Australia), and other cultivars. It was used in Brazil as a replacement for *B. decumbens* because of the devastating damage done by spittlebug. *B. humidicola* recovers more rapidly from spittlebug damage than does *B. decumbens*. It is very competitive with weeds and with associated legumes and can withstand heavy grazing pressure, but has the lowest nutritive value of the *Brachiaria* species discussed here. The CP concentrations of from 4 to about 7% may limit its value, although digestibility is reported to range from about 50 to 70%. The cultivar 'Tully' has been used successfully in the humid tropics in Australia. Because of its aggressiveness, *B. humidicola* is being used extensively in spite of its low nutritive value. Growth rate varies from about 150 to 500 g steer^{-1} day^{-1}, however, because of its high carrying capacity, beef gain per hectare is good.

Brachiaria mutica (Forssk.) Stapf (paragrass) is a highly palatable grass that is widely distributed in tropical and subtropical areas with annual rainfall of about 1500 mm or more, or in swampy areas receiving less rainfall. It has long stolons with short rhizomes and ascending or decumbent stems. It grows well in flooded areas (to less than 1 m of water), but is very sensitive to frost and cool temperatures. Nutritive value of foliage decreases as leaf-to-stem ratios decrease with plant development to maximum heights of 1 to 2 m. The CP ranges from about 10 to 14% and 3 to 6% for leaves and stems, respectively, and whole plant digestibility is from about 40 to 65%. Because of competitiveness of the grass, association with legumes is difficult. Paragrass normally is planted vegetatively because seed yields are low. In Australia, however, seed are combine harvested,[54] and seed was recently harvested in Florida. Variability of the species is believed to be slight. *Eriochloa puctata* (L.) Desv. ex Hamillt. (*Eriochloa polystachya* Kunth, caribgrass) and paragrass have similar growth habits and areas of adaptation and are difficult to distinguish without the presence of inflorescences.

Brachiaria ruziziensis Germain & Evrard (ruzigrass, congograss) has its origin in the Ruzizi Valley in eastern Zaire and Burundi. Although cultivar 'Kennedy' is a palatable, leafy perennial, it has not been used to any great extent in its area of adaptation, possibly because it does not tolerate low fertility soils.

Other *Brachiaria* species include *B. arrecta* (Hack. ex T. Durand and Schinz) Stapf (syn. B. radicans), tannergrass, from tropical Africa. It has been planted in the past in Latin America, but is out of favor because of possible toxicity to grazing animals and the availability of newer species. *B. distachya* (L.) Stapf originated in the Indian subcontinent, southeast Asia, Australia, and the Pacific Islands. *B. distachya* is one of the few annuals of the species being used for grazing. *B. subquadripara* (Trin.) Hitchc., corigrass, of tropical Asian origin, is used as a forage in coconut plantations.

Chloris gayana Kunth (Rhodes grass) is a variable species of diploid and tetraploid genotypes that has been very successful in the subtropics, yet has not been widely used in Latin America. Rhodes grass has been cultivated since the 1890s in southern Africa, where seed of what was to become cultivar 'Pioneer' was distributed to New South Wales, Australian farmers in 1902. By 1905, it had spread to Queensland, Australia. This diploid cultivar, or close relatives (it crosses easily), is now naturalized in much of subhumid Queensland. It is very persistent but does not have many of the more favorable attributes of later cultivars. 'Callide,' 'Samford,' and 'Boma' are tetraploids, and 'Pioneer,' 'Katambora,' and 'Bell' are diploids. All are tolerant of drought. Callide is very popular with growers in Australia at the present time. Cultivar Bell (Rhodes grass scale tolerant) has been used in Texas for many years.

Callide Rhodes grass stems can extend from 0.5 to 2 m high. It is a perennial, stoloniferous, sometimes tufted grass that has more cold tolerance than many other tropical grass species. The available cultivars respond very well to N fertilization. Rhodes grass can survive dry seasons of up to about 6 months, yet can tolerate periodic flooding or waterlogging. Also, it is tolerant of fire, tolerates high saline conditions,[55] and grows on most fertile soils with soil pH above about 4.5. Introduced to southern Florida in 1989, Callide was found to be the fastest growing grass between about November and March when adequately supplied with N fertilizer. Also, it has a good balance between CP and digestibility, with CP up to about 18 to 20% and digestibility up to above 70%. Its regrowth, even after foliage-killing frosts, is faster than other warm-season perennial grasses. Callide will not compete well with Bahia grass or common Bermuda grass. A variety of legumes will associate well with Callide. Seed production, seedling vigor, and seedling recruitment characteristics are good.

Cynodon dactylon (L.) Pers. (Bermuda grass) is a common weed with worldwide distribution. This rhizome-developing species is better adapted to moderately drained soils than wetter environments. It is being used for grazing in the higher latitude subtropics, for example, the U.S. Gulf Coast states. Besides the common type, there have been releases of similar, but higher yielding types. In the 1940s a hybrid cultivar, 'Coastal,' was developed and has been used in many areas of the world along with many other hybrids, such as 'Coastcross 1,' (1967). Of the more than 20 cultivars officially released in the U.S., the most widely used varieties are common, 'Coastal,' 'Tifton 44,' 'Tifton 78,' and 'Tifton 85.'[54]

Cynodon aethiopicus Clayton & Harlan, *C. nlemfuensis* Vanderyst variety *nlemfuensis* and variety *robustus* Clayton & Harlan, and *C. plectostachyus* (K. Schum.) Pilger are of east African origin. All have been called star grass or African star grass.[56,57] They differ morphologically from Bermuda grass because they lack rhizomes. They are stoloniferous, vigorous, and respond well to high soil fertility. They are planted vegetatively, are drought tolerant, competitive, and widely adapted to tropical and subtropical regions with more than about 650 mm annual rainfall. Much of the literature naming *C. plectostachyus*, prior to about 1970, refers to *C. nlemfuensis*.

C. nlemfuensis was introduced to Puerto Rico and distributed throughout Latin America. It is used for grazing, particularly in Mexico and Central America, and in Florida in the U.S. The University of Florida has released cultivars 'McCaleb' (*C. aethiopicus*),[58] 'Ona,'[59] 'Florico,'[60] and 'Florona'[61] (all *C. nlemfuensis*). Star grass also has been used successfully in South Africa under high fertilizer inputs and intense grazing regimes. Nutritive value of the star grasses depends on the age of regrowth since leaf-to-stem ratios decrease rapidly in the summer after about 10 weeks of growth. Thus, stocking to utilize only the upper leafy canopy can produce excellent animal performance, while grazing management producing utilization of mature stems results in lower animal responses.[62]

Digitaria eriantha Steud. (digitgrass) is a recently expanded species from taxonomic revisions in 1981[63] and 1984.[64] It now includes the former important stoloniferous, subtropical species *D. decumbens* Stent with cultivars 'Pangola,'[65] 'Transvala,'[66] and 'Slenderstem,'[67] and *D. pentzi* Stent. cultivar 'Taiwan.'[68] A hybrid cultivar, 'Survenola,' was developed in Florida by Schank et al.[69] A tufted type, *D. smutsii* Stent, is represented by Australian cultivars 'Premier'[70] and 'Advance.'[44] Premier and Advance produce viable seed, while the others are planted vegetatively. Because of the high quality of these grasses, an effort was made to develop a stoloniferous digitgrass that produced viable seed. Using *D. milanjiana* (Rendle) Stapf, an erect or sometimes stoloniferous or pseudorhizomatous type, cultivar 'Mardi,' was released in Malasia; however, it failed to produce adequate amounts of seed. In 1993, a planting of cultivar 'Jarra,' another genotype from Australia, was successfully harvested for seed. Jarra has some of the characteristics of Pangola, as do cultivars 'Arnhem' and 'Strickland.'[46,54] Because of the variability of the species, the potential for developing other genotypes suitable for diverse environmental regions is strong. At the present time, most of the digitgrasses in Latin America and elsewhere in the tropics, except for small holdings, are hardly used.

Echinochloa polystachya (H.B.K.) Hitch. (alemangrass) is a robust, perennial subaquatic or aquatic species that is vegetatively planted. It is native of Louisiana and Texas, through the West Indies to Buenos Aires, Argentina. Alemangrass was grazed primarily as a naturalized species until the last decade or so when isolated plantings were made. It has coarse, erect, stoloniferous culms 1 to 2 m long. Alemangrass can form extensive colonies in flooded or swampy areas, but also can survive better-drained soils in areas with 1000 mm or more of well-distributed annual rainfall. It has been grazed in Florida and used experimentally to accelerate drying of colloidal phosphate settling ponds from central Florida's phosphate industry. A cultivar, 'Amity,' was released in Australia and is one of two grasses replacing or complementing paragrass in the ponded pasture management system in Queensland, Australia.[41,71] Little variation exists within alemangrass. *E. colona* (L.) Link (jungle rice) and *E. crus-galli* (L.) Beauv. (barnyard millet) are used, but these can become weed pests in rice fields.

Hemarthria altissima (Poir.) Stapf & Hubbard (limpograss) is a perennial, stoloniferous/rhizomatous, and variable species believed to have originated in southern Africa. It can be found (native/naturalized) from the Rio Grande river area of Texas, through Mexico down to Paraguay, Bolivia, and Brazil. It is also found in Italy, Turkey, and Australia. Introductions from collections made in 1964 and 1971 by Oakes were widely distributed throughout Latin America.[72]

Depending on genotype, limpograss can reach a height of about 1 m or more. It is tolerant of waterlogging, yet survives droughts. It is well suited for the subtropics because of good growth in cool weather and survival in areas where several annual frosts occur. In 1978, three cultivars, 'Redalta,' 'Greenalta,' and 'Bigalta,' were released by the University of Florida,[73,74] and in 1984 cultivar 'Floralta' was released.[75] Of these, Redalta is the shortest with the lowest nutritive value and the most cold tolerance. Greenalta was not very vigorous and was seldom planted. Bigalta, a tetraploid, is the tallest but has highest digestibility. Bigalta did not persist well under the grazing management typically used. Floralta is preferred at the present time because it persists and competes better than Bigalta, however, its nutritive value is considerably less.[74,76] The main area of adaptation in Florida is in the peninsula. The *Hemarthria* genus is comprised of 12 species[39,77] of which species such as *H. compressa* (L. F.) R. Brown and *H. siberica* are better adapted to temperate climates than is *H. altissima*. This is a genus that warrants further evaluation for the subtropics.

Hymenachne amplexicaulis (Rudge) Nees (hymenachne) is widely distributed in tropical America (and naturalized in south Florida). Hymenachne cultivar 'Olive' was released in 1988 in Australia for use in ponded pasture systems. This robust rhizomatous perennial has erect or ascending stems up to about 2 m. It produces harvestable seed but is not as acceptable to cattle as is Amity alemangrass. It is reputed to be less drought tolerant than paragrass.[41,71]

Hyparrhenia rufa (Nees) Stapf (jaraguágrass) is a tufted perennial species with some annual varieties. Stems grow to about 2.5 m high. It has been planted in most Latin American countries by seed. It is adapted to seasonally dry climates (up to about 5 months) with annual rainfall of about 1000 mm or more. Digestibility and CP concentrations drop drastically as the dry season progresses. Presently, it is being replaced by more productive grasses of higher nutritive value.

Panicum maximum Jacq. (guinea grass) is a very diverse, seed-producing species with about a dozen botanical varieties. It is native to fertile soils of Africa. Guinea grass is widely distributed in the tropics and subtropics with annual rainfall of about 900 mm or more. Guinea grass can survive long droughts and grows well when soil pH is above 5.0, even on soils high in Al and Mn. Guinea grass can be used for "cut and carry" feed and most cultivars can also be grazed successfully, if not overstocked. It is better adapted to fertile, non-waterlogged conditions. Guinea grass can become a weed in citrus. Its high quality makes it useful as a "fattening" grass in the tropics. 'Hamil,' 'Riverside,' 'Makueni,' and 'Gatton' are Australian cultivars, while 'Colonao' and 'Vencedor' are Brazilian releases.[54]

'Petrie' (green panic), (*P. maximum* var. *trichoglume* Eyles), is a shorter bunch type that is excellent for grazing and can survive annual rainfall as low as 560 mm. Green panic survives light frosts and grows well on all but deep sand and heavy clay soils, and it can also tolerate heavy grazing pressures.

Paspalum atratum (atra paspalum) is a newly evaluated species from moist or low areas of Mato Grosso and surrounding states of Brazil into Bolivia. It is a perennial and grows up to 2 m high. It has recently been introduced to Florida and has been evaluated in Brazil.[42,43,78] This apomict tolerates waterlogged soils and temporary flooding, but does not grow rapidly during the cool season (December to March in south Florida). Foliage is damaged by frost. The Florida introduction has been described[42] and was released as cultivar 'Suerte.'[43] There are several closely related species that may permit development of other cultivars for diverse environments.

Paspalum notatum Fluegge (Bahia grass) is one of the few widely used grasses native to Latin America. It is now found in almost all tropical and subtropical regions and even in some warm temperate areas. It is primarily used as a pasture or lawn grass. Bahia grass is low in forage quality, rhizomatous, creeping, and very persistent, even under intense grazing pressure and on low fertility soils. It can also respond to nitrogen fertilization and retain substantial amounts of nitrogen longer than do most highly productive tropical grasses.[79] Because of low animal preference and low palatability, it has become a weed in better quality pastures in Florida and elsewhere. In spite of its lower quality, Bahia grass comprises about 85% of the improved pastures in Florida because of its long-term persistence. It is a satisfactory grass for mature cattle, but is less desirable for growing animals. After about 4 to 5 months into the growing season, digestibility can fall to 40% or less, even though CP concentrations may remain above 5 to 7%. The decline appears to begin and to be associated with the advent of seed production. The most widely distributed cultivar is 'Pensacola' from Florida, with 'Argentine' and 'Paraguayan' being more recently planted.[21] More productive selections of Pensacola Bahia grass have been made in Georgia with the cultivar 'Tifton 9' resulting.[80] Several tropical legumes can associate with Bahia grass; however, in Florida the most persistent legume has been 'Florida' carpon desmodium [*Desmodium heterocarpon* (L.) DC.], a perennial that has persisted about two decades in Bahia grass pastures in south Florida.[81,82] *Vigna parkeri* Baker has also persisted well in Bahia grass pastures.[83] Persistence of the annuals, *Aeschynomene americana* and *Alysicarpus vaginalis* (L.) DC., has been less certain.[79]

Pennisetum ciliare (L.) Lam (synonym, *Cenchrus ciliaris* L.), buffelgrass, is a deep-rooted, drought-tolerant perennial that is not suitable to wetter soil types. It has been an important forage grass in the southwestern U.S., northern Mexico, and Australia since the late 1940s and early 1950s. There are about 20 strains or genotypes that have been released by these and other countries. The most widely used cultivars presently are: common buffelgrass, which is called American in Australia, 'Biloela' (1956, Australia), 'Gayndah' (1930s, Australia), 'Molopo' (1950s, South Africa), 'Nueces' (1977, U.S.), 'Nunbank' (1961, Australia), and 'Western Australian' (1910, Australia).[84]

Pennisetum clandestinum Hochst. ex Chiov. (Kikuyu grass) is a subtropical grass, naturally occurring across the plateau of East and Central Africa. It has spread to most of the humid, 1000 to 3000 m highlands of the world, where it can become a weed in vegetable growing areas. It is a low-growing stoloniferous/rhizomatous grass. The Australian cultivars 'Breakwell,' 'Crofts,' 'Noonan,' and 'Whittet' produce viable seed. The common type has been planted in the largest area in Australia. There is little demand for this grass in Latin America. Kikuyu grass is not adapted to the lowland tropics including peninsular Florida. This might be expected since its optimum growth range is from about 25°C day and 20°C night temperatures.

Pennisetum purpureum Schumach (elephant grass, napier grass) is a robust African grass that has been introduced to all tropical areas of the world. It has naturalized in many of them. There are about 25 cultivars and strains and about 16 hybrids [*P. purpureum x P. glaucum* (L.) R. Br.] selected for forage value. The typical erect type (i.e., cultivar 'Merker'), with short rhizomes, has stems that can reach a height up to about 7 m with a stem diameter up to about 3 cm. This grass probably has been the most important grass in small farm "cut-and-carry" management systems primarily for dairy cattle. Persistence is good under this system if annual rainfall is about 1000 mm or more (without a long dry season). This grass is difficult to manage under grazing. Napier grass is usually planted vegetatively because of low or no seed production. Because of its deep root system, grass vigor can be markedly reduced under prolonged waterlogging. Napier grass grows vigorously in the subtropics and tropics in high fertility soils. The quality of napier grass, regardless of fertilizer input, depends upon the leaf-to-stem ratio, which decreases as the regrowth interval increases. It is not uncommon to have 12 to 20% CP in leaves and a digestibility up to 70%, but digestibility of stems is low. Early assessments of intensive management of napier grass were presented by Vincente-Chandler et al.[85,86]

The University of Florida developed a dwarf elephant grass cultivar, 'Mott,' from a Georgia selfed (self fertilized) progeny of cultivar 'Merkeron.' It has many of the desired attributes of a perennial grass. Because of its relatively short height, leaf-to-stem ratios are good even after 10 weeks of regrowth. Unfortunately, it does not produce seed, so it must be planted vegetatively.[87] It also can be difficult to maintain stands under grazing.

Setaria sphacelata (Schumach.) Stapf & C. E. Hubb. ex Chipp. variety *sericea* (Stapf) Clayton (setaria, golden timothygrass) is from tropical and subtropical Africa. It was first planted as a pasture grass in Kenya and later in most of the humid tropics of the world. Additional subspecies of this highly variable genus are *Setaria sphacelata* (Schumach.) Stapf & C. E. Hubb. ex Chipp. variety splendida (Stapf) Clayton (giant setaria) and *Setaria sphacelata* (Schumach.) Stapf & C. E. Hubb. ex Chipp., variety sericea x variety splendida hybrid (splendagrass, tetraploid). Propagation of this tufted (rarely rhizomatous) grass is by seed or vegetatively. It has erect stems to 3 m high. Setaria tolerates temporary waterlogging and light frost, is compatible with numerous legumes, and persists with annual rainfall above about 1000 mm. It is not well adapted to acid soils below about pH 5.0 or to alkaline soils. Prominent Australian cultivars of setaria are 'Nandi' (diploid), 'Narok,' 'Kazungula,' and 'Solander' (tetraploids).[54] High oxalate concentrations may reduce animal intake. Kazangula appears to be the popular cultivar in more moist areas of Brazil. Setaria tolerates temporary high stocking rates but rotational grazing is recommended.[88] In Florida, setaria does not form a dense sward or persist well.

V. SUMMARY AND CONCLUSIONS

Commercially useful tropical grass cultivars are now widely distributed and readily available in most areas. Even widely distributed commercial varieties can become unavailable rather quickly. The present sophisticated global resource preservation and enhancement system has the necessary technological base to maintain the existing collections of more than 15,000 tropical grass and 50,000 tropical legume accessions. Recent trends suggest that the value of these resources cannot be taken for granted, and continuing support of international research and development interests must be actively maintained. Many of the available genotypes have not been adequately evaluated. Thus, there is less need for general collection of tropical forage germplasm, although the threat of complete loss of some genotypes makes some specific collection rather urgent. Because the available variability within species has increased as collections increased, many species should be reevaluated using the broader germplasm base. Emphasis on further collection should generally be restricted to species in genera that have already yielded viable cultivars, or to genera with large numbers of uncollected species. In a few instances, additional collections should be made of species that have already been developed into cultivars.

When large numbers of different species are being evaluated simultaneously, or even with individual species where within-species variability is large, it is imperative that grazing animals be utilized for evaluation as early as possible to determine acceptability of the various plant genotypes to grazing animals and plant response to grazing. Newer evaluation methods can be used to reduce the cost and time of cultivar development.

Substantial international support was generated for germplasm collection and development of suitable storage facilities during the 1970s and 1980s. Along with the lagging evaluation of the extensive germplasm collections currently available, a need exists for widespread international commitment to the effective maintenance of both the genetic material and its availability. Some equitable means of supporting these efforts must be developed, especially considering the current financial constraints on national and international research organizations. Otherwise, the potentially useful materials that have been collected and stored will become unavailable for evaluation, especially in many tropical areas where economic constraints may restrict even modest investment in uncertain germplasm. Seed multiplication, international correspondence, and especially free availability of small quantities of seed of the diverse germplasm are essential for continuing progress in evaluation and development of forages throughout the tropics. Unfortunately, these assets of existing germplasm programs were taken for granted during the past few decades of relatively good financing. Means of continuing such policies now merit international attention. Considering the successes attained through evaluation of only a small portion of tropical grass and legume germplasm under only a portion of the available environments and management levels, the potential for pasture development in the tropics and subtropics is substantial.

REFERENCES

1. Hanson, J., Getting the most out of ILCA's forage gene bank, *ILCA Newsletter*, 10, 4, 4, 1991.
2. Clements, R. J. and Cameron, D. G., Eds., *Collecting and Testing Tropical Forage Plants*, CSIRO, Melbourne, Australia, 1980.
3. Reid, R., Collection and use of climatic data in pasture plant introduction, in *Collecting and Testing Tropical Forage Plants*, Clements, R. J. and Cameron, D. G., Eds., CSIRO, Melbourne, Australia, 1980.
4. Gibbens, A., Saving seeds for future generations, *Science*, 254, 804, 1991.
5. Davies, W. E., A global plan for forage genetic resources, *Proc. XV Int. Grassland Congr.*, Kyoto, Japan, 1985, 107.
6. Anon., Sexta reunion del comite de la red internacional de evaluacion de pastos tropicales (RIEPT) en Veracruz, Mexico, *Pasturas Tropicales*, 11, 1, 29, 1989 (in Spanish).
7. Anon., *Seeds for Our Future*, The U.S. National Plant Germplasm System, Program Aid 1470, USDA, ARS, Washington, D.C., 1990.
8. Asay, K. H., Contributions of introduced germplasm in the development of grass cultivars, in *Use of Plant Introductions in Cultivar Development* (Part 1), Shands, H. L. and Wiesner, L. E., Eds., Crop Science Society of America, Madison, WI, 1991.
9. Brown, A. H. D., Core collections: A practical approach to genetic resources management, *Genome*, 31, 818, 1989.
10. Clayton, W. D., Tropical grasses, in *Genetic Resources of Forage Plants*, McIvor, J. G. and Bray, R. A., Eds., CSIRO, Melbourne, Australia, 1983, 39.
11. Cohen, J. I., Williams, J. T., Plucknett, D. L., and Shands, H., *Ex situ* conservation of plant genetic resources: global development and environmental concerns, *Science*, 253, 965, 1991.
12. Jones, R. J. and Walker, B., Strategies for evaluating forage plants, in *Genetic Resources of Forage Plants*, McIvor, J. G. and Bray, R. A., Eds., CSIRO, Melbourne, Australia, 1983, 185.
13. Hacker, J. B., Priorities and activities of the Australian Tropical Forages Genetic Resource Center, *Trop. Grassl.*, 31, 243, 1997.
14. National Research Council, *Managing Global Genetic Resources, The U. S. National Plant Germplasm System*, National Academy Press, Washington, D.C., 1991.
15. Stalker, H. T. and Chapman, C., Eds., *Scientific Management of Germplasm: Characterization, Evaluation, and Enhancement*, IBPGR Training Course: Lecture Series 2, IBPGR, Rome, 1989.
16. Reid, R., Forage genetic resources — their national and international importance, *Trop. Grassl.*, 31, 251, 1997.

17. Anon., Forage genetic resources: Meeting the requirements of industry, *Trop. Grassl.*, 31, 241, 1997.
18. Gonzalez, B., Generalidades, in *FUSAGRI Pastos, Serie Petroleo y Agricultura*, Maraven, Petroleos de Venezuela, 1986, 9 (in Spanish).
19. Mott, G. O., Ed., *Handbook for the Collection, Preservation, and Characterization of Tropical Forage Germplasm Resources*, University of Florida and CIAT, Cali, Colombia, 1979.
20. Serrao, E. A. S. and Neto, M. L. S., The adaptation of tropical forages in the Amazon region, in *Tropical Forage in Livestock Production Systems*, Doll, E. C. and Mott, G. O., Eds., Special Pub. No. 24, American Society of Agronomy, Madison, WI, 1975, 31.
21. Kretschmer, A. E., Jr. and Hood, N. C., *Paspalum notatum* in Florida, USA, in *Forage Seed Production.* Vol. 2. *Tropical and Subtropical Species*, Loch. D. S. and Ferguson, J. E., Eds., CABI Publishing, Wallingford, Oxon, U.K., 1999, 335.
22. Blumenstock, D. I., Distribution and characteristics of tropical climates, *Proc. 9th Pacific Sci. Congr.*, 20, 3, 1957.
23. McWilliam, J. R., Response of pasture plants to temperature, in *Plant Relations in Pastures*, Wilson, J. R., Ed., CSIRO, Melbourne, Australia, 1978, 17.
24. Reid, R. and Strickland, R. W., Forage plant collection in practice, in *Genetic Resources of Forage Plants*, McIvor, J. G. and Bray, R. A., Eds., CSIRO, Melbourne, Australia, 1983, 149.
25. Vanderborght, T., *Wild Phaseoleae — Phaseolinae Collection: IBPGR Base Collection for Wild Phaseolus and Vigna Species*, B-1860, Natural Botanic Garden, Belgium, (undated).
26. Anon., *Forage and Browse Plants for Arid and Semi-Arid Africa*, IBPGR, Rome, 1984.
27. Loch, D. S. and Ferguson, J. E., Eds., *Forage Seed Production* Vol. 2. *Tropical and Subtropical Species*, CABI Publishing, Wallingford, Oxon, U.K., 1999.
28. Bogdan, A. V., *Tropical Pasture and Fodder Plants*, Longman, New York, 1977.
29. Cooper, J. P., Potential production and energy conversion in temperate and tropical grasses, *Herb. Abstr.*, 40, 1, 1970.
30. Eyles, A. G., Cameron, D. G., and Hacker, J. B., *Pasture Research in Northern Australia — Its History, Achievements and Future Emphasis*, Res. Rep. 4, CSIRO Div. Tropical Crops and Pasture, CSIRO, Brisbane, Australia, 1985.
31. Humphreys, L. R. and Riveros, F., *Tropical Pasture Seed Producion*, FAO Plant Production and Protection Paper 8, FAO, Rome, 1986.
32. Kretschmer, A. E., Jr. and Pitman, W. D., Tropical and subtropical forages, in *Forages, the Science of Grassland Agriculture*, 5th ed., Barnes, R. F., Miller, D. A., and Nelson, C. J., Eds., Iowa State University Press, Ames, IA, 1994, 283.
33. Loch, D. S., Tropical herbage and seed production — origins, progress, and opportunities, *J. Appl. Seed Prod.*, 9, 14, 1991.
34. 't Mannetje, L. and Jones, R. M., Eds., *Plant Resources of South-East Asia*, Purdoc Sci. Publ., Wageningen, The Netherlands, 1992.
35. Minson, D. J., Nutritional differences between tropical and temperate pastures, in *Grazing Animals*, Morley, F. H. W., Ed., Elsevier Sci. Publ., Amsterdam, 1980, 143.
36. Minson, D. J., *Forage in Ruminant Nutrition*, Academic Press, London, 1990.
37. O'Reilly, M. V., *Better Pastures for the Tropics*, Arthur Yates, Rockhampton, Queensland, Australia, 1975.
38. Skerman, P. J. and Riveros, F., Eds., *Tropical Grasses*, FAO, Rome, 1990.
39. Watson, L. and Dallwitz, M. J., *The Grass Genera of the World*, CAB International, Wallingford, Oxon, U.K., 1992.
40. Zeven, A. C., *Dictionary of Cultivated Plants and their Regions of Diversity Excluding Most Ornamentals, Forest Trees and Lower Plants*, Center for Agricultural Publ. and Documentation, Wageningen, The Netherlands, 1982.
41. Anning, P. and Kernot, J., Eds., *North Queensland Beef Production Series 1. Ponded Pastures*, Queensland Dep. Primary Industries, Townsville, Queensland, Australia, 1989.
42. Kretschmer, A. E., Jr., Kalmbacher, R. A., and Wilson, T. C., Preliminary evaluation of atra paspalum: a high quality, seed producing perennial forage grass for Florida, *Soil Crop Sci. Soc. Florida Proc.*, 53, 60, 1994.
43. Kalmbacher, R. S., Brown, W. F., Colvin, D. L., Dunavin, L. S., Kretschmer, A. E., Jr., Martin, F. G., Mullahey, J. J., and Rechcigl, J. E., *'Suerte' Atra Paspalum, Its Management and Utilization.* Florida Agric. Exp. Stn. Circ. S-397, Gainesville, FL, 1997.

44. Hacker, J. B., Breeding tropical grasses for ease of vegetative propagation and for improved seed production, *Proc. XV Int. Grassland Congr.*, Kyoto, Japan, 1985, 251.
45. Oram, R. N., A. Grasses, 23. *Digitaria*, (b). *Digitaria milanjiana* (Rendle) Staph. (finger grass) cv. Jarra. Queensland Dep. Primary Industries, Australia, *Trop. Grassl.*, 27, 55, 1993.
46. Oram, R. N., A. Grasses, 23. *Digitaria*, (b). *Digitaria milanjiana* (Rendle) Staph. (finger grass), cv. Arnhem. Northern Territory Dep. Primary Industries, Australia, *Trop. Grassl.*, 31, 376, 1997.
47. Chippendahl, L. K. A., A guide to the identification of grasses in South Africa, in *The Grasses and Pastures of South Africa,* Meredith, D., Ed., Cape Times Ltd., Parow, C. P., South Africa, 1959, 7.
48. Toledo, J. M., Vera, R., Lascano, C., and Lenne, J. M., *Andropogon gayanus Kunth: A Grass for Tropical Acid Soils*, CIAT, Cali, Colombia, 1990.
49. Mejia, M. *Andropogon gayanus Kunth: Bibliografia Analitica*, CIAT, Cali, Colombia, 1984 (in Spanish).
50. Ferguson, J. E. and de Andrade, R. P., *Andropogon gayanus* in Latin America, in *Forage Seed Production.* Vol. 2. *Tropical and Subtropical Species*, Loch, D. S. and Ferguson, J. E., Eds., CABI Publishing, Wallingford, Oxon, U.K., 1999, 381.
51. Heering, J. H., Initial evaluation of *Brachiaria* species, *Germplasm Newsletter*, No. 20, ILCA, Addis Ababa, Ethiopia, 1989.
52. De Souza, F. H. D., *Brachiaria* spp. in Brazil, in *Forage Seed Production.* Vol. 2. *Tropical and Subtropical Species*, Loch, D. S. and Ferguson, J. E., Eds., CABI Publishing, Wallingford, Oxon, U.K., 1999, 371.
53. Loch, D. S., *Brachiara decumbens* (signal grass) — a review with particular reference to Australia, *Trop. Grassl.*, 11, 141, 1977.
54. Loch, D. S. and Ferguson, J. E., Tropical and Subtropical Seed Production: an Overview, in *Forage Seed Production.* Vol. 2. *Tropical and Subtropical Species*, Loch, D. S. and Ferguson, J. E., Eds., CABI Publishing, Wallingford, Oxon, U.K., 1999, 1.
55. Taleisnik, E., Peyrano, G., and Arias, C., Response of *Chloris gayana* cultivars to salinity. 1. Germination and early vegetative growth, *Trop. Grassl.*, 31, 232, 1997.
56. Clayton, W. D. and Harlan, J. R., The genus *Cynodon* L. C. Rich. in tropical Africa, *Kew Bull.*, 24, 185, 1970.
57. Harlan, J. R., Cynodon species and their value for grazing and hay, *Herbage Abst.*, 40, 233, 1970.
58. Hodges, E. M., Boyd, F. T., Dunavin, L. S., Kretschmer, A. E., Jr., Mislevy, P., and Stanley, R. L., Jr., *'McCaleb' Stargrass*, Florida Agric. Exp. Stn. Circ. S-231, Gainesville, FL, 1975.
59. Hodges, E. M., Mislevy, P., Dunavin, L. S., Ruelke, O. C., and Stanley, R. L., Jr., *'Ona' Stargrass*, Florida Agric. Exp. Stn. Circ. S-268A, Gainesville, FL, 1984.
60. Mislevy, P., Brown, W. F., Caro-Costas, R., Vincente-Chandler, J., Dunavin, L. S., Hall, D. W., Kalmbacher, R. S., Overman, A. J., Ruelke, O. C., Sonoda, R. M., Sotomayor-Ríos, A., Stanley, R. L., Jr., and Williams, M. J., *Florico Stargrass*, Florida Agric. Exp. Stn. Circ. S-361, Gainesville, FL, 1989.
61. Mislevy, P., Brown, W. F., Dunavin, L. S., Hall, D. W., Kalmbacher, R. S., Overman, A. J., Ruelke, O. C., Sonoda, R. M., Stanley, R. L., Jr., and Williams, M. J., *Florona Stargrass*, Florida Agric. Exp. Stn. Circ. S-362, Gainesville, FL, 1989.
62. Pitman, W. D., *Management of Stargrass Pastures for Growing Cattle Using Visual Pasture Characteristics*, Florida Agric. Exp. Sta. Bull. 884, Gainesville, FL, 1991.
63. Kok, P. F. D., Notes on *Digitaria* in South Africa, *Bothalia*, 13, 457, 1981.
64. Kok, P. F. D., Studies on *Digitaria* (Poaceae) I. Enumeration of species and synonymy, *South African J. Bot.* 3, 184, 1984.
65. Hodges, E. M., Killinger, G. B., McCaleb, J. E., Ruelke, O. C., and Allen, R. J., Jr., *Pangola Digitgrass*, Florida Agric. Exp. Stn. Bull. 718A, Gainesville, FL, 1975.
66. Boyd, F. T., Schank, S. C., Smith, R. L., Hodges, E. M., West, S. H., Kretschmer, A. E., Jr., Brolmann, J. B., and Moore, J. E., *Transvala Digitgrass: A Tropical Forage Resistant to 1. Sting Nematode 2. Pangola Stunt Virus*, Florida Agric. Exp. Stn. Circ. S-222, Gainesville, FL, 1973.
67. McCaleb, J. E. and Hodges, E. M., *Slenderstem Digitgrass*, Florida Agric. Exp. Stn. Circ. S-201, Gainesville, FL, 1969.
68. Kretschmer, A. E., Jr., Allen, R. J., and Hodges, E. M., *Taiwan Digitgrass*, Florida Agric. Exp. Stn. Circ. S-258, Gainesville, FL, 1978.

69. Schank, S. C., Ruelke, O. C., Ocumpaugh, W. R., Moore, J. E., and Hall, D. W., *Survenola Digitgrass: A Tropical Forage Grass*, Florida Agric. Exp. Stn. Circ. S-292, Gainesville, FL, 1982.
70. Oram, R. J., Ed., *Register of Australian Herbage Plant Cultivars*, CSIRO, Queensland, Australia, 1990.
71. Wilden, J. H. and Chapman, D. G., *Ponded Pasture Systems — Capitalizing on Available Water*, Queensland Dep. of Primary Industries Bull. RQR 87006, Rockhampton, Queensland, Australia, 1987.
72. Oakes, A. J., Hemarthria collection from South Africa, *Turrialba*, 23, 37, 1973.
73. Quesenberry, K. H., Dunavin, L. S., Hodges, E. M., Killinger, G. B., Kretschmer, A. E., Jr., Ocumpaugh, W. R., Roush, R. D., Ruelke, O. C., Schank, S. C., Smith, D. C., Snyder, G. H., and Stanley, R. L., Jr., *Redalta, Greenalta, and Bigalta Limpograss, Hemarthria altissima, Promising Forages for Florida*, Florida Agric. Exp. Stn. Bull. 802, Gainesville, FL, 1978.
74. Christiansen, S., Ruelke, O. C., Ocumpaugh, W. R., Quesenberry, K. H., and Moore, J. E., Seasonal yield and quality of Bigalta, Redalta, and Floralta limpograss, *Trop. Agric. (Trinidad)*, 65, 49, 1988.
75. Quesenberry, K. H., Ocumpaugh, W. R., Ruelke, O. C., Dunavin, L. S., and Mislevy, P., *Floralta — A Limpograss Selected for Yield and Persistence in Pastures*, Florida Agric. Exp. Stn. Circ. S-312, Gainesville, FL, 1984.
76. Pitman, W. D., Machen, R. V., and Pond, K. R., Grazing evaluation of Bigalta and Floralta limpograss, *Crop Sci.*, 34, 210, 1994.
77. Clayton, W. D. and Renvoize, S. A., *Genera Graminum Grasses of the World*, Kew Bull. Addit. Series 13, London, 1986, chap. 1.
78. Quarin, C. L., Valls, J. F. M., and Urbani, M. H., Cytological and reproductive behaviour of *Paspalum atratum*, a promising forage grass for the tropics, *Trop. Grassl.*, 31, 114, 1997.
79. Pitman, W. D., Portier, K. M., Chambliss, C. G., and Krestchmer, A. E., Jr., Performance of yearling steers grazing bahiagrass pastures with summer annual legumes or nitrogen fertilization in subtropical Florida, *Trop. Grassl.*, 26, 206, 1992.
80. Burton, G. W., Registration of 'Tifton 9' Pensacola bahiagrass, *Crop Sci.*, 29, 1326, 1989.
81. Krestchmer, A. E., Jr., Brolmann, J. B., Snyder, G. H., and Coleman, S. W., *'Florida' Carpon Desmodium [Desmodium heterocarpon (L.) DC.], a Perennial Tropical Forage Legume for Use in South Florida*, Florida Agric. Exp. Stn. Circ. S-260, Gainesville, FL, 1979.
82. Pitman, W. D. and Kretschmer, A. E., Jr., *Carpon Desmodium for Peninsular Florida Pastures: Considerations for Establishment and Use*, Florida Agric. Exp. Stn. Circ. S-385, Gainesville, FL, 1993.
83. Pitman, W. D., Grass-legume interactions as affected by diverse tropical legume germplasm in the USA Gulf Coast States, in *Identifying Germplasm for Successful Forage Legume-Grass Interactions*, Springer, T. L. and Pittman, R. N., Eds., United States Department of Agriculture, Agricultural Research Service, Washington, D.C., 1996, 28.
84. Loch, D. S., *Cenchrus ciliaris* in Australia, in *Forage Seed Production*. Vol. 2. *Tropical and Subtropical Species*, Loch, D. S. and Ferguson, J. E., Eds., CABI Publishing, Wallingford, Oxon, U.K., 1999, 351.
85. Vincente-Chandler, J., Silva, S., and Figarella, J., The effect of nitrogen fertilization and frequency of cutting on the yield of: I, Napiergrass, II, Guineagrass, and III, Paragrass, *J. Agric. Univ. Puerto Rico*, 43, 215, 1959.
86. Vincente-Chandler, J., Caro-Costas, R., Pearson, R. W., Abruna, F., Figarella, J., and Silva, S., *El Maneja Intensivo de Forrajeras Tropicales en Puerto Rico*, Puerto Rico Agric. Exp. Stn. Bull. 202, Mayaguez, Puerto Rico, 1967 (in Spanish).
87. Sollenberger, L. E., Prine, G. M., Ocumpaugh, W. R., Hanna, W. W., Jones, C. S., Schank, S. C., and Kalmbacher, R. S., *Mott Dwarf Elephantgrass: A High Quality Forage with Potential in Florida and in the Tropics*, Florida Agric. Exp. Stn. Circ. S-356, Gainesville, FL, 1988.
88. Jones, R. J. and Jones, R. M., Liveweight gain from rotationally and continuously grazed pastures of Narok setaria and Samford rhodesgrass fertilised with nitrogen in southeast Queensland, *Trop. Grassl.*, 23, 135, 1989.

4 Germplasm Resources of Tropical Forage Legumes

Albert E. Kretschmer, Jr. and W.D. Pitman

CONTENTS

I. INTRODUCTION

The greatest diversity of tropical forage legume germplasm is in the American tropics. While native legume populations have undoubtedly contributed to forage for livestock, the common situation is for natural legume populations in tropical America to be more prevalent outside pastures than within them. Beginning with commercial use of the genus *Stylosanthes* in the 1930s,[1] Australia has led in the initial adoption and, somewhat in reverse, subsequently has led in collection, assessment, and evaluation of diverse tropical legume germplasm. During the last half of the 20th century, numerous tropical forage legume cultivars were developed. The initial important cultivars, which were subsequently distributed throughout tropical and subtropical regions of the world, were developed in Australia. This tropical forage legume cultivar development consisted primarily of the selection and commercialization of naturally occurring genotypes with very limited contribution from plant breeding.[2]

Early tropical forage legume cultivars were often selected from plot evaluations, where excessive emphasis was placed on forage production potential. Resulting insufficient tolerance to commercial grazing management, limited competitive ability with aggressive tropical grasses, and narrow areas of adaptation led to many failures with initial tropical pasture legume evaluations and commercial plantings. Nonetheless, vigorous early growth of the widely distributed cultivars repeatedly demonstrated the tremendous potential of tropical forage legumes. This potential continues to propel the search for tropical forage legumes that will hopefully be as widely useful as the tropical forage grass cultivars.

II. UNIQUE ASPECTS OF TROPICAL LEGUME GERMPLASM

In contrast to the coarse, unpalatable roughage typical of mature tropical grasses, many tropical legumes retain leaves of high forage quality throughout the growing season. Tropical legumes can enhance dietary protein, rate of passage, intake, and subsequent animal performance when added to tropical grass pastures. Thus, the primary objective of the incorporation of legumes into tropical

0-8493-2318-5/01/$0.00+$.50

grass pastures is to provide increased CP (crude protein) in the diet, since CP concentration of tropical grasses is often lower than that required for animal maintenance. As little as 10% legume in a 4% CP grass diet of sheep was found to overcome CP deficiency, and led to large increases in total herbage intake.[3] Increased rates of gain by young growing cattle have also been reported in response to low proportions of tropical legumes in grass pastures.[4] Also, legumes supply additional nitrogen to the pasture through recycling, thus increasing forage production and carrying capacity. Well-managed tropical grass/legume pastures can be highly productive and can be stocked at 3 to 4 animal units per hectare during the growing season when moisture is adequate.

Tropical legumes are a much more diverse group of plants than are the tropical grasses or the temperate legumes. Tropical legumes range from low-growing, prostrate types to trees and vines, which can grow to several meters in height. They also vary from herbaceous to woody growth forms. Many, especially some woody species, have limited usefulness as forage due to presence of thorns, accumulation of toxic secondary metabolic products, or simply production of foliage at a height beyond reach of livestock. Some tropical legumes produce little forage, with a high proportion of photosynthate translocated to extensive underground root and storage systems. Other species produce high proportions of lush top growth and fail to persist when repeatedly defoliated. Some tropical legumes thrive in waterlogged soils and others require well drained soils for survival. Some thrive primarily in arid locations. Some species of tropical legumes possess competitive advantages over other plants on highly infertile soils and even on soils with typically toxic levels of some elements. Some contrasting species of tropical legumes are productive and persistent only on fertile sites. Some are competitive with pasture grasses, and others rapidly succumb to such competition. Likewise, there are grazing-tolerant and susceptible types.

It has become evident that, in general, tropical legumes are less persistent than associated grasses under typical continuous grazing practices, and they are less widely adapted than the grasses. The common recommendation of inoculating the legume seed with rhizobial bacteria at initial planting to ensure effective nitrogen fixation can be a substantial complication to the planting process in remote areas. Fortunately, not all plantings fail due to lack of inoculation, since many tropical legumes can be effectively nodulated with a broad spectrum of rhizobial bacteria common to tropical soils where native legumes are present.

In general, the collection, storage, and initial evaluation of tropical legume germplasm is similar to that for the grasses. The above-mentioned association with nitrogen-fixing bacteria adds the consideration of collecting and assessing effectiveness of the legume-bacterial strain associations. The greater specificity of the legumes to edaphic and climatic conditions provides greater restrictions on suitability of sites for preliminary assessment and initial seed increase. Additional considerations with particular application to tropical legume germplasm collection, storage, and assessment have been presented by Bray,[5] Stace and Edye,[6] Kretschmer,[7] Sonoda et al.,[8] and Schultze-Kraft and Benavides.[9]

III. PRESENT AVAILABILITY OF GERMPLASM

There are about 18,000 species of *Leguminosae* with subfamilies *Caesalpinioideae, Mimosoideae,* and *Papilionoideae*. They are separated into 42 tribes and 651 genera. Although the primary site of evolution of legumes may be Africa prior to continental shifts, the majority of tropical legume genera and species are currently found in Latin America.[10,11]

The tropical forage legumes that have been released for commercial use are included in about 6 tribes. Many tropical legume species are not suited as forages because of their toxicity, antiquality, or woody unpalatable attributes. Almost all of the commercial tropical legumes are found in the subfamily *Papilionoideae*. Exceptions include leucaena [*Leucaena leucocephala* (Lam.) de Wit.] and desmanthus (*Desmanthus virgatus* Willd.) in *Mimosoideae*, and *Chamaechrista rotundifolia* (Pers.) Greene in *Caesalpinioideae*.

The diversity of tropical legume species found to be suitable as forage is much greater than that of temperate legumes, but their successful incorporation into grazing management systems is still not perfected. For example, Australian growers have had very good success with several tropical legumes that are very persistent. In contrast, there is almost no present demand for tropical legume seed for use in Brazil and elsewhere in Latin America in spite of the large number of native American species. Many native tropical legumes in these countries are inadvertently grazed because of their natural occurrence. Early evaluations under clipping or short periods of grazing indicated adaptability of many tropical legumes. However, under the grazing management approaches typically used, lack of legume survival has resulted in lack of demand for seed. It appears that the recent interest and use of leucaena in Brazil may stimulate more use of this and other legume species in tropical America.

Lack of long-term persistence is the reason for lack of success of tropical legumes in many areas. The reasons for lack of survival cannot be easily defined, but must include the grazing animal in most instances, although droughts, flooding, infertile soils, grass competition, diseases, etc. may weaken or kill the legume in grass-legume pastures. The predictability of legume adaptation to a given environment from clipping evaluations is poor. A high percentage of legumes released as cultivars from clipping experiments have failed to survive when grazed. For this reason evaluation of the response to grazing should be initiated as soon as possible in the evaluation scheme.[7] The philosophy of tropical legume evaluation is beginning to change from the standard "temperate forage" field-plot method of cut and analyze. Part of this change results from the large backlog of untested species and genotypes of different morphological and agronomic characteristics that are available from germplasm banks.

The present number of tropical legume collections of six germplasm banks is at least 52,000 (Table 4.1). Many of these are duplicate holdings because of the formal and informal interaction among collectors, evaluators, and holders. The number would be reduced if human edible *Vigna* species were not listed from SRPIS, which includes more than half of the total tropical legume collection at that institution. Data for CENARGEN are incomplete, and there are many smaller collections not reported here. Aside from the food legumes, *Vigna* and *Phaseolus*, the largest collection numbers are in *Stylosanthes, Desmodium,* and *Centrosema.*

The selected list and brief description of tropical legumes presented should be supplemented by the following references: Anonymous,[12] and 1984,[13] Blumenstock,[14] Bogdan, [15] Bray,[5] Eyles and Cameron,[16] Gonzales,[17] Gramshaw et al.,[18] Gutteridge and Shelton,[19] Hague et al.,[20] Hanson,[21] Kretschmer,[7] Loch,[22] 't Mannetje and Jones,[23] McWilliam,[24] Minson,[25] Peoples and Herridge,[26] Polhill et al.,[10] Polhill and Raven,[11] and 1981,[27] Reid,[28] Skerman,[29] Stinton,[30] Summerfield and Bunting,[31] Whiteman,[32] Whyte et al.,[33] Williams,[34] and Zeven.[35]

IV. IMPORTANT LEGUME SPECIES

Aeschynomene is a primarily tropical genus with about 160 species.[36-40] *Aeschynomene* is native to all American tropical areas between about 30°N and 30°S latitudes up to about 2500 m; however, some *Aeschynomene* species are found on the Atlantic coast in the U.S. to 40°N and in South America to about 35°S latitude. The range on the Pacific coast is from 28°N to 17°S. The Old World distribution occurs principally in Africa with a few species occurring in Asia and the Pacific Islands. About half of the species are xeric and the remainder are hydrophytes. General morphology ranges from herbs several centimeters in height to erect or prostrate subshrubs and shrubs, to tree-like plants up to 8 m high. About 40% of the species are perennial.[37] No species are known to be toxic to cattle, although at least one [*A. elaphroxylon* (Guill. & Perr.) Taub.] is armed with sharp thorns.

Aeschynomene americana L. (aeschynomene, American jointvetch) is a self-regenerating annual herbaceous legume (when plants are young), erect/branching with plant height reaching 1 to 2 m.

TABLE 4.1
Approximate Number of Genera and Species of Selected Tropical Herbaceous and Browse Forage Legumes

	Germplasm Bank											
	SRPIS		IRFL		CSIRO		CIAT		ILCA		CENARGEN	
Genus	Spp.[a]	Acc.	Spp.	Acc.	Spp.	Acc.	Spp.	Acc.	Spp.	Acc.	Spp.	Acc.
Acacia	2	4	1	1	24	50	31	983	67	178	—	—
Aeschynomene	43	675	48	767	33	457	9	256	9	175	4	346
Alysicarpus	5	17	7	84	14	353	—	—	—	—	—	—
Cajanus	4	125	1	29	2	209	—	—	—	—	—	—
Calopogonium	2	12	3	67	5	144	4	526	—	—	—	—
Centrosema	6	51	17	443	32	1231	33	2376	12	326	4	469
Crotalaria	20	233	18	83	71	376	25	272	24	187	—	—
Desmanthus	4	65	6	139	9	313	—	—	—	—	—	—
Desmodium	37	276	39	780	98	1531	47	2777	27	166	3	350
Galactia	3	12	11	109	15	253	12	548	—	—	—	—
Indigofera	36	118	29	94	80	455	17	221	25	176	—	—
Lablab	1	101	1	44	1	153	—	—	1	184	—	—
Leucaena	13	696	10	338	12	688	13	193	17	174	—	—
Lotononis	2	22	22	87	20	107	—	—	—	—	—	—
Macroptilium	5	205	11	570	16	688	10	601	—	—	2	136
Macrotyloma	3	19	8	92	10	163	—	—	—	—	—	—
Mucuna	3	48	4	33	14	57	—	—	—	—	—	—
Neonotonia	2	187	1	115	1	291	—	—	1	259	—	—
Phaseolus[b]	[c]	—	11	75	19	280	—	—	6	283	—	—
Pueraria	—	—	—	—	—	—	4	237	—	—	—	—
Rhyncosia	6	35	12	78	40	367	14	453	12	140	—	—
Sesbania	17	50	17	50	38	325	—	—	18	305	—	—
Stylosanthes	10	121	19	536	51	2277	25	3564	14	1127	7	1678
Teramnus	4	25	19	536	10	310	4	373	—	—	—	—
Vigna[b]	38	8255	38	362	55	688	33	728	21	417	2	61
Zornia	5	29	8	104	22	198	17	1025	9	256	3	514
Total	271	114[d]	361	56	692	120	298	151	263	44	25	36

[a] Spp. = number of species; Acc. = number of accessions. Total numbers of species and accessions may be somewhat misleading because of considerable duplication among the germplasm banks listed. Some collections have also increased to some extent since information was made available.

[b] Includes mostly human food species except data from CSIRO, which are only forage types.

[c] None included in forage collection.

[d] Total numbers (rounded) of accessions in 100s.

Note: Appreciation is extended to M. Spinks, USDA, SRPIS, Griffin, GA; B. C. Pengelly, CSIRO-ATFGRC, Brisbane, Australia; J. Hansen, ILCA, Addis Ababa, Ethiopia; and R. Schultze-Kraft, Universitat Hohenheim, Stuttgart, Germany for assisting with the compilations.

It is used successfully in Florida (common) and in Australia (cultivars 'Glenn' and 'Lee').[41] Hodges et al.[42] and Kalmbacher et al.[43] have described the use of aeschynomene in Florida. It is best suited to moist or waterlogged soils that temporarily flood.

Lack of persistence is partially due to erratic early season rainfall, which often is sufficient for aeschynomene germination. A subsequent dry period can often be fatal to the slow-developing seedlings. In Bahia grass and other grass pastures, establishment can be increased markedly by lightly disking or chopping the grass sod in early spring, and particularly by burning the pasture prior to germination of aeschynomene.

The CP and digestibility of leaves range from 15 to 25 and 60 to 70%, respectively, during the growing season, however, at the onset of seed formation and plant maturation, quality and intake by grazing livestock begin to decrease until plant death. Since the portion of grazed aeschynomene plants is restricted to the pinnately compound leaves and tender stems, there is little relationship between total plant yield or quality and animal performance. Aeschynomene is also used as green manure. It is similar in appearance to some genotypes of *A. villosa* Poir.

Aeschynomene evenia C. Wright (evenia aeschynomene, evenia jointvetch) is a newly described species not previously used for grazing.[44] A native of Texas, Cuba, and northern South America into Brazil, evenia aeschynomene is a short-lived perennial (annual where severe frosts occur) that has many of the attributes and similar appearance of American jointvetch. Compared with American jointvetch, however, evenia jointvetch has several agronomic advantages. It persists through mild winters in subtropical climates and, being day-length neutral, produces seed during all seasons. Possibly, it can be better used as an annual in colder climates than can American jointvetch whose flowering is triggered only by short days, and thus sometimes does not produce seed before frost. Seed weight and seedling vigor of evenia jointvetch are greater than for American jointvetch, which should help it compete better in Bahia grass and in other sods. Seed germination and moderate seedling growth occur continuously throughout the mild period of winter in south Florida, while American jointvetch seeds usually germinate only in the spring. Rapid growth of American jointvetch seedlings does not occur until late May or June. With proper management of evenia jointvetch, at least two seed harvests can be made annually.

Aeschynomene falcata (Pois.) DC. (jointvetch) is a perennial native to altitudes of 1800 m in northwestern and east-central South America. Prostrate and herbaceous with stems to about 1 m, *A. falcata* is drought tolerant and adapted to the subtropics, but not to the humid tropics. An Australian cultivar, 'Bargoo,' has been used since 1973 and persists well where adapted.[45]

Aeschynomene villosa Poir. (villosa jointvetch) is a variable, annual/perennial prostrate/erect legume species found from Mexico through most of Latin America. Several gentoypes have been selected in Australia from which the cultivars 'Kretschmer' and 'Reid' were released in 1995.[41] Evaluations in Florida have shown that this species can persist under a variety of grazing management systems, however, it appears that it will not tolerate prolonged waterlogging.

Alysicarpus vaginalis (L.) DC. [alyceclover, buffalo clover (Australia)], is one of about 30 species native to tropical Asia. It is naturalized in most countries of tropical America. There are two distinct forms of alyceclover, one being predominately erect and annual (cultivated form) and the other, prostrate, stoloniferous with some genotypes perennial. Most of the Latin American collections are the prostrate type. The cultivated form has been used in tropical America with some success, although regeneration from seed is not always assured. Stems can reach a height of 1.5 m and more. It is an excellent, if stemmy, hay crop, but does not tolerate as much waterlogging as American jointvetch. The commercial types are damaged by root-knot nematodes (*Meloidogyne* species). It requires about 1000 mm annual rainfall and a soil pH above about 5.0 for best growth. Several of the prostrate introductions overwinter in peninsular Florida, but also appear to be less tolerant of excess soil water than American jointvetch. Both forms are palatable to cattle.[46]

Arachis glabrata Benth. (rhizoma peanut) is one of more than 70 species in this genus.[47] It is native to Brazil, Argentina, and Paraguay.[48,49] Rhizoma peanut is highly persistent and rhizomatous. The cultivars 'Florigraze,'[50] 'Arb,' 'Arblick,' and 'Arbrook' have been released by the University of Florida. Rhizoma peanut must be planted vegetatively by using rhizomes and may require 2 to 3 years to become well established. Like most *Arachis* species, it is of very high quality. It is used as a pure stand for hay, and in association with commonly used grasses for grazing. Its growth is reduced when grown on waterlogged sites, with best growth occurring on well-drained sites receiving 1000 mm or more annual, well-distributed rainfall. It can survive severe winters in north Florida and southern Georgia.

Arachis pintoi Krap. & Greg., nom. nud. (pinto peanut) is native to central Brazil. It was first collected in 1954 and since has been distributed throughout the tropics. The cultivar 'Amarillo,' released in Australia in 1987, was also released as 'Mani forragero' (Panama, 1997) and 'Povenir' (Costa Rica, 1998). The single superior genotype is in limited use, with different cultivar names, in other Latin American countries. Pinto peanut is a vigorous stoloniferous species that has prostrate to erect stems when growing in dense swards. It can produce 3 to 5 tons ha^{-1} of seed, although until recently, most plantings were vegetative.[51-58] Establishment by seed is more rapid than by vegetative planting.[59] Seedling vigor and acceptability to animals are excellent for this yellow-flowered wild peanut.[60,61] The CP and digestibility of pinto peanut plants range from about 15 to 20 and 60 to 75%, respectively.

Because of its shade tolerance, it has been suggested that pinto peanut could be used for grazing in tree plantations, or simply as ground cover in orchards.[62] In Florida, Amarillo pinto peanut has not been as vigorous as other seed-producing pinto genotypes, as evidenced by the light green color of foliage and slow stolon development. Other *A. pintoi* introductions have had a dark green leaf color. Further evaluation of seed-producing wild peanuts is in progress. Present published and unpublished Florida data indicate that *A. stenosperma* produces large quantities of nuts and is adapted as a cover crop in citrus, while *A. kretschmeri* is an excellent wildlife feed.[63] Only a few of the estimated 420 accessions (in the U.S.) have been thoroughly evaluated.[54]

Calopogonium mucunoides Desv. (calopo) is an annual or short-lived perennial that is indigenous to tropical America. It has been grazed inadvertently in many tropical pastures, where it has invaded or become naturalized. One of eight species in the genus, it is now widespread in Africa and Asia. A short-day flowering response limits the usefulness of calopo in the subtropics because of poor or no seed production. Similar to tropical kudzu [*Pueraria phasioloides* (Roxb.) Benth.], it has been used for more than 70 years as a cover crop in plantation agriculture, in rotation with rice as a green manure/grazing crop in Brazil, and for grazing to a limited extent in Australia. In the tropics, it persists well with annual, well-distributed or seasonal rainfall above 1000 mm. It is not very palatable. There appears to be little variation within the species. Another species, *C. caeruleum* (Benth.) Sauv. is very unpalatable and, therefore, very persistent when grass mixtures are heavily grazed.

Centrosema is a primarily tropical genus of 32 perennial and annual named species. Collection and evaluation of this genus has been intense during the past two or more decades.[64,65] About 3700 accessions comprise the world's germplasm collection.[64] Most species are twining perennials with trifoliolate leaves. Flower color ranges from white to purple to red. In spite of the large research effort, use of *Centrosema* species has remained limited to small environmental niches. There has been very little or no use for grazing in Latin America, some use as cover in plantation crops, and for grazing primarily in Australia.

Centrosema acutifolium Benth. is a perennial, twining species that has only recently been developed in Colombia (cultivar 'Vichada') for use in the hot tropics with annual rainfall of 1000 to 2500 mm.[66] It has been evaluated for use as a protein bank for dairy operations.[67] It is similar to *C. pubescens* Benth. in appearance and nutritional quality, but can tolerate acid (pH down to about 4.3) soils and soils high in Al and Mn.

Centrosema macrocarpon Benth. is found from Mexico through the northern latitudes of Brazil and is similar in appearance and adaptation to *C. acutifolium*.[68] It is reported to be the most disease resistant of the *Centrosema* species evaluated to date, but, like the others, is susceptible to damage by leaf-cutting ants.

Centrosema pascuorum Mart. ex Benth. (centurion) is a herbaceous twining annual found mainly in semi-arid areas of Central and South America that have a long dry season. It is utilized in the Northern Territory of Australia for grazing. Cultivars 'Cavalcade' (1984)[69] and 'Bundey' (1986)[41] have been released in Australia. It requires a reliable 4 to 6 month wet season, can tolerate prolonged flooding, and is also drought tolerant. A very palatable species, it is a heavy seed producer when good moisture conditions exist.

Centrosema pubescens Benth. (centro) is very widely distributed in the semi-humid areas of the Latin American tropics. It will not grow well on soils with high Al and Mn contents or when pH is below about 5.0. It can persist with as little as 750 mm annual rainfall, but is more productive in areas receiving 1000 mm or more. In Australia, the commercial genotype has been successfully used for several decades in a small coastal area of Queensland.[70] Because of its late flowering and lack of seed production in south Florida, it does not persist well. The cultivar 'Belalto' (1971), originally thought to be *C. pubescens,* but since identified as *C. schiedeanum* L., has better cool-season growth and stronger stoloniferous growth, but is later flowering than common centro.

Centrosema virginianum Benth. has many of the same characteristics as *C. pubescens*, but its flowering period is not influenced by day length. It is native to Maryland (possibly New Jersey) to Oklahoma and south into Argentina, and it is often found in the Gulf Coast states and in Florida in the U.S. In spite of the evaluation of many genotypes and a few hybrids, there is no commercial use.[71] Natural out crossing was estimated to be about 18%.[72]

Chamaecrista rotundifolia (Pers.) Greene (formerly *Cassia rotundifolia* Pers.) is called round-leafed cassia. It is an annual or short-lived perennial semi-erect to prostrate herb with bifoliolate leaves. It is native to Mexico, the Caribbean, and into Brazil and Uruguay. It has been very successful in the 700 to 1400 mm annual rainfall areas of Queensland, Australia where cultivar 'Wynn' was developed.[73] An estimated natural out crossing of 13% has been reported.[72] It is moderately grazed by cattle especially when associated grass availability is low, and it can contribute to pasture quality, production, and animal gains.[74-77] Its persistence depends primarily on sufficient soil seed reserves, which reflect primarily rainfall patterns and grazing management.[75,78,79] Wynn and other types may be susceptible to leaf diseases.[80]

Clitoria ternatea L. (butterfly pea) is a perennial twining legume with a woody crown that can survive about 500 mm annual rainfall but grows best with about 1500 mm. It is not tolerant of flooding. It has been evaluated in Mexico (cultivar 'Tehuana,' 1988) and Brazil (common), and elsewhere in Latin America. In Australia, cultivar 'Milgarra' (1990) was released; however, it has not been widely used there or in tropical Asia despite apparent adaptation.[81]

Desmanthus virgatus (L.) Willd. (desmanthus) probably originated in Mexico and is now widespread in the Americas from Arizona, Texas, and Florida to Argentina. The genus includes erect to prostrate perennial and annual herbs with some browse subshrub or shrub growth forms. Its leaves are bipinnate. From the botanical variety *virgatus* came 1991 Australian releases of cultivars 'Marc' (early), 'Bayamo' (mid-season), and 'Yuman' (late flowering).[41,82,83] A physical mixture of the three is being used in Australia and elsewhere in the Pacific region. Collected from areas with annual rainfall of 250 to 2000 mm and up to 2000 m elevation, most cultivars grow best in soils with pH above about 6.0. Most will persist on sand as well as clay soils.[84] Desmanthus is a nontoxic browse legume (and a substitute for leucaena in some areas) especially when mature. The CP in stems can be about 6.5%, and CP in leaves is often more than 20%. Crowns can survive frosts, although foliage is susceptible to damage. There is a large collection of genotypes, and there is a good possibility that the use of this species will increase.

Desmodium is a genus with a large number of tropical to temperate species. Several have been developed into successful cultivars.[85]

Desmodium heterocarpon (L.) DC. (carpon desmodium) is native to subtropical and tropical areas of India, southeast and eastern Asia, the Pacific Islands, and Australia.[85] This is a long-lived perennial prostrate to erect (depending on grazing pressure) herb with a woody crown upon maturation. The cultivar 'Florida' (variety *heterocarpon*) has persisted when grown in association with Bahia grass and other commonly used grasses in peninsular Florida.[86,87] Other genotypes that are more erect become prostrate when heavily grazed. Most accessions are short-day-length plants that flower in Florida in September through October with seed ready for harvest in late October to November. Florida carpon desmodium is better adapted to the humid subtropics than to more tropical areas. It will survive moderate droughts and temporary flooding, but best growth occurs on moderately drained, moist soils. Initial seedling growth is slow, but long-term persistence of individual plants and excellent seedling recruitment make it a more desirable species than annuals such as *Aeschynomene americana*. Nutrient requirements for carpon desmodium, as for most other tropical legumes, generally are less than those for temperate legumes. Soil pH should be maintained above about 5.0. It is not adapted to low pH soils or soils high in Al and Mn. Carpon desmodium is equally palatable in mixtures with most associated grasses in spite of a 2 to 3% tannin concentration. There is an estimated natural out crossing of 4% within the species.[72]

Desmodium heterocarpon (L.) DC. subspecies *ovalifolium* (Prain) Ohashi (formerly *D. ovalifolium* Wallich ex Gagnep.) is a perennial, creeping, stoloniferous legume that is well adapted to the humid and subhumid tropics.[9,88] It has been used in southeast Asia as a cover crop in plantation agriculture and for grazing for many years.[9] Compared with Florida carpon desmodium, cultivar 'Itabela' (Brazil, 1989) is not as palatable.

Desmodium intortum (Mill.) Urb. (greenleaf desmodium) is indigenous to the highlands from 800 to 2500 m, from Mexico to southern Brazil. Trailing but not twining, this large perennial can grow in the lower elevations in the subtropics, but is not adapted to the humid or semi-humid tropics. Requiring about 1000 to 1500 mm annual rainfall and a soil pH of above about 5.0, cultivar 'Greenleaf' (Australia, 1963) is used in isolated areas of the Australian subtropics. Its range in the subtropics is limited because seed production is too late for seedling recruitment to occur.

Desmodium uncinatum (Jacq.) DC. (silverleaf desmodium) is native to Central and South America. It is not as common as Greenleaf, but has a similar growth habit. Cultivar 'Silverleaf' (Australia, 1962) is earlier flowering and its area of use is less in Australia than that of Greenleaf, however, it is being used in Kenya.

Gliricidia sepium (Jacq.) Kunth ex wap. (gliricidia) is a small deciduous tree (to 12 m). It is native to the seasonally dry Pacific coast of Central America, and is now naturalized from Mexico to South America, the Caribbean, West Africa, and Asia. Used primarily in cut-and-carry forage systems, gliricidia has recently been evaluated in several countries.[89-93] In areas where both gliricidia and leucaena are adapted, leucaena is generally superior.

Lablab purpureus (L.) Sweet (lablab bean, hyacinth bean) is from the Old World tropics and has been used for grazing in Australia, and for human consumption in the Caribbean and elsewhere. An erect to climbing, annual or short-lived perennial, it grows well in slightly acid to alkaline soils, but is not tolerant of waterlogging. Consumption of this legume by dairy cattle passes an undesirable flavor into unpasteurized milk. Cultivars from Australia are 'Rongai' (1962), 'Highworth' (1973), and 'Koala' (1996). It is marketed in Texas as 'Tecomate' primarily as wildlife feed.[41] Because it can grow as an annual in temperate areas, additional germplasm evaluation is warranted.[94]

Leucaena leucocephala (Lamb.) de Wit. (leucaena) is the most widely evaluated and used species of the browse legumes. Native or naturalized in most tropical areas with moderately drained soils and a pH greater than 5.5 to 6.0, leucaena is a very long-lived perennial that can reach a height of about 16 m (cultivar 'El Salvador').[95-99] The bipinnate leaves and smaller stems are readily consumed by cattle. Cultivation of leucaena for grazing began in Australia about three decades ago, with some of the original plantings in Australia still existing. Although many other browse shrubs and trees have been evaluated, beef gains from leucaena have been the highest. Its major drawback, however, has been the adverse effect of mimosine on the thyroid of the consuming animal. Mimosine is an amino acid found in small quantities in the legume. Excess consumption of the legume can lead to the loss of hair and the formation of goiters. After identification of the cause and a correction for mimosine toxicity in ruminants (it is still toxic to nonruminants), there has been increased evaluation and use of leucaena in Brazil, Florida, Asia, and most other tropical areas.[100,101] Cattle are inoculated with ruminal bacteria to detoxify the effect of mimosine.

Normally, leucaena is planted in single to four rows, 1 to 5 m apart and separated by grass strips of up to about 5 m. It can survive temperatures of about –7°C or less, which increases its area of adaptation into warm temperate areas. Because of slow seedling growth, grazing normally is delayed until the end of the first growing season. The leucaena psyllid (*Heteropsylla cubana*) devastated foliage several years ago in Australia and elsewhere,[102-104] however, this damage has not diminished the enthusiasm of growers using leucaena. Grazing management and possibly parasitic insects appear to have reduced the problem. Establishment difficulties in south Florida have prevented its use commercially. At present, the most popular cultivar is 'Cunningham' (Australia, 1977).

Macroptilium atropurpureum (DC.) Urb. (siratro) occurs naturally from southern Texas to Argentina and northern Brazil. It has been evaluated in almost all tropical and subtropical areas of the world. Its popularity increased from about 1967 in Australia with cultivar 'Siratro' until the mid-1970s. Demand for seed declined slowly because of its lack of persistence under poor grazing management. 'Siratro' leaves and stems are susceptible to *Uromyces appendiculatus* (rust), which can be very damaging to forage yields. Rather recent germplasm collections in Mexico and subsequent selection and breeding[105] led to the Australian release of the cultivar 'Aztec,' a line bred with rust resistance. A rust-resistant germplasm has also been registered (U.S., 1992).[106] Siratro is also susceptible to foliar leaf blight (*Rhizoctonia solani*).[107] Under grazing, success of this herbaceous, twining legume has not been very good, especially in the humid tropics. The cultivar 'Siratro' in Australia has been used primarily in coastal Queensland.[108,109] It persists well in south Florida under cutting every 45 to 60 days, but does not survive under the commonly used grazing practices.[110] In addition to not persisting well in Florida, Brazil, and elsewhere, it can become a costly weed in citrus because of its viney growth habit and difficulty of eradication.

Macroptilium lathyroides (L.) Urb., (phasey bean) a short-lived perennial, has been used as a pioneer crop in Australia, Florida, and elsewhere.[111] Although possessing many desirable forage characteristics, limited stand life greatly reduces the use of this legume. It can become a difficult weed to control in citrus because of its twining habit during plant maturation.

Other *Macroptilium* species are being evaluated. *M. gracile* (Poepp. Ex Benth.) Urb. is an example of one of the prostrate types that may withstand heavier grazing pressure than does Siratro.

Neonotonia wightii (Wight and Arn.) Lacky (glycine, perennial soybean) has a twining growth habit similar to that of Siratro. Originally classified as *Glycine javanica* L., it is less palatable and has been more persistent in tropical America than Siratro, possibly because of its tolerance to rust. Cultivars 'Tinaroo,' 'Cooper,' 'Clarence' (all released in 1962) and 'Malawi' (1975) were released in Australia, where they have been most successful at higher elevations in the tropical zone of Queensland.[41,112] Glycine is tolerant of drought and temporary waterlogging. It has persisted well in some areas of Latin America. Seed production is not consistent in south Florida because of its late flowering habit.

Pueraria phaseoloides (Roxb.) Benth. (tropical kudzu, puero) is a robust trailing, perennial species. It is native to East and Southeast Asia, and has been used throughout the tropical world beginning more than 80 years ago as a cover crop (sometimes grazed) in rubber and oil palm plantations. It is adapted and widely distributed in the high rainfall areas of the tropics, where it is tolerant of flooding and waterlogging. It can also survive droughts. It is not adapted to the subtropics because of lack of seed production in such environments and poor winter survival. One of the commercially used genotypes is not very palatable, and this genotype persists well since cattle consume it only as a last resort. Other genotypes are reported to be moderately palatable. Although tropical kudzu produces seed, almost all plantings in tropical America have been made vegetatively.

Sesbania sesban (L.) Merrill (sesbania) is one of two groups of different types of about 50 species of the genus.[113] Genotypes are annual or short-lived perennial shrubs or trees. Sesbania is cultivated in Africa, Asia, and tropical America, being adapted to annual rainfall areas of 500 to 2000 mm and up to about 2300 m elevation. Sesbania tolerates waterlogging and can be grazed, but normally is used in a cut-and-carry system. Its quality and palatability are less than those of leucaena.[114-116] Cultivar 'Mount Cotton' is a recent Australian release.[117]

Stylosanthes is a trifoliolate genus of 30 species. More species of this annual/perennial genus have been successful under grazing management systems than species from any other tropical legume genus. Most successful have been Townsville stylo (*S. humilis* Kunth), an annual; Caribbean stylo [*S. hamata* (L.) Taub.], primarily annual or short-lived perennial; shrubby stylo (*S. scabra* Vog.), perennial; and stylo [*S. guianensis* (Aub. Sw.], perennial. Susceptibility to anthracnose disease caused by *Colletotrichum gloeosporioides* has been a problem with many of the original cultivars.[118,119] The genus is characterized by small yellowish flowers. Most species of *Stylosanthes* are not tolerant of prolonged waterlogging. As a genus, *Stylosanthes* is known for its ability to extract P from soils where it is not available to other species. Most genotypes can significantly contribute to animal diets.[120] For a thorough overview of research on the genus *Stylosanthes*, the reader is referred to Stace and Eyde.[6]

Stylosanthes guianensis (stylo) is comprised of seven botanical varieties with variety *guianensis* being used most in developing cultivars. The origin of the species is Brazil. This erect, sometimes prostrate perennial is naturalized in most Latin American countries, but not in the U.S. or the Caribbean Islands. Cultivars represented by the variety *guianensis* are 'Schofield' (1930s), 'Cook' (1971), 'Endeavour' (1971), and 'Graham' (1979) from Australia; 'Deodora,' 'Deodora II,' 'IR-1022' (early 1970s), and 'Mineirao' (1993) from Brazil;[121] 'Pucallpa' (Peru, 1985, also released as 'Reyan' in China, 1991), and 'Savanna' (Florida, U.S., 1992). The variety *intermedia* (Vogel) Hassl. (fine stem stylo) is represented by common (Australia, 1965) and 'Oxley' (Australia, 1969). From Brazil (1983) came cultivar 'Bendeirante' of variety *pauciflora* M. B. Ferreira & S. Costa (tardio stylo).[41,120] As a group they are moderately drought tolerant, cannot withstand continuous waterlogging, survive light frosts, and require a soil pH of > 5.0. This species is now seldom used in Australia and elsewhere because of anthracnose, but in southern Florida the disease has not been severe enough to kill mature plants. It has reduced seed production drastically at times. Stylo is competitive with associated grasses and has good palatability and adequate quality.

Stylosanthes hamata (L.) Taub. (Caribbean stylo) is native to south Florida, the Caribbean into Colombia, Venezuela, and Brazil. It is an annual to short-lived perennial. The Australian tetraploid cultivars 'Verano' (1973) and 'Amiga' (1997) are adapted to northern Queensland, Australia. Being more tolerant to anthracnose, in Australia, Caribbean stylo has replaced large areas of *S. humilis*, which had been killed by the disease. Caribbean stylo has also been used successfully in Thailand, India, and several west African countries. Verano is moderately drought tolerant but has not persisted well in south Florida, possibly a result of periodic high water tables.

Stylosanthes humilis Kunth (Townsville stylo) is an annual, similar in appearance to *S. hamata*, that became naturalized in northern Australia at the beginning of the 20th century. Later it was seeded in large areas, where it became the major legume used. More recently, other species replaced Townsville stylo after large areas had been devastated by anthracnose. Although anthracnose did not damage commercial plantings of *S. humilis* in the 1970s in south Florida, these stands did not have long-term persistence. Several cultivars from Australia, Thailand, and Brazil are available. *Stylosanthes macrocephala* M. B. Ferreira & S. Costa cultivar 'Pioneiro' from Brazil was released in 1983 before its species name change.

Stylosanthes scabra Vogel (shrubby stylo) is strongly perennial, growing up to 2 m high. It is the most drought tolerant of the commonly used stylos. Because of its low leaf-to-stem ratio (as high as 96% by weight of stems) shrubby stylo is the least nutritious species of the genus. It is adapted to areas with annual rainfall of 500 to 2000 mm. In northern Australia, cultivars 'Seca' (1976), 'Fitzroy' (1979), and 'Siran' (1990) replaced *S. hamata* in the drier climates, although Fitzroy is more susceptible to anthracnose than Seca and Siran. Seca is the most commonly used cultivar.[122] One cultivar (Q10042) has been released in India. Some cross-pollination and species integration have occurred, which has led to hybrids in the general population that are difficult to distinguish. In 1996, cultivars 'Primar' and 'Unica' were released in Australia.[123] They were selections made from a *S. scabra* collection. Although they are taxonomically unique and are being reclassified, they closely resemble *S. scabra*. These cultivars are superior to Seca in several traits.[122] Seca is used in a cut-and-carry system in India. A resume of the world collections of shrubby stylo was recently reported.[124]

Vigna adenantha (G. Mey.) Marechal, Marcherpa, and Stainer is native to Mexico, Central America, and through tropical areas of Argentina. It is found in tropical Africa, Taiwan, and other islands. It is a perennial, viney legume with stems to about 4 m long. It is one of about 150 species in the genus. Its growth in pastures is similar to that of Siratro and Neonotonia, but it is much more tolerant of waterlogging than these species and is not attacked by rust. It has potential in seasonally dry to humid subtropics, but mechanical seed harvesting is difficult.

Vigna Parkeri Baker subspecies *maranguensis* (creeping vigna) originated in Africa and has been naturalized in the highlands of Papua, New Guinea. There are three subspecies of *V. parkeri*. This legume is short, viney, or stoloniferous, moisture loving, and tolerant of infertile low pH (> 5.0) soils. The cultivar 'Shaw' (Australia, 1984) flowers indeterminately, thus mechanically harvested seed yields are low. Shaw seems to be adapted to subtropical humid climates. Creeping vigna appears promising in south Florida, however, it can be eliminated by burning.[125] It is also successful in some areas in Australia.[71,126] Because little is known of the genotypic variation of the species, further collections may provide an opportunity to develop genotypes with a wider range of adaptation. Earlier flowering would contribute to its success in the subtropics.

Zornia latifolia Smith (zornia) is one of about 80 species of the genus primarily native to the Americas and Africa. This heterogeneous species has a wide adaptation to soils and climate, and includes erect to prostrate genotypes with leaves primarily bifoliolate. An accession being evaluated in Florida is daylength neutral, cannot tolerate long-term flooding, but withstands high grazing pressure and appears to associate well with Bahia grass. This genus has been insufficiently evaluated in spite of availability of over 2000 accessions.

There are other tropical legumes of less current importance that have been successfully used in very limited areas. Included are: *Cajanas cajan* (pigeon pea, primarily for human consumption); *Clitoria ternatea*,[127] *Crotalaria* species (mostly toxic); *Galactia striata* (Jacq.) Urb.; *Indigofera hirsuta* L. (hairy indigo, which was used in Florida and elsewhere when the search for tropical legumes began and is still used sparingly; many other species of *Indigofera* are toxic); *Lotononis bainesii* Bak. (used successfully in Australia and Argentina, where it is adapted to near temperate climates, being frost tolerant); *Macrotyloma* species[128] (used primarily in Australia); *Mucuna* species (used for grazing or hay); *Rhynchosia* species (not used commercially; generally unpalatable); *Teramnus labialis* (L.F.) Spreng. (grazed as an invader in improved or natural pastures[129] where it persists well, but has not been used commercially); and *Teramnus uncinatus* (L.) Sw. (grazed as a naturally occurring legume along roadsides, but does not persist when seeded for pasture because of overgrazing).

V. SUMMARY AND CONCLUSIONS

After World War II, the value and use of naturalized tropical legumes stimulated the search for types with better attributes. Selection of superior genotypes extended the range of environments where tropical legumes contribute to productivity of native and improved pastures. A slow change from evaluation of centrosema, pueraria, and the early stylo and desmodium genotypes gradually shifted to assessment of additional new species and other genera in search of genotypes with attributes necessary for successful persistence and quality enhancement. The main centers of tropical legume diversity, Brazil and Mexico, were traversed with collectors, primarily Australians, and later collectors from CIAT and ILCA. Concurrently, botanists renewed their interest in the taxonomy of Papilionoidea, from which come most of the legumes described here. Evaluation in other areas began and seed exchanges as well as collections increased. Recently, tropical forage legumes began to be investigated for use by small landholders and for use in integrated agricultural systems. The trends have evolved to a widespread emphasis on the evaluation of leucaena and other shrub legumes (gliricidia and desmanthus) and wild peanuts. In addition to the enhanced forage quality initially emphasized, nitrogen fixation levels up to 290 kg ha^{-1} annually from herbaceous species and 575 kg ha^{-1} annually from shrub legumes have been reported.[130]

In the late 1980s and continuing today, support for the work has subsided. Fortunately, the germplasm banks hold a treasure of untested germplasm. Several recent Australian conferences have provided information on the future tropical pasture and forage trends in that country and elsewhere.[131-133] Tropical pasture research has followed the pattern set by Australia in the past. Currently, this source of leadership and research continuity appears to be in jeopardy.

Sufficient tropical forage legume germplasm evaluations have been conducted over the past few decades to provide a considerable body of information. Broad classifications of adaptation and potential uses can now be assigned to many species. In the past, entries in germplasm evaluations were, to some extent, dependent upon timely availability. Future assessments must be more carefully planned to make more efficient use of limited resources and to increase the probability of successful results. Many previous evaluations have failed to produce even potentially useful genotypes. The narrow range of adaptation of most tropical pasture legume species has become apparent. Thorough preliminary study can, in many cases, narrow the pool of realistically probable species to a rather small group. Any predetermined management constraints or existing utilization systems may further limit the likely prospects. Identification of the agronomic and social limitations and thorough assessment of the edaphic and climatic factors will most often result in a rather small set of species to consider. Assessment of core collections of the identified species can further increase efficiency. The available information base, tremendous germplasm resource assembled, and potential benefits provide a compelling opportunity for continued evaluation and development of tropical forage legume germplasm.

REFERENCES

1. Humphreys, L. R., Townsville lucerne: history and prospect, *J. Aust. Inst. Agric. Sci.*, 33, 3, 1967.
2. Clements, R. J., Developing persistent pasture legume cultivars for Australia, in *Persistence of Forage Legumes*, Marten, G. C., Matches, A. G., Barnes, R. F., Brougham, R. W., Clements, R. J. and Sheath, G. W., Eds., American Society of Agronomy, Madison, Wisconsin, 1989, 505.
3. Minson, D. J. and Milford, R., The energy values and nutritive value indices of *Digitaria decumbens, Sorgum almum,* and *Phaseolus atropurpureum*, *Australian J. Agric. Res.*, 17, 411, 1966.
4. Aiken, G. E., Pitman, W. D., Chambliss, C. G., and Portier, K. M., Responses of yearling steers to different stocking rates on a subtropical grass-legume pasture, *J. Anim. Sci.*, 69, 3348, 1991.
5. Bray, R. A., Germplasm sources for genetic improvement of forage legumes in Australia and New Zealand, in *Forage Legumes for Energy Efficient Animal Production*, Barnes, R. F., Ball, P. R., Brougham, R. W., Marten, G. C., and Minson, D. J., Eds., United States Department of Agriculture, Agricultural Research Service, Washington, D.C., 1985, 278.
6. Stace, H. M. and Edye, L. A., Eds., *The Biology and Agronomy of Stylosanthes*, Academic Press, Sydney, Australia, 1984.
7. Kretschmer, A. E., Jr., Tropical forage legume development, diversity, and methodology for determining persistence, in *Persistence of Forage Legumes*, Marten, G. C., Matches, A. G., Barnes, R. F., Brougham, R. W., Clements, R. J., and Sheath, G. W., Eds., American Society of Agronomy, Madison, Wisconsin, 1989, 117.
8. Sonoda, R. M., Kretschmer, A. E., Jr., and Wilson, T. C., Evaluation of *Macroptilium atropurpureum* (DC) Urb. germplasm for reaction to foliar diseases, *Soil Crop Sci. Soc. Florida Proc.*, 51, 25, 1991.
9. Schultze-Kraft, R. and Benavides, G., *Germplasm Collection and Preliminary Evaluation of Desmodium ovalifolium Wall.*, Genetic Resources Communication 12, CSIRO, Brisbane, Australia, 1988.
10. Polhill, R. M., Raven, P. H., and Sirton, C. H., Evolution and systematics of the Leguminosae, in *Advances in Legume Systematics* (Part 1), Polhill, R. M. and Raven, P. H., Eds., Royal Botanic Gardens, Kew, Surry, England, 1981.
11. Polhill, R. M. and Raven, P. H., Eds., *Advances in Legume Systematics* (Part 1), Royal Botanic Gardens, Kew, Surry, England, 1981.
12. Anon., *Tropical Legumes: Resources for the Future*, National Academy Press, Washington, D.C., 1979.
13. Anon., *Forage and Browse Plants for Arid and Semi-Arid Africa*, IBPGR, Rome, 1984.
14. Blumenstock, D. I., Distribution and characteristics of tropical climates, *Proc. 9th Pacific Sci. Congr.*, 20, 3, 1957.
15. Bogdan, A. V., *Tropical Pasture and Fodder Plants*, Longman, New York, 1977.

16. Eyles, A. G., Cameron, D. G., and Hacker, J. B., *Pasture Research in Northern Australia — Its History, Achievements and Future Emphasis*, Res. Rep. 4, CSIRO Div. Tropical Crops and Pasture, CSIRO, Brisbane, Australia, 1985.
17. Gonzalez, B., Generalidades, in *FUSAGRI Pastos, Serie Petroleo y Agricultura*, Maraven, Petroleos de Venezuela, 1986, 9 (in Spanish).
18. Gramshaw, D., Reed, J. W., Collins, W. J., and Carter, E. D., Sown pastures and legume persistence: An Australian overview, in *Persistence of Forage Legumes*, Marten, G. C., Matches, A. G., Barnes, R. F., Brougham, R. W., Clements, R. J., and Sheath, G. W., Eds., American Society of Agronomy, Madison, Wisconsin, 1989, 1.
19. Gutteridge, R. C. and Shelton, H. M., Eds., *Forage Tree Legumes in Tropical Agriculture*, CAB International, Tucson, Arizona, 1993.
20. Hague, I., Jutzi, S., and Neate, P. J. H., Eds., *Potential of Forage Legumes in Farming Systems of Sub-Saharan Africa*, ILCA, Addis Ababa, Ethiopia, 1986.
21. Hanson, J., Getting the most out of ILCA's forage gene bank, *ILCA Newsletter*, 10 (4), 4, 1991.
22. Loch, D. S., Tropical herbage and seed production — origins, progress, and opportunities, *J. Appl. Seed Prod.*, 9, 14, 1991.
23. 't Mannetje, L. and Jones, R. M., Eds., *Plant Resources of South-East Asia*, Purdoc Sci. Publ., Wageningen, The Netherlands, 1992.
24. McWilliam, J. R., Response of pasture plants to temperature, in *Plant Relations in Pastures*, Wilson, J. R., Ed., CSIRO, Melbourne, Australia, 1978, 17.
25. Minson, D. J., Nutritional differences between tropical and temperate pastures, in *Grazing Animals*, Morley, F. H. W., Ed., Elsevier Sci. Publ., Amsterdam, 1980, 143.
26. Peoples, M. B. and Herridge, D. G., Nitrogen fixation by legumes in tropical and subtropical agriculture, *Adv. Agron.*, 44, 155, 1990.
27. Polhill, R. M. and Raven, P. H., Eds., *Advances in Legume Systematics* (Part 2), Royal Botanic Gardens, Kew, Surry, England, 1981.
28. Reid, R., Collection and use of climatic data in pasture plant introduction, in *Collecting and Testing Tropical Forage Plants*, Clements, R. J. and Cameron, D. G., Eds., CSIRO, Melbourne, Australia, 1980.
29. Skerman, P. J., *Subtropical Forage Legumes*, Plant Production and Protection Series 2, FAO, Rome, 1977.
30. Stinton, C. H., Ed., *Advances in Legume Systematics* (Part 3), Royal Botanic Gardens, Kew, Surry, England, 1987.
31. Summerfield, R. J. and Bunting, A. H., Eds., *Advances in Legume Science*, Royal Botanic Gardens, Kew, Surry, England, 1980.
32. Whiteman, P. C., *Tropical Pasture Science*, Oxford University Press, New York, 1980.
33. Whyte, R. O., Nilsson-Leissner, G., and Trumble, H. C., *Legumes in Agriculture*, FAO Agric. Studies No. 21, FAO, Rome, 1953.
34. Williams, R. J., Tropical legumes, in *Genetic Resources of Forage Plants*, McIvor, J. G. and Bray, R. A., Eds., CSIRO, Melbourne, Australia, 1983, 17.
35. Zeven, A. C., *Dictionary of Cultivated Plants and Their Regions of Diversity Excluding Most Ornamental, Forest Trees and Lower Plants*, Center for Agricultural Publ. and Documentation, Wageningen, The Netherlands, 1982.
36. Bishop, H. G., Pengelly, B. C., and Ludke, D. H., Classification and description of a collection of the legume genus *Aeschynomene, Trop. Grassl.*, 22, 160, 1988.
37. Kretschmer, A. E., Jr. and Bullock, R. C., *Aeschynomene* spp.: distribution and potential use, *Soil Crop Sci. Soc. Florida Proc.*, 39, 145, 1980.
38. Kretschmer, A. E., Jr. and Snyder, G. H., Potential of *Aeschynomene* species for pastures in the tropics, *Proc. XIV Int. Grassland Congr.*, Lexington, Kentucky, 1983, 783.
39. Kretschmer, A. E., Jr., Snyder, G. H., and Wilson, T. C., Productivity and persistence of selected *Aeschynomene* ssp., *Soil Crop Sci. Soc. Florida Proc.*, 45, 174, 1986.
40. Rudd, V. E., American species of *Aeschynomene*, *U.S. National Herb.*, 32, 1, 1955.
41. Loch, D. S. and Ferguson, J. E., Tropical and subtropical seed production: an overview, in *Forage Seed Production.* Vol. 2. *Tropical and Subtropical Species*, Loch, D. S. and Ferguson, J. E., Eds., CABI Publishing, Wallingford, Oxon, U.K., 1999, 1.

42. Hodges, E. M., Kretschmer, A. E., Jr., Mislevy, P., Roush, R. D., Ruelke, O. C., and Snyder, G. H., *Production and Utilization of the Tropical Legume Aeschynomene*, Florida Agric. Exp. Stn. Circ. S-290, Gainesville, Florida, 1982.
43. Kalmbacher, R. S., Mislevy, P., Martin, F. G., Prevatt, J. W., Chambliss, C. G., and Kidder, G., *Establishment of Aeschynomene in Bahiagrass Sod*, Florida Agric. Exp. Stn. Circ. S-355, Gainesville, Florida, 1988.
44. Kretschmer, A. E., Jr., Pitman, W. D., and Wilson, T. C., *Aeschynomene evenia* (evenia aeschynomene), a perennial legume for grazing in south Florida, *Soil Crop Sci. Soc. Florida Proc.*, 53, 52, 1994.
45. Wilson, G. P. M., Jones, R. M., and Cook, B. G., Persistence of jointvetch (*Aeschynomene falcata*) in experimental sowings in the Australian subtropics, *Trop. Grassl.*, 16, 155, 1982.
46. Gramshaw, D., Pengelly, B. C., Muller, F. W., Harding, W. A. T., and Williams, R. J., Classification of a collection of the legume *Alysicarpus* using morphological and preliminary agronomic attributes, *Australian J. Agric. Res.*, 38, 355, 1987.
47. Krapovickas, A., A proposed taxonomic summary of the genus *Arachis*, in *Report of a Workshop on the Genetic Resources of Wild Arachis Species*, IBPGR and ICRISAT, Eds., CIAT, Cali, Colombia, 1990, 9.
48. IBPGR and ICRISAT, Eds., *Report of a Workshop on the Genetic Resources of Wild Arachis Species*, CIAT, Cali, Colombia, 1990.
49. IBPGR and ICRISAT, Eds., *Descriptors for Groundnuts*, IBPGR, Rome, 1992.
50. Prine, G. M., Dunavin, L. S., Moore, J. E., and Roush, R. D., *'Florigraze' Rhizoma Peanut, a Perennial Forage Legume*, Florida Agric. Exp. Stn. Circ. S-275, Gainesville, Florida, 1981.
51. Grof., B., Forage attributes of the perennial groundnut *Arachis pintoi* in a tropical savanna environment in Colombia, *Proc. XV Int. Grassland Congr.*, Kyoto, Japan, 1985, 168.
52. Cook, B. G. and Franklin, T. G., Crop management and seed harvesting of *Arachis pintoi* Krap. et Greg. nom. nud., *J. Appl. Seed Prod.*, 6, 26, 1988.
53. Oram, R. J., B. Legumes 21. Arachis (a) *Arachis pintoi* Krap et Greg nom nud. (Pinto peanut) cv. Amarillo, *Australian J. Exp. Agric.*, 30, 445, 1990.
54. Kerridge, P. C. and Hardy, B., Eds., *Biology and Agronomy of Forage Arachis*. Internatl. Center Trop. Agric. (CIAT), Apartado Aereo 6713, Cali, Colombia, 1994.
55. Smith, P., Pastures for prosperity — Seeds forum., 6. What we want from the seed industry in the future — merchant's viewpoint, *Trop. Grassl.*, 30, 88, 1996.
56. Kerridge, P. C. and Loch, D. S., *Arachis pintoi* in Australia and Latin America, in *Forage Seed Production*. Vol. 2. *Tropical and Subtropical Species,* Loch, D. S. and Ferguson, J. E., Eds., CABI Publishers, Wallingford, Oxon, U.K., 1999, 407.
57. Kretschmer, A. E., Jr., Simpson, C. E., Wilson, T. C., and Pitman, W. D., Evaluation of wild nut-producing *Arachis* species for forage, *Proc. XVII Int. Grassland Congr.*, Rockhampton, Australia, 1994, 2122.
58. Bowman, A. M. and Wilson, G. P., Persistence and yield of forage peanuts (*Arachis* spp.) on the New South Wales north coast, *Trop. Grassl.*, 30, 402, 1996.
59. Zdravko, B. and Fisher, M. J., Effect of planting method and soil texture on the growth and development of *Arachis pintoi, Trop. Grassl.*, 30, 395, 1996.
60. Hernandez, M., Argel, P. J., Ibrahim, M. A., and 't Mannetje, L., Pasture production, diet selection and liveweight gains of cattle grazing *Brachiaria brizantha* with or without *Arachis pintoi* at two stocking rates in the Atlantic Zone of Costa Rica, *Trop. Grassl.*, 29, 134, 1995.
61. Jones, R. M., Persistence of *Arachis pintoi* cv. Amarillo on three soil types at Samford, south-eastern Queensland, *Trop. Grassl.*, 27, 11, 1993.
62. Firth, D. J. and Wilson, G. P. M., Preliminary evaluation of species for use as permanent ground cover in orchards on the north coast of New South Wales, *Trop. Grassl.*, 29, 18, 1995.
63. Kretschmer, A. E., Jr. and Wilson, T. C., Use of the wild peanut, *Arachis kretschmeri*, as wildlife feed, *Soil Crop Sci. Soc. Florida Proc.*, 57, 108, 1998.
64. Schultze-Kraft, R., Williams, R. J., Coradin, L., Lazier, J. R., and Kretschmer, A. E., Jr., *1989 World Catalog of Centrosema Germplasm*, CIAT and IBPGR, Cali, Colombia, 1989.
65. Schultze-Kraft, R. and Clements, R. J., *Centrosema: Biology, Agronomy, and Utilization*, CIAT, Cali, Colombia, 1990.

66. Schultze-Kraft, R., Benavides, G., and Arias, A., Collection of germplasm and preliminary evaluation of *Centrosema acutifolium, Pasturas Tropicales,* 9, 12, 1987 (in Spanish).
67. Mosquera, P. and Lascano, C., Produccion de leche de vaca en pasturas de *Brachiaria decumbens* sin y con acceso controlado a bancos de proteins, *Pasturas Tropicales*, 14, 2, 1992 (in Spanish).
68. Schultze-Kraft, R., Natural distribution and germplasm collection of the tropical pasture legume *Centrosema macrocarpum* Benth., *Angew Botanik*, 60, 407, 1986.
69. Oram, R. J., b. *Centrosema pascuorum* Mart. ex Benth. (centurion), in *Register of Australian Herbage Plant Cultivars*, CSIRO, Queensland, Australia, 1990, 272.
70. Teitzel, J. K. and Burt, R. L., *Centrosema pubescens* in Australia, *Trop. Grassl.*, 10, 5, 1976.
71. Jones, R. M. and Clements, R. J., Persistence and productivity of *Centrosema virginianum* and *Vigna parkeri* cv. Shaw under grazing on the coastal lowlands of south-east Queensland, *Trop. Grassl.*, 21, 55, 1987.
72. Maass, B. L. and Torres, A. M., Off-types indicate natural out crossing in five tropical legumes in Colombia, *Trop. Grassl.*, 32, 124, 1998.
73. Strickland, R. W., Greenfield, R. G., Wilson, G. P. M., and Harvey, G. L., Morphological and agronomic attributes of *Cassia rotundifolia* Pers., *C. pilosa* L. and *C. trichopoda* Benth., potential forage legumes for northern Australia, *Australian J. Exp. Agric.*, 25, 100, 1985.
74. Jones, R. M., Bunch, G. A., and McDonald, C. K., Ecological and agronomic studies on *Chamaecrista rotundifolia* cv. Wynn related to modelling of persistence, *Trop. Grassl.*, 32, 153, 1998.
75. Clements, R. J., Jones, R. M., Valdes, L. R., and Bunch, G. A., Selection of *Chamaecrista rotundifolium* by cattle, *Trop. Grassl.*, 30, 389, 1996.
76. Tarawali, S. A., Evaluation of *Chamaecrista rotundifolia* accessions as a fodder resource in subhumid Nigeria, *Trop. Grassl.*, 29, 129, 1995.
77. Cook, B. G., Persistent new legumes for heavy grazing. 2. Wynn round-leafed Cassia, *Queensland Agric. J.*, 114, 119, 1988.
78. Jones, R. M. and Bunch, G. A., Yield and population dynamics of *Chamaechrista rotundifolia* cv. Wynn in coastal south-eastern Queensland as affected by stocking rate and rainfall, *Trop. Grassl.*, 29, 65, 1995.
79. Partridge, I. N. and Wright, J. W., The value of round-leafed cassia (*Cassia rotundifolia* cv. Wynn) in a native pasture grazed with steers in south-east Queensland, *Trop. Grassl.*, 26, 269, 1992.
80. Chakraborty, S., Charudattan, R., and DeValerio, J. T., Reaction of selected accessions of forage *Cassia* spp. to some fungal pathogens, *Trop. Grassl.*, 27, 32, 1994.
81. Hall, T. J., Adaptation and agronomy of *Clitoria ternatea* L. in northern Australia, *Trop. Grassl.*, 19, 156, 1985.
82. Jones, R. M. and Brandon, N. J., Persistence and productivity of eight accessions of *Desmanthus virgatus* under a range of grazing pressures in subtropical Queensland, *Trop. Grassl.*, 32, 145, 1998.
83. Gardiner, C. P. and Burt, R. L., Performance characteristics of *Desmanthus virgatus* in contrasting tropical environments, *Trop. Grassl.*, 29, 183, 1995.
84. Jones, R. M., Jones, R. J., and Rees, M. C., Evaluation of tropical legumes on clay soils at four sites in southern inland Queensland, *Trop. Grassl.*, 31, 95, 1997.
85. Ohashi, H., *Ginkoana — Contributions to the Flora of Asia and the Pacific Region*, No. 1, Tokyo Academia Scientific Books, Tokyo, Japan, 1973.
86. Kretschmer, A. E., Jr., Brolmann, J. B., Snyder, G. H., and Coleman, S. W., *'Florida' Carpon Desmodium [Desmodium heterocarpon (L.) DC], a Perennial Tropical Forage Legume for Use in South Florida*, Florida Agric. Exp. Stn. Circ. S-260, Gainesville, Florida, 1979.
87. Pitman, W. D. and Kretschmer, A. E., Jr., *Carpon Desmodium for Peninsular Florida Pastures: Considerations for Establishment and Use*, Florida Agric. Exp. Stn. Circ. S- 385, Gainesville, Florida, 1993.
88. Ohashi, H., Taxonomic studies in *Desmodium heterocarpon* (L.) DC. (Leguminosae), *J. Japan Bot.*, 66, 14, 1991.
89. Withington, D., Glover, N., and Brewbaker, J., *Gliricidia sepium (Jacq.) Walp.: Management and Improvement*, NFTA Special Publ. 87-01, Nitrogen-Fixing Tree Association, Wiaimanalo, Hawaii, 1987.
90. Bray, R. A., Palmer, B., and Ibrahim, T. M., Performance of shrub legumes at four sites in Indonesia and Australia, *Trop. Grassl.*, 31, 31, 1997.

91. Bray, R. A., Ibrahim, T., Palmer, B., and Schlink, A. C., Yield and quality of *Gliricidia sepium* accessions at two sites in the tropics, *Trop. Grassl.*, 27, 30, 1993.
92. Adejumo, J. O., Effect of plant age and harvest date in the dry season on yield and quality of *Gliricidia sepium* in southern Nigeria, *Trop. Grassl.*, 26, 21, 1992.
93. Cobbina, J. and Atta-Krah, A. N., Forage productivity of *Gliricidia* accessions on a tropical Alfisol in Nigeria, *Trop. Grassl.*, 26, 248, 1992.
94. Fribourg, H. A., Overton, J. R., McNeill, W. W., Culverhouse, E. W., Montgomery, M. J., Smith, M., Carlisle, R. J., and Robinson, N. W., Evaluations of the potential of hyacinth bean as an annual warm-season forage in the mid-South, *Agron J.*, 76, 905, 1984.
95. Anon., *Leucaena Research in the Asian-Pacific Region*, Proc. Workshop NFTA and IDRC, IDRC, Ottawa, Canada, 1983.
96. Bray, R. A., Breeding leucaena, in *Forage Legumes for Energy Efficient Animal Production*, Barnes, R. F., Ball, P. R., Brougham, R. W., Marten, G. C., and Minson, D. J., Eds., USDA, ARS, Washington, D.C., 1985, 317.
97. Brewbaker, J. L., Species in the genus Leucaena, *Leucaena Res. Rep.*, 7, 6, 1987.
98. Jones, R. M. and Bunch, G. A., Long-term records of legume persistence and animal production from pastures based on Safari Kenya clover and leucaena in subtropical coastal Queensland, *Trop. Grassl.*, 29, 74, 1995.
99. Jones, R. J., The value of *Leucaena leucocephala* as a feed for ruminants in the tropics, *World Animal Rev.*, 31, 13, 1979.
100. Jones, R. J., Leucaena toxicity and the ruminant degradation of mimosine, in *Plant Toxicology*, Seawright, A. A., Hegarty, M. P., James, L. F., and Keeler, R. F., Eds., CSIRO, Brisbane, Australia, 1985.
101. Jones, R. J. and Megarrity, R. G., Successful transfer of DHP-degrading bacteria from Hawaiian goats to Australian ruminants to overcome the toxicity of leucaena, *Australian Vet. J.*, 63, 69, 1986.
102. Bray, R. A. and Woodroffe, T. D., Effect of leucaena psyllid on yield of *Leucaena leucocephala* cv. Cunningham in southeast Queensland, *Trop. Grassl.*, 25, 356, 1991.
103. Castillo, A. C., Cuyugan, O. C., Fogarty, S., and Shelton, H. M., Growth, psyllid resistance and forage quality of *Leucaena leucocephala* x *L. pallida*, *Trop. Grassl.*, 31, 188, 1997.
104. Austin, M. T., Williams, M. J., Hammond, A. C., Frank, J. H., and Chambliss, C. G., Psyllid population dynamics and plant resistance of *Leucaena* selections in Florida, *Trop. Grassl.*, 30, 223, 1996.
105. Bray, R. A., Sonoda, R. M., and Kretschmer, A. E., Jr., Pathotype variability of rust caused by *Uromyces appendiculatus* on *Macroptilium atropurpureum, Plant Disease*, 75, 430, 1991.
106. Kretschmer, A. E., Jr., Sonoda, R. M., Bullock, R. C., and Wilson, T. C., Registration of IRFL 4655 *Macroptilium atropurpureum* germplasm, *Crop Sci.*, 32, 836, 1992.
107. Sonoda, R. M., Kretschmer, A. E., Jr., and Wilson, T. C., Evaluation of *Macroptilium atropurpureum* (DC) Urb. germplasm for reaction to foliar diseases, *Soil Crop Sci. Soc. Florida Proc.*, 51, 25, 1991.
108. Jones, R. J. and Jones, R. M., The ecology of Siratro based pastures, in *Plant Relations in Pastures*, Wilson, J. R., Ed., CSIRO, Melbourne, Australia, 1976, 353.
109. English, B. H., Macroptilium atropurpureum in Australia, in *Forage Seed Production.* Vol. 2. *Tropical and Subtropical Species*, Loch, D. S. and Ferguson, J. E., Eds., CABI Publishing , Wallingford, Oxon, U.K., 1999, 407.
110. Kretschmer, A. E., Jr., Sonoda, R. M., Bullock, R. C., Snyder, G. H., Wilson, T. C., Reid, R., and Brolmann, J. B., Diversity in *Macroptilium atropurpureum* (DC) Urb., *Proc. XV Int. Grassland Congr.*, Kyoto, Japan, 1985, 155.
111. Pitman, W. D., Kretschmer, A. E., Jr., and Chambliss, C. G., *Phaseybean, a Summer Legume with Forage Potential for Florida Flatwoods*, Florida Agric. Exp. Stn. Circ. S- 330, Gainesville, Florida, 1986.
112. Cameron, A. G., Tropical and subtropical pasture legumes 4. Glycine (*Neonotonia wightii*): An outstanding but soil specific legume, *Queensland Agric. J.*, 10, 311, 1984.
113. Heering, J. H., Nokoe, S., and Mohammed, J., The classification of a *Sesbania sesban* (ssp. *sesban*) collection. I. Morphological attributes and their taxonomic significance, *Trop. Grassl.*, 30, 206, 1996.
114. Heering, J. H., Nokoe, S., and Mohammed, J., The classification of a *Sesbania sesban* (ssp. *sesban*) collection. II. Agronomic attributes and their relation to biomass estimation, *Trop. Grassl.*, 30, 215, 1996.

115. Evans, D. O. and Rotar, P. P., *Sesbania in Agriculture*, Westview Press, London, 1987.
116. Evans, D. O. and Macklin, B., Eds., *Perennial Sesbania Production and Use*, (A manual of practical information for extension agents and development workers), Nitrogen Fixing Tree Assoc., Waimanalo, Hawaii, 1990.
117. Anon., B. Legumes., 24. *Sesban*, (a)., *Sesbania sesban* (L.) Merril (sesban) cv. Mount Cotton, *Trop. Grassl.*, 29, 188, 1995.
118. Lenne, J. M. and Trutman, P., *Diseases of Tropical Pasture Plants*, CABI Publishing, Wallingford, Oxon, U.K., 1994.
119. Irwin, J. A., Cameron, D. G., and Lenne, J. M., Responses of *Stylosanthes* to anthracnose, in *The Biology and Agronomy of Stylosanthes*, Stace, H. M. and Edye, L. A., Eds., Academic Press, New York, 1984, 295.
120. Gardener, C. J. and Ash, A. M., Diet selection in six *Stylosanthes*-grass pastures and its implications for pasture stability, *Trop. Grassl.*, 28, 109, 1994.
121. Anon., *Stylosanthes guianensis* cv. Mineirao, Nova leguminosa para agropecuaria dos Cerrados, *Pasturas Tropicales*, 15, 32, 1993 (in Portuguese).
122. Edye, L. A., Hall, T. J., Clem, R. L., Graham, T. W. G., Messer, W. B., and Rebgetz, R. H., Sward evaluation of eleven "*Stylosanthes seabrana*" and *S. scabra* cv. Seca at five subtropical sites, *Trop. Grassl.*, 32, 243, 1998.
123. Anon., Caatinga stylo *Stylosanthes*-'Pimar' and 'Unica', *Plant Varieties J.*, 9, 19, 1996.
124. Maass, B. L., Die troposche weideleguminose *Stylosanthes scabra* Vog. variabilitat, leistungstand und moglichkeiten zuchterischer verbesserung, *Landbauforschung Volkenrode, Sonderheft*, 1989, 97 (in German).
125. Pitman, W. D. and Adjei, M. B., Response of Shaw creeping vigna and a perennial alyceclover accession to burning, *Trop. Grassl.*, 28, 53, 1994.
126. Cook, B. G. and Jones, R. M., Persistent new legumes for intensive grazing. 1. Shaw creeping vigna, *Queensland Agric. J.*, 113, 89, 1987.
127. Reid, R. and Sinclair, D. F., *An Evaluation of a Collection of Clitoria ternatea for Forage and Grain Production*, Genetic Resources Comm. No. 1, CSIRO, Brisbane, Australia, 1980.
128. Blumenthal, M. J. and Staples, L. B., Origin, Evaluation and use of *Macroptyloma* as forage — a review, *Trop. Grassl.*, 27, 16, 1993.
129. Pengelly, B. C. and Eagles, D. A., Diversity in the tropical legume genus *Teramnus*, *Trop. Grassl.*, 30, 298, 1996.
130. 't Mannetje, L., Potential and prospects of legume-based pastures in the tropics, *Trop. Grassl.*, 31, 81, 1997.
131. Anon., Pastures for prosperity, Proceedings of the fifth tropical pasture conference, *Trop. Grassl.*, 30, 1, 1996.
132. Anon., Tropical pasture establishment, *Trop. Grassl.*, 27, 257, 1993.
133. Anon., Northern dairy feed base 2001, *Trop. Grassl.*, 27, 129, 1993.

5 Breeding and Improvement of Tropical Grasses

Byron L. Burson and Bruce A. Young

CONTENTS

I. INTRODUCTION

Based on modern taxonomic treatments, the grass family (Poaceae or Gramineae) is the fourth largest group of flowering plants. It consists of about 10,000 species and, depending on the type of classification, has from 651[1] to 785[2] genera. Tropical grasses comprise a large portion of the family. Four of the eight largest genera, *Eragrostis*, *Panicum*, *Paspalum,* and *Digitaria*, consist of many tropical species.[2]

Much of the world's land mass is used for livestock production, which is dependent on perennial grasses. Unfortunately, their importance is not recognized because their economic contribution to providing food for mankind is indirect. Consequently, grasses receive less attention and funding in comparison to food and fiber crops. Limited funding and the difficulty in breeding grasses have impeded the development of improved cultivars. Progress is further restricted in tropical grasses because they grow primarily in areas of the world where monetary resources for research are limited or unavailable.

II. FLORAL CHARACTERISTICS

A. Flower

Grass flowers are small, unobtrusive, and enclosed inside several bract-like structures (Figure 5.1A). Only during or shortly after anthesis are the uppermost structures visible. Grass flowers lack a large, colorful perianth and are reduced to essentially the reproductive organs. A flower typically consists of two small lodicules (generally considered the reduced perianth), three stamen (each with a two-lobed anther attached to a filament), and a pistil composed of a uniocular ovary bearing two stigmas, which are feathery extensions of the styles (Figure 5.1B).

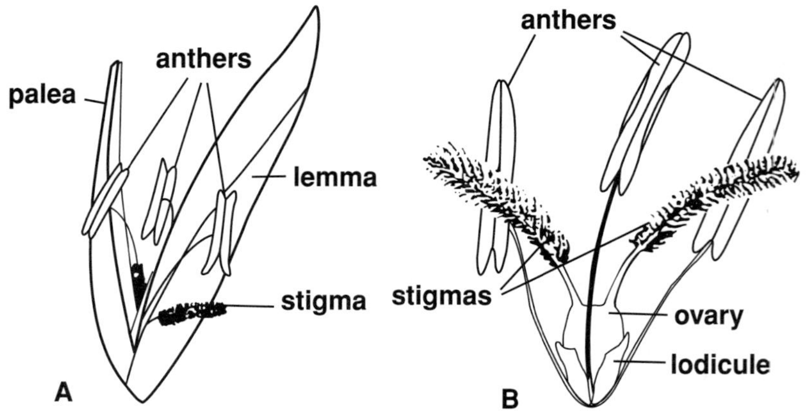

FIGURE 5.1 Typical grass floret and flower. (A) Grass floret at anthesis with the lemma and palea forced apart exposing the stigmas and anthers of the enclosed flower. (B) Parts of the grass flower with the lemma and palea removed. (From Pohlman, J. M., *Breeding Field Crops*, AVI Publishing Co., Inc., Westport, Conn., 1979. With permission from Chapman & Hall, New York.)

B. Floret

The flower is enclosed within two bracts, the lemma and palea. Collectively, the lemma, palea, and enclosed flower make up the floret (Figure 5.1A). The lemma is attached to a short axis or rachilla and, when more than one floret is present, they are attached to the rachilla in an alternating arrangement.

C. Spikelet

A spikelet typically consists of two glumes, one or more florets, and the associated rachilla, and is often referred to as the basic unit of a grass inflorescence. The glumes are empty bracts located immediately below the lowermost floret at the base of the spikelet and they vary in size, texture, and other characteristics. Some grasses have only one glume (some taxa in the Paniceae tribe), whereas others have no glumes (some taxa of the Oryzeae tribe). The number of florets in a spikelet varies from one to many depending on the taxa.

There is considerable variation in floral structure, but grass flowers are classified as either perfect or imperfect. A perfect flower has both the stamens and the pistil, whereas an imperfect flower has either the stamens or the pistil. Most grasses have perfect flowers.

D. Monoecism

Monoecious plants have imperfect flowers, but both the staminate and pistillate flowers are on the same plant. The location of the pistillate and staminate flowers may vary depending on the species. Both maize (*Zea mays* L.) and eastern gama grass (*Tripsacum dactyloides* [L.] L.) are monecious plants, but their flowers are arranged differently. In maize, the flowers are located on separate inflorescences, but in gama grass, the staminate and pistillate flowers are on the same inflorescence.

E. Dioecism

Dioecious plants have the staminate and pistillate flowers on different plants as in pampas grass (*Cortaderia selloana* [Schultes & Schultes] Asch. & Graebner) and buffalo grass (*Buchloë dactyloides* [Nutt] Engelm.).

Both of these floral anomalies, monoecism and dioecism, influence the breeding protocol and obviously are useful for controlled hybridization.

III. PRELIMINARY RESEARCH ACTIVITIES

Before initiating a plant improvement program, a breeder should establish priorities and conduct preliminary research regarding the target species. The first priority is to identify the characters that need improvement and then determine what is necessary for their alteration. For a listing and discussion of some of the more common traits that frequently need improvement, see Sleper.[3]

A thorough knowledge of the target species is essential. This begins with understanding the taxonomy of the species and its close relatives. Obviously, genetic variability for the trait under consideration also must exist in the taxon for a breeding program to succeed.

A. Germplasm Acquisition

A wide array of germplasm should be assembled. Plant material can be obtained from a number of sources. Many countries have an agency that collects, increases, preserves, and distributes germplasm. These agencies usually provide seed upon request. The recent global interest in the decline of biodiversity has encouraged many countries to collect and preserve their native germplasm. Some of the better-known national and international agencies that maintain and distribute germplasm of tropical and subtropical forages are USDA-ARS, CSIRO, EMBRAPA-CENARGEN, CIAT, ILCA, and other CGIAR centers involved with tropical forages.

Additional sources of germplasm usually can be obtained from other grass breeding programs working with the same target species or close relatives. However, many tropical grasses may not be available through these sources because of the limited number of breeding programs investigating them. Another means of obtaining additional germplasm is by plant exploration and collection in the center of origin of the species or genus of interest. These collection trips are made in the area considered to be the center of origin because that is where the greatest amount of genetic diversity exists for that particular genus or species. Tropical forage grasses are native to the warmer regions of the world with a majority originating in Africa.

B. Germplasm Evaluation

After germplasm is acquired, it should be closely evaluated both in the field and the laboratory. Initially, the germplasm should be screened to determine the amount of diversity that exists for the traits the breeding program is addressing. The more promising accessions should be selected and evaluated for method of pollination, fertility, chromosome number, method of reproduction, and flowering behavior.

1. Method of Pollination

In most grasses, the pollen is transported by the wind, but in some tropical grasses, insects transport the pollen.[4] Regardless of how the pollen is transported, the plants are either self- or cross-pollinated or a combination of both.

Self-pollination is determined by growing an accession in isolation or by enclosing its inflorescences in a pollinating bag prior to anthesis. Cross-pollination is determined by growing several accessions of the same species in an area where they will interpollinate. It is advantageous if the pollen source has a recognizable dominant trait that is expressed in the offspring when cross-pollination occurs. The percent seed set under open- and self-pollinated conditions provides a general indication as to the mode of pollination as well as the plant's fertility. It is not unusual for a cross-pollinated species to produce some self-pollinated seed and for self-pollinated species to produce some open-pollinated seed. Knowing the method of pollination is essential in selecting the appropriate breeding protocol. Although most perennial tropical grasses are predominately cross-pollinated, there are exceptions. For example, many *Paspalum* species, with the exception of Pensacola Bahia grass (*P. notatum* var. *saurae* Parodi),[5] are highly self-pollinated.[6]

In some species, the method of pollination can be determined by closely observing the floral organs at anthesis. Eastern gama grass and many *Pennisetum* species are protogynous in that the stigmas are exserted from the floret one or more days prior to anther exsertion. This permits the stigmas to be pollinated by adjacent plants and ensures cross-pollination. In buffelgrass (*Pennisetum ciliare* [L.] Lam syn *Cenchrus ciliaris* L.), the stigmas are receptive when they are exserted from the floret, which can be 72 hours prior to anther exsertion.[7] Protandry is the opposite of protogyny in that anthers are exserted prior to stigma exsertion; however, this behavior has been reported only in cool-season grasses.[8] Both protogyny and protandry ensure cross-pollination and can be used in a hybridization program.

2. Self-Incompatibility and Male Sterility

Self-incompatibility and different forms of male sterility are two other means of preventing self-pollination and can be used in hybridization programs. Self-sterile accessions were discovered in sexual, diploid *Paspalum* species.[6,9] Burson[6] determined that the self-incompatibility in *P. intermedium* Munro. ex Morong. resulted from the failure of its pollen tubes to grow beyond the stigma. This self-incompatible germplasm has been used to produce interspecific *Paspalum* hybrids without tedious hand emasculations. Male sterile pearl millet (*Pennisetum glaucum* [L.] R. Br.) has been a valuable tool in improving that species and in producing a number of interspecific *Pennisetum* hybrids.[10]

It is apparent that naturally occurring morphological or developmental variations have evolved to enforce a specific mode of pollination. However, the method of pollination of a plant is rarely obligate and can be influenced by the environment and/or through genetic changes. It is important that all germplasm under evaluation be screened closely for method of pollination and fertility, including pollen viability.

3. Method of Reproduction

Knowing how a species reproduces is extremely important and its method of reproduction should be understood before a breeding program is initiated. This is of paramount importance in tropical grasses because apomixis is prevalent in many of the genera. Apomixis is an asexual form of reproduction whereby seed are produced in the absence of fertilization. An unreduced egg in the embryo sac develops into an embryo without fertilization. Thus, there is no variability among the progeny because all are genetically identical to the maternal parent. Obviously, sexual and apomictic species require different breeding methodologies.

4. Cytogenetics

Because of the cytogenetic diversity of tropical grasses, an understanding of the cytology and chromosome number of a species is essential before initiating a breeding program. The chromosome number varies widely in most tropical genera. For example, in the genus *Paspalum*, it ranges from a low of $2n = 2x = 12$ in *P. hexastachyum* Parodi[11] to a high of $2n = 16x = 160$ in *P. floridanum* Michx.[12] Some genera have species with different base chromosome numbers. *Paspalum* has base numbers of 6 and 10; *Panicum*, 8, 9, and 10; and *Pennisetum*, 5, 7, 8, and 9. Polyploidy is common in tropical grasses. Many species have cytotypes with different chromosome numbers. For example, dallis grass (*Paspalum dilatatum* Poir.) biotypes have chromosome numbers of $2n = 4x = 40$, $5x = 50$, and $6x = 60$[13,14] and Bahia grass (*P. notatum* Flugge) cytotypes have 20, 30, 40, and 50 chromosomes.[12,15] In dallis grass, biotypes with the same chromosome number have different genomic compositions and, consequently, different meiotic chromosome pairing behaviors.[13,14]

In addition to differences in ploidy levels, there are also different types of polyploidy. Allopolyploids, autopolyploids, and segmental allopolyploids all occur in tropical grasses and influence the expression of genetic traits.

Irregular meiosis often occurs in polyploids and produces gametes without the full complement of chromosomes. This reduces pollen viability and causes sterility. Fertility is also reduced in aneuploid plants. Therefore, it is important to know the chromosome number, meiotic chromosome behavior, and pollen viability of germplasm used in a breeding program. This is especially true for tropical grasses because cytological irregularities and chromosomal variations are common.

If interspecific hybridization is an approach being contemplated, an understanding of the cytogenetic relationships between the different species involved will greatly facilitate the program.

5. Flowering Behavior

Knowledge of a species' photoperiod response is important. Although most tropical grasses are either day neutral or short-day,[4] a few tend to be long-day in their flowering response. In the southern U.S., *Paspalum intermedium* flowers in May and June, whereas *P. conspersum* Schrad. ex Schult. flowers in September and October, and dallis grass flowers continuously from May to November. The time of day when anthesis occurs should be ascertained because species tend to vary for this characteristic.[16] A knowledge of these two events is essential when it is necessary to synchronize flowering to cross different plants or species.

IV. APOMIXIS

Because apomixis is prevalent in tropical and subtropical grasses, especially in the Panicoideae and Chloridoideae subfamilies, this reproductive phenomenon merits special attention in discussing grass breeding. Watson[17] lists the genera in which apomixis is reported. Because the cytology and reproductive behavior of many tropical grasses have not been studied in depth, numerous apomictic species probably remain unidentified. If apomixis is present in a target species, it is imperative that the breeder understand this reproductive phenomenon before establishing a breeding protocol. This begins with an understanding of the different apomictic mechanisms and extends to selecting and using the appropriate apomictic breeding scheme for genetic improvement.

Before discussing the apomictic mechanisms, a brief review of the development of a sexual embryo sac is in order. A single cell, known as the archesporial or megaspore mother cell, in the hypodermal layer of a young ovule, enlarges and undergoes meiosis to produce a linear tetrad of megaspores. Because these four cells are the products of meiosis, their chromosome number is haploid or half the number of the adjacent somatic nucellar cells of the ovule, and the linear order of the chromosomes is altered because crossing over occurred during meiosis. The three megaspores nearest to the micropyle degenerate, while the remaining chalazal megaspore (the one most distant from the micropyle) enlarges and undergoes three mitotic divisions to produce an eight-nucleate embryo sac (female gametophyte) of the *Polygonum* classification. The nuclei in the mature embryo sac are all reduced in chromosome number and differentiate as an egg cell, two synergid cells, two polar nuclei that form the central cell, and three or more antipodal cells. Immediately after anthesis, the pollen grains come in contact with the stigmas and usually germinate upon contact with the stigma papilla. Following germination, the pollen tubes grow through the stigma, style, and ovary, and a tube enters into the embryo sac and releases two nuclei. One nucleus fertilizes the egg cell to produce the embryo, and the other sperm nucleus unites with the two polar nuclei to produce the endosperm. These structures, along with other tissues in the ovule, develop into a mature seed.

A. Apomictic Mechanisms

Three apomictic mechanisms are generally recognized: apospory, diplospory, and adventitious embryony. Apospory and diplospory are sometimes referred to as gametophytic apomixis. The classification of the three types is based on the origin and development of the end product. To determine which mechanism is involved, it is necessary to microscopically observe the development inside an ovule. This is usually accomplished by dehydrating the pistil, embedding it in paraffin, sectioning it with a microtome, staining it with the appropriate dyes, and examining megasporogenesis and subsequent embryo sac development in the serial sections of the ovule with light microscopy. Another method is to dehydrate the pistil, clear it with methyl salicylate, and observe the development inside the intact ovule using interference microscopy.[18]

1. Apospory

In most aposporous species, the archesporial cell or megaspore mother cell normally enlarges and undergoes meiosis to produce a linear tetrad of megaspores. However, at this stage or shortly thereafter, one or more of the adjacent nucellar cells in the ovule begin to enlarge and their nuclei become more dense. These enlarged nucellar cells begin to divide mitotically. By this time, the functional megaspore, which had developed from meiosis, usually aborts and the enlarging nucellar cells occupy its area. As the nucellar cells continue to enlarge, they become vacuolated and their nuclei divide mitotically to produce aposporous embryo sacs with a varying number of nuclei. The difference in nuclei number and arrangement is a function of the number of the mitotic divisions that occur and the orientation of the spindle during the first mitotic division. Because of these developmental differences, aposporous embryo sacs are subdivided into the bipolar "*Hieracium* type" and the monopolar "*Panicum* type."[19-21] Most aposporous tropical and subtropical grasses have the four-nucleate *Panicum* type of embryo sac development, and most aposporous temperate grasses have the eight-nucleate *Hieracium* type of embryo sac.[21]

For the *Hieracium* type, the unreduced nucellar cell undergoes three mitotic divisions to produce an eight-nucleate embryo sac that is organized similar to the reduced *Polygonum* type.[20] During the first mitosis in the enlarged nucellar cell, the polarization and orientation of the spindle are usually parallel to the long axis of the cell and the resultant nuclei are at opposite ends of the cell. The subsequent two mitotic divisions produce four nuclei at each end of the sac and a large vacuole develops in the center. Thus, the sac is bipolar. These unreduced nuclei differentiate as an egg cell, two synergid cells, two polar nuclei, and three antipodal cells and appear similar to a reduced *Polygonum* sac.

In the *Panicum* type sac, there is no initial polarization and only two mitotic divisions occur. During the first mitotic division, the spindle is oriented crosswise to the long axis of the cell and the two nuclei produced are located in the micropylar end of the developing sac. At the same time, a vacuole forms in the chalazal end. These two nuclei divide again producing a mature four-nucleate embryo sac with all nuclei located in the micropylar end and a large vacuole in the chalazal end. In most warm-season grasses with this type of sac, the nuclei usually differentiate as an egg cell, one polar nucleus, and two synergids.[22-24] However, in buffelgrass, they sometimes differentiate as an egg cell, two polar nuclei, and one synergid.[24] Occasionally, only three nuclei are present and they differentiate as an egg and two polar nuclei.[24] Contrary to popular belief, all apomictic Panicoideae grasses do not have the *Panicum* type embryo sac. Many *Paspalum* species have bipolar development because the first mitotic spindle is parallel to the long axis of the cell and nuclei develop at each end of the sac, similar to the *Hieracium* type of development. However, only four to six nuclei are usually produced but none differentiate into antipodals. Since most aposporous embryo sacs in tropical and subtropical grasses do not have antipodal cells, the absence of antipodals is an important criterion used to cytologically distinguish between mature aposporous sacs and meiotically derived sexual sacs. Apospory is the most common form of apomixis in tropical grasses.

2. Diplospory

A diplosporous embryo sac originates from a generative cell (archesporial cell or megaspore mother cell) in an ovule, whereas aposporous embryo sacs develop from somatic cells. Unreduced diplosporous embryo sacs are produced by at least three different mechanisms, but only two are reported to occur in apomictic tropical grasses. The most common is the *Antennaria* type. Initially, the megaspore mother cell differentiates similar to processes in ovules of sexual plants. Instead of undergoing meiosis, it usually begins to elongate and becomes vacuolated, but its nucleus never initiates meiosis. Eventually, the nucleus divides mitotically, producing an unreduced, two-nucleate embryo sac. The number of additional mitotic divisions varies for different species and even within species. Weeping love grass (*Eragrostis curvula* [Schrad.] Nees),[25] lehmann love grass (*E. lehmanniana* Nees),[26] and eastern gama grass[27,28] are diplosporous apomicts with the *Antennaria* type of development, and the number of nuclei in their mature embryo sacs differs. In weeping love grass, mature diplosporous sacs with four, six, and eight nuclei were reported by Voigt and Bashaw.[25] Only sacs with four nuclei were observed in lehmann love grass and the nuclei remained in the micropylar end of the sac. The nuclei differentiated as an egg, two synergids, and one polar nucleus. At maturity the synergids disintegrated quickly and two-nucleate sacs were frequently observed.[26] In eastern gama grass, the sacs had eight nuclei (egg, two synergids, two polars, and three antipodals) and appeared similar to a *Polygonum* type sac.[27,28] To distinguish between eight-nucleate diplosporous sacs and meiotically derived sexual sacs, the early developmental stages in the ovule have to be carefully observed. The best evidences of the *Antennaria* type of diplospory are the absence of (1) meiosis, (2) a dyad, and (3) linear tetrad or any remains of the three micropylar members of the linear tetrad. The presence of a single, elongated megaspore mother cell extending from just above the micropyle toward the center of the ovule also suggests *Antennaria* type development.

The second type of diplospory reported in grasses is the *Taraxacum* type. In this case the megaspore mother cell actually initiates meiosis, but the chromosomes do not pair with their homologues because of asynapsis. These unpaired chromosomes (univalents) remain scattered throughout the metaphase plate during metaphase I and do not migrate to the poles at anaphase I. They remain congregated near the center of the cell and eventually are enclosed by a nuclear membrane forming a single nucleus (restitution nucleus) with an unreduced chromosome number. This restitution nucleus undergoes a second division that is analogous to meiosis II followed by cytokinesis to produce two unreduced cells. The chalazal member of this dyad usually divides mitotically to produce an unreduced eight-nucleate embryo sac and the micropylar cell deteriorates.[19] Chao[29] reported the *Taraxacum* type of diplospory in asynaptic plants of two *Paspalum* species. Even though the form of diplospory in eastern gama grass is the *Antennaria* type, Leblanc et al.[28] reported a very low incidence of the *Taraxacum* type of development in a few accessions of eastern gama grass, *T. dactyloides hispidum* Hitchc., *T. bravum* Gray, and *T. zopilotense* Hern.-Xol. and Randolph. This type of diplospory is rare in tropical grasses.

The third type of diplospory is the *Ixeris* type. It has not been reported in tropical grasses.

3. Adventitious Embryony

In adventitious embryony, embryos develop directly from unreduced somatic cells in the ovule by mitotic division, but an embryo sac is not formed. The embryos appear more like developing buds. However, for the mature unreduced embryo to develop into a seedling, a sexual embryo sac has to develop in the same ovule. Adventitious embryos lack a food source and are dependent upon the endosperm in the mature sexual embryo sac for their nutrition. This mechanism has not been reported in tropical grasses, but it is common in *Citrus*.

B. Pseudogamy

Pseudogamy is a term associated with apomixis and refers to the endosperm development in an apomictic sac (aposporous and diplosporous). It has been demonstrated in some apomictic grasses that the polar nuclei and a sperm nucleus fuse to produce endosperm.[19,20,22] However, in other apomictic grasses, it is not known if fertilization is necessary or if only pollination is required for endosperm development and perhaps for the maturation of a precocious embryo. In either case, seed set in an apomict is influenced by pollen viability.

C. Parthenogenesis

Parthenogenesis is the autonomous development of an embryo without pollination. When mature ovules of apomictic plants are examined prior to anthesis, a young embryo is often observed instead of the expected egg cell, indicating the embryo developed without fertilization.[24]

D. Obligate Apomicts

Both aposporous and diplosporous apomicts are also classified by the degree or level of apomixis expressed. Plants that reproduce entirely by apomixis are obligate apomicts. Common dallis grass is an example. This grass is a natural pentaploid hybrid with an unbalanced chromosome constitution and is apparently completely apomictic because aneuploid progeny, which would result from an occasional meiotic event, have never been reported.

E. Facultative Apomicts

Species that reproduce by both apomixis and sexuality are facultative apomicts, and most apomictic species are facultative. Even though apomixis is genetically controlled, there is evidence that different environmental stimuli, primarily photoperiod, alter the level of apomixis expressed in facultative apomicts.[30-32]

The breeder should know if a species is an obligate or a facultative apomict. The level of apomixis and sexuality in facultative apomicts influences the strategies used in breeding apomictic species.

F. Indicators of Apomixis

Besides cytologically observing megasporogenesis and embryo sac development as described above, there are other methods of detecting apomixis. A practical method for a breeder is a progeny test. Because apomixis is an asexual cloning process whereby the seeds are genetic copies of the maternal parent, the resulting progeny from an obligate apomict or a highly apomictic plant are uniform and appear identical to that parent. Thus, uniform progeny from a heterozygous or cross-pollinated plant suggests apomixis. However, uniform progeny from a self-pollinated, homozygous plant can be mistaken for apomixis and the breeder should be diligent in avoiding such misclassifications by knowing the mode of pollination of the species in question. The most reliable way to determine apomixis is to combine a cytological analysis with a progeny test.

Detecting apomixis in a facultative apomict is more difficult. Both of the above-mentioned methods can be used, but a more thorough evaluation is required. The presence of a high number of identical progeny along with an occasional off-type plant indicates facultative apomixis. However, difficulty arises when more off-type plants are present in a progeny row than uniform types resembling the maternal parent, or when approximately equal numbers of uniform and off-type plants occur. In these cases, it cannot be determined whether the maternal parent reproduces sexually or by facultative apomixis. In both situations, the embryo sac development in all of the progeny should be cytologically examined.

Additional tools are now available to assist in identifying apomictic plants and determining mechanisms; however, these have not eliminated the necessity of cytologically evaluating the germplasm, but only serve to aid in addressing problems. The presence or absence of callose in the walls of megaspore mother cells has been used to determine if a plant reproduces sexually or by diplospory.[33] Callose is absent in the walls of the megaspore mother cells in diplosporous plants, but present in the sexual species. Molecular markers linked to the genomic region controlling apomixis have been used to identify apomictic plants in a population segregating for method of reproduction.[34] Unfortunately, obligate and facultative apomicts cannot be distinguished from one another using this approach, and cytological studies are necessary to identify the two reproductive types.

A grass breeder should be aware of additional characteristics that may indicate apomictic reproduction. These are only "indicators" and should not be used as a means for predicting or classifying the reproductive behavior of a plant or population. In hybridization programs where the emasculation technique is very reliable, the recovery of uniform progeny resembling the maternal parent suggests that the female parent may be apomictic. Another indication of apomixis occurs when a paternal parent in a cross has a dominant genetic marker that is not expressed in the offspring. Other potential indicators of apomixis include a maternal parent with (1) irregular meiosis, (2) aneuploid chromosome number, or (3) unbalanced ploidy level (3X, 5X, 7X, etc.) while the chromosome number of its progeny remain constant and have good seed set. Seed set should not be used to classify the reproductive behavior of plants with an unbalanced ploidy level. This approach has been advocated, but it is not reliable because not all apomictic plants have good seed set and sexual plants can be more fertile than their apomictic counterparts. Another problem with using seed set is that facultative apomicts cannot be identified. The emergence of twin seedlings from germinating seed is another indicator of apomixis. Twin seedlings are much more likely to occur in seed from an aposporous apomict because multiple embryo sacs frequently develop in their ovules. Caution should be used in employing this characteristic to predict apomixis because twin seedlings can develop in sexual plants because of polyembryony and can be mistaken for apomixis.

More comprehensive information on apomixis is available elsewhere.[19-21,35,36]

V. BREEDING APOMICTIC GRASSES

Because apomixis is prevalent in tropical grasses, there is a high probability that it will be encountered in a breeding program. Apomixis can be either an impediment or a valuable tool to genetic improvement. Traditionally, apomictic breeding was merely the selection and increase of superior, naturally occurring, apomictic ecotypes. Most cultivars of apomictic tropical grasses used today were obtained in this manner. An excellent example is guinea grass (*Panicum maximum* Jacq.), an important forage grass throughout the tropics, especially the West Indies and tropical South America. Different apomictic ecotypes exist in nature and the more vigorous forage types were selected and released as cultivars. A number of apomictic guinea grass cultivars and strains have been obtained in this manner. Two recent examples are the cultivars 'Tanzania-1' and 'Mombaça.' Tanzania-1 was selected and released in 1971 in the Ivory Coast by the French research organization ORSTOM. This cultivar was introduced into Brazil where it is being promoted by EMBRAPA. Using African germplasm obtained from ORSTOM, EMBRAPA breeders recently selected and released another apomictic ecotype as the cultivar Mombaça. Both of these apomictic cultivars are superior to the widely grown cultivar 'Colonião' (Colonial) for several traits. The selection and release of superior apomictic ecotypes as cultivars is common in other warm-season grasses such as buffelgrass, weeping love grass, Bahia grass, Old World bluestems (*Bothriochloa-Dichanthium* complex), dallis grass, *Paspalum plicatulum* Michx., *Brachiaria* species and others.[4,37]

The actual breeding of apomictic grasses using hybridization is a relatively recent occurrence and progress is promising. A major breakthrough occurred when it was determined that apomixis is genetically controlled and the method of reproduction can be manipulated similar to any other genetically controlled trait. Several researchers have investigated the genetic control of apomixis in various tropical and subtropical grasses and considerable controversy has resulted. Bashaw and Hanna[36] and Savidan[21] have summarized the different findings and theories. However, indications are that apomixis is controlled by one to a few genes. Molecular mapping in *Brachiaria*, maize-*Tripsacum* hybrids, buffelgrass, and *Pennisetum squamulatum* supports the concept that the gene(s) controlling apomictic reproduction are under simple genetic control.[38-41] However, Carman[42] recently hypothesized that apomixis may result from asynchronously expressed duplicate genes that control female development and not from specific apomictic genes.

The success of breeding apomicts depends on the availability of sexual germplasm that is cross-compatible with its apomictic counterparts. Sexual plants usually exist in predominately apomictic species, but they are difficult to discover. Most have been found in large germplasm collections from the species' center of origin or in large breeding populations. Typically, apomixis occurs only in polyploids, while diploids are sexual. If the only source of sexuality is at the diploid level, it is necessary to double the diploid's chromosomes to produce a sexual tetraploid to cross with a tetraploid apomict. Burton and Forbes[43] and Savidan[44] used this approach to produce sexual tetraploid Bahia grass and guinea grass germplasm, respectively. However, others[25,45-47] have discovered naturally occurring sexual tetraploids in buffelgrass, guinea grass, and weeping love grass.

In apomictic breeding, the sexual plant is used as the female parent and is pollinated with the apomict. Pollen development is not influenced by apomixis and the pollen of most apomicts is sufficiently viable. Hybridization of a sexual and an apomict results in new genotypes that usually segregate for method of reproduction. The apomictic F_1 hybrids are true-breeding and the heterosis expressed is captured and remains unchanged over generations. Seed set in the apomictic hybrids usually is adequate; however, poor seed set can be a problem. This breeding system is very efficient because seed of the fertile, superior, apomictic hybrids can be increased and the germplasm evaluated immediately. It also allows for the rapid release of new cultivars since the lengthy evaluation procedures encountered in breeding sexual cultivars are eliminated. Another advantage is that isolation is not necessary for seed production.

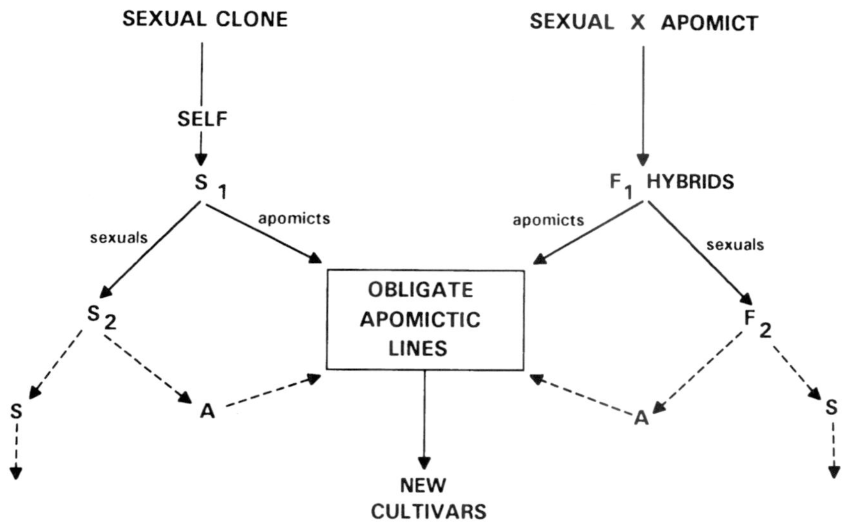

FIGURE 5.2 Diagram of a breeding scheme used to produce true breeding apomictic cultivars using obligate apomicts. Sexual parent is heterozygous for method of reproduction. (From Taliaferro, C. M. and Bashaw, E. C., *Crop Sci.*, 6, 473, 1966. With permission from the Crop Science Society of America.)

A. Obligate Apomicts

For efficient breeding, the optimal situation is to cross an obligate sexual female with an obligate apomict and their progeny segregate for method of reproduction. It is also advantageous to produce more apomictic than sexual progeny in the F_1 generation because every true-breeding apomictic hybrid is a potential cultivar. Many naturally occurring apomicts are the products of wide crosses and are highly heterozygous. Because they are propagated vegetatively, this genetic diversity is locked in and maintained indefinitely. When a sexual plant is crossed with an apomict, the contact with sexuality releases the heterozygosity stored in the apomict and a myriad of different phenotypes is expressed in the F_1 progeny. If the female parent is heterozygous for sexuality, the F_1 hybrids will segregate for method of reproduction.

Taliaferro and Bashaw[48] crossed a sexual tetraploid plant, discovered in a field of apomictic common buffelgrass,[45] with different apomictic buffelgrass ecotypes and reported the method of reproduction was controlled by two epistatic genes. They proposed a breeding scheme to produce true-breeding apomictic F_1 cultivars (Figure 5.2). The sexual plant, a presumed mutant, was heterozygous for method of reproduction. This plant was unique because when self-pollinated, its progeny segregated for method of reproduction (Figure 5.2). Unfortunately, that plant is no longer available. When the sexual buffelgrass plants available today are self-pollinated, the progeny do not segregate for method of reproduction, but are sexual. Because of this, the selfing of a sexual plant to produce S_1 progeny segregating for mode of reproduction, illustrated in Figure 5.2, is not applicable today. However, the first apomictic buffelgrass cultivar developed and released using the breeding scheme in Figure 5.2 was 'Higgins,' a superior S_1 strain selected from selfed progeny from the sexual plant.[49] Two true-breeding, apomictic F_1 buffelgrass hybrids were selected from approximately 100 hybrids recovered from controlled sexual x apomictic crosses and were released as 'Nueces' and 'Llano' buffelgrass in 1977.[50] Contrary to that reported by Savidan,[21] these were the first apomictic cultivars developed and released that were not merely ecotype selections. The basic hybridization scheme outlined in Figure 5.2 is also being used in breeding *Brachiaria* species in Brazil[51] and Colombia[85] and guinea grass in Japan.[23,52]

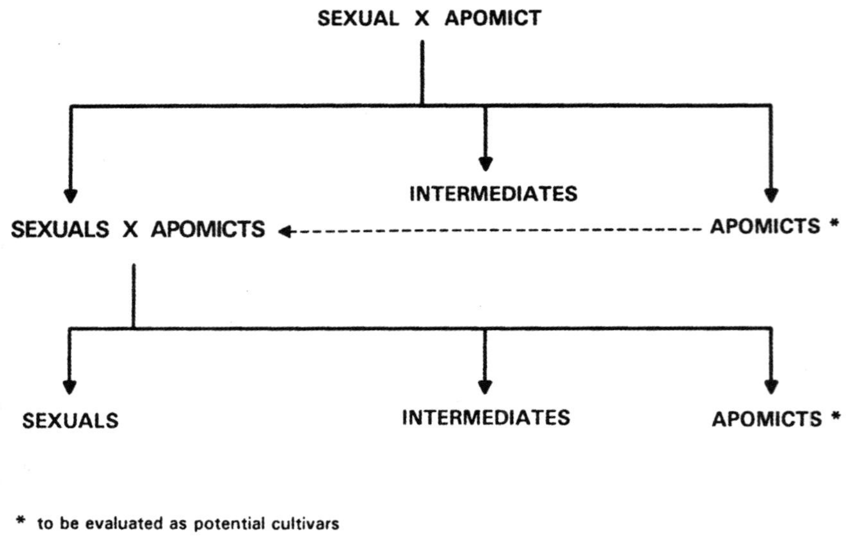

FIGURE 5.3 Diagram of a breeding scheme used to produce new apomictic cultivars using facultative apomicts. (From Voigt, P. W. and Burson, B. L., *Proc. XIV Int. Grassl. Congr.*, 1983, 160. With permission from the authors.)

B. Facultative Apomicts

Because facultative apomicts have the capacity for sexual reproduction, some variability is expressed in their progeny. The major problem in breeding facultative apomicts is the inability to control this variation.[53] If an obligate sexual plant is available to use as the female parent, facultative apomicts, usually the highly apomictic, are crossed onto the sexual female parent. Facultative apomicts are more difficult to use in a breeding program than obligate apomicts because the majority of the hybrids are also facultative rather than obligately apomictic or obligately sexual. It is usually difficult to determine the level of sexuality and apomixis by progeny testing. Extensive progeny testing is necessary to identify the more highly apomictic hybrids, but cytological analysis of embryo sac development is required. One advantage of breeding forage grasses compared to other crops is that a high level of uniformity in the end product is not always necessary. Therefore, if variability is not excessive, facultative apomicts can be used to develop improved highly apomictic forage cultivars.

This approach was used to develop improved weeping love grass hybrid germplasm.[54] A sexual plant was crossed with a highly apomictic accession, and the highly sexual hybrids were identified and crossed with the superior highly apomictic hybrids. This procedure was repeated until desirable, highly apomictic hybrids were recovered for testing. A diagram outlining this apomictic breeding scheme is shown in Figure 5.3.

The breeding schemes outlined above provide the basic principles of breeding both obligate and facultative apomicts when a sexual source is available. For simplicity purposes, the models are presented on the assumption that apomixis is controlled by a single gene; however, in reality, both models have been successful in developing improved germplasm. Hanna[55] discusses these basic breeding schemes in more detail by addressing the outcome depending on the genetic composition of the parents. Savidan[21] discusses the models on the basis of using the sexual hybrids to increase the diversity in the sexual pool.

C. Fertilization of Unreduced Gametes

A rare phenomenon that provides a means for gene transfer in obligate apomicts without sexual counterparts is the fertilization of an unreduced egg (2n + n).[56] This process is also referred to as

B_{III} hybridization, and the hybrids are called B_{III} hybrids.[57] Apparently, the unreduced egg in an apomictic embryo sac does not initiate cell division prior to pollination and subsequently is fertilized with a sperm nucleus from a pollen tube that has entered the embryo sac. The resultant B_{III} hybrid has the normal somatic chromosomes from the apomictic maternal parent plus a haploid set from the paternal parent. This produces a plant with a higher chromosome number genetically different from its apomictic maternal parent.

This event occurs infrequently but has played a major role in polyploidization and speciation of tropical grasses. Bashaw et al.[57] crossed two facultative apomicts, *Pennisetum flaccidum* Griseb. and *P. mezianum* Leeke, and recovered 92 B_{III} hybrids and 109 normal (n + n) B_{II} hybrids. Many of the B_{III} hybrids were more fertile than the B_{II} hybrids. These findings demonstrate that the frequency of B_{III} hybridization may be sufficiently high so that this approach might be used to improve some apomictic species.

Little is known about what initiates this phenomenon; however, Martinez et al.[58] provided some insight. They reported that the frequency of B_{III} hybrids increased in apomictic Bahia grass (*Paspalum notatum* Flugge) when the stigmas were pollinated two to three days prior to anthesis. Vielle et al.[24] investigated the ultrastructure of egg cells in both apomictic and sexual embryo sacs of buffelgrass and determined that a cell wall developed around and enclosed the apomictic egg cell prior to anthesis. However, it did not form around the sexual egg cell until after fertilization. This may explain the findings of Martinez et al.[58] because the cell wall may prevent a pollen tube from penetrating the apomictic egg cell at the time of anthesis, but when pollinated early, the tube enters the unreduced egg cell because a cell wall has not formed.

VI. BREEDING SEXUAL GRASSES

Genetic improvement of forage grass species that reproduce sexually is usually practiced using one of two methodologies — ecotypic selection or recurrent selection. Ecotypic selection is the evaluation and release of superior accessions. Accessions are populations of plants grown from seed collected from one or more plants. Ecotypic selection involves little breeding per se, but does require the field collection and screening of numerous accessions. Recurrent selection is any form of plant improvement in which the breeding methodology is cyclical in nature. Starting with a base population, superior plants are identified and recombined using one of several protocols to concentrate the desirable genes and form the population for the next cycle of selection. The process is then repeated until the desired level of response is achieved or there is no more genetic variation to exploit.

The success of recurrent selection depends on the degree of inheritance; i.e., heritability of the characteristic for which selection is being practiced. Recurrent selection can enhance the agronomic performance of a given population, while the success of ecotypic selection is directly dependent on the superiority of the initial collection of plants and focuses on thorough testing to identify the "best" accessions and to give assurance of their genetic superiority. No genetic improvement of the original accessions is anticipated from ecotypic selection, although culling of "poor" plants is often practiced in the early stages of evaluation.

A. Ecotypic Selection

Most tropical cultivars are either superior accessions selected from natural populations or superior intraspecific hybrids. Field collections are made of open-pollinated seed from individual plants. In some cases, bulked seed from several plants is planted, usually in long rows in one or more environments (locations) and evaluated for agronomic potential including forage and seed production and persistence. Sometimes inferior plants are culled at this stage. Vegetative increases of individually collected plants that are sterile or produce little seed are also a form of ecotypic selection. Seed (or vegetative material) is increased from promising accessions, tested in larger plantings, and released as cultivars, if warranted. Pensacola Bahia grass, a widely grown, seed-propagated species, is a prominent example of a cultivar developed through ecotypic selection. Several Bahia grass plants were collected near Pensacola, Florida. Seed from these plants was increased, tested, and released as Pensacola Bahia grass. 'Redalta,' 'Greenalta,' and 'Bigalta' limpograss (*Hemarthria altissima* [Poir.] Stapf and C. E. Hubbard)[59] are increases of three superior accessions collected in South Africa. Although they reproduce sexually, these cultivars produce little seed and, therefore are vegetatively propagated. 'Transvala' digitgrass (Pangolagrass) (*Digitaria decumbens* Stent) is another example of a vegetatively propagated cultivar that originated as an increase of a single plant collected in South Africa.[60]

A variation of ecotypic selection has been used to develop many tropical cultivars. Intraspecific hybridization is used in generating progeny from which the selection of superior F_1 genotypes can be made. 'Merkeron' napier grass (*Pennisetum purpureum* Schumach.) is a vegetatively propagated F_1 clone resulting from a cross between a tall genotype and a leafy, disease-resistant dwarf genotype.[61] 'Mott' napier grass (elephant grass) is a vegetative increase of a selfed progeny of Merkeron napier grass.[62]

B. Recurrent Selection

Two basic forms of recurrent selection are used in forage grass breeding — mass selection and the polycross method.

1. Mass Selection

Mass selection, although the term is not descriptive, begins with the selection of superior plants (for one or more characters) from a base population. Seed is collected from the selected plants and bulked to form the population for the next selection cycle. The cycle is usually, but not necessarily, repeated several times. The number of cycles depends on the extent of improvement desired and whether the gain per cycle has diminished below the cost-benefit threshold. Because each cycle contains a genetic contribution from unselected plants (open-pollinated seed is the result of fertilization by pollen from all plants, selected and unselected), the potential genetic gain in each cycle is reduced. Therefore, mass selection is usually reserved for characteristics with very high heritability. However, mass selection is easier and the amount of time for each selection cycle is less than for other recurrent selection methods.

2. Polycross Method

The polycross method is widely used in forage grass breeding, although it has been underutilized in the improvement of tropical species. Although the cycle interval is longer than for mass selection, the polycross method is more efficient for those agronomic characters that are of low to moderate heritability. Because even high heritability estimates often have large standard errors, the polycross method is generally preferred by forage breeders because it gives the best probability that significant breeding progress will be made. Genetic studies that have identified the pattern and degree of inheritance of a particular character will often suggest an alteration of the basic procedure to increase the efficiency of the polycross method. However, all variations have one step in common. Once superior plants are identified, they are recombined in isolation so that only the genes from the selected plants contribute to the next cycle of selection. With mass selection, only one half of the genes from the selected plants contribute to the next generation.

The basic procedure of the polycross method starts with the identification of superior plants from the base population. These selected plants are vegetatively propagated (cloned) into several identical plants (ramets) and recombined by planting the "replicated" genotypes in a randomized block design (to insure random pollen distribution) in an isolated crossing block. Seed is then collected from the polycross and the cycle is repeated.

3. Variations of the Polycross Method

a. *Single plant selection vs. progeny testing*

If the heritability of a character is relatively high, acceptable breeding progress can be made using "single-plant" selection. Individual superior plants are selected from a population, cloned and recombined in isolation. Seed is bulked from the polycross and a random sample of plants grown from this seed is used for the next selection cycle.

If heritabilities are low, the polycross method with progeny testing is usually recommended. This increases the probability that truly genetically superior plants are chosen for the next cycle, because the "best" plants are identified on the basis of the actual performance of their progeny. Performance is then confirmed before the next cycle of selection. In this variation of the polycross method, plants are selected on the basis of their phenotype and recombined in isolation as in single plant selection. At this point, seed are collected from individual plants. Seed from each genotype are bulked over replicates to form half-sib families. The resulting progeny are evaluated in family rows in a replicated nursery. Superior families are identified from the mean performance of their progeny. The original maternal plants that produced the superior progeny are vegetatively propagated and recombined in an isolated crossing block to produce plants for the next cycle of selection. The disadvantage of progeny testing is that it increases the generation time for each cycle.

b. *Restricted recurrent phenotypic selection (RRPS)*

This variation of the polycross method has been very successful in increasing the forage yield of Pensacola Bahia grass.[63,64] It imposes many restrictions or methodological changes on the basic concept of the polycross method to improve breeding efficiency. The efficiency of RRPS is based on its very rigorous protocol and tightly controlled cultural practices. Nine cycles of selection resulted in the release of 'Tifton 9' Bahia grass.[60] Yield was increased 47% with concomitant increases in seed production and no deterioration in forage quality.[65] Details of RRPS are also outlined in Sleper.[3]

C. Hybridization

Hybridization is the exchange of genetic material between two plants. The taxonomic relationship between the parents dictates if the cross is intraspecific, interspecific, or intergeneric hybridization. More hybrids are usually produced from intraspecific than from intergeneric crosses. Crosses between sexual species as well as those between sexual and apomictic plants fit into these three hybridization categories. A knowledge of the parents' method of pollination, chromosome number, meiotic behavior, mode of reproduction, pollen viability, fertility, and flowering behavior is essential before attempting any crosses because of potential differences.

The success of a hybridization program depends largely on the pollination system of the species. Obviously, naturally cross-pollinated species are easier to hybridize than self-pollinated species. Fortunately, most perennial forage grasses are cross-pollinated, but many are capable of some self-pollination.

Species with imperfect flowers require less preparation because emasculation of the florets of the female parent is not necessary. This is also true for plants with perfect flowers that have protogynous flowering behavior (early exsertion of the stigmas) as well as those that are self-incompatible or male sterile. Self-fertilization is low enough in some species that crosses can be made by enclosing an inflorescence from each parent in the same pollinating bag and allowing them to interpollinate.[54] Care should be taken to insure that both inflorescences are similar in maturity and flower at the same time. The inflorescence of the male parent should be removed from the bag within a few days after anthesis to prevent its seed from falling in the bag and contaminating the seed lot. Crossing blocks with the female parent in the center surrounded by ramets of the male parent with adequate isolation from other plants can also be used to produce hybrids.

When self-pollinated species are crossed, it is necessary to either emasculate the florets of the female parent or use a procedure to delay anther dehiscence until the anthers are removed. Because grass flowers are very small and difficult to emasculate, mist chambers are frequently used to delay dehiscence until the anthers can be removed from the floret using sharp-pointed forceps. However, the temperature should be 21°C or less for this technique to work. Chemical gametocides that render the pollen nonviable have been used in *Pennisetum* and *Cenchrus* hybridization.[56] For more information regarding emasculation and pollination techniques, see Burson.[16]

Regardless of the taxonomic relationship between the parents, the most successful F_1 hybrids involving tropical grasses reproduce asexually, either by apomixis or vegetatively through stolons and/or rhizomes. Poor seed set is the major limitation in developing tropical forage grass cultivars using interspecific and intergeneric hybridization.

1. Intraspecific Hybridization

Provided there is ample variation within a species, intraspecific hybridization has the most potential for developing improved cultivars and has been the most successful approach because of the close relationship of the parents. They are normally easy to cross, and a large number of hybrids are usually produced. Contrary to interspecific hybrids, most intraspecific hybrids have sufficient fertility that they can be propagated by seed. The greatest success in developing F_1 hybrid cultivars has been with Bermuda grass, *Cynodon dactylon* (L.) Pers. var. *dactylon*. It is highly cross-pollinated, large numbers of hybrids are easily obtained, and the F_1 hybrids can be propagated vegetatively. The cultivar 'Coastal' was selected from 5000 F_1 hybrids produced from natural crosses between 'Tift' Bermuda grass and an introduction from Africa.[66] Because it is propagated vegetatively by stolons, Coastal retains its hybrid vigor indefinitely similar to apomictic cultivars. Coastal has many desirable traits and is widely grown throughout the southern U.S.

Burton[67] crossed two self-sterile selections of Pensacola Bahia grass and selected two F_1 hybrids that were superior to Pensacola in forage production. The hybrids were released as 'Tifhi 1' and 'Tifhi 2'; however, seed production problems prevented their widespread use.

2. Wide Hybridization

The term wide hybridization includes both interspecific and intergeneric hybridization. In tropical grasses, numerous interspecific and some intergeneric hybrids have been produced. Some were produced for the genetic improvement of a species, but most were for phylogenetic and biosystematic studies. From an improvement standpoint, wide hybridization is used to transfer easily recognized traits from one species to another and to capitalize on heterosis in F_1 hybrids. However, poor seed set in the F_1 hybrids and inability to vegetatively propagate them have essentially prevented all wide hybrids from being released as improved forage grass cultivars. This sterility problem can sometimes be circumvented by producing hybrids that are apomictic or propagated by stolons or rhizomes.

Interspecific *Cynodon* hybrids have been produced for both cultivar development and biosystematic studies. The tropical species *C. nlemfuensis* Vanderyst var. *robustus* was crossed with *C. dactylon* var. *dactylon* and produced sterile F_1 hybrids that could be propagated vegetatively. The best of these F_1 hybrids produced more dry matter and had higher IVDMD than Coastal Bermuda grass and was released as the cultivar 'Tifton 78.'[68] Sterile, triploid F_1 *C. transvaalensis* Burtt-Davy x *C. dactylon* hybrids were produced and released as vegetatively propagated turfgrasses.[69] Harlan et al.[70,71] produced more than 700 interspecific hybrids between several different *Cynodon* species from Africa and Asia. These were used for biosystematic studies that resulted in a taxonomic revision of the genus. Even though most *Cynodon* species are highly self-sterile, the hybrids produced for this biosystematic study were made using hand emasculations and pollinations.

More interspecific hybrids probably have been made between more different *Paspalum* species than in any of the other tropical grasses. Essentially all the hybrids were made by hand emasculation and pollination because most sexual *Paspalum* species are highly self-pollinated. A majority of the hybrids were used for a phylogenetic investigation of different taxa with emphasis on determining the progenitors of dallis grass and Bahia grass.[72-74] Some of these hybrids were excellent forage types, but, unfortunately, they were sterile and could not be propagated by seed.

Pennisetum-Cenchrus interspecific and intergeneric hybrids also have been produced. Even though most of the species are protogynous and highly cross-pollinated, controlled emasculations and gametocides were used to produce the hybrids. They were used to investigate apomixis in the *Pennisetum-Cenchrus* agamic complex.[57,75] Much of this research focused on buffelgrass. Some of the hybrids are desirable forage types with adequate seed fertility and are being evaluated as forage grasses.

Pearl millet is a *Pennisetum* species that has been used extensively in interspecific hybridization programs for both germplasm enhancement and basic genetic studies. Triploid hybrids from crosses between diploid pearl millet and tetraploid napier grass produced high yields of high-quality forage but were sterile. The chromosomes of these sterile F_1 hybrids were doubled using colchicine, but the fertile amphihexaploids were not as desirable forage types as the sterile triploids.[76] An interspecific hybridization program was initiated in 1978 at Tifton, GA, with the objective of transferring the gene(s) controlling apomixis from wild *Pennisetum* species to cultivated pearl millet. Hybrids were produced between pearl millet and four wild species.[77,78] Crosses and backcrosses between tetraploid pearl millet ($2n = 4x = 28$) and hexaploid, apomictic *P. squamulatum* ($2n = 6x = 54$) have produced the most encouraging results in the interspecific transfer of the genes controlling apomixis.[77]

Interspecific and intergeneric hybrids were made among numerous Old World bluestems for a biosystematic treatment of the subtribe Bothriochloininae.[79] Interspecific hybrids have been made in *Brachiaria*,[51] *Digitaria*,[80] *Eragrostis*,[25] and *Setaria*[81] primarily to develop improved germplasm; however, some of the *Brachiaria* and *Eragrostis* hybrids were made to investigate the inheritance of apomixis.

For years, intergeneric hybrids were produced between maize and eastern gama grass to elucidate the progenitors of maize. Because eastern gama grass is a wild relative of maize, several programs were initiated to transfer the gene(s) for apomixis from *Tripsacum* to maize. The first program to attempt this challenge was in the former USSR.[82] Today, there are two programs pursuing this goal. One is the IRD-CIMMYT (formerly ORSTOM-CIMMYT) program in Mexico,[83] and the other effort is by USDA-ARS at Woodward, Oklahoma.[84] These programs have produced numerous hybrids in an attempt to transfer apomixis to maize. Commercially acceptable apomictic maize has not been produced, but progress has been made and considerable basic information regarding the genetics of apomixis has resulted.

The main point to remember in using wide hybridization for germplasm enhancement is that chances of success are very limited because the F_1 hybrids will probably be sterile. Only those that reproduce by apomixis or are vegetatively propagated will have any chance of being released as an improved cultivar.

From this overview of the breeding and improvement of tropical and subtropical grasses, it is obvious that an extremely large number of improved forage grass cultivars have been developed and released throughout the world. Because of the number of cultivars, diversity of genera/species and geographical scope of the different breeding programs, it was impossible to cover every program. Fortunately, Loch and Ferguson[37] have published a very comprehensive list identifying most of the tropical grass cultivars that have been released from breeding programs throughout the warmer regions of the world. Readers interested in specific cultivars should refer to that publication.

VII. CONCLUSION

When considering the complexity and diversity of most tropical grasses (i.e., ploidy levels, meiotic irregularities, methods of reproduction, modes of pollination, and a host of other traits that impact genetic improvement), it is apparent that grass breeders have made remarkable progress in the selection and development of improved cultivars. The prevalence of apomixis in tropical grasses, especially members of the Panicoideae subfamily, has impacted the improvement of many species. However, apomixis has been a mixed blessing considering (1) the number of superior apomictic ecotypes that have been selected in the wild and released as cultivars and (2) how it can be used as a breeding tool to fix permanent heterosis in new apomictic F_1 hybrids when sexual germplasm is available. Within the past few years, much has been learned regarding the control and manipulation of apomixis. This basic information has led to the development of different breeding methods where apomixis is used as a dynamic breeding tool in developing improved, true breeding, apomictic cultivars in some tropical grasses. As new sources of sexual germplasm are discovered in other apomictic grasses, apomictic breeding will be used to develop additional improved apomictic cultivars in the future. Just as grass breeders have made significant progress in the selection and development of improved cultivars in the past, the future appears even more promising. With the recent development of molecular techniques such as breeding tools, grass breeders and geneticists are beginning to use these techniques in their programs. These new developments will undoubtedly contribute to the breeding of improved tropical grass cultivars in the future.

REFERENCES

1. Clayton, W. D. and Renvoize, S. A., *Genera Graminum Grasses of the World*, Kew Bulletin Additional Series XIII, Her Majesty's Stationery Office, London, 1986, chap. 1.
2. Watson, L. and Dallwitz, M. J., *The Grass Genera of the World*, C.A.B. International, Wallingford, U.K., 1992, chap. 3.
3. Sleper, D. A., Forage grasses, in *Principles of Cultivar Development*, Vol. 2, Fehr, W. R., Ed., Macmillan Publishing Company, New York, 1987, chap. 6.
4. Bogdan, A. V., *Tropical Pasture and Fodder Plants*, Longman Inc., New York, 1977, 20.
5. Burton, G. W., Breeding Pensacola bahiagrass, *Paspalum notatum*: I. method of reproduction, *Agron. J.*, 47, 311, 1955.
6. Burson, B. L., Pollen germination, pollen tube growth and fertilization following self and interspecific pollination of *Paspalum* species, *Euphytica*, 36, 641, 1987.
7. Shafer, G. S., Burson, B. L., and Hussey, M. A., Stigma receptivity and seed set in protogynous buffelgrass, *Crop Sci.*, 40, 391, 2000.
8. Connor, H. E., Breeding systems in the grasses: a survey, *N. Z. J. Bot.*, 17, 547, 1979.
9. Norrmann, G. A., Quarin, C. L., and Burson, B. L., Cytogenetics and reproductive behavior of different chromosome races in six *Paspalum* species, *J. Heredity* 80, 24, 1989.
10. Burton, G. W. and Powell, J. B., Pearl millet breeding and cytogenetics, in *Advances in Agronomy*, Vol. 20, Norman, A. G., Ed., Academic Press, New York, 1968, 49.
11. Quarin, C. L., Relaciones citotaxonomicas entre *Paspalum almum* Chase y *P. hexastachyum* Parodi (Gramineae), *Bonplandia* 3, 115, 1974.
12. Gould, F. W., *The Grasses of Texas*, Texas A&M University Press, College Station, 1975, 500.
13. Bashaw, E. C. and Forbes, I., Jr., Chromosome numbers and microsporogenesis in dallisgrass, *Paspalum dilatatum* Poir., *Agron. J.*, 50, 441, 1958.
14. Burson, B. L., Voigt, P. W., and Evers, G. W., Cytology, reproductive behavior, and forage potential of hexaploid dallisgrass biotypes, *Crop Sci.*, 31, 636, 1991.
15. Tischler, C. R. and Burson, B. L., Evaluating different bahiagrass cytotypes for heat tolerance and leaf epicuticular wax content, *Euphytica* 84, 229, 1995.
16. Burson, B. L., Warm-season grasses, in *Hybridization of Crop Plants*, Fehr, W. A. and Hadley, H. H., Eds., Am. Soc. Agron. and Crop Sci. Soc. Am., Madison, 1980, chap. 50.

17. Watson, L., The grass family, Poaceae, in *Reproductive Versatility in the Grasses*, Chapman, G. P., Ed., Cambridge University Press, Cambridge, 1990, chap. 1.
18. Young, B. A., Sherwood, R. T., and Bashaw, E. C., Cleared-pistil and thick-sectioning techniques for detecting aposporous apomixis in grasses, *Can. J. Bot.*, 57, 1668, 1979.
19. Nogler, G. A., Gametophytic apomixis, in *Embryology of Angiosperms*, Joshi, B. M., Ed., Springer-Verlag, Berlin, 1984, chap. 10.
20. Asker, S. E. and Jerling, L., *Apomixis in Plants*, CRC Press, Boca Raton, 1992, 49.
21. Savidan, Y., Apomixis: Genetics and Breeding, in *Plant Breeding Reviews*, Vol. 18, Janick, J., Ed., John Wiley & Sons, Inc., New York, 2000, chap. 2.
22. Warmke, H. E., Apomixis in *Panicum maximum* Jacq., *Am. J. Bot.*, 41, 5, 1954.
23. Nakagawa, H., Embryo sac analysis and crossing procedure for breeding apomictic guineagrass (*Panicum maximum* Jacq.), *JARQ*, 24, 163, 1990.
24. Vielle, J.-Ph., Burson, B. L., Bashaw, E. C., and Hussey, M. A., Early fertilization events in the sexual and aposporous egg apparatus of *Pennisetum ciliare* (L.) Link, *Plant J.*, 8, 309, 1995.
25. Voigt, P. W. and Bashaw, E. C., Apomixis and sexuality in *Eragrostis curvula*, *Crop Sci.*, 12, 843, 1972.
26. Voigt, P. W., Burson, B. L., and Sherman, R. A., Mode of reproduction of lehmann lovegrass, *Crop Sci.*, 32, 118, 1992.
27. Burson, B. L., Voigt, P. W., Sherman, R. A., and Dewald, C. L., Apomixis and sexuality in eastern gamagrass, *Crop Sci.*, 30, 86, 1990.
28. Leblanc, O., Peel, M. D., Carman, J. G., and Savidan, Y., Megasporogenesis and megagametogenesis in several *Tripsacum* species (Poaceae), *Am. J. Bot.*, 82, 57, 1995.
29. Chao, C-Y, Megasporogenesis and megagametogenesis in *Paspalum commersonii* and *P. longifolium* at two ploidy levels, *Bot. Notister*, 127, 267, 1974.
30. Knox, R. B. and Heslop-Harrison, J., Experimental control of aposporous apomixis in a grass of the Andropogoneae, *Bot. Notiser*, 116, 127, 1963.
31. Quarin, C. L., Seasonal changes in the incidence of apomixis of diploid, triploid, and tetraploid plants of *Paspalum cromyorrhizon*, *Euphytica*, 35, 515, 1986.
32. Hussey, M. A., Bashaw, E. C., Hignight, K. W., and Dalmer, M. L., Influence of photoperiod on the frequency of sexual embryo sacs in facultative apomictic buffelgrass, *Euphytica*, 54, 141, 1991.
33. Peel, M. D., Carman, J. G., and Leblanc, O., Megasporocyte callose in apomictic buffelgrass, Kentucky bluegrass, *Pennisetum squamulatum* Fresen., *Tripsacum* L., and weeping lovegrass, *Crop Sci.*, 37, 724, 1997.
34. Ozias-Akins, P., Lubbers, E. L., Hanna, W. W., and McNay, J. W., Transmission of the apomictic mode of reproduction in *Pennisetum*: co-inheritance of the trait and molecular markers, *Theor. Appl. Genet.*, 85, 632, 1993.
35. Gustafsson, A., *Apomixis in Higher Plants, II. The Causal Aspect of Apomixis*, Lunds Univ. Arsskr., N.F. Avd. 2, 43, 1947, 71.
36. Bashaw, E. C. and Hanna, W. W., Apomictic reproduction, in *Reproduction Versatility in the Grasses*, Chapman, G. P., Ed., Cambridge University Press, Cambridge, 1990, chap. 5.
37. Loch, D. S. and Ferguson, J. E., Tropical and subtropical forage seed production: an overview, in *Forage Seed Production Vol. 2: Tropical and Subtropical Species*, Loch, D. S. and Ferguson, J. E., Eds., CAB International, Wallingford, U.K., 1999, chap. 1.
38. Pessino, S. C., Evans, C., Ortiz, J. P. A., Armstead, I., Valle, C. B. do and Hayward, M. D., A genetic map of the apospory-region in *Brachiaria* hybrids: identification of two markers closely associated with the trait, *Hereditas*, 128, 153, 1998.
39. Leblanc, O., Grimanelli, D., Gonzalez-de-Leon, D., and Savidan, Y., Detection of the apomictic mode of reproduction in maize-*Tripsacum* hybrids using maize RFLP markers, *Theor. Appl. Genet.*, 90, 1198, 1995.
40. Gustine, D. L., Sherwood, R. T., Gounaris, Y., and Huff, D., Isozyme, protein and RAPD markers within a half-sib family of buffelgrass segregating for apospory, *Crop Sci.*, 36, 723, 1996.
41. Ozias-Akins, P., Roche, D., and Hanna, W. W., Tight clustering and hemizygosity of apomixis linked molecular markers in *Pennisetum squamulatum* implies genetic control of apospory by divergent locus that may have no allelic form in sexual genotypes, *Proc. Natl. Acad. Sci.*, 95, 5127, 1998.
42. Carman, J. G., Asynchronous expression of duplicate genes in angiosperms may cause apomixis, bispory, tetraspory, and polyembryony (review), *Biol. J. Linn. Soc.*, 61, 51, 1997.

43. Burton, G. W. and Forbes, I., Jr., The genetics and manipulation of obligate apomixis in common bahiagrass (*Paspalum notatum* Flugge), in *Proc. VIII Int. Grassl. Congr.*, Reading, (England), 1960, 66.
44. Savidan, Y. H., Genetics and utilization of apomixis for the improvement of guineagrass (*Panicum maximum* Jacq.), in *Proc. XIV Int. Grassl. Congr.*, Smith, J. A. and Hayes, V. W., Eds., Westview Press, Inc., Boulder, 1983, 182.
45. Bashaw, E. C., Apomixis and sexuality in buffelgrass, *Crop Sci.*, 2, 412, 1962.
46. Smith, R. L., Sexual reproduction in *Panicum maximum, Crop Sci.*, 12, 624, 1972.
47. Hanna, W. W., Powell, J. B., Millot, J. C., and Burton, G. W., Cytology of obligate sexual plants in *Panicum maximum* Jacq. and their use in controlled hybrids, *Crop Sci.*, 13, 695, 1973.
48. Taliaferro, C. M. and Bashaw, E. C., Inheritance and control of obligate apomixis in breeding buffelgrass, *Pennisetum ciliare, Crop Sci.*, 6, 473, 1966.
49. Bashaw, E. C., Registration of Higgins buffelgrass, *Crop Sci.*, 8, 397, 1968.
50. Bashaw, E. C., Registration of Nueces and Llano buffelgrass, *Crop Sci.*, 20, 112, 1980.
51. Valle, C. B. do, Glienke, C., and Leguizamon, G. O. C., Breeding of apomictic *Brachiaria* through interspecific hybridisation, in *Proc. XVII Int. Grassl. Congr.*, Palmerston North, (New Zealand), 1993, 427.
52. Nakajima, K., Apomixis and its application to plant breeding, in *Gamma Field Symposia No. 29*, Institute of Radiation Breeding, NIAR, MAFF, Ohmiya-machi, Ibaraki-ken, (Japan), 1990, 71.
53. Bashaw, E. C., Problems and possibilities of apomixis in the improvement of tropical forage grasses, in *Tropical Forages in Livestock Production Systems*, Doll, E. C. and Mott, G. O., Eds., ASA Special Publication No. 24, Am. Soc. Agron., Crop Sci. Soc. Am., Soil Sci. Soc. Am., Madison, WI, 1975, 23.
54. Voigt, P. W. and Burson, B. L., Breeding of apomictic *Eragrostis curvula*, in *Proc. XIV Int. Grassl. Congr.*, Smith, J. A. and Hayes, V. W., Eds., Westview Press, Inc., Boulder, CO, 1983, 160.
55. Hanna, W. W., Use of apomixis in cultivar development, in *Advances in Agronomy*, Vol. 54, Sparks, D. L., Ed., Academic Press, Inc., New York, 1995, 333.
56. Bashaw, E. C. and Hignight, K. W., Gene transfer in apomictic buffelgrass through fertilization of an unreduced egg, *Crop Sci.*, 30, 571, 1990.
57. Bashaw, E. C., Hussey, M. A., and Hignight, K. W., Hybridization (N + N and 2N + N) of facultative apomictic species in the *Pennisetum* agamic complex, *Int. J. Plant Sci.*, 153, 466, 1992.
58. Martinez, E. J., Espinoza, F., and Quarin, C. L., B_{III} Progeny (2n+n) from apomictic *Paspalum notatum* obtained through early pollination, *J. Heredity*, 85, 295, 1994.
59. Quesenberry, K. H., Dunavin, L. S., Jr., Hodges, E. M., Killinger, G. B., Kretschmer, A. E., Jr., Ocumpaugh, W. R., Roush, R. D., Ruelke, O. C., Schank, S. C., Smith, D. C., Snyder, G. H., and Stanley, R. L., Registration of Redalta, Greenalta, and Bigalta limpograss, *Crop Sci.* 19, 294, 1979.
60. Boyd, F. T., Schank, S. C., Smith, R. L., Hodges, E. M., West, S. H., Kretschmer, A. E., Jr., Brolman, J. B., and Moore, J. E., Transvala digitgrass: A tropical forage resistant to 1. sting nematode 2. pangola stunt virus, *Fla. Agric. Exp. Stn. Cir.* No. S-222, 1973.
61. Burton, G. W., Registration of 'Merkeron' napiergrass, *Crop Sci.*, 29, 1327, 1989.
62. Sollenberger, L. E., Prine, G. M., Ocumpaugh, W. R., Hanna, W. W., Jones, C. S., Jr., Schank, S. C., and Kalmbacher, R. S., Registration of 'Mott' dwarf elephantgrass, *Crop Sci.*, 29, 827, 1989.
63. Burton, G. W., Improved recurrent restricted phenotypic selection increases bahiagrass yields, *Crop Sci.*, 22, 1058, 1982.
64. Burton, G. W., Recurrent restricted phenotypic selection, in *Plant Breeding Reviews*, Vol. 9, Janick, J., Ed., John Wiley & Sons, New York, 1992, chap. 5.
65. Burton, G. W., Registration of 'Tifton 9' Pensacola bahiagrass, *Crop Sci.*, 29, 1326, 1989.
66. Burton, G. W., Breeding bermudagrass for the southeastern United States, *J. Am. Soc. Agron.*, 39, 551, 1947.
67. Burton, G. W., Plant breeding 1910-1984, in *Gene Manipulation in Plant Improvement*, 16th Stadler Genet. Symp., Gustafson, J. P., Ed., Plenum Press, New York, 1984, chap. 1.
68. Burton, G. W. and Monson, W. G., Registration of 'Tifton 78' bermudagrass, *Crop Sci.*, 28, 187, 1988.
69. Burton, G. W., Breeding better bermudagrasses, in *Proc IX Int. Grassl. Congr.*, Sao Paulo, Brazil, 1965, 93.
70. Harlan, J. R., de Wet, J. M. J., and Richardson, W. L., Hybridization studies with species of *Cynodon* from East Africa and Malagasy, *Am. J. Bot.*, 56, 944, 1969.

71. Harlan, J. R., de Wet, J. M. J., Rawal, K. M., Felder, M. R., and Richardson, W. L., Cytogenetic studies in *Cynodon* L. C. Rich. (Gramineae), *Crop Sci.*, 10, 288, 1970.
72. Burson, B. L., Phylogenetic investigations of *Paspalum dilatatum* and related species, in *Proc. XIV Int. Grassl. Congr.*, Smith, J. A. and Hayes, V. W., Eds., Westview Press, Inc., Boulder, CO, 1983, 170.
73. Burson, B. L., Phylogenetics of apomictic *Paspalum dilatatum*, in *Proc XVI Int. Grassl. Congr.*, Nice, (France), 1989, 413.
74. Quarin, C. L., Burson, B. L., and Burton, G. W., Cytology of intra- and interspecific hybrids between two cytotypes of *Paspalum notatum* and *P. cromyorrhizon*, *Bot. Gaz.*, 145, 420, 1984.
75. Hussey, M. A., Bashaw, E. C., Hignight, K. W., Wipff, J., and Hatch, S. L., Fertilisation of unreduced female gametes: A technique for genetic enhancement within the *Cenchrus-Pennisetum* agamic complex, in *Proc. XVII Int. Grassl. Congr.*, Palmerston North, (New Zealand), 1993, 404.
76. Hanna, W. W., Gaines, T. P., Gonzalez, B. and Monson, W. G., Effect of ploidy on yield and quality of pearl millet x napiergrass hybrids, *Agron. J.*, 76, 969, 1984.
77. Dujardin, M. and Hanna, W. W., Cytogenetics of double cross hybrids between *Pennisetum americanum-P. purpureum* amphiploids and *P. americanum* x *P. squamulatum* interspecific hybrids, *Theor. Appl. Genet.*, 69, 97, 1984.
78. Hanna, W. W. and Dujardin, M., Role of apomixis in building and maintaining genome combinations, in *Proc. 2nd Int. Symp. on Chromosome Engineering in Plants*, Kimber, G., Ed., College of Agriculture, Univ. Extension, Univ. Missouri, Columbia, 1990, 112.
79. Harlan, J. R., Chheda, H. R., and Richardson, W. L., Range of hybridization with *Bothriochloa intermedia* (R. Br.) A. Camus, *Crop Sci.*, 2, 480, 1962.
80. van Heemert, C. and Schank, S. C., New breeding techniques: Terylene bags for *Digitaria* breeding, in *Proc. Soil Crop Sci. Soc. Fla.*, 28, 150, 1968.
81. Hacker, J. B., Cytology of species hybrids in the *Setaria sphacelata* complex, *Aust. J. Bot.*, 16, 533, 1968.
82. Petrov, D. F., Fokina, E. S., and Belovsova, N. I., Experimental production of apomixis in corn, *Doklady Akad. Nauk SSSR*, 281, 509, *(Doklady Bot. Sci.*, 281, 31), 1985.
83. Savidan, Y., LeBlanc, O., and Berthaud, J., Transfer of apomixis through wide crosses, in *Apomixis: Exploiting Hybrid Vigor in Rice*, Khush, G. S., Ed., IRRI, Los Banos, Phlippines, 1994, 23.
84. Kindiger, B., Sokolov, V., and Khatypova, I. V., Evaluation of apomictic reproduction in a set of 39 chromosome maize-*Tripsacum* backcross hybrids, *Crop Sci.*, 36, 1108, 1996.
85. Miles, J. W. and do Valle, C. B., Manipulation of apomixis in *Brachiaria* breeding, in *Brachiaria: Biology, Agronomy, and Improvement,* Miles, J. W., Maass, B. L., and do Valle, C. B., Eds., CIAT-EMBRAPA, Cali, Colombia, 1996, 164.

6 Tropical Forage Legume Breeding

Kenneth H. Quesenberry and David S. Wofford

CONTENTS

I. INTRODUCTION

Leguminous plants are represented by approximately 19,000 species in over 750 genera, and are economically second only to the grass family. Legumes are a valuable source of forestry products (timber, resins, etc.), various commercial chemical products (dyes, tannins, insecticides, etc.), edible

0-8493-2318-5/01/$0.00+$.50

grain (soybean among others), forage consumed by grazing animals or as hay, and plants in conservation systems for erosion control. Historically, forage legumes have played an important role in agriculture, and over 200 genera are considered to have commercial potential. Prior to the commercial production of inorganic fertilizers, they were widely grown for their contribution to soil fertility, either as part of a general soil improvement program or in specific crop rotations. The replenishing of organic matter in agricultural systems is crucial to soil conservation, particularly in areas of the tropics where fragile soil conditions and extremes of rainfall exist. The enhanced awareness of soil fertility maintenance and/or improvement as a component of ecologically based production systems should increase the use of legumes in many countries. As environmental concerns about the use of inorganic N fertilizer become more important worldwide, the use of forage legumes should once again become a significant and integral factor even in highly mechanized agricultural systems.

Tropical forage legumes are grown in numerous countries, not simply in those with a defined tropical climate. Collectively they represent a diverse pool of germplasm with a wide range of phenotypic types. Growth habits range from viney to erect and from low growing to large trees. Some species are well-adapted to short-drought regions, whereas others are useful in areas that experience longer drought periods. Because of the large number of available species, it is likely that, with adequate research, a species could be identified for almost any specific location. The increased use of systematic grazing systems in tropical areas typically entails improved pastures composed of a grass-legume mixture. In these systems, the N supplied by the legume via biological N_2 fixation is used by the grass species, greatly improving both production and quality of forage. The overall effect of increased forage quality is improved livestock performance and animal reproduction. Since more native grasslands are being used in livestock systems, there should be an ever-increasing use of forage legume species in these plant-animal management programs.

II. LIMITATIONS OF TROPICAL FORAGE LEGUME BREEDING

A. Current Knowledge Base

A principal factor limiting the genetic improvement of tropical forage legumes is the scarcity of information in the literature for many of them, since investigations have concentrated on only a few genera. Overall, the published information is poorly distributed across species, however, a reasonable amount of information does exist for a few species that were targeted for investigation by individual countries. The ability to successfully manipulate genetic systems is related to the amount of information available for the genetic mechanisms operating in a species. Until there is a substantial increase in genetic investigations of tropical legumes, this aspect will continue to be a hindrance to improvement through either classical breeding or biotechnology methods.

B. Past and Current Breeding Objectives

A major effort to utilize tropical forages has been undertaken in Australia due to the low number of native legume species. Much of the work has been focused on species evaluation and direct selection of introduced lines. Hutton[1] described the characteristics that a tropical legume must possess in order to be successfully used in pasture systems: (1) efficient N_2-fixation, producing sufficient nitrogen and protein for productive plant-animal systems, (2) sufficient plant response to phosphate so that it is capable of exploiting areas with poor soils, (3) the ability to persist despite competition from accompanying grass species and typical climatic stresses, (4) tolerance to variable soil conditions, (5) economic seed production and suitable plant structure for mechanized harvesting procedures, (6) resistance to diseases and insects, (7) suitable year-round quality for animal feed, and (8) lack of antiquality chemical constituents. Virtually all of these desirable attributes have been or currently are breeding and selection objectives in tropical forage legume improvement programs.

C. Taxonomic Problems

The systematic classification of tropical forage legume species is much less well understood than that of temperate species. In several tropical forage legume genera, e.g., *Arachis*, comprehensive taxonomic treatments of all species are not, or have only recently, become available.[2] For others, only portions of the genus have been treated, e.g., *Desmodium*.[3] In some genera, the classical taxonomic treatments do not correspond well with findings regarding intra- and interspecific crossing compatibility.[4]

D. Reproductive Behavior

Most tropical forage legume species whose reproductive behavior has been well characterized are either partially or completely self-pollinated.[4,5] This is in stark contrast to the major temperate forage legumes, e.g., alfalfa *Medicago sativa* L., red clover *Trifolium pratense* L., and white clover *Trifolium repens* L., which are predominately cross-pollinated. Research with *Aeschynomene americana* L. (American joint vetch) has shown that although the species is predominately self-pollinated, as much as 25% outcrossing may occur when outcrossing conditions are maximized.[5] Other work with *Stylosanthes guianensis* (Aublet) Sw. has also shown a mean outcrossing rate of 15% varying from 2 to 33% among individuals.[6] The high percentage of outcrossing in predominately self-pollinated species will influence the choice of breeding methods.

E. Problems with Animal Assessment

Although some intrinsic value may be allocated to tropical legumes for use in crop rotation systems and some value may accrue through hay sales, most of the value of a tropical legume must be realized through increased livestock production. Thus, the ultimate test of worth of any new tropical forage legume cultivar must be determined by its impact on improved livestock production. Laboratory estimates of forage quality combined with clipping study data may be used initially to extrapolate potential gains, but ultimately animal assessment under field conditions will be required. Persistence and productivity under grazing of selected lines may be very different than under a cutting regime, but extensive animal evaluations of large numbers of introductions are not practical. Nevertheless, most researchers agree that evaluation of plant response to grazing defoliation is important and should be determined relatively early in a selection program. Methods such as "mob grazing" of rows or small plots of large numbers of plant introductions to determine plant response to grazing defoliation may help overcome some of these limitations.[7,8] Large-scale field experiments to measure animal gain or reproductive response to the addition of tropical forage legumes to animal diets generally are expensive to conduct and subject to bias created by differential forage availability. Such experiments are generally conducted only on advanced lines nearing release as cultivars.

III. GENERAL CONSIDERATIONS FOR TROPICAL LEGUME IMPROVEMENT

A. Species Selection

The selection of species is the most important factor in the potential use of a tropical legume. Most of the species considered for use originated in Central and/or South America, with lesser numbers from various African countries and Southeast Asia, but these species are often considered for use outside of these regions. Although there have been published procedures for selecting and evaluating species, a standardized method for introduction and evaluation cannot be used in all situations. The highly specific ecogeographic characteristics of a region where a species will be used have to be considered in this process.

In addition to considering the general features of an environment when selecting species for evaluation, the size of the intended region of use can be a significant factor. The larger a geographical zone over which one intends to use a species, the more locations must be involved in evaluating the plant material. This will usually result in variable responses for a particular species, and not lead to a single line or even species that will perform well throughout the region. Several countries have developed a basic plan to evaluate new species, the most common theme being one of introducing large numbers of lines followed by extensive screening of the materials at multiple locations. Although this format has proven reasonably useful, it is not realistic for countries without the infrastructure to undertake such activities. Additionally, factors other than agronomic performance may need to be accessed during an evaluation process.

A model stressing the efficiency of evaluation of forage legumes has been proposed for subtropical Japan.[9] Within this model, the region was subdivided into several zones based on climatic and edaphic factors (temperature, soil pH, and drought stress) and economic factors (maximum profits and minimum cost of production). This categorization can then be used to limit zonal evaluation to those species with a reasonable potential to succeed. Growth chamber evaluations could be used to determine species response to such basic factors as temperature and photoperiod. Based on the zonal classifications, species can then be selected for small-plot field trials to quantify productivity, quality, and animal performance. Promising lines identified in these evaluations would undergo further investigation under commercial conditions with subsequent release of those that would benefit the producers of this zone. In this program, the probability of attaining success would hopefully be enhanced in an efficient manner. The economic factors would balance the cost of production with the benefit of the legume species. In countries with relatively high land prices and/or other location- or region-specific economic parameters, it may prove judicious to examine the economic results as closely as the agronomic results. Regardless of the approach adopted, a systematic evaluation method should be used to identify species that have potential in a geographic area. This evaluation should be viewed, however, as a first step since most, if not all, identified species will still have deficiencies that need to be addressed in breeding programs.

B. Targeting Environment and Production Systems

The first step for any program of plant improvement is identification of general adaptation of species to a region. Factors such as total annual precipitation and seasonal rainfall distribution must be considered. Most tropical regions experience at least some dry periods during the year. In some regions with pronounced severe dry seasons, only reseeding annual legumes may be feasible. In other less extreme environments, perennial legumes that can defoliate during a short dry season and regrow from root reserves at the onset of a rainy period may be well suited. Adaptation to periodic flooding may be essential for legume survival in some areas, whereas long-term soil waterlogging may be problematic for a legume in other areas.[10]

General soil type, including texture, structure, and pH will influence species selection. Many tropical legumes produce satisfactory yields at pH ranges lower than those of temperate forage legumes, but with some species extensive germplasm evaluation and/or plant breeding have been required to identify genotypes within a species that will tolerate low pH.[11] The primary factor limiting growth in low-pH soils is generally Al toxicity coupled with low P availability. Both germplasm evaluation and plant breeding have been effective in identifying legume species tolerant to high levels of soil Al, and determining variability for this trait within species.

Extremes of temperatures in a region may also impact species selection. Low temperatures are generally not a problem in the tropics, except at high altitudes. However, in subtropical regions of Australia, New Zealand, and the southeastern United States, tolerance of a species to frost and freezes often becomes the primary factor affecting species selection. For some tropical grasses, tissue tolerance to mild frost has been observed,[12] but less variability in this trait appears to be observed in tropical legumes. Variability for winter survival within and among legume species has been observed after moderate freezing temperatures. Research has shown that in some species, e.g., *Centrosema*, this trait may be associated with height of the cotyledonary node above the soil surface.

Legume species with large tap roots and low crowns tend to be superior to viney species for winter hardiness, and those with extensive rhizome systems generally will be superior to either. Defoliation management during late fall can also dramatically affect winter survival.[13]

Species selection must not only target the environment, but also the livestock production system in which the legume will be used. Tropical forage legumes will most often be used as a component of a mixed grass-legume sward for grazing, but numerous other utilization schemes may need to be evaluated. For example, certain shrubby or tree legumes cannot be readily adapted for extensive planting for general cow-calf grazing. These same species may, however, be quite suitable for fodder bank deferred grazing or for intensive "cut-and-carry" small holder production systems. Other examples of production systems that might utilize a legume include deferred dry season grazing, grazing management practices favoring legume competition, and use of legumes in pure stand for high-quality hay production. Any of these types of variable management systems may cause a particular legume species, which performed poorly under continuous grazing, to excel.

The nutritional needs of the class of livestock in the production system are an additional aspect to consider in species selection. Mature shrubby legumes may meet the protein and energy needs of a dry, pregnant beef cow, but will likely be inadequate for young calves or dairy cows. Likewise, goats and browsing livestock may prefer shrubby or small tree legumes, whereas cattle and sheep would not. Beef cows with young calves will require high levels of protein and energy. Often one of the most pronounced effects of including a forage legume in a beef production system is an increase in annual conception and calving rates. The identification and selection of a legume that produces high-quality feed during the breeding season may be a major factor in species selection.

C. Plant Collection vs. Breeding

Active research programs involving tropical legumes have been in place for over 40 years in some locations. Nevertheless, crop improvement is still in its infancy for many of these plants. Thus, when a scientist identifies a legume species that appears to have potential in a region, an immediate question becomes, "Should a program of plant breeding and selection be initiated, or should more germplasm be sought for evaluation?" It has been suggested that 70 to 80% of the ultimate long-term value of introducing a new species will be obtained by the first release. Knowing that for most tropical forage legume species, broad genetic variability exists for many traits, the wise course at the onset appears to be evaluation of a broadly based collection of plant introductions. In the past 30 years, extensive efforts in plant collecting have been carried out by the CIAT Tropical Pastures Program, Cali, Columbia; the CSIRO, Tropical Crops Division, Brisbane, Australia; the USDA National Plant Germplasm System, Beltsville, MD (U.S.), and various other national and international agencies. Available introductions of a species should be thoroughly evaluated before additional collection trips or plant breeding programs are proposed. Even in circumstances where a released cultivar is known to have a particular weakness, e.g., disease or nematode susceptibility, etc., germplasm evaluation should be considered first. Nevertheless, circumstances will surely arise where specific traits from an otherwise poorly adapted introduction appear desirable for transfer to an improved cultivar. Specific breeding programs and objectives for the major tropical forage legumes will be discussed later in this chapter.

D. General Breeding Methodology

Generally, the first step in most tropical forage legume improvement programs has been mass selection or pure line selection out of mixed populations, either plant introductions or landraces. Where breeders have made controlled hybridizations of tropical forage legumes, a pedigree selection system or some modification of it has been used most often. Modifications have included F_2 to F_3 bulk advances, natural selection of a reseeding F_2 bulk, and development of homogenous $F_5 - F_6$ blends.[7] Some workers have suggested that the single-seed descent method may be well adapted to tropical forage legume breeding.[14] In an effort to increase efficiency of multiple cycles of selection in predominantly self-pollinated species, Miles[6] proposed a scheme of using selected marker genes, e.g., those for flower color, to identify natural outcrosses. Using this scheme, a cycle of recurrent selection can be completed in three years, and a new cycle initiated in year four. Reviews of plant breeding and selection programs generally show that additional cycles are more valuable than slight increases in improvement per cycle, thus any scheme that shortens the number of years per cycle appears to have merit in breeding tropical forage legumes. Identification of simply inherited marker genes is the key to using the scheme proposed by Miles,[6] but this is often easy in many species.

E. Biotechnology Applications

Biotechnology applications utilizing genetic transformation have the most potential in crops where the desired trait cannot be identified by germplasm evaluation or manipulated by plant breeding. As already suggested, both of these avenues of research appear to remain fruitful in tropical forage legumes and should be exhausted before resorting to genetic transformation approaches. A requirement for most methods of genetic transformation is that an individual transformed cell must be regenerated into a complete plant. Although *in vitro* regeneration is well documented for many crop species, there are few reports of success in tropical forage legumes. Whole plant regeneration of *S. guianensis* from several different explant tissues has been reported.[15] It has since been found that protoplasts of this genus are also capable of *in vitro* regeneration.[16] In general, *Stylosanthes* seems to be quite amenable to *in vitro* regeneration unlike many, if not most, tropical forage legume species. Useful tissue-culture protocols for other tropical species have only recently been reported.[17-20] Like many other species, regeneration has been shown to be genotype specific.

Genetic transformation of tropical forage legumes has only been reported for *Stylosanthes*.[21] Transgenic plants of *S. guianensis*, CIAT 184, were developed using standard procedures for an *Agrobacterium* system. Transmission of the novel DNA through sexual generation was also reported. *Agrobacterium*-mediated transformation studies have been conducted with *Alysicarpus vaginalis* (L.) DC. genotypes in Florida, however, no transgenic plants have been recovered to date (unpublished results from our laboratory). Additional research both on *in vitro* regeneration protocol and efficient transformation techniques will likely be needed on each species being considered for recombinant DNA applications.

IV. BREEDING OBJECTIVES

Identification of a clearly defined objective must be the first step when the decision is made to initiate a program of plant breeding for a tropical forage legume. The objective must address the general considerations discussed earlier, and should be clearly focused on the intended use of the plant material. Plant breeding objectives for tropical forage legumes can be grouped into six categories: (1) adaptation, (2) forage quality, (3) plant growth and development, (4) improved establishment, (5) pest resistance, and (6) persistence.

A. Adaptation

Selection objectives with an adaptation component usually relate to climatic, edaphic, or grazing adaptation. Some examples of climatic adaptation selection objectives include winterhardiness in subtropical zones, ability to withstand prolonged drought or flooding, and altered photoperiod response for improved seed production. A problem encountered with selection for winterhardiness is the unreliability of natural "cold enough, but not too cold" winter temperatures. Solutions include multiple plantings across different coldhardiness zones and screening over several years. Although growth chambers or greenhouses may be used to supplement selection programs in the field, they generally cannot substitute for them. However, greenhouse selection schemes may be very effective to evaluate flooding or drought tolerance. Likewise, photoperiod responses of tropical forage legumes are generally highly heritable and quite amenable to selection under greenhouse or field conditions.

Edaphic selection objectives with tropical forage legumes are usually related to growth in low pH soils, which often have high Al saturation, low available P, and low levels of certain micronutrients. This complex soil environment, coupled with an interaction between *Bradyrhizobium* strains and soil pH levels, create difficulty in maintaining uniformity of selection environments over years and locations. Nevertheless, reasonably heritable genetic variability for good growth in low pH, high Al saturation soils has been identified in several tropical forage legumes.[22,23] Greenhouse screening procedures may be useful for preliminary evaluation of a large number of lines,[24] but field testing will ultimately be needed.

Selection objectives related to grazing adaptation usually involve modification of plant structure or selection for persistence under grazing. Most plant germplasm or plant breeding evaluation schemes include a component of testing advanced lines under grazing. Research in Florida[8,9] has suggested benefits from using grazing animals earlier in the selection scheme when persistence under grazing is an objective of the selection program. If grazing animals are used as a selection tool in the early stage of an evaluation program when many lines with variable plant growth types are present, care must be taken to avoid the effects of selective grazing. Even when a "mob grazing" approach is used in an early stage evaluation, differences in residual dry matter remaining after grazing should be determined. Modified plant types such as lower plant crowns, improved leaf-to-stem ratio, and increased branching are usually highly heritable and amenable to selection under field conditions.

B. Forage Quality

The three major aspects of quality in tropical forage legumes that plant breeders have attempted to modify are digestibility, crude protein (CP) concentration, and antiquality factors. The widespread use of *in vitro* procedures for estimating forage digestibility and, more recently, the application of infrared techniques to estimate digestibility allow large numbers of samples to be screened in a relatively short period of time. Although legumes are generally thought to have "good" digestibility, the presence of antiquality factors such as tannins often result in low *in vitro* digestibility values and reduced animal performance in *in vivo* studies.[25] Care must be taken to correlate *in vitro* estimations of digestibility with *in vivo* results. CP concentration of well-nodulated tropical forage legume leaves is generally not a limiting factor in animal nutrition; however, as these species mature they may become stemy with reduced total CP concentration. Breeding objectives related to forage quality generally attempt to modify quality parameters by modification of plant morphological characteristics such as leaf-to-stem ratio.

Selection for reduced levels of antiquality components may be feasible. Research has shown variation in the tannin levels in *Desmodium heterocarpon* (L.) DC.[26] and for mimosine levels in *Leucaena leucocephala* (Lam) Dewit.[1] These traits usually have relatively high heritability and can be selected under field conditions. It is important to sample plant tissue of the same age, as the concentration of antiquality factors usually varies with age of tissue, and often increases with age.

C. Plant Growth and Development

Determining total forage dry matter yield (DMY) and seasonal distribution of yield is almost always a part of general breeding objectives. These traits often have only moderate or low heritability, can be evaluated only under field conditions, and usually must be evaluated over multiple years. Thus, they may be among the more difficult and expensive traits to modify in a breeding and selection program. Nevertheless, their overriding importance for livestock production demands that they remain a high-priority objective. An additional complication imposed by selection for DMY is that the entire plant must be harvested to estimate yield, often necessitating an additional year to intercross selected high-yielding parents.

Selection for altered partitioning of assimilates between above- and belowground plant components may increase total harvested DMY. Research in Florida has suggested that "Florigraze" rhizoma perennial peanut (*Arachis glabrata* Benth.) partitions more than 50% of its assimilates to belowground rhizomes.[27] However, other research has shown that this rhizome pool of carbohydrate is vital for long-term persistence of rhizoma perennial peanut.[14] Thus, drastic shifts in partitioning may be detrimental in the long term.

D. Establishment

Breeding and selection objectives related to establishment must first consider the method of propagation of the species. Most tropical forage legumes are seed propagated, but some, such as rhizoma perennial peanut, are propagated by digging and planting rhizomes.[27] Selection for lines that spread rapidly from planted rhizomes is a major selection objective in this species. Alternatively, for some stoloniferous species, planting of vegetative tops may improve establishment. Thus, selection for adventitious rooting could become an objective.

Obtaining a suitable compromise between forage production and seed production is a long-standing problem of all forage breeders. While maximum forage dry matter production is desirable, a new cultivar is not likely to succeed without good seed yield. Quantifying seed yield in the same trial used to evaluate forage yield is not usually possible. Nevertheless, selection for seed yield must be an important objective of most tropical forage legume breeding programs. Seed shattering is a problem with many tropical forage legume species; thus, selection for reduced shattering may greatly improve harvested seed yield. Specialized seed harvesting and/or processing equipment may need to be developed to improve harvested seed yield and quality.

Seedling vigor is often poor in tropical forage legumes. However, the general trait of seedling vigor has been suggested to consist of at least five components: (1) speed of germination, (2) speed of elongation, (3) emergence force, (4) leaf area expansion rate, and (5) relative growth rate.[28] Several of these components are related to seed size, which in some species, e.g., *Trifolium subterranean* L.,[29] has been positively correlated with seedling vigor. Any breeding program to improve seedling vigor in tropical forage legumes should start by assessing the variability for seed size within a species. The relationship of seed size to the components of seedling vigor, especially relative growth rate, should also be determined.

E. Disease and Insect Resistance

Of the various objectives considered for manipulation by plant breeding, selection for pest resistance or transfer of pest resistance from an unproductive introduction into a productive cultivar are among the most common. These traits are often qualitatively inherited, and relatively easy to screen for in a greenhouse.[30] Incorporation of resistance to a particular pest may greatly broaden the area of adaptation of a species.

The primary considerations for choosing pest resistance as a breeding objective are accessible sources of variability for the trait, cooperator support (plant pathologist, entomologist, or nematologist), and ease of screening for resistance on relatively young plants. Additional considerations include the variability of the pathogens. In some diseases, such as rust on *Macroptilium atropurpureum* (DC.) Urb., multiple races have been identified and breeding methods that pyramid genes using blends of pure lines with multiple genes for resistance to different races of the pathogen[31] may be needed.

F. Persistence

The general opinion of producers and many researchers is that tropical C-4 grasses are highly competitive and persistent for long periods, whereas tropical forage legumes are less aggressive and less persistent. This general conclusion may be true for some species combinations, but long-term, highly persistent grass-legume associations are possible. If lack of persistence is a problem with a tropical forage legume, any program selected for improved persistence must first identify the apparent cause(s) of legume loss. Persistence of a species as a sward component may be achieved by either long-term perenniality of original plants or recruitment of new individuals. Unless a species is rhizomatous or stoloniferous, it is unlikely that original individual legume plants will persist for many years. Under this scenario, selection for natural reseeding ability may become an objective. Such an objective should include management practices likely to be adopted by typical producers. Conversely, improved persistence of rhizomatous or stoloniferous species may be desirable. In such circumstances, selection for increased rhizome size or improved adventitious rooting may achieve the objective. These types of selection objectives generally can only be carried out under field conditions. They will usually also need animal defoliation to approximate persistence in the "real world." As such, selection for persistence may be among the more difficult selection objectives.

Other factors related to persistence include general competitive ability with grasses and quantity of N fixed by the legume. Competition for light between small tropical forage legume seedlings and tropical grasses is often extreme. Failure to control grass growth in periods when new legume seedlings are being recruited can greatly alter the legume composition of a sward.[33] Viney-type legumes may compete better for light when grown with tall tropical grasses, but such legumes may be more susceptible to over grazing. Presence of factors such as tannins in leaves, which may make the legume less acceptable to the grazing animals, can improve the overall competitive position of the legume.

V. SPECIFIC TROPICAL FORAGE LEGUME BREEDING PROGRAMS

Several institutions around the world are conducting research on selection and improvement of tropical forage legumes. The three primary locations discussed in this review, based on frequency of published articles, are the CSIRO Division of Tropical Crops and Pastures in cooperation with various Australian state Divisions of Plant Industry, the CIAT Tropical Pastures Program, and the University of Florida's Institute of Food and Agricultural Science. Research at each of these locations was active through the 1980s and early 1990s, but the number of scientists working on tropical forage legumes at each location has decreased markedly since 1990. This reduction in plant breeding research will likely result in a long-term decrease in development of improved cultivars and utilization of these species in developing countries. Although a number of genera have been evaluated by these institutions, the principal ones of focus, and the ones reviewed in this chapter, are *Aeschynomene, Arachis, Centrosema, Desmodium, Leucaena, Macroptilium,* and *Stylosanthes*. This review attempts only to document the selection or breeding history of named cultivars of these genera, with a brief summary of the attributes of each.

A. Breeding and Selection in *Aeschynomene*

The genus name has also become the most widely used common name for the species *A. americana*, although some research publications refer to this species as American joint vetch to distinguish it from the cultivar "Bargoo" of *A. falcata* released in 1973.[34] The known released cultivars of *A. americana* are "Glenn" and "Lee," developed in Australia, and released by the Queensland Division of Primary Industry.[35-37] Glenn was derived from CPI 58491, which was originally collected near the coast 45 km South of Tampico, Mexico. This cultivar was produced by mass selection of superior plants in CPI 58491. According to Bishop, et al.,[35] the superior attributes of Glenn are its tolerance of wet soils and waterlogging, superior regeneration from seed, persistence, and ability to spread from seed under heavy grazing. Plants of the cultivar Lee, which was registered in 1994,[37] have been reported to persist for up to four years under grazing in coastal Australia with 25 to 50% of plants surviving for more than one year.[36] This cultivar remained green longer in the growing season, and had higher DMY and improved persistence with competitive companion grasses. In north-central Florida, Lee is often frosted before flowering or at least before seed maturity, and plants did not perennate in most winters.

In Florida, a common ecotype of *A. americana* was identified in the 1950s as a forage legume with good potential. The general lots of Florida common can be traced to seed harvested in the 1950s from native stands by various private individuals and seed companies, and the USDA Soil Conservation Service.[38] More seed of this ecotype are sold in Florida than any other tropical legume. It is well adapted to seasonally wet flatwoods soils in peninsular Florida (Figure 6.1A).

More recently, a broad range of *Aeschynomene* germplasm has been evaluated in south[39,40] and north[41] Florida. Evaluation of P.I.s showed broad variability for flowering, plant height, leafiness, and winter live-over. Additional greenhouse screening showed variability for response to root-knot nematodes (RKN).[42] Advanced selections have been evaluated from this work and a germplasm of *Aeschynomene evenia* C.Wright was released in 1997.[43]

Through studies of the pollination biology, a method of emasculating and hybridizing *A. americana* was identified.[44] Using this technique, hybrids among lines varying in photoperiod response were produced. This work led to the identification of a major gene for day-neutral response[45] and elucidation of the genetics of photoperiod response in three photoperiod responsive lines.[46] Additional research identified a single dominant gene controlling the glabrous stem trait.[47] A field study maximizing potential for outcrossing using the glabrous stem gene as a marker showed an average of 27% outcrossing when numerous pollinators were present.[5] F_7 breeding selections from hybrids among the glabrous stem accession and Florida common were similar to Florida common in vigor and regrowth, but one is seven to ten days later in flowering. Seed production has been comparable to Florida common in north Florida.

In Australia, two cultivars of villose joint vetch (*A. villosa* Poir.), "Reid" (formerly CPI 91209) and "Kretschmer" (formerly CPI 93621), were released in Queensland in 1995.[48] These CPI numbers were collected in Mexico at altitudes of 1250 m and 2420 m, respectively. Characteristics of this species include high seed yields, prostrate to semierect growth habit, adaptation to a wide range of soil types in medium rainfall areas, and tolerance of light frosts.

B. Breeding and Selection in *Arachis*

Although the common peanut, a.k.a. groundnut, (*Arachis hypogea* L.) has been used for both hay and grazing in the past, current economic conditions limit its use for forage. This discussion will focus only on the perennial species of the genus *Arachis*. The genus has been divided into nine taxonomic sections, and the two perennial species of importance for forage cultivation, *A. glabrata* Benth. and *A. pintoi* Krapov. and W.C. Gregory, have been assigned to sections *Rhizomatosae* and *Caulorrhizae*, respectively.[2]

The common name rhizoma perennial peanut has been adopted for cultivars of *A. glabrata*, and will be used in this discussion. Most research with rhizoma perennial peanut has been conducted in the southeastern United States. The original research, in Florida, focused on germplasm evaluation. This research showed that the species was strongly perennial, with established stands now persisting for about 40 years under pasture management. Established stands may yield up to 10,000 kg ha^{-1} with crude protein ranging from 140 to 180 g kg^{-1}. New fields of rhizoma peanut are normally established by planting dormant rhizomes that have been dug from established plantings during the winter. This relatively low vegetative propagation ratio (ha of source : ha planted) is the principal limitation to more extensive use of this species, as no serious disease or nematode pest has been reported.[27,49,50]

In the early 1960s, a rapidly spreading plant was observed between plots of PI 118457 and PI 151982 on the Agronomy Farm at the University of Florida at Gainesville. Vegetative material of this plant, a suspected seedling from PI 118457, was collected, propagated, tested extensively, and released as the cultivar "Florigraze"[27] (Figure 6.1B). The main superior attributes of Florigraze

FIGURE 6.1 Tropical forage legumes: (A) Common Florida ecotype of *Aeschynomene americana* growing in association with limpograss (*Hemarthria altissima*) on wet flatwoods soil in south Florida; (B) "Florigraze" perennial peanut (*Arachis glabrata*) being grazed by dairy cattle in north central Florida; (C) Centro, (*Centrosema pubescens*) growing in experimental plots, with flower in center of photo; (D) "Florida" carpon desmodium (*Desmodium heterocarpon*) growing in association with Bahia grass (*Paspalum notatum*) in south Florida, note flower in center and prominent leafmarking on leaves; (E) plant introduction of *Leucaena leucocephala* growing in experimental plots, note immature and mature flowers and seed pods; (F) beef cattle grazing "Savanna" stylo (*Stylosanthes guianensis*) in central Florida.

were improved rate of vegetative establishment, spread, and good competitiveness with associated perennial tropical grasses. A second cultivar "Arbrook" selected from PI 262817 was released in 1986.[51] Other research with rhizoma perennial peanut has shown that seed production is limited by various physiological factors, but is not related to pollen viability or failure to germinate.[52,53] Additional germplasm evaluation has shown that most accessions of *A. glabrata* are highly resistant to early (*Cercospera arachidicola*) and late (*Cercosporidium personatum*) leaf spot disease, leaf rust (*Puccinia arachidis*), and peanut RKN (*Meloidogyne arenaria*).[54] Research in Louisiana evaluated 300 naturally occurring seedlings of Florigraze and found that none outperformed Florigraze for all forage traits, suggesting that this chance seedling is in fact a chance hybrid expressing hybrid vigor.[55] Recently the cultivar "Prine" rhizoma peanut (formerly CPI93483) has been proposed for release in subtropical Australia.[56]

Genetic improvement research and cultivar development of *Arachis pintoi* was initially carried out by the Tropical Pastures Program of CIAT, Cali, Colombia. Research in Colombia has shown that this legume has high leaf CP content (17%) and *in vitro* dry matter digestibility (IVDMD) (62%) in both the wet and dry season. The legume was preferentially selected by grazing animals (selection index of 0.56 to associated grasses), showing that this legume was well accepted by grazing cattle throughout the year.[57] Additional research has shown that this legume produces large soil seed reserves and is compatible with productive stoloniferous grasses such as *Bracharia humidicola* (Rendle) Schweickt.[58] Based on these and other results, germplasm of various perennial *Arachis* species was introduced and evaluated in Australia. From these evaluations, CPI 58113 was released as the cultivar "Amarillo" in Australia and has been a commercial success.[59] Additional evaluations of this germplasm are continuing in Colombia and Australia. The proceedings of a workshop held at CIAT in 1993 summarize the biology and agronomy of forage *Arachis* on a worldwide basis.[60]

C. Breeding and Selection in *Centrosema*

This genus consists of 35 species of herbaceous legumes, the majority of which are viney perennials. There is variation among species with a few being annuals and/or small shrub types. Recent investigations on mode of reproduction using flower color genetic markers,[61] isozymes, and DNA markers[62] have demonstrated that accessions in some species have relatively high levels of outcrossing. These results are important for germplasm collection and maintenance and for cultivar development programs. Chromosome counts for this genus indicate that all species are 2n = 22, except for *C. virginianum* (L.) Benth, which has 18 chromosomes. *Centrosema* species are found naturally in a tremendous range of climates. Four of the most scrutinized species include *C. brasilianum* (L.) Benth., *C. pubescens* Benth., *C. pascuorum* C. Martius ex Benth., and *C. virginianum* (L.) Benth. Of these, only *C. pubescens* has attained importance as a forage crop and/or as a cover crop. Regardless of species, most *Centrosema* cultivars are simply selected plant introductions. Some researchers have considered that at least a third of *Centrosema* species have some agricultural potential even though only limited germplasm of several species has been collected and evaluated.

C. brasilianum is a short-lived perennial species that has recently gained more attention as a forage crop. There are currently over 250 accessions in the world collection and these exhibit variation for several agronomic traits. Investigations in Nigeria[63] lead to the identification of two groups for further research. One group, consisting of 10 accessions, exhibited fast establishment, high production during the establishment year, and good drought tolerance. Eleven accessions were categorized into the second group and these produced high yield under grazing, had high competitive ability, and demonstrated good drought tolerance. One cultivar of this species has been released, "Oolloo."[64] Oolloo is not a bred cultivar, but a direct selection of CPI 55696, which was collected near the Petrolina Airport in Brazil. This perennial produces from 3 to 4 t/ha of dry matter under rainfed conditions with 30 to 100% of the plants surviving from one season to the next.

C. pubescens is a self-pollinated species that has a vigorous, twining growth habit (Figure 6.1C). A native of tropical South America, this species has a longer history of domestication than most other tropical legume species. It was introduced into Southeast Asia no later than the 19th century and has been widely used as a cover crop on rubber plantations during the 20th century. Except for two released cultivars, "Belalto" [65] and "Cardillo,"[66] all other commercial types are either ecotypes or common centrosema, usually referred to as Centro. The released cultivars have both been simply plant introduction selections and not bred cultivars. Belalto was a plant introduction from the San Jose Plateau of Costa Rica. Cardillo was selected as one of ten promising accessions from 396 evaluated in southeastern Queensland.

C. pascuorum is a prostrate annual species native to arid tropical regions of South America. It may be found in areas with less than 1000 mm of annual precipitation and is, therefore, very drought tolerant. It is quite susceptible to RKN, although damage may be less in a grass-legume pasture than in experimental plantings. As an annual, regeneration in a pasture must occur by reseeding, therefore, it is crucial to allow for proper seed production and maturation. Released cultivars include "Cavalcade" and "Bundey," both of which were developed in Australia. Cavalcade was derived from a cross between two Brazilian plant introductions, but is superior to its parents in seed and forage yield. Pedigree selection was used in the F_2, F_4, and F_6 generations with single-seed descent practiced in the F_3 and F_5. The selection criterion in the F_2 and F_4 was for vigorous plants that flowered prolifically at the optimum time for seed production (late March) and showed no visible symptoms of root-knot nematodes or other pathogens.[67] The F_6 was planted in small field plots and evaluated for two years, with subsequent regional testing prior to release. Bundey is an introduction from Soledade in the Brazilian province of Paraiba. Evaluations began in 1977, and it was released in 1987. It flowers later than Cavalcade, has smaller seed, and has hairy stems and petioles unlike Cavalcade's glabrous habit. Bundey outperforms Cavalcade in areas with a longer growing season due to either a higher total rainfall or a better distribution of rain.[68]

Centrosema virginianum is a climbing perennial vine found from Uruguay to the northeastern United States. Although this species has the widest area of adaptation of any of the *Centrosema* species, it is not normally found in wetter areas. It is generally thought to be the most genetically diverse species with a wide range of variation for many characters including vigor, flowering date, frost resistance, and drought resistance.

D. Breeding and Selection in *Desmodium*

The genus *Desmodium* contains a number of herbaceous and shrubby legumes thought to have forage potential. Shrub type *Desmodium* such as *D. discolor* V. Vogel., *D. distortum* (Aublet) J. F. Macbr., and *D. gyroides* (Roxb. ex Link) DC. are used as forage material in areas of Central and South America. This chapter will focus only on the species *D. intortum* (Mill.) Urb. Greenleaf desmodium, *D. uncinatum* (Jacq.) DC. Silverleaf Desmodium, *D. heterocarpon* (L.) DC. Carpon Desmodium, and *D. heterocarpon DC.* subsp. *ovalifolium* Ohashi. Ohashi[3] reduced *D. ovalifolium* Wall. to a subspecies of *D. heterocarpon*, but, as suggested by Schultze-Kraft and Benavides,[69] we have used the earlier and more widely recognized name of *D. ovalifolium* based on agronomic distinctiveness. A comprehensive review of research in the Caribbean region on all species of the genus *Desmodium*, thought to have forage potential, was presented by Thro and Quesenberry.[70]

Several excellent review articles have summarized the available research information on *D. intortum* and *D. uncinatum*.[71-73] Thus this review will focus primarily on germplasm selection and cultivar development. These two species have similar characteristics in that both are perennial with trailing viney stems that root freely at the nodes. The two species appear to be relatively closely related and will hybridize, although some F_1 seedlings are weak with poor survival.[74,75] Other species that have been hybridized with these species include *D. sandwicense* E. Meyer, *D. canum* Schinz and Thell., *D. aparines* (Link) DC., and *D. sericophyllum* Schldl. Viability and fertility of these hybrids have varied greatly.[72] One cultivar each has been released in the species *D. intortum* and *D. uncinatum*. "Silverleaf" *D. uncinatum* originated from CPI 8990, which was introduced to Australia from Brazil. "Greenleaf" *D. intortum* is a mixture of three introductions, CPI 17916 from El Salvador, CPI 18009 from Guatemala, and CPI 23189 of unknown origin. Cameron[73] presents a handy reference table of distinguishing characteristics of these cultivars and species. The primary limitation of these species appears to be poor persistence under grazing with reasons for this only partly understood.[72]

Seeds of *D. heterocarpon* were introduced into the U.S. for research purposes in 1954. The original accession (P.I. 217910) was received from the Forest Botanist, Forest Research Institute,

Dehra Dun, Uttar Pradesh, India.[25] After testing at Tifton, GA, seeds were sent to the AREC, Ft. Pierce, FL. The common name carpon desmodium was adopted for this species, and the cultivar "Florida" was released from a seed increase of P.I. 217910 (Figure 6.1D). This species is adapted to areas with high summer rainfall, but will not tolerate prolonged flooding. In subtropical Florida this cultivar will persist in pastures for several years and will also regenerate from abundant seed production. In the northern part of peninsular Florida and the panhandle, it may suffer winterkill in some years. Once established, it associates well with Bahia grass (*Paspalum notatum* Flugge.) and other tropical grasses. The primary limitations to more widespread use appear to be difficulties with establishment due to marginal seedling vigor and susceptibility to RKN, where this pest is a problem. Recent research has revealed variability for resistance to various RKN species among introductions of *D. heterocarpon* and *D. ovalifolium*.[30] Other research demonstrated broad variability among plant introductions of *D. heterocarpon* and *D. ovalifolium* for a number of morphological and agronomic traits.[4,69]

Although no named cultivars have been released from breeding and selection efforts among these two species, research has shown that hybrids can be successfully produced.[4] Due to problems with premature abscission of emasculated flowers, these workers used variability for leaf mark and flower color as markers to aid in identification of hybrid progeny from crosses made in the early morning (prior to 10.00 h) without emasculation. Rate of hybrid production per flower crossed was low (2%), but over 100 F_1 hybrids from 20 different parent line combinations were identified. Purple flower color was shown to be dominant to white and controlled by a single gene. Date of flowering in an F_2 population between an early and a late flowering parent segregated over the range of the parents with a mode near the midpoint. Pollen fertility of F_1 hybrids was slightly less than that of the parents, but generally was above 90%.[76] Two advanced breeding lines from these hybrids had superior dry matter yield, root-knot nematode tolerance, seed production, and persistence compared to Florida carpon desmodium,[77] but these lines require additional regional testing before recommendation for release.

Local ecotypes of *D. ovalifolium* have been utilized for some time as cover crops in Southeast Asian plantations, however, its use has declined because of susceptibility to pink disease caused by the fungus *Sclerotium salmonicolor*. Research in the CIAT Tropical Pastures Program indicated this species has good potential on acid, low-fertility soils in high-rainfall regions in the American tropics. Countries where it has been suggested to have potential include Bolivia, Brazil, Colombia, Costa Rica, Cuba, Ecuador, Honduras, Mexico, Nicaragua, Panama, Peru, and Venezuela.[78] In addition to the possible problems with RKN mentioned above, susceptibility of early introductions to stem gall nematode (*Pterotylenchus cecidogenus*) was reported in Colombia.[79,80] This early research was conducted using a very limited number of germplasm accessions. Several additional collections were made in various Southeast Asian countries, resulting in dramatic broadening of the variability for most traits evaluated.[69,81,82] Germplasm of *D. ovalifolium* has been widely evaluated in the Caribbean region.[70] A core collection of 18 accessions of *D. ovalifolium* has been established based on (1) representativeness of the geographical distribution of the species, (2) environmental conditions at the collection sites, (3) agronomic performance, and (4) quality parameters of the accessions. This core collection is being evaluated at six contrasting environments in Colombia to establish genotype x environment interactions for agronomic and quality parameters.[83] Natural outcrossing in *D. ovalifolium* based on purple flowered progeny from a white flowered strain was 4%, which was lower than for *Centrosema virginianum* (18%) and *Galactia striata* (13%).[61] No known cultivars of this species have been released to date, although considerable germplasm evaluation continues at a number of locations.

E. Breeding and Selection in *Leucaena*

Because of the multipurpose uses of *Leucaena*, numerous reviews have been written thoroughly documenting the varied uses in such fields as agronomy, agroforestry, and soil conservation.

Germplasm collections have increased over the past three decades and taxonomic relationships have been elucidated by both classical methods and by studies of sexual compatibility from crosses.[84-88] The above works recognized only 16 valid species descriptions, however, a recent workshop suggested that taxonomists now recognize 22 species.[89] This workshop reviewed adaptation, quality, and use of *Leucaena* in farming systems. This review will focus only on applications of breeding and selection for forage types of *L. leucocephala* (Figure 6.1E) and crosses with related species to incorporate specific resistance to pests or acid soil tolerance.

Early breeding efforts in Australia concentrated on locating and exploiting variability for various forage characteristics in *L. leucocephala*. Two cultivars were released in 1962 by the Queensland Pasture Liasion Committee: "Peru" (CPI 18614) and "El Salvador" (CPI 18623). Each of these cultivars resulted from increase of plant introductions made to Australia in 1954.[87] Flowering and pollination biology and methods for making controlled crosses have been described,[88] revised, and improved.[90] Although the species appears to be primarily self-pollinated, the above methods made the production of intra- and interspecific hybrids relatively easy.

A breeding program to develop types adapted to grazing with high basal branching density and high forage yield was initiated at Samford, Australia in 1956–57.[91] In a series of papers, Gray[92-94] showed that erect habit was dominant over bushy habit and absence of strong basal branching was dominant over its presence. In this breeding effort, selection for yield in progenies was delayed until the F_4 generation when lines were nearing uniformity. Due to repeated frost damage of plant material in the dry season, the research was transferred to Landsdown near Townsville, where three lines were selected from among 28 previously identified lines. These three lines were further evaluated along with Peru for yield of edible dry matter in two additional experiments. Line 3, an F_4 selection from the cross of "Guatemala" X Peru, had 49% greater yield than Peru in an experiment at Landsdown and was subsequently released as "Cunningham."[91]

In early research with *L. leucocephala,* the presence of the amino acid mimosine was shown to cause health problems (thyroid malfunction, loss of hair, and goiter development) in animals consuming diets containing a high percentage of this legume. During the 1970s, breeding efforts were focused on selecting types with reduced levels of mimosine from interspecific hybrids of Cunningham and Peru with low mimosine lines of *L. pulverulenta* (Schldl.) Benth. Problems with fertility and seed production were encountered initially, but fertile, low mimosine lines were eventually selected. However, these lines generally were lower yielding than "Cunningham." In the early 1980s, studies comparing the toxicity of *Leucaena* in goats in Australia and Hawaii showed that rumen bacteria in the Hawaii goats were detoxifying DHP (the toxic metabolite of mimosine) with no observed health problems in goats on a high *Leucaena* diet.[95] Later research purified the bacteria and cultures were introduced into Australia. Goats and cattle in Australia dosed with DHP-degrading bacteria acquired the ability to degrade DHP,[96] and increased average daily gain per head from 0.52 kg per day in the undosed group to 1.03 kg in the treated group.[97] Similar results have been obtained in Florida by dosing cattle with DHP degrading bacteria.[98] In both experiments, the control groups eventually acquired the DHP degrading bacteria from incidental contact, indicating that the bacterium is relatively easy to transmit. In Florida, it was shown to persist over a winter season when animals were not grazing *Leucaena*. These findings have resulted in elimination of low-mimosine content as a breeding objective in *L. leucocephala*.

Breeding for tolerance to acid soils has been a major selection objective to improve the adaptability of *L. leucocephala* to acidic Oxisols and Ultisols in tropical America and Southeast Asia, which have high Al and low Ca saturations. Germplasm evaluation at CIAT showed that *L. diversifolia* (Schldl.) Benth. (2n = 52) and hybrids of this species with *L. leucocephala* (2n = 104) had tolerance to acid soil conditions.[22] The F_1 hybrids had the expected 78 chromosomes, with generally from 20 to 24 univalents. Nevertheless, some 30 to 40% of F_1 trees set open-pollinated seed. After additional selection at Planaltina, Brazil, F_3 and F_4 selections from hybrids were identified with acid tolerance and either the branching and large leaflets needed in a grazing type, or apical dominance and small leaflets suitable for forestry types. The research further indicated that, due to the variability among F_1 trees (apparently due to the heterozygous nature of the outcrossing *L. diversifolia*), F_2 and subsequent generations should be raised from only the most vigorous trees in the previous generation, rather than from a bulk of all trees showing some acid tolerance.[11] Additional research in Southeast Asia demonstrated that F_3 lines selected in Brazil were 35 to 79% taller than Cunningham when grown in acid sulfate soils.[99] However, reports from the 1998 Leucaena workshop indicated that no accessions were specifically adapted to strongly acid-infertile soils.[100] Of six environmental variables identified as limiting growth in *Leucaena*, soil acidity was the most important across 19 diverse sites in Australia, New Guinea, Indonesia, Vietnam, Phillippines, Laos, and Kenya over a 2.5-year period.[101] These workers reported that the KX2 F_1 hybrid accession of *L. pallida* K748 X *L. leucocephala* K636 was broadly adapted and relatively high yielding across all environments including cool temperature environments.

Beginning in the mid 1980s, damage to leucaena from feeding and oviposition by the psyllid *Heteropsylla cubana* Crawford began to be observed. Most of the widely planted selections of *L. leucocephala* were susceptible to this insect. Other species including *L. esculenta* (Moncino and Sesse ex DC.) Benth., *L. pallida* Britton and Rose, and *L. retusa* Benth have been reported to have some degree of psyllid resistance.[102] Additionally, Hutton and Chen[99] reported that F_3 lines from crosses of *L. leucocephala X L. diversifolia* were more tolerant of a severe psyllid attack than was the control Cunningham. Recent reports from India[103] and Florida[104] confirm high levels of psyllid resistance in *L. esculenta* and *L. pallida*; however, in Florida, the psyllid susceptible *L. leucocephala* lines yielded greater biomass than the less vigorous *L. esculenta* and *L. pallida*.

Genetic improvement research is continuing in the genus. The *L. leucocephala* cultivar Tarramba was released in Australia in 1997. It is reported to have superior seedling vigor, growth characteristics, and plant height (242 cm at 13 months vs. 109 and 126 cm for Cunningham and Peru, respectively).[105] In addition to the programs in Australia, active genetic improvement programs are underway in Hawaii and Brazil. Austin, et al.[106] reported that both programs are based on tetraploid level crosses using *L. leucocephala*, *L. pallida*, and *L. diversifolia* germplasm in various combinations with the aim of capitalizing on heterosis for growth, psyllid resistance, cold tolerance, and acid soil tolerance. In reviewing a career of breeding experiences with *Leucaena*, Brewbaker[107]concluded that success in genetic improvement of *Leucaena* was a function of harnessing five types of genetic systems: self-incompatibility and male sterility, hybrid vigor, interspecific hybrid diversity, quantitative trait loci (QTL) for tree form, and QTL for stress and pest tolerance. Other research from Hawaii[85] has shown that interspecific hybrid barriers between 16 *Leucaena* species are relatively weak with 77% of 118 attempted hybrid combinations being successful.

F. Breeding and Selection in *Macroptilium*

Two species of this genus have been extensively evaluated, *M. atropurpureum* (DC.) Urb. and *M. lathyroides* (L.) Urb. The former is commonly referred to as Siratro, due to the release of a cultivar of the same name in 1960.[108] "Siratro" was the first tropical legume cultivar bred and released in Australia. It was derived from a cross between two Mexican introduction lines, CPI 16877 and CPI 16879, collected in Vera Cruz State and Matlopa in the state of San Luis Potosi, respectively. Both parents were considered to be relatively persistent types. A cross was made and elite individuals selected in the F_2 generation. These selections were subjected to cattle grazing for two subsequent generations, at which time three promising F_4 families were identified. These three families were bulked in equal amounts to form Siratro. The plant is a creeping, viney perennial that will, under adequate conditions, form a dense mat of vegetative growth. Siratro is a deep-rooted species, well adapted to areas with a minimum of 800 mm annual precipitation, but performs adequately in areas receiving as little as 650 mm. It is diploid, $2n = 2x = 22$, and is self-pollinated. Siratro is tolerant of a soil pH range from 4.5 to 8.0, but an adequate supply of P is necessary for growth and stand maintenance.

Work on this species partially originated due to pest problems in phasey bean, *M. lathyroides*, a related species that is quite susceptible to RKN. In this respect, Siratro has been a successful alternative as it has proven resistant to these soil-borne pathogens. Siratro, like most viney forage legumes, is more sensitive to grazing pressure, and has poor persistence if more than moderate grazing pressure is utilized.

Siratro is very sensitive to rust (*Uromyces appendiculatus*), which can cause massive leaf drop and subsequent stand decline. Bray has reported on genetic control of rust resistance in *M. atropurpureum*.[109,110] In an evaluation of 179 accessions, 59 exhibited resistance to an Australian isolate of the fungus. The results of crosses made between these resistant lines and Siratro indicated several different genetic mechanisms conferring resistance depending upon the geographic source of the other parent. He found resistance could be due to either a single dominant gene, two dominant genes, or a major dominant gene with associated modifying genes. As rust species typically develop several physiological races through their sexual life cycle, resistance to one isolate may not necessarily confer resistance to a different isolate. To evaluate this possibility, 10 near-isogenic lines developed in a backcross program with Siratro were exposed to 10 rust isolates: one from Australia, two from Florida, and seven from Mexico.[31,32] Each of these lines was backcrossed to Siratro for four generations with selection for rust-resistant types followed by two generations of selfing to identify individuals that were homozygous for the rust-resistance genes. Although six lines were resistant to all rust isolates, four lines were susceptible to one or more of the rust isolates. Differential isolates of this pathogen suggest that new cultivars should contain several different genes for resistance to counter the natural variation in the rust populations. Four of these lines were selected and mixed to produce the multiline cultivar "Aztec." This cultivar is morphologically very similar to Siratro except for rust-resistance. In field trials where the pathogen was present, Aztec plants remained rust-free and had 30% higher leaf production than did Siratro. The rust resistant genes in Aztec trace to four distinct sources (Colombia; El Salvador; Oaxaca, Mexico; and Sonora, Mexico) which should provide a wide geographic area of adaptation where it exhibits a high level of rust resistance.

Another characteristic that has hindered Siratro is pod shattering. This is a common phenomenon in many tropical legumes, and results in poor commercial seed production, hence a high seed cost to producers. Unless seed production can be improved either through breeding or by the use of management practices, high seed cost will reduce the impact any cultivar may have in the agricultural community. Other breeding work has continued on *M. atropurpureum* and several late-generation families are under evaluation. A primary objective of this work has been increased production as compared to Siratro. Although preliminary indications suggest the development of lines that yield significantly more than Siratro, grazing evaluations to assess their relative persistence must be performed before material can be released.

G. Breeding and Selection in *Stylosanthes*

One of the most desirable features of *Stylosanthes* is its suberect to erect growth habit, which makes it more suitable for grazing systems than viney-type legumes. Because of the natural diversity among and within its species, *Stylosanthes* has yielded more cultivars for use in tropical pastures than any other genus of tropical forage legumes. In their area of origin, these species have long been appreciated as valuable pasture plants; however, they have only been utilized in other regions in the last 100 years. For example, *S. guianensis* is regarded as the first domesticated species in the genus and was used as a cover crop on Malaysian plantations in the late 1800s. *S. humilis* Kunth was the first species of the genus evaluated for its forage potential and was grown for this purpose in the early part of this century in Australia. Although there is considerable variation within some species, it is not difficult to identify plants with potential in the intended area of establishment based on climate and environment. *S. guianensis* and *S. humilis* have become increasingly important partly due to their being adaptable to wet and dry climates, respectively. The use of *Stylosanthes* has been reduced in Australia because of extreme susceptibility to anthracnose caused by *Colletotrichum gloeosporioides*. The extent of the problem is exacerbated by the presence of pathotypes and races within pathotypes of this fungus. The incidence of this disease has greatly reduced the use of certain species of *Stylosanthes* and lead to a search for sources of resistance in germplasm collections.[111] Genetic resistance to this pathogen in tetraploid *S. hamata* has been reported to be quantitative in nature and primarily a function of additive genetic variance.[112] To overcome the genetic diversity within this fungus, a coordinated, multidisciplinary approach should be adapted using broad-based germplasm to develop cultivars with stable resistance to anthracnose.[113]

S. humilis has been widely grown in areas that have between 600 and 1200 mm of annual precipitation with dry seasons of more than four months. This annual or short-lived perennial species is self-pollinated with a chromosome count of $2n = 2x = 20$. The primary limitation of growth is temperature as it will not develop if the mean temperature falls below 23°C. This legume has a long history of importance in Australia. After being accidently introduced early in this century, its value as a pasture crop was "discovered" near the city of Townsville, hence it is commonly referred to as Townsville stylo.[114] Townsville is a widely variable population that has become naturalized in regions of Australia. In addition to Townsville stylo, three cultivars of this species have been released, which differ with respect to flowering. These cultivars were developed by mass selection from naturalized pastures of Townsville stylo and selected for high yield and maturity differences. These cultivars, "Gordon," "Lawson," and "Paterson," are late-, mid-, and early-season flowering types, respectively.[115]

S. guianensis is a prominent species of *Stylosanthes* grown in many countries throughout the world (Figure 6.1F). In general, it is best adapted to wetter areas of the tropics and subtropics. The literature for this species can be confusing as it has also been identified as *S. guyanensis* or *gracilis*. Attempts have been made to clarify the taxonomy[116,117] and this highly polymorphic species has been partitioned into the following varieties: *guianensis*, *gracilis*, *intermedia*, *robusta*, *dissitiflora*, and *longiseta*. Of these six varieties, cultivars have been developed in two, *guianensis* and *intermedia*. *S. guianensis* var *guianensis* germplasm was the source for the Australian cultivars "Schofield," "Cook," "Endeavour," and "Graham." In general, this variety is a rather large group of self-pollinated diploids ($2n = 20$) that differ for several agronomic traits, including DMY and seed production. Materials within this group also differ with respect to the climate of their site of origin. Schofield was introduced into Australia from Brazil in the early 1930s and was the most widely grown cultivar in Australia until the late 1970s when susceptibility to anthracnose all but terminated its use.[118] Cook, selected in Colombia, is earlier flowering and slightly more tolerant of anthracnose than Schofield. In Australia, Endeavour, a Guatemalan selection, has proven to be as susceptible to anthracnose as Schofield. Graham was released in 1979 and is more persistent than the other cultivars of this variety. Other cultivars in this group have been released in Brazil. "Savanna" stylo was developed in Florida and released in 1987.[119] This cultivar performs well in subtropical Florida, particularly in autumn when it produces more forage than most tropical legumes. It was developed by mass selection after a 12-year period of natural selection in a field originally planted with 22 accessions from South America. Obviously, some natural outcrossing occurred in this material as it is distinct from the original parental sources. As area of this cultivar has increased in Florida, anthracnose has become a limiting factor to seed production. Evaluation for disease resistance in *S. guianensis* has suggested that resistance to anthracnose is available, and that a program of recurrent selection would enhance resistance to this fungus.

Other species of *Stylosanthes* that are increasing in importance include *S. capitata* J. Vogel, *S. hamata* (L.) Taubert, and *S. scabra* J. Vogel. *S. capitata* performs well under extremely acidic soil conditions, and is tolerant to a wide range of *Colletotrichum gloeosporioides* pathotypes. Most collections were obtained from areas of Brazil and Venezuela, as it seems to be rather limited and sporadic in its natural habitat. One cultivar has been released, "Capica," which is simply a bulk of five accessions, CIAT 1315, 1342, 1693, 1728, and 1943.[120] Further breeding work on this species has involved a pedigree selection program.[121] Progeny from crosses between CIAT 1019 and CIAT

1097 were evaluated for dry matter production, vigor, nutrient concentrations of N, Ca, Mg, K, and P, and seed yield in a soil of pH 4.7. Transgressive segregates were noted for these traits in a large F_2 population that was subjected to grazing. Further pedigree selection resulted in the identification of seven F_7 families that yielded significantly more dry matter than the best parent, with a range of 6 to 44% more forage produced. Four F_7 families had significantly higher seed yield than the best parent. These advanced lines had approximately the same anthracnose resistance in the field as the most resistant parent. These results indicate that the development of an improved cultivar of this low-acid tolerant species is quite possible.

S. hamata is found in the Caribbean and along the coastlines of South and Central America as well as Florida in the United States. This drought-tolerant species contains both diploid (2n = 20) and tetraploid (2n = 40) types, most of which have an annual growth habit. The tetraploid forms seem to be more tolerant of acidic soil conditions than the diploids. There is substantial evidence that diploid *S. hamata* and diploid *S. humilis* are the parents of the tetraploids in this species.[122,123] "Verano," a self-pollinating tetraploid type, was released in 1973 in Australia, but can be traced to a collection made at the Maracaibo Airport, Venezuela in 1965 (CPI 38842).[118] This cultivar is very much a tropical plant and performs poorly even in the subtropics. It has adequate anthracnose tolerance and is competitive in some native grass swards. "Amiga" is another tetraploid cultivar with attributes similar to Verano.[124] These have become increasingly important in areas where common ecotypes have largely succumbed to anthracnose.

S. scabra is native to Bolivia, Brazil, Colombia, Ecuador, and Venezuela. It is a strong perennial with a small shrub growth habit reaching 2 m in height. Like *S. hamata*, this species is also a tetraploid with 2n = 40, and these two species are known to naturally intercross. *S. scabra* is reported to be an allotetraploid having *S. viscosa* as one parental species with the other parent unknown.[125] Isozyme analysis[126] of several *Stylosanthes* species indicated a close genetic relationship between tested cultivars of *S. hamata* and *scabra*. A very drought-tolerant species having a pronounced taproot, *S. scabra* is well adapted to semiarid tropical regions. Released cultivars include "Fitzroy," "Seca," and "Siran."[118,127] Fitzroy, derived from CPI 40205, is a bushy, mid-season flowering type that has proven to be very susceptible to anthracnose. Consequently, its use is limited to regions of low rainfall or at higher latitudes. Seca, a late-flowering, erect cultivar with moderate anthracnose resistance, was derived from CPI 40292. Siran is a composite of three lines developed by breeding/selection from four different sources resistant to anthracnose. Cameron et al.,[128] reported on the response to recurrent selection for anthracnose resistance in *S. scabra* crosses between Fitzroy and 12 accessions. They showed significant improvement in resistance through three cycles of selection. Their use of numerous, genetically diverse sources should confer broad-based resistance on the subsequent populations developed. Other important traits that appear to be amenable to genetic improvement in this species include time of flowering, frost tolerance, seed yield, seedling vigor, dry matter production, plant height, and plant growth habit.[124,129]

VI. CONCLUSION

Our review of the potential of various tropical forage legume species and plant breeding efforts to date with these species suggests that no species fulfills all characteristics desired in a successful forage legume. All have one or more deficiencies for some environments. Only through organized plant improvement programs will new cultivars be developed that contain those characteristics in commercially acceptable cultivars. By necessity, this involves the same type of long-term effort accorded cultivated temperate legume species such as alfalfa, red and white clover, etc. Due to the immense attention these temperate crops have received, more economical cultivars are available than those existing prior to these efforts. Breeding programs on many tropical legumes are still in their infancy, but should yield great rewards if continued or expanded.

In addition to traditional breeding approaches, biotechnology should be considered in terms of potential improvement in tropical legumes. As the first genetically engineered cultivars are released, this technology should be assessed for tropical forage legumes and utilized where appropriate. The ability to incorporate novel genes through transformation procedures could greatly impact the suitability of a legume crop. The use of both traditional and modern techniques in organized programs of plant improvement of tropical forage legumes should be beneficial to producers as they strive to meet market demands for animal products.

REFERENCES

1. Hutton, E. M., Selection and Breeding of Tropical Pasture Legumes, in *Tropical Forage Legumes*, Skerman, P. J., Cameron, D. G., and Riveros, F., Eds., FAO, Rome, 173, 1988.
2. Krapovickas, A., and Gregory, W. C., Taxonomy of the genus *Arachis* (*Leguminosae*), *Bonplandia*, 8, 1, 1994 (in Spanish).
3. Ohashi, H., *The Asiatic Species of Desmodium and its Allied Genera (Leguminosae), Ginkgoana — Contributions to the Flora of Asia and the Pacific Region No. 1,* Academia Scientific Book, Inc., Tokyo, 1973.
4. Quesenberry, K. H., McKellar, M. A., and Moon, D. E., Evaluation and hybridization of germplasm in the *Desmodium heterocarpon-D. ovalifolium* species complex, in *Proc. XVI Intl Grassland Congress*, Nice, France, 251, 1989.
5. McKellar, M. A., Deren, C. W., and Quesenberry, K. H., Outcrossing in Aeschynomene, *Crop Sci.*, 31, 476, 1991.
6. CIAT, Plant breeding, in *Annual Report Tropical Pastures Program*, Centro Internacional de Agricultura Tropical, Cali, Colombia, 1984, 21.
7. Quesenberry, K. H., Smith, R. L., Schank, S. C., and Ocumpaugh, W. R., Tropical grass breeding and early generation testing with grazing animals, in *Proceedings of the 34th Southern Pasture and Forage Crop Improvement Conference*, Auburn University, Auburn, Alabama, 1977.
8. Gildersleeve, R. R., Ocumpaugh, W. R., Quesenberry, K. H., and Moore, J. E., Mob-grazing of morphologically different *Aeschynomene* species, *Trop. Grassl.*, 21, 123, 1987.
9. Kitamura, Y., Introduction of tropical legumes and development of legume-based pastures in subtropical Japan, *Tropical Agr. Research Series No. 18*, Tropical Agr. Res. Ctr., Japan., 173, 1985.
10. Albrecht, S. L., Bennett, J. M., and Quesenberry, K. H., Growth and nitrogen fixation of *Aeschynomene* under water stressed conditions, *Plant and Soil*, 60, 309, 1981.
11. Hutton, E. M., Field selection of acid-soil tolerant leucaena from *L. leucocephala X L. diserfolia* crosses, *Trop. Agric.* (Trinidad), 67, 2, 1989.
12. Quesenberry, K. H., Ocumpaugh, W. R., and Ruelke, O. C., *Hemarthria altissima*: a pasture grass for the tropics, in *Proc. XIV Intl. Grasslands Congress*, Smith, J. A., and Hays, V. W., Eds., Lexington, KY, Westview Press, Boulder, Colorado, 788, 1983.
13. Ortega-S, J. A., Sollenberger, L. E., Quesenberry, K. H., Cornell, J. A., and Jones, Jr., C. S., Productivity and persistence of rhizoma peanut pastures under different grazing managements, *Agron J.*, 84, 799, 1992.
14. Baltensperger, D. D., Breeding self-pollinated forage legumes, in *Southern Pasture and Forage Crop Improvement Conference*, Lexington, KY, 22, 1988.
15. Meijer, E. G. M., and Broughton, W. J., Regeneration of whole plants from hypocotyl-, root-, and leaf-derived tissue cultures of the pasture legume *Stylosanthes guianensis*, *Physiol. Plant.*, 52, 280, 1981.
16. Meijer, E. G. M., and Steinbiss, H. H., Plantlet regeneration from suspension and protoplast cultures of the tropical pasture legume *Stylosanthes guianensis* (Aubl.), *Sw. Annals of Botany*, 52, 305, 1983.
17. Angeloni, P. N., Rey, H. Y., and Mroginski, L. A., Regeneration of plants from callus tissue of the pasture legume *Centrosema brasilianum*, *Plant Cell Rep.* 11, 519, 1992.
18. Wofford, D. S., Baltensperger, D. D., and Quesenberry, K. H., *In vitro* culture response of alyceclover genotypes on four media systems, *Crop Sci.*, 32, 261, 1992.
19. Wofford, D. S., Quesenberry, K. H., and Baltensperger, D. D., Tissue culture regeneration of *Desmodium*, *Crop Sci.*, 32, 265, 1992.
20. Rey, H. Y., and Mroginski, L. A., Regeneration of plants from callus tissue of *Aeschynomene* spp., *Plant, Cell, Tissue and Organ Cult.*, 45, 185, 1996.

21. Sarria, R., Calderon, A., Thro, A. M., Torres, E., Mayer, J., and Roca, W. M., Agrobacterium-mediated transformation of *Stylosanthes guianensis* and production of transgenic plants, *Plant Sci.*, 96, 119, 1994.
22. Hutton, E. M., Acid soil tolerant hybrids in the tree legume *Leucaena*, in *Proc. XV Intl. Grassland Congress*, Kyoto, Japan, 199, 1985.
23. Keller-Grein, G., and Passoni, F., Agronomic evaluation of *Centrosema* germplasm on an acid ultisol in the humid tropics of Peru. III. Selected *Centrosema pubescens* Benth. accessions, in *Proc. XVII Intl. Grassland Congress*, Palmerston North, Hamilton, Lincoln, New Zealand, and Rockhampton, Australia, 2145, 1993.
24. Soffes, A. R., and Quesenberry, K. H., Tolerance to soil acidity and resistance to the root-knot nematode, *Meloidogyne* spp. in Leucaena, in *Agronomy Abstracts*, ASA, Madison, WI, 1982, 180.
25. Kretschmer, A. E., Jr., Brolmann, J. B., Snyder, G. H., and Coleman, S. W., *Florida Carpon Desmodium, a Perennial Tropical Forage Legume for Use in South Florida*, Circular S-260, University of Florida, Florida Agric. Exp. Stn., Gainesville, FL, 1979.
26. Albrecht, K. A., and Quesenberry, K. H., Genetic, environmental and morphological effects on tannin concentrations of tropical forage legumes, in *Agronomy Abstracts*, ASA, Madison, WI, 1987, 140.
27. Prine, G. M., Dunavin, L. S., Moore, J. E., and Rousch, R. D., *'Florigraze' Rhizoma Peanut a Perennial Forage Legume*, Circular S-275, University of Florida, Florida Agric. Exp. Stn., Gainesville, FL, 1981.
28. Carelton, A. E., and Cooper, C. S., Seed size effects upon seedling vigor of three forage legumes, *Crop Sci.*, 12, 183, 1972.
29. Black, J. N., The influence of seed size and depth of sowing on emergence and early vegetative growth of subterranean clover (*Trifolium subterraneum* L.), *Austr. J. Ag. Res.* 7, 98, 1956.
30. Quesenberry, K. H., and Dunn, R. A., Variability among *Desmodium* species for response to root-knot nematodes, *Crop Sci.*, 27, 1234, 1987.
31. Bray, R. A., Sonoda, R. M., and Kretschmer, A. E., Jr., Pathotype variability of rust caused by *Uromyces appendiculatus* on *Macroptilium atropurpureum*, *Plant Dis.* 75, 430, 1991.
32. Bray, R. A., and Woodroffe, T. D., Macroptilium atropurpureum (DC) Urban (atro) cv. Aztec, *Aust. J. Exp. Agr.,* 35, 121, 1995.
33. Sollenberger, L. E., Quesenberry, K. H., and Moore, J. E., Effects of grazing management on establishment and productivity of aeschynomene overseeded in limpograss pastures, *Agron. J.*, 79, 78, 1987.
34. Oram, R. N., Register of Australian herbage plant cultivars, CSIRO Division of Plant Industry, Melbourne, 276, 1990.
35. Bishop, H. G., Ludke, D. H., and Rutherford, M. T., Glen joint vetch: a new pasture legume for Queensland coastal areas, *Queensland Agric. J.*, 111, 241, 1985.
36. Bishop, H. G., and Hilder, T. B., The *Aeschynomene* genus as a source of pasture legumes for tropical and subtropical Australia, in *Proc. XVII Intl. Grassland Congress*, Palmerston North, Hamilton, Lincoln, New Zealand, and Rockhampton, Australia, 2152, 1993.
37. Bishop, H. G., Cook, B. G., Hopkinson, J. M., and Hilder, T. B. *Aeschynomene americana* L. (American jointvetch) cv. Lee. *Aust. J. Exp. Agric.* 35:122, 1996.
38. Hodges, E. M., Kretschmer, A. E., Jr., Mislevy, P., Rousch, R. D., Ruelke, O. C., and Snyder, G. H., *Production and Utilization of the Tropical Legume Aeschynomene,* Circular S-290, University of Florida, Institute of Food and Agricultural Sciences, 1982.
39. Kretschmer, A. E., Jr., and Bullock, R. C., *Aeschynomene* spp.: Distribution and potential use, *Soil Crop Sci. Soc. Fla. Proc.*, 39, 145, 1980.
40. Kretschmer, A. E., Jr., and Snyder, G. H., Potential of *Aeschynomene* spp. for pastures in the tropics, in *Proc. XIV Intl. Grassland Congress*, Smith, J. Allen, and Hays, Virgil W., Eds., Lexington, KY, Westview Press, Boulder, Colorado, 1981, 783.
41. Quesenberry, K. H., and Ocumpaugh, W. R., Forage potential of *Aeschynomene* species in North Central Florida, *Soil Crop Sci. Soc. Fla. Proc.*, 40, 159, 1981.
42. Quesenberry, K. H., Hardy, S. R., and Dunn, R. A., Evaluating *Aeschynomene americana* L. germplasm for response to *Meloidogyne* spp., in *Proc. XV Intl. Grassland Congress*, Kyoto, Japan, 1985, 158.
43. Kretschmer, A. E., Jr., Pitman, W. D., Wilson, T. C., and Bullock, R. C. Registration of evenia *aeschynomene* IRFL 1645 germplasm. *Crop Sci.* 16, 1675, 1996.
44. Hardy, S. R., and Quesenberry, K. H., Artificial hybridization of *Aeschynomene americana* L. (A tropical forage legume), *Soil Crop Sci. Soc. Fla. Proc.*, 43, 174, 1984.

45. Deren, C. E., and Quesenberry, K. H., Inheritance of photoperiodic vs. day neutral flowering in aeschynomene, *J. Hered.*, 80, 327, 1989.
46. Deren, C. E., and Quesenberry, K. H., Inheritance of photoperiod induced flowering in three photoperiodic lines of *Aeschynomene americana* L., *Theoret. Appl. Gen.*, 78, 825, 1989.
47. Deren, C. W., and Quesenberry, K. H., Glabrous stem: a seedling marker gene in *Aeschynomene americana*, *Trop. Grassl.* 21, 188, 1987.
48. Bishop, H. G., Cook, B. G., English, B. H., Bushell, J. J., and Hilder, T. B., More Aeschynomene pasture legumes for the tropics and sub-tropics. *Proc. XVIII Intl. Grassland Congress*, Winnipeg and Saskatoon, Canada, 1997, 1-1.
49. Baltensperger, D. D., Prine, G. M., and Dunn, R. A., Root-knot nematode resistance in *Arachis glabrata*, *Peanut Sci.* 13, 78, 1986.
50. Ruttinger, A. E., Quesenberry, K. H., Prine, G. M., and Moore, G. A., Germplasm evaluation of wild perennial *Arachis* introductions, in *Agronomy Abstracts*, ASA, Madison, WI, 1988, 94.
51. Prine, G. M., Dunavin, L. S., Glennon, R. J., and Rousch, R. D., *'Arbrook' Rhizoma Peanut a Perennial Forage Legume*, Circular S-332, University of Florida, Florida Agric. Exp. Stn., Gainesville, FL, 1986.
52. Niles, W. L., *Factors Inhibiting Seed Production in Florigraze Rhizoma Peanut* (*Arachis glabrata* Benth)., Ph.D. dissertation, Univ. of Florida, Gainesville, Florida, 1989.
53. Niles, W. L. and Quesenberry, K. H., Pollen germination of rhizoma peanut cv. 'Florigraze', *Peanut Sci.*, 19, 105, 1992.
54. Quesenberry, K. H., Ruttinger-Lamperti, A., and Kelly, C. A., Perennial *Arachis* germplasm for the subtropics and tropics, in *Proc. XVII Intl. Grassland Congress*, Palmerston North, Hamilton, Lincoln, New Zealand, and Rockhampton, Australia, 1993, 2138.
55. Venuto, B. C., Elkins, W. M., Hintz, R. W., and Reed, R. L., Comparison of seed-derived lines from 'Florigraze' rhizoma perennial peanut. *Crop Sci.* 37, 1098, 1997.
56. Cook, B. G., personal communication, 1999.
57. Lascano, C. E., and Thomas, D., Forage quality and animal selection of *Arachis pintoi* in association with tropical grasses in the eastern plains of Colombia, *Grass & Forage Sci.*, 43, 433, 1988.
58. Grof, B., Forage attributes of the perennial groundnut *Arachis pintoi* in a tropical savanna environment in Colombia, in *Proc XV Intl. Grassland Congress*, Kyoto, Japan, 1985, 168.
59. Cook, B. G., and Loch, D. S., Commercialisation of *Arachis pintoi* cv. Amarillo in northern Australia, *Proc. XVII Intl. Grassland Congress*, Palmerston North, Hamilton, Lincoln, New Zealand and Rockhampton, Australia, 1993, 2140.
60. Kerridge, P. C., and Hardy, B., *Biology and Agronomy of Forage Arachis, CIAT Publication* No. 240, Centro Internacional de Agricultura Tropical, Cali, Colombia, 1994.
61. Maass, B. L., and Torres, A. M.. Off types indicate natural outcrossing in five tropical forage legumes in Colombia, *Tropical Grasslands*, 32, 124, 1998.
62. Penteado, M. I., Saenz-de-Miera, L. E., and Perez-de-la-Vega, M., Genetic resources of Centrosema spp: genetic changes associated to the handling of an active collection, *Genetic Resources and Crop Evolution,* 43, 85, 1996.
63. Peters, M., Kramer, H., Starawali, S. A., and Schultze-Kraft, R., Characterization of a germplasm collection of the tropical pasture legume *Centrosema brasilianum* in subhumid West Africa, *J. Agr. Sci.,* 130, 139, 1998.
64. Cameron, A. G., and Lemke, B. G., *Centrosema brasilianum* (L.) Benth. (Centro) cv. Oolloo, *Aust. J. Exp. Agr.,* 37, 717, 1997.
65. Clements, R. J., Williams, R. J., Grof, B., and Hacker, J. B., Centrosema, in *The Role of Centrosema, Desmodium, and Stylosanthes in Improving Tropical Pastures*, Burt, R. L., Rotar, P. P., Walker, J. L., and Silvey, M. W., Eds., Westview Tropical Agriculture Series No. 6. Westview Press, Boulder, CO, 1983, 96.
66. Queensland Department of Primary Industries, Variety: 'Cardillo' syn. Q 252561/CPI 43197. *Plant Varieties J.,* 10, 17, 1997.
67. Clements, R. J., Winter, W. H., and Thomson, G. J., Breeding *Centrosema pascuorum* for northern Australia, *Trop. Grassl.*, 20, 59, 1986.
68. Oram, R. N., Register of Australian herbage plant cultivars. b. legumes. 15. Centrosema. b. *Centrosema pascuorum* Mart. ex Benth. (centurion) cv. Bundy., *J. Aust Inst. Agric. Sci.*, 53, 125, 1987.

69. Schultze-Kraft, R., and Benavides, G., Germplasm collection and preliminary evaluation of *Desmodium ovalifolium* Wall., *Gen. Resources Comm.*, 12, 1, 1988.
70. Thro, A. M., and Quesenberry, K. H., Characterization and use of *Desmodium* species in the Caribbean tropics and subtropics, *Proc. Caribbean Food Crops Soc.*, 26, 678, 1990.
71. Bryan, W. W., *Desmodium intortum* and *Desmodium uncinatum*, *Herbage Abstracts* 39, 183, 1969.
72. Imrie, B. C., Jones, R. M., and Kerridge, P. C., Desmodium, in *The Role of Centrosema, Desmodium, and Stylosanthes in Improving Tropical Pastures*, Burt, R. L., Rotar, P. R., Walker, J. L., and Silvey, M. W., Eds., Westview Tropical Agricultural Series No. 6, Westview Press, Boulder, CO, 1983, 97.
73. Cameron, D. G., Tropical and subtropical pasture legumes 3. Desmodiums (*Desmodium intortum* and *D. uncinatum*): legumes for the cool moist tropics and subtropics, *Queensland Agric. J.*, 110, 271, 1984.
74. Hutton, E. M., and Gray, S. G., Hybridization between the legumes *Desmodium intortum, D. uncinatum*, and *D. sandwicense*, *J. Aust. Inst. Agric. Sci.*, 33, 122, 1967.
75. Chow, K. H., and Crowder, L. V., Hybridization of *Desmodium* species, *Euphytica*, 22, 399, 1973.
76. McKellar, M. A., and Quesenberry, K. H. Chromosome pairing and pollen viability in *Desmodium ovalifolium* Wall. X *Desmodium heterocarpon* (L.) DC. hybrids. *Aust J. Bot.,* 40, 243, 1992.
77. Quesenberry, K. H., and Moon, D. E., Evaluation of *Desmodium* species hybrids for dry matter and seed yield. *Soil Crop Sci. Soc. Florida Proc.*, 55, 103, 1996.
78. Pizarro, E.A. (Ed.), *Resultados 1982–1985. 3a.Reunion de la RIEPT*, Red Internacional de Evaluacion de Pastos Tropicales, Centro International de Agricultura Tropical, Cali, Colombia, Vols I and II, 1985.
79. Lenne, J. M., Stem gall nematode on *Desmodium ovalifolium* in Colombia, *Plant Disease*, 67, 557, 1983.
80. Siddiqi, M. R., and Lenne, J. M., *Pterotylenchus cecidogenus*, a new stem gall nematode parasitizing *Desmodium ovalifolium* in Colombia, *J. Nematol.*, 16, 62, 1984.
81. Pattanavibul, S., and Schultze-Kraft, R., Collecting germplasm of *Desmodium* and *Pueraria phaseoloides* in Thailand and peninsular Malaysia, in *Proc. XV Intl. Grassland Congress*, Kyoto, Japan, 1985, 112.
82. Schultze-Kraft, R., Tuan, H. D., and Ha, N. P., Collection of native forage legume germplasm in Vietnam, in *Proc. XVII Intl. Grassland Congress*, Palmerston North, Hamilton, Lincoln, New Zealand and Rockhampton, Australia, 1993, 233.
83. Schmidt, A., Lascano, C. E., Maas, B. L., and Schultze-Kraft, R., An approach to define genotype x environment interaction in a core collection of *Desmodium ovalifolium. Proc. XVIII Intl. Grassland Congress*, Winnipeg and Saskatoon, Canada, 1997, 1-59.
84. Brewbaker, J. L., Species in the genus *Leucaena*, *Leucaena Research Reports* 7, 6, 1987.
85. Sorrensson, C. T., and Brewbaker, J. L., Interspecific compatibility among 15 *Leucaena* species (Leguminosae: Mimosoideae) via artificial hybridizations, *Am. J. Bot.*, 81, 240, 1994.
86. Zarate, S. P., Taxonomic revision of the genus *Leucaena* Benth. from Mexico, *Bulletin of the International Group for the Study of the Mimosiodeae*, 12, 24, 1984.
87. Gray, S. G., A review of research on *Leucaena leucocephala*, *Trop. Grassl.*, 2, 19, 1968.
88. Gray, S. G., A study of variation, hybridization, and the F_1 generation of inter-strain crosses in *Leucaena glauca*, *J. Aust. Inst. of Agric. Sci.*, 26, 287, 1960.
89. Shelton, H. M., The Leucaena genus: new opportunities for agriculture, in Shelton, H. M., Gutteridge, R. C., Mullen, B. F., and Bray, R. A., Eds., *Leucaena — Adaptation, Quality and Farming Systems*, ACIAR Proceedings Series. No. 86., Australian Council for International Agricultural Research, Canberra, Australia, 1998, 15.
90. Sorrensson, C. T., Pollinating and emasculating techniques for *Leucaena* species, *Leucaena Research Reports*, 9, 127, 1988.
91. Hutton, E. M., and Beattie, W. M., Yield characteristics in three bred lines of the legume *Leucaena leucocephala*, *Trop. Grassl.*, 10, 187, 1976.
92. Gray, S. G., Inheritance of growth habit and quantitative characters in intervarietal crosses in *Leucaena leucocephala* (Lam.) de Wit., *Aust. J. Agric. Res.*, 18, 63, 1967.
93. Gray, S. G., General and specific combining ability in varieties of *Leucaena leucocephala* (Lam.) de Wit, *Aust. J. Agric. Res.*, 18, 71, 1967.
94. Gray, S. G., The components of variation in an intervarietal cross in *Leucaena leucocephala* (Lam.) de Wit, *Aust. J. Agric. Res.*, 18, 77, 1967.

95. Jones, R. J., and Megarrity, R. G., Comparative toxicity responses of goats fed on *Leucaena leucocephala* in Australia and Hawaii, *Aust. J. Agric. Res.*, 34, 781, 1983.
96. Jones, R. J., and Megarrity, R. G., Successful transfer of DHP-degrading bacteria from Hawaii goats to Australian ruminants to overcome the toxicity of leucaena, *Aust. Vet.*, 63, 259, 1986.
97. Quirk, M. F., Bushell, J. J., Jones, R. J., Megarrity, R. G., and Butler, K. L., Live weight gains on leucaena and native grass pastures after dosing cattle with rumen bacteria capable of degrading DHP, a ruminal metabolite from leucaena, *J. Argic. Sci., Camb.*, 111, 156, 1988.
98. Hammond, A. C., Allison, M. J., Williams, M. J., Prine, G. M., and Bates, D. B., Prevention of *leucaena* toxicosis of cattle in Florida by ruminal inoculation with 3-hydroxy-4-pyridone-degrading bacteria, *Am. J. Vet. Res.*, 50, 2176, 1989.
99. Hutton, E. M., and Chen, C. P., Meeting the challenge of adapting *Leucaena leucocephala* to the acid oxisols and ultisols of South America and southeast Asia, in *Proc. XVII Intl. Grassland Congress*, Palmerston North, Hamilton, Lincoln, New Zealand, and Rockhampton, Australia, 1993, 2124.
100. Mullen, B. F., Shelton, H. M., Basford, K. E., Castillo, A. C., Bino, B., Victorio, E. E., Acasio, R. N., Tarabu, J., Komolong, M. K., Galga, K. K., Khoa, L. V., Co, H. X., Wandera, F. P., Ibrahim, T. M., Clem, R. L., Jones, R. J., Middleton, C. H., Bolam, M. J. M., Gabunada, F., Stur, W. W., Horne, P. M., Utachak, K., and Khanh, T. T., Agronomic adaptation to environmental challenges in the genus *Leucaena*, in Shelton, H. M., Gutteridge, R. C., Mullen, B. F., and Bray, R. A., Eds., *Leucaena — Adaptation, Quality and Farming Systems, ACIAR Proceedings Series*, No. 86., Australian Council for International Agricultural Research, Canberra, Australia, 1998, 39.
101. Mullen, B. F., Castillo, A., Shelton, H. M., Wong, C. C., Wandera, F. P., Middleton, C., Clem, R. L., Bino, B., Khoa, L. V., Ibrahim, T. M., Horne, P. M., Gutteridge, R. C., Castillo, A. C., and Middleton, C. H., Low temperature and acid soil tolerance in Leucaena, in Shelton, H. M., Gutteridge, R. C., Mullen, B. F., and Bray, R. A., Eds., *Leucaena — Adaptation, Quality and Farming Systems, ACIAR Proceedings Series,* No. 86., Australian Council for International Agricultural Research, Canberra, Australia, 1998, 71.
102. Wheeler, R. A., and Brewbaker, J. L., Results from the international *Leucaena* psyllid trial network, *Leucaena Res. Rep.*, 10, 11, 1989.
103. Kahn, A. K. F., Devarathinam, A. A., Sudhakar, D., and Dorairaj, M. S., Variability of *Leucaena* species for psyllid infestation, *Madras Agric. J.,* 83, 743, 1996.
104. Austin, M. T., Williams, M. J., Hammond, A. C., Frank, J. R., and Chambliss, C. G., Psyllid population dynamics and plant resistance of *Leucaena* selections in Florida. *Tropical Grasslands* 30, 223, 1996.
105. Queensland Department of Primary Industries, Variety: 'Tarramba" syn K636, *Plant Varieties J.,* 1, 19, 1997.
106. Austin, M. T., Sun, W., Brewbaker, J. L., and Schifino-Wittmann, M. T., Developing Leucaena hybrids for commercial use, in Shelton, H. M., Gutteridge, R. C., Mullen, B. F., and Bray, R. A., Eds., *Leucaena — Adaptation, Quality and Farming Systems, ACIAR Proceedings Series,* No. 86., Australian Council for International Agricultural Research, Canberra, Australia, 1998, 82.
107. Brewbaker, J. L., Wei-Guo-Sun, Sun, W. G., and Dieters, M. J., Improvements in nitrogen fixing trees for enhanced site quality, in Matheson, A. C., Nikles, D. G., Harwood, C. E., and Walker, S. M., *Tree Improvement for Sustainable Tropical Forestry, QFRi-IUFRO,* Caloundra, Queensland, Australia, 1996, 437.
108. Hutton, E. M., Siratro — a tropical pasture legume bred from *Phaseolus atropurpureus*, *Aust. J. Exp. Agr. Anim. Husb.*, 2(5), 117, 1962.
109. Bray, R. A., Inheritance of rust resistance in *Macroptilium atropurpureum*, *Plant Path.*, 37, 88, 1988.
110. Bray, R. A., Breeding for rust resistance in *Macroptilium atropurpureum*, in *Proc. XVII Int. Grassl. Congr.*, 1993, 2128.
111. Cameron, D. F., Miller, C. P., Edye, L. A., and Miles, J. W., Advances in research and development with stylosanthes and other tropical pasture legumes, in *Proc. XVII Intl. Grasslands Congress*, Palmerston North, Hamilton, Lincoln, New Zealand, and Rockhampton, Australia, 1993, 2109.
112. Iamsupasit, N., Cameron, D. F., Cooper, M., Chakraborty, S., and Edye, L. A., Inheritance of anthracnose resistance in the tropical pasture legume *Stylosanthes hamata, Aust. J. Agr. Res.,* 46, 1353, 1995.
113. Cameron, D. F., Chakraborty, S., Davis, R. D., Edye, L. A., Irwin, J. A. G., Manners, J. M., and Staples, I. B., A multi-disciplinary approach to anthracnose disease of stylosanthes in Australia, in *Proc XVI Intl Grassland Congress*, Nice, France, 1989, 719.

114. Humphreys, L. R., Townsville lucerne: history and prospect, *J. Aust. Inst. Agr. Sci.*, 33, 3, 1967.
115. Gillard, P., and Fisher, M. J., The ecology of Townsville stylo-based pastures in Northern Australia, in *Plant Relations in Pastures*, Wilson, J. R., Ed., CSIRO, Melbourne, 1978, 340.
116. Mannetje, L. 't, A revision of varieties of *Stylosanthes guianensis* (Aubl.) Sw, *Aust. J. Bot.*, 25, 347, 1977.
117. Mannetje, L. 't, Considerations on the taxonomy of the genus *Stylosanthes*, in *The Biology and Agronomy of Stylosanthes*, Stace, H. M., and Edye, L. A., Eds., Academic Press., North Ryde, N. S. W., Australia, 1984, 1.
118. Skerman, P. J., Cameron, D. G., and Riveros, F., *Tropical Forage Legumes*, FAO, Rome, 1988, 391.
119. Brolmann, J. B., Registration of FP-8400 pencilflower germplasm, *Crop Sci.* 27, 153, 1987.
120. Edye, L. A., Grof, B., and Walker, B., Agronomic variation and potential utilization of *Stylosanthes*, in *The Biology and Agronomy of Stylosanthes*, Stace, H. M., and Edye, L. A., Eds., Academic Press, North Ryde, N. S. W., Australia, 1984, 547.
121. Hutton, E. M., and Grof, B., Increased yield from transgressive segregates of a *Stylosanthes capitata* cross in a tropical oxisol, *Trop. Agric.* (Trinidad), 70, 345, 1993.
122. Liu, C. J., Musial, J. M., and Smith, F. W., Evidence for a low level of genomic specificity of sequence-tagged sites in *Stylosanthes, Theor. Appl. Gen.*, 93, 864, 1996.
123. Curtis, M. D., Cameron, D. F., and Manners, J. M., Molecular evidence that diploid Stylosanthes humilis and diploid Stylosanthes hamata are progenitors of allotetraploid *Stylosanthes hamata* cv. Verano, *Genome,* 38, 344, 1995.
124. Edye, L. A., and Hall, T. J., Development of new *Stylosanthes* cultivars for Australia from naturally occurring genotypes, in *Proc. XVII Intl. Grassland Congress*, Palmerston North, Hamilton, Lincoln, New Zealand and Rockhampton, Australia, 1993, 2159.
125. Stace, H. M., and Cameron, D. F., Cytogenetics and evolution, in *The Biology and Agronomy of Stylosanthes*, Stace, H. M., and Edye, L. A., Eds., Academic Press, North Ryde, N. S. W., Australia, 1984, 49.
126. Vithanage, V. and Chakraborty, S., Identification of selected cultivars and accessions of *Stylosanthes* by isozyme analysis, *Tropical Grassl.*, 26, 156, 1992.
127. Hopkinson, J. M., and Walker, B., Seed production of *Stylosanthes* cultivars in Australia, in *The Biology and Agronomy of Stylosanthes*, Stace, H. M., and Edye, L. A., Eds., Academic Press, North Ryde, N. S. W., Australia, 1984, 433.
128. Cameron, D. F., Boland, R. A., Chakraborty, S., Jamieson, B., and Irwin, J. A. G., Recurrent selection for partial resistance to anthracnose disease in shrubby stylo (*Stylosanthes scabra*), in *Proc. XVII Intl. Grassland Congress*, Palmerston North, Hamilton, Lincoln, New Zealand, and Rockhampton, Australia, 1993, 2137.
129. Maass, B. L., and Schultze-Kraft, R., Characterization and preliminary evaluation of a large germplasm collection of the tropical forage legume *Stylosanthes scabra* Vog., in *Proc. XVII Intl. Grassland Congress*, Palmerston North, Hamilton, Lincoln, New Zealand, and Rockhampton, Australia, 1993, 2151.

7 Constraints and Developments in the Enhancement of Tropical Forage Grasses of Economic Importance

Antonio Sotomayor-Ríos and Stan C. Schank (deceased)

CONTENTS

I. INTRODUCTION

Breeding constraints such as differences in chromosome number and genetic irregularities in meiosis and consequent lack of seed production are some of the probable causes of the slow development of new superior hybrids or varieties of forage crops in the tropics. However, when deviations from the normal plant life cycle (e.g., apomixis or asexual reproduction through seed) occur in important tropical forage species, these can be utilized effectively for the improvement of such grasses as *Brachiaria*, *Cenchrus*, and *Panicum*. Basic aspects of apomixis and methods of its use for the improvement of tropical grasses are discussed in the chapter by Burson and Young.

In this chapter, emphasis is given to selected tropical grass improvement programs at several locations on five of the most important tropical forage grass genera during the last decades. An initial section addresses a basic component of a variety enhancement program, availability of genetic variability.

II. GERMPLASM RESOURCES

The keystone for the success of any grass breeding and improvement program is the utilization of genetic variation within the species being studied.[1-3] The Germplasm Resources Information

Network (GRIN), a computerized network monitored by the National Plant Germplasm System (NPGS) of the U. S. Department of Agriculture (USDA), is a centralized information system designed to aid in the preservation and distribution of germplasm available within its regional centers. The GRIN system is a direct aid to plant breeding programs in general, not only within the United States but also worldwide.

Once new genetic resources are obtained through donations and explorations, they become part of GRIN and are accessible to all plant breeders. There is a need to search for useful tropical grass and legume germplasm and to incorporate it into inventories that are available to tropical forage breeders. This new germplasm should not only have high yield potential, disease and pest resistance, and good adaptation to different soil and environmental conditions, but should also be of good quality. As of September, 1999 (Gilbert Lovell, personal communication), there were 4218 accessions of six important tropical grass genera stored in three locations of the NPGS: the Genetic Resources Unit (GRU), Griffin, Georgia; the National Seed Storage Laboratory (NSSL), Ft. Collins, Colorado; and the North Central Regional Plant Introduction Station (NCRPIS), Ames, Iowa; as follows:

Genus	Site	Total
Brachiaria	GRU	15
Cenchrus	NSSL	4
	GRU	831
Cynodon	NSSL	2
	GRU	308
Digitaria	GRU	525
Panicum	NCRPIS	882
	NSSL	162
	GRU	616
Pennisetum	NSSL	27
	GRU	846
Total		4,218

Most tropical pasture selection and improvement programs have relied on the use of introductions, and excellent collections of the most important tropical and subtropical forage grasses have been established at CENARGEN (Centro Nacional de Recursos Geneticos e Biotecnologia) of EMBRAPA (Empresa Brasileira de Pesquisa Agropecuaria), Campo Grande, Brazil, and at CIAT (Centro Internacional de Agricultura Tropical), Cali, Colombia. Also, there are extensive collections at the USDA-ARS and at agricultural research stations in Florida, Georgia, and Puerto Rico, where much time and effort have been devoted to improving tropical and subtropical grasses through breeding.

III. *BRACHIARIA*

The genus *Brachiaria*, native to Africa, has been found in Australia, Asia, and tropical America.[4] It comprises about 100 species of diverse habitats ranging from shaded to open areas and from semidesert to swampy terrain. The basic chromosome number of the genus is either x = 7 or 9. Plants are bisexual, having bisexual spikelets with hermaphroditic florets.[5] Usually Brachiarias are apomictic, but sexual biotypes have been utilized in breeding improvement programs.[6,7] The natural distribution of the species and germplasm collection sites have been described by Keller-Grein et al.,[8] who summarized the geographic, climatic, and edaphic characteristics of collection sites of seven *Brachiaria* species native to Africa (i.e., *brizantha, decumbens, dictyoneura, humidicola, jubata, nigropedata*, and *ruziziensis*) that are important for tropical pasture development or are represented by a large number of accessions in germplasm collections. The Brachiarias are probably the most important forage grasses utilized in tropical America. There were over 75 million hectares of *Brachiaria* pasture, mostly in the Brazilian tropics, in 1996.[9] Two of the most extensively used cultivars are 'Basilik' of *B. decumbens* Stapf and 'Marandu' of *B. brizantha* (A. Rich.) Stapf.[10] Both species are commonly known as signalgrass. According to Lapointe and Miles,[11] probably the first introduction of a *Brachiaria* into the New World from Africa was paragrass [*B. mutica* (Forsk.) Stapf or *purpurascens* Raddi] during the 18th and 19th centuries. Paragrass or 'malojillo' as it is known in Spanish is commonly found throughout the tropics and is capable of producing excellent forage yields under intensive management.[12] It also produces abundant quantities of seed in the tropics and is tolerant of drought and flooding.[13] In Puerto Rico, much research has been conducted on *Brachiaria* species. Warmke[14] found paragrass to be an allotetraploid (2n = 36) and reported 36.3% germination in the seeds produced. In 1960 at the Agricultural Experiment Station of the University of Puerto Rico (AES-UPR) in Rio Piedras, a *B. brizantha* introduction from Ceylon was determined to be a tetraploid (2n = 36) with irregular microsporogenesis. It showed lagging univalents at anaphase I and micronuclei in the spore quartets, which might have contributed to its lack of seed set.[15,16] Two additional Brachiarias, congo, *B. ruziziensis* (Germain and Evrard), and tanner, *B. arrecta* (Dur. and Schinz) Stent, the former introduced into Puerto Rico by seed from Australia and the latter, vegetatively from Africa, were also studied by the AES-UPR in the early 1970s. Congograss was found to be a diploid (2n = 18) with normal meiosis. Tanner is a tetraploid with lagging univalents observed in meiosis, an aspect which probably is related to its lack of seed set. An interspecific, high-yielding hybrid was developed at the USDA-ARS Tropical Agricultural Research Station (TARS) in Mayagüez in 1997 by crossing paragrass (female parent) x kleingrass (*Panicum coloratum* L.) PI 410177. The *Brachiaria* x *Panicum* interspecific hybrid (BPIH 104) was evaluated under field conditions in a comparison with five *Panicum maximum* genotypes (three guinea grass hybrids, common guinea grass, and 'Tobiata,' a cultivar developed in Brazil). The BPIH and 'Tobiata' were the top dry matter producers, with annual yields of over 40 t/ha.[7] In 1988 scientists at CIAT and EMBRAPA initiated formal grass breeding programs for the improvement of Brachiarias, but genetic recombination was not possible due to the presence of apomixis in almost all of the species of this genus tested. The cultivar 'Basilisk,' a widely grown *decumbens*, is very susceptible to several species and genera of spittlebug (*Homoptera: Cercopidae*). These insects can cause serious losses on millions of hectares of improved *Brachiaria* pastures.[17] But, with the development of a sexual tetraploid biotype of *B. ruziziensis,* which is cross-compatible with *B. decumbens* and *B. brizantha,*[18] applied breeding programs have been initiated, and broad-based sexual tetraploid populations showing resistance to the spittlebug are being developed at CIAT and EMBRAPA.[6]

With the expanding use of *Brachiaria* as a pasture grass in tropical America, production of *Brachiaria* seed has received increased attention, especially in Brazil.[9] This country is growing sufficient seed to supply the local demand and export markets.

IV. *CYNODON*

The genus *Cynodon* (tribe Chlorideae) native to Africa, comprises some of the world's most important grasses. Its basic chromosome number is x = 9 or 10.[5] Probably the best known taxa of the genus is *C. dactylon* (L.) Pers., Bermuda grass or couch grass. This species is well distributed in the tropics and subtropics between latitudes 45°N and 45°S. A revised classification of the genus *Cynodon* was proposed by Clayton and Harlan.[19] These authors stated, "Some compromise is necessary in order to arrive at a system whose units are easy to recognize in practice and at the same time constitute a fair summary of cytogenetic relationships." Their proposed classification is based mainly on a series of morphologic and taxonomic traits: 1. *C. plectostachyus*; 2. *C. dactylon;* 3. *C. aethiopicus*, and 4. *C. nlemfuensis* with two varieties, *robustus* and *nlemfuensis*. According to Harlan,[20] *C. nlemfuensis* differs from *C. dactylon* by the absence of underground rhizomes. Harlan pointed out that, "In general, the term star grass has been used for the robust, non-rhizomatous taxa and Bermuda grass for the rhizomatous forms."

Var. *robustus* is a more vigorous plant, with long, slender racemes; but the growth habit of var. *nlemfuensis* is very similar to that of *C. dactylon*, and the species are often more difficult to separate if the basal parts are missing from the specimen. Both varieties are diploid (2n = 18) with tetraploid strains known, so far, only under cultivation.

In order to illustrate the diversity of the material as forage, Harlan[20] made reference to a selection of *C. nlemfuensis* var. *nlemfuensis* made by H. R. Chheda of the University of Ibadan, Nigeria, from material collected in the Lake Manyanara area of Tanzania. After testing, it proved to be more productive than the local *Cynodon* and was released as cultivar 'IB-8.' The selection is a tetraploid and was identified as *C. nlemfuensis* var. *nlemfuensis*, a star grass. The development of IB-8 demonstrates that robust tropical species are as capable of being improved through plant breeding as *C. dactylon*.

A significant contribution to the improvement of the genus was made by Dr. Glenn Burton, USDA-ARS, Tifton, Georgia in the early 1940s with the development of Coastal Bermuda grass,[21] a hybrid derivative between a local Georgia strain of *Cynodon dactylon* and an introduction from South Africa. This hybrid was a landmark not only in forage development in the southeastern U.S. but also in the annals of plant breeding.

C. nlemfuensis var *nlemfuensis* is an example of a star grass introduction which has made an important contribution to agriculture in tropical and subtropical America. For the last two decades, this star grass has been recognized as an outstanding forage grass in Puerto Rico and Florida;[12, 22] and, due to its vigor, aggressiveness, and resistance to insects and diseases, it has replaced pangola grass (*Digitaria decumbens* Stent.) in many areas of the tropics. In 1957, the AES-UPR introduced two strains of star grass from Kenya. They were originally thought to be *C. plectostachyus,* but, in 1972, were reclassified as *C. nlemfuensis* var. *nlemfuensis* by J. M. J. de Wet of the University of Illinois (personal communication). The PR PI numbers assigned to the star grass strain introductions were 2341 and 2342. Field tests with seven grasses conducted at the Gurabo Substation of the AES-UPR during 1958-60 showed that dry matter and protein yields of PR PI 2341 were superior to those of most of the other grasses tested (Sotomayor-Ríos et al., unpublished work). The excellent potential of star grass under grazing management was later demonstrated on a private dairy farm in Orocovis, Puerto Rico.[23] Field tests conducted in 1973–1975 at the Corozal Substation[24] demonstrated the agronomic potential of 10 *Cynodon* introductions. In that study, the two highest yielding introductions were Coastcross-1 Bermuda grass, *C. dactylon*, USDA PI 293611, developed by Burton in 1967[25] and *C. plectostachyus*, USDA PI 341817. The hydrocyanic acid potential (HCN-p) of the introductions ranged from 0 ppm for a Bermuda grass to a maximum of 333 ppm for a star grass (*C. plectostachyus*) for the 30-day cutting interval. Two hundred ppm HCN-p is considered to be the "threshold of danger" to ruminant animals and can have adverse effects on their health. Generally, HCN-p content decreased as the age of the forage increased. In 1972, star grass strain 2341 was brought from Puerto Rico to the Agricultural Research and Education Center (AREC) of the University of Florida, Ona, and released in 1989 as 'Florico' star grass.[26] In 1973, this star grass was also introduced into Honduras, where it is now an important component of pastures. According to Mislevy and Martin[22] the increased popularity of the star grasses has been attributed to their considerable drought tolerance, ease of establishment, good persistence, and resistance to most pests.

No reports or studies are available on the cytology and chromosome number of star grass to explain its lack of seed formation. In unpublished work by the authors, the chromosome number of star grass was found to be 2n = 4x = 36. Some cells showed evidence of quadrivalent associations; others showed quadrivalents and univalents, indicating a lack of homology between the chromosomes, a possible cause of the absence of seed formation in the grass.

V. *DIGITARIA*

The genus *Digitaria* (tribe Panicoideae) contains a number of pasture grasses of varying economic importance. It has over 200 species, mainly of tropical and subtropical regions of the world. Its basic chromosome number is x = 9, 15, or 17.[5] The most well-known *Digitaria* of the pasture species is pangola grass (*D. decumbens* Stent), which became an important pasture grass in Puerto Rico and other tropical countries in the early 1950s. It has excellent adaptation to the tropics but is susceptible to damaging attack by the yellow sugarcane aphid (*Sipha flava* Forbes) especially during cool, dry months, and to other insects, particularly the chinchbug (*Blissus leucopterus* Say) and the two-line spittlebug (*Prosapia bicincta* Say). Pangola grass is difficult to improve by breeding since it is sterile, an aspect which was of concern to pasture specialists and farmers three decades ago. The findings of Dirven and van Hoof[27] that a stunting virus on pangola grass was spreading in Surinam and that it was susceptible to attack by such insects as the sugarcane aphid prompted grass specialists and breeders in Puerto Rico and elsewhere in the tropics to look for a replacement within the genus. In May 1966 over 250 clones belonging to various species of *Digitaria* were introduced into the AES-UPR at Rio Piedras. Most of these accessions had been collected by A. J. Oakes in southern Africa during his trip in 1964.[28] One of the first *Digitaria* studies in Puerto Rico was conducted in the late 1950s at the AES-UPR in Río Piedras.[29] The taxonomy and reproductive behavior of giant pangola, *D. valida* Stent, were studied to determine the grass's genetic potential in a breeding program. The study found that giant pangola had 42 somatic chromosomes, while other research reported 24, 30, and 36 for this species. The process of microsporogenesis was highly irregular, characterized by grouping of multivalents, lagging univalents, and micronuclei. The degeneration of the megaspore mother cell was thought to be the main reason for the low degree of fertility obtained.

Thirty selections from the Oakes *Digitaria* collection were evaluated for seed set at the Corozal Substation.[30] Of these, *D. umfolozi* had the highest caryopsis formation (60%). It proved to have an excellent dry forage yield, but was susceptible to attack by the yellow sugarcane aphid. Meiotic behavior of the B chromosomes in *Digitaria* was reported by Shambulingappa.[31] He also conducted detailed meiotic studies of 12 accessions of *Digitaria* comprising five species. Ten were found to be tetraploids; one, a hexaploid; and one had 2n = 34 chromosomes.[32] In Florida, with the cooperation of researchers in the Caribbean area, two new cultivars in the genus were released with tolerance to pangola stunt virus. These new cultivars were 'Transvala'[33] and 'Survenola.'[34] 'Transvala' is a triploid (2n = 3x = 21) and 'Survenola', a hexaploid (2n = 6x = 42).

At the Corozal Substation of the UPR-AES, the *in vitro* organic matter digestibility (IVOMD), crude protein content, and mineral composition of 11 grasses were compared in two seasons and evaluated in relation to grazing ruminant requirements.[35] A selection of *D. eriantha* was the highest in IVOMD for both seasons, being superior to star grass strain 2341 (69.7 vs. 58.2%) during the long-day season and (62.2 vs. 50.9%) during the short-day season. This selection is susceptible to rust (*Puccinia oahuensis* Ell and Ev.), but is a good candidate to be improved by breeding due to its high forage quality.

VI. *PANICUM*

The genus *Panicum* (tribe Paniceae) is native to Africa and has over 370 species of diverse habitats distributed throughout the tropics and warm temperate regions of the world. Plants are bisexual and have bisexual spikelets with hermaphroditic flowers.[5] Probably the most well-known species of the genus in the tropics is *P. maximum* Jacq. (guinea grass), which is apomictic, although the occurrence of sexual plants has been reported.[36] Based on progeny tests with *P. maximum*, Warmke[37] concluded that the off-type plants he obtained (1.3 to 4.7%) had arisen by sexual recombination. Smith[38] completely isolated sexual plants of *P. maximum* by testing off-type plants among formerly identified apomictic plants. Hutton[39] at EMBRAPA in Brazil reported the development of promising acid-tolerant lines of *P. maximum* from crosses between an apomictic male parent and cv. 'Tift 49,' a sexual cultivar developed by W. W. Hanna, USDA-ARS, Tifton, Georgia.[36] The sexual potential of six guinea grass varieties determined through embryo sac analysis was reported by Javier,[40] although no sexual plants were isolated by him.

The basic chromosome number of the genus is x = 7, 8, 9, or 10.[5] The somatic chromosome number of the genus *Panicum* reported prior to the work of Warmke[37] was 18, 36, 54, or 72, suggesting a polyploid series built upon a basic number of nine. Warmke's studies determined that the somatic chromosome number in four types of guinea grass he had identified in Puerto Rico — common, gramalote, broad-leaf, and fine-leaf — was 32, although they differed significantly in size and in certain morphological characteristics. He also found two types of cultivar 'Borinquen,' one with 32 and one with 48 somatic chromosomes and paragrass, a related species, with 36. The chromosome number of five sexual plants of *P. maximum* was reported by Hanna et al. to be 2n = 4x = 32.[36] An excellent review of the genetic resources, modes of reproduction, and breeding procedures of the genus was made by Savidan et al.[41] who suggested a breeding scheme to cross sexual x apomictic *Panicums*.

Guinea grass is a very diverse species suitable for grazing, green soilage, hay, and silage. It is one of the most widely utilized tropical and subtropical grasses due to its high yields and adaptation to almost all types of soil and climatic conditions. Probably the greatest weakness of guinea grass is its indeterminate flowering and seed shattering. Seed harvesting is difficult since the spikelets drop 7 to 11 days after pollination. Young[42] and Burson et al.[43] studied the anatomical basis for seed shattering in kleingrass and guinea grass and found that two different abscission layers were present. A primary layer was located in the pedicel near the base of the glumes, and a secondary layer extended across the rachilla. They concluded that the major contributor to seed shattering was the primary abscission layer. In Puerto Rico, many unsuccessful attempts have been made to search for guinea grass ecotypes with inflorescences that retain seed for longer periods. A possible way of developing a guinea grass without the seed shattering characteristic is by transferring the genes for such a trait from other sources such as *P. fasciculatum* Swartz or *P. coloratum* (2n = 4x = 32). An additional alternative for obtaining these genes in *P. maximum* is by searching progenies in advanced generations where this specific trait can be found. The only source of resistance available at this time is in kleingrass. Unfortunately, local common guinea grass (2n = 4x = 32) cannot be utilized in crosses with kleingrass due to the difference in chromosome number of the two species.

A good possibility for studying the inheritance of resistance to seed shattering in the *Panicums* is by utilizing previously mentioned interspecific hybrid BPIH 104 developed at TARS.[7] This F_1 is a vigorous, leafy plant with an inflorescence similar to most of the Panicums. Apparently, the resistance to seed shattering is controlled by recessive genes since the F_1s all had seed shattering. Present research indicates that the search for seed shattering resistant genes should be done in advanced generations.

Numerous F_1 *P. maximum* hybrids have been developed at TARS, Mayagüez[44,45] using the sexual lines of 'Tifton 49' ('Tift 49') and 'SPM 92' (sexual *P. maximum* 92) as female parents.[46] 'Tifton 49' is a population having sexual and facultative apomicts, while 'SPM 92' is a seed-propagated population consisting of only sexual progeny. Crosses were made in the field using controlled pollen dehiscence following the technique of Schertz and Clark[47] with modifications. The female flowers were covered with plastic bags at about 6 P.M. in the evening prior to making the crosses. This late-day bagging must be done in order to avoid the high temperatures that develop inside the plastic bags during the daylight hours. The next morning, the anthers were removed and pollen was applied from the selected male parent. The procedure was repeated for two to three days until enough flowers were pollinated. Male parents were selected on the basis of their agronomic superiority (yield data) and performance in other countries. Field data obtained during a one-year period showed that the *in vitro* dry matter digestibility (IVDMD) and dry matter yield of the F_1

hybrids were higher than those of either parent.[7] In Cuba, high yielding *P. maximum* hybrids were developed by Segui et al.[48]

Guinea grasses are usually susceptible to ergot caused by *Claviceps maximensis* Theis[49] especially during the rainy season. According to Vicente-Chandler et al.,[12] ergot infection does not pose a health threat to animals feeding on guinea grass although it can affect the quality of the seed produced.

VII. *PENNISETUM*

The genus *Pennisetum* (tribe Panicoideae) consists of about 80 species (mostly perennial) that are of great economic importance in the tropics and subtropics.[50] Elephant or napier grass (*Pennisetum purpureum* Schum.) is the highest dry matter-producing pasture species in the tropics and subtropics. It is a robust perennial requiring high levels of nitrogen fertilization to maximize its forage production.[51] In Puerto Rico, dry matter yields of 62 t/ha/yr have been reported under intensive management (at 90-day harvest intervals).[12] This grass has a wide adaptation to various soil conditions from acid with low fertility to slightly alkaline. There are many varieties and hybrids within the tetraploid (2n = 4x = 28) *P. purpureum* species,[52] the most important being napier or common elephant grass, kinggrass (PI 300086), Merkeron, and a series of Taiwan selections.[53] Merkeron (PI 531087) is an F_1 hybrid developed by Burton[54] between a very leafy dwarf and a tall selection of napier grass. This hybrid is considered to be an excellent example of valuable germplasm concentrated and preserved in a single perennial clone. Dwarf napier grasses (N75, N114, N127, and N128) were selected from among selfed progeny of Merkeron.[55] They have shorter internodes than the tall napier grass and do not need to be cut after grazing since no long, dry stems are left on the plant. The first comparison of tall and dwarf napier grass cultivars was made by Hanna et al.[56] These authors found that the dwarf cultivars yielded one-third as much plant material as the tall cultivars. Sollenberger et al.[57] conducted three experiments between 1987 and 1991 to determine the factors underlying the high quality of 'Tifton N75,'[55] a cultivar named 'Mott,'[57,58] and to assess the quality potential of napier grass x pearl millet [*Pennisetum glaucum* (L.) R. Br.] hybrids. These authors concluded that the superior forage quality of Mott is due to its high leaf-to-stem ratio and to the proportion and distribution of tissues in the leaf lamina.

Kinggrass is also a vigorous accession of elephant grass and was introduced into Panama in 1970 from the United States. It has also been distributed to many areas in the tropics.[53] For many years, it was believed that kinggrass was a hybrid between *P. glaucum* (formerly *P. americanum*) and *P. purpureum;* however, chromosome studies proved it to be an intraspecific hybrid within the *P. purpureum* complex.[59]

Napier grass reproduction is asexual, although the grass produces sexual seed which is extremely small, for use in commercial propagation. An alternative to obtain seed from napier grass was suggested by Powell and Burton.[60] This method consists of developing triploid seed utilizing a dwarf cytoplasmic male-sterile pearl millet inbred ('Tifton 23DA') (2n = 14) crossed with elephant grass (2n = 28). The F_1 plants have excellent vigor, but, when mature, they are sterile. At the present time, no commercial triploid hybrids are available.

Considerable work on the development of *Pennisetum* hexaploids (2n = 6x = 42) that produce viable seed has been conducted at the Institute of Food and Agricultural Sciences, University of Florida, Gainesville. Additional information on this aspect of plant breeding can be obtained from the Institute. The hexaploid hybrids originated from a series of crosses involving 'Tift 23DA' and 'Mott' dwarf elephant grass. Triploid progeny from this cross were grown in tissue culture as described by Rajasekaran et al.[61] and two hexaploid plants were obtained from the tissue culture. The seeds produced on the hexaploids were smaller than pearl millet, but possessed good seedling vigor. The nutritive value of these hybrids is similar to that of 'Mott,'[62] but some problems have been encountered in the long-time persistence of the hybrids.[63,64] Genetic improvement of the hexaploid hybrids as well as transferring the good quality genes of the dwarf plants to them should have priority in *Pennisetum* breeding programs.

VIII. FUTURE PROSPECTS FOR THE IMPROVEMENT OF TROPICAL FORAGE GRASSES BY BREEDING

As the world population continues to increase at an alarming rate, the demand for food will increase, especially in the developing countries. Forage grasses and forage legumes will play an important role as a substitute for the cereal grains normally used for feeding beef cattle and other livestock, which provide the main protein and food sources for humans. It will be imperative to see that scientists throughout the tropics, where the population explosion is highest, devote more effort to the development of superior forages having the quality and quantity surpassing those presently utilized. Of prime importance will be the development of forages capable of growing in vast areas of the tropics where soil conditions such as low fertility, high aluminum content, and drought exist.

Desirable traits now lacking in many forages such as improved quality and disease and pest resistance will need to be incorporated into future pasture material. Also superior methods of propagation by mechanical means should be developed. Breeders must continue to look for new sources of genetic diversity for the most important tropical forage grasses and legumes. Apomixis and other important mechanisms for fixing traits must receive more attention from breeders.

The many collections of important forage species around the world will need to be re-evaluated, and genes useful for breeding should be recognized and maintained under the core collection concept. These core collections should be available to breeders throughout the world for the final development of superior forages needed by the world population in the years ahead when food availability will be essential for the survival of the human race.

The environmental and population changes that have been predicted throughout the world, such as high increases in CO_2, methane, and the rupture of the ozone layer, will require that temperate zone forages be replaced by adaptable tropical and subtropical forages. There will be a need to search for physiological mechanisms similar to those found in the C_3 plants that have digestibility superior to those found in C_4 plants.

REFERENCES

1. Burson, B. L., Warm-season grasses, in *Hybridization of Crop Plants*, Fehr, W. R. and Hadley, H. H., Eds., ASA-CSSA, Madison, Wisconsin, 1980, 695.
2. Sleper, D. A., Forage grasses, in *Principles of Cultivar Development*, Fehr, W. R., Ed., Macmillan Pub. Co., New York, 1987, 161.
3. Vogel, K. P. and Sleper, D. A., Alteration of plants via genetics and plant breeding, in *Forage Quality, Evaluation and Utilization*, Fahey, G. C., Ed., American Society of Agronomy, Madison, Wisconsin, 1994, 891.
4. Renvoize, S. A., Clayton, W. D., and Kabuye, C. H. S., Morphology, taxonomy, and natural distribution of *Brachiaria* (Trin.) Griseb., in *Brachiaria: Biology, Agronomy, and Improvement*, Miles, J. W., Maass, B. L., and do Valle, C. B., Eds., Centro Internacional de Agricultura Tropical (CIAT), Cali, Colombia, 1996.
5. Watson, L. and Dallwitz, M. J., *The Grass Genera of the World*, CAB International, Wallingford, Oxon, U.K., 1992.
6. Miles, J. W. and do Valle, C. B., Advances in breeding apomictic *Brachiaria* in tropical America, in *Proc. XVIII Int. Grassl. Congress*, Vol. 1, 1997, 4–63.
7. Sotomayor-Ríos, A., Torres-Cardona, S., and Quiles-Belén, A., Agronomic comparison of four *Panicum* hybrids and two cultivars at three cutting intervals in Puerto Rico, *J. Agric. Univ. P. R.*, 82, 141, 1988.
8. Keller-Grein, G., Maass, B. L., and Hanson, J., Natural variation in *Brachiaria* and existing germplasm collections, in *Brachiaria: Biology, Agronomy and Improvement*, CIAT, Cali, Colombia, 1996, 16.
9. Santos-Filho, L. F., Seed production: perspective from the Brazilian private sector, in *Brachiaria: Biology, Agronomy, and Improvement*, Miles, J. W., Maass, B. L., and do Valle, C. B., Eds., CIAT, Cali, Colombia, 1996.

10. Miles, J. W., Lapointe, S. L., Escandon, M. L., and Sotelo, G., Inheritance of resistance to spittlebug (*Homoptera:Cercopidae*), in interspecific *Brachiaria* spp. hybrids, *J. Econ. Entomol.*, 88, 1477, 1995.
11. Lapointe, S. L. and Miles, J. M., Germplasm case study: *Brachiaria* species, in *Pastures for the Tropical Lowlands. CIAT's Contribution*, CIAT, Cali, Colombia, 1992.
12. Vicente-Chandler, J. Caro-Costas, R., Abruña, F., and Silva, S., *Producción y Utilización Intensiva de las Forrajeras en Puerto Rico*, Bol. 271, Est. Exp. Agric., Univ. P. R., 1983.
13. Skerman, P. J. and Riveras, F., *Tropical Grasses*, FAO, Rome, 1990.
14. Warmke, H. E., Cytotaxonomic investigations of some varieties of *Panicum maximum* and of *P. purpurascens* in Puerto Rico, *Agron. J.*, 43, 143, 1951.
15. Sotomayor-Ríos, A., Vélez-Fortuño, J., Woodbury, R., Schertz, K. F., and Sierra–Bracero, A., Description and cytology of a form of signal grass (*Brachiaria brizantha* Stapf) and its agronomic behavior compared to guinea grass (*Panicum maximum* Jacq.), *J. Agric. Univ. P. R.*, 44, 208, 1960.
16. Schank, S. C. and Sotomayor-Ríos, A., Cytological studies on *Brachiaria* species, *Proc. Soil Crop Sci. Soc. Florida*, 28, 156, 1968.
17. Cardona, C., Miles, J. M., and Sotelo, G., An improved methodology for massive screening of *Brachiaria* spp. genotypes for resistance to *Aeneolamia varia (Homoptera:Cercopidae)*, *J. Econ. Entomol.*, 92, 490, 1999.
18. Swenne, A., Louant, B. P., and Dujardin, M., Induction par la colchicine de formes autotetraploids chez *Brachiaria* (Graminée) (*Homoptera*:Cercopidae), *Agron. Trop.*, 36, 134, 1981.
19. Clayton, W. D. and Harlan, J. R., The genus *Cynodon*, L. C. Rich. in tropical Africa, *Kew Bull.*, 24, 185, 1970.
20. Harlan, J. R., *Cynodon* species and their value for grazing and hay, *Herbage Abst.*, 230, 1970.
21. Burton, G. W., *Coastal Bermuda Grass*, Circ. 10 (revised), Georgia Coastal Plain Exp. Stn., Tifton, Georgia, 1948.
22. Mislevy, P. and Martin, F. G., Comparison of Tifton 85 and other *Cynodon* species for production and nutritive value under grazing, *Proc. Soil Crop Sci. Soc. Florida*, 57, 77, 1998.
23. Caro-Costas, R., Vicente-Chandler, J., and Abruña, F., Comparison of heavily fertilized Congo, Star, and Pangola grass pastures in the humid mountain region of Puerto Rico, *J. Agric. Univ. P. R.*, 60, 179, 1976.
24. Vélez-Santiago, J., Sotomayor-Ríos, A., and Torres-Rivera, S., Effect of three harvest intervals and two fertilizer rates on the yield and HCN content of ten *Cynodon* cultivars, *J. Agric. Univ. P. R.*, 63, 35, 1979.
25. Burton, G. W., Registration of Coastcross-1 bermudagrass, *Crop Sci.*, 12, 125, 1972.
26. Mislevy, P., Brown, W. F., Caro-Costas, R., Vicente-Chandler, J., Dunavin, L. S., Hall, D. W., Kalmbacher, R. S., Overman, A. J., Ruelke, O. C., Sonoda, R. M., Sotomayor-Ríos, A., Stanley, R. L., and Williams, M. J., Registration of 'Florico' stargrass, *Crop Sci.*, 33, 358, 1983.
27. Dirven, J. G. P. and van Hoof, H. A., *Tijdschr. PlZiekt.*, 66, 344, 1960.
28. Oakes, A. J., *Digitaria* collection in South Africa, *Trop. Agric.*, 42, 323, 1965.
29. Sotomayor-Ríos, A., Schertz, K. F., Woodbury, R., and Vélez-Fortuño, J., Description and reproduction behavior of giant pangola (*Digitaria valida* Stent.), *J. Agric. Univ. P. R.*, 44, 53, 1960.
30. Sotomayor-Ríos, A., Vélez-Fortuño, J., and Spain, G., Forage yields and plant character correlations in 30 *Digitaria* selections, *J. Agric. Univ. P. R.*, 55, 53, 1971.
31. Shambulingappa, K. G., Occurrence and meiotic behaviour of B chromosomes in *Digitaria, Cytologia*, 33, 539, 1968.
32. Shambulingappa, K. G., Cytological studies in *Digitaria pentzii* Stent., *J. Agric. Univ. P. R.*, 54, 401, 1970.
33. Schank, S. C., Boyd, F. T., Smith, R. L. Hodges, E. M., West, S. H., Kretschmer, Jr., A. E., Brolmann, J. B., and Moore, J. E., Registration of 'Transvala' digitgrass, *Crop Sci.*, 30, 1368, 1990.
34. Schank, S. C., Ruelke, O. C., Ocumpaugh, W. R., Moore, J. E., and Hall, D. W., Registration of 'Survenola' digitgrass, *Crop Sci.*, 30, 1369, 1990.
35. Ramos-Santana, R. and McDowell, L. R., *In vitro* digestibility, crude protein content, and mineral concentrations of *Cynodon, Brachiaria,* and *Digitaria* accessions in a humid tropical region of Puerto Rico, *Communications in Soil Sci. and Plant Analysis*, 27, 2687, 1996.
36. Hanna, W. W., Powell, J. B., Millot, J. C., and Burton, G. W., Cytology of obligate sexual plants in *Panicum maximum* Jacq. and their use in controlled hybrids, *Crop Sci.*, 13, 695, 1973.

37. Warmke, H. E., Apomixis in *Panicum maximum, Amer. J. Bot.*, 41, 5, 1954.
38. Smith, R. L., Sexual reproduction in *Panicum maximum* Jacq., *Crop Sci.*, 12, 624, 1972.
39. Hutton, E. M., Breeding acid-tolerant lines of the tropical grass *Panicum maximum* Jacq., in *Proc. XVI Int. Grassl. Congr.*, 1989, 355.
40. Javier, E. Q., The flowering habits and mode of reproduction of guineagrass (*Panicum maximum* Jacq.), in *Proc. XI Int. Grassl. Congr.*, 1970, 284.
41. Savidan, Y. H., Jank, L., Costa, J. C. G., and do Valle, C. B., Breeding *Panicum maximum* in Brazil. 1. Genetic resources, modes of reproduction and breeding procedures, *Euphytica*, 41, 107, 1989.
42. Young, B. A., Variation for seed shattering in kleingrass (*Panicum coloratum*), in *Agronomy Abstr.*, American Society of Agronomy, Madison, Wisconsin, 1981, 77.
43. Burson, B. L., Correa, J., and Potts, H. C., Anatomical basis for seed shattering in kleingrass and guineagrass, *Crop Sci.*, 23, 747, 1983.
44. Sotomayor-Ríos, A., Rivera-Amador, E., and Quiles-Belén, A., Breeding for seed shattering resistance in guineagrass, in *Proc. XXIX Ann. Meeting, Caribbean Food Crops Soc.*, Fort-de-France, Martinique, 1993.
45. Sotomayor-Ríos, A., Quiles-Belén, A., and Johnson, E., Identification of *Panicum* hybrids derived from sexual and apomictic plants, in *Agron. Abstr.*, American Society of Agronomy, Madison, Wisconsin, 1994, 124.
46. Hanna, W. W. and Nakagawa, H., Registration of 'Tift SPM 92' sexual germplasm, *Crop Sci.*, 34, 547, 1984.
47. Schertz, K. F. and Clark, L. E., Controlling dehiscence with plastic bags for hand crosses in sorghum, *Crop Sci.*, 7, 540, 1967.
48. Sequí, E., Mendoza, F., y Machado, H., Nuevos híbridos de *Panicum maximum* con altos rendimientos de ms y elevada calidad. Est. Exp. "Indio-Hatuey," Matanzas, Cuba, *Pastos y Forrajes*, 18, 29, 1995.
49. Theis, T., *Some Diseases of Puerto Rican Forage Crops*, Bull. 51, Fed. Exp. Stn., Mayagüez, Puerto Rico, 1953.
50. Brunken, J.N., A systematic study of *Pennisetum* sect. *Pennisetum (Gramineae), Amer. J. Bot.*, 64, 161, 1977.
51. Bogdan, A. V., *Tropical Pasture and Fodder Plants (Grasses and Legumes)*, Longman, London, 1977.
52. Pereira, A. V., Escolha de variedades de capim elefante, in *Anais do 10E simposi* o sobre manejo da pastagem, Peixoto, A. M., de Moura, J. C., and Faria, V. P., Eds., FEALQ, Piracicaba, Sao Paulo, Brazil, 1992, 47.
53. Mendoza, P. E. and Schank, S. C., Production and utilization of kinggrass and other Pennisetums for meat and milk production, in *Proc. Int. Conf. on Livestock and Poultry in the Tropics*, University of Florida, Gainesville, Florida, 1987, 35.
54. Burton, G. W., Registration of 'Merkeron' napiergrass, *Crop. Sci.*, 29, 1327, 1989.
55. Hanna, W. W. and Monson, W. G., Registration of dwarf 'Tift N75' napiergrass germplasm, *Crop Sci.*, 28, 870, 1988.
56. Hanna, W. W., Monson, W. G., and Hill, G. M., Evaluation of napiergrass, in *Proc. XVII Int. Grassl. Congress*, 1993, 402.
57. Sollenberger, L. E., Moore, J. E., Flores-C., J. A., Chaparro, C. J., and Macoon, B., Forage quality determinants of Mott elephantgrass and *Pennisetum* hybrids, in *Proc. XVII Int. Grassl. Congress*, 1993, 201.
58. Sollenberger, L. E. and Jones, Jr., C. S., Beef production from nitrogen-fertilized Mott dwarf elephantgrass and Pensacola bahiagrass pastures, *Trop. Grassl.*, 23, 129, 1989.
59. Martínez, R. O., Monzoate, M., Herrera, R. S., Cruz, R., and Torres, V., Obtention of king grass (*Pennisetum purpureum*) clones from tissue culture. Selection and evaluation of mutants, in *Proc. XVI Int. Grassl. Cong.*, Vol. 1, 1989, 439.
60. Powell, J. B. and Burton, G. W., A suggested commercial method of producing an interspecific hybrid forage in *Pennisetum, Crop Sci.*, 6, 378, 1966.
61. Rajasekaran, K., Schank, S. C., and Vasil, I. K., Characterization of biomass production, cytology, and phenotypes of plants regenerated from embryogenic callus cultures of *Pennisetum americanum* X *P. purpureum* (hybrid triploid napiergrass), *Theor. Appl. Genet.*, 73, 4, 1986.

62. Schank, S. C., Smith, R. L., and Seib, J., Evaluation of triploid and hexaploid derivatives from *Pennisetum glaucum* X *P. purpureum* crosses at the beef research unit, in *Proc. XXVI Caribbean Food Crops Soc. Ann. Meeting*, Mayagüez, Puerto Rico, 1991, 354.
63. Spitaleri, R., *Agronomic Performance and Ensiling Characteristics of Seeded Pennisetum Hexaploid Hybrids*, M. S. Thesis, University of Florida, Gainesville, Florida, 1992.
64. Macoon, B., Defoliation effects on yield, persistence, and quality-related characteristics of four *Pennisetum* forage genotypes, M. S. Thesis, University of Florida, Gainesville, Florida, 1992.

8 Biotechnology in Tropical Forage Crops

Rex L. Smith

CONTENTS

I. INTRODUCTION

The world's crop production is entering the age of biotechnology. All of the world's major crop species have been transformed and have expressed foreign genes. Transgenic crops, or "genetically modified" (GM) crops, as the media refers to them, are now being grown on a large scale. In the United States nearly 60 million acres of GM crops were grown in 1998, up tenfold since 1996.[1] In 1998 over 25% of the maize, 38% of the soybeans, 45% of the cotton, and 35% of the canola grown in the U.S. were "genetically modified."[2] Most of those transgenic crops express herbicide tolerance (mainly Roundup™) and/or insect resistance, mainly the *Bacillus thuringiensis* toxic protein (Bt). Farmers like the transgenic crops because of their reduced pesticide needs and higher subsequent profits. However, the rapid increase in scope of transgenic crops and their control in

0-8493-2318-5/01/$0.00+$.50

the hands of just a few multinational corporations has caused concern among consumers, especially in Europe. Consumers are also uneasy about health safety concerns based mainly on fear of the unknown and mistrust of governmental regulating agencies. There is no data supporting the detrimental health effects of GM foods. Consumer groups are pushing legislation to label and/or restrict transgenic food products. While these concerns are being worked out, growth in the use of transgenic crops will undoubtedly be slowed. In the long run, biotechnology will make very significant contributions to making crop production more sustainable and more environmentally friendly, and in improving nutritional quality and productivity, especially in marginal crop environments.

In addition to the traits mentioned above, numerous other traits are suitable for genetic engineering. Some are commercially available, such as plants with virus resistance and others with modified fatty acids. Other traits at various stages in the developmental pipeline include disease resistance, other fatty acid modifications, improved nutritional composition, more efficient nitrogen utilization, and stress tolerance. Other more exotic useful traits are being researched, such as using plants to produce human antibodies, hormones, and vaccines. Long-term goals of plant genetic engineering include production of industrial feedstocks to be used in new products such as plastics, fibers, and designer oils.[3] Leading authorities believe that biotechnology will play a vital role in the 21st century in providing food and industrial products to sustain the world's rapidly growing population.

Biotechnological research, in both the public and private sectors, is in a state of technological revolution stimulated by the potential for profit. Large private and public expenditures are being made in the race to control these emerging technologies, genes and traits, as well as the final products. Structural (mapping and sequencing of genomes) and functional (isolation and characterization of genes) genomics are being done using high speed ("high throughput") technologies that churn out very large volumes of data that are being managed, analyzed, and summarized by "Bioinformatics" (special computer programs on high speed/volume computers). High throughput gene isolation and characterization is rapidly changing biotechnology from being gene poor, that is, not having sufficient genes for good progress, to having a surplus of genes. However, in both the public and private domains, genes, traits, and processes are being patented as rapidly as they are defined making them no longer freely available for general usage. Also the data assembled by the private sector is proprietary and is unavailable for public use. In addition to the above-mentioned safety concern, the ownership of genes and genomic data is raising serious concerns among the public about corporate control of crop genetics, cultivar development, and crop (food) production.

Since biotechnological developments are expensive and costs must be recovered from their products, biotechnological developments are "economics" driven, and, therefore, have been aimed at major food and cash crops rather than at forage crops. This is not because forages are not important, as they are essential to world agricultural systems, but because they have lower visibility and profit/funding potential and receive lower research priorities. In this report the limited biotechnology research on tropical forages that has been reported will be described, but since it is limited, useful methodology and research approaches will sometimes be discussed with nontropical forage examples. The publication by the Crop Science Society of America entitled *Molecular and Cellular Technologies for Forage Improvement* (1998)[4] reviews progress in temperate forage species *Medicago, Festuca, Lolium, Agrostis*, and *Dactylis,* so that will not be covered here.

II. GENETIC ENGINEERING

Although a good deal of crop production is now in the age of biotechnology, this is not true for most forages, especially tropical and subtropical species. Genetic engineering of plants, including forages, involves the stable integration into the plant nuclear genome of foreign DNA (genes) along with promoters and other sequences needed for gene activity via gene transfer technologies. Gene transfer technologies have been developed to work for all plants. However, since stable integration of transformed genes onto the host genome is a rare event, powerful selection systems involving tissue culture and plant regeneration must be used to recover the transformed plant. Often tissue/cell culture and plant regeneration problems restrict the species that can be transformed. These technologies are now sufficiently advanced to provide unique crop improvement opportunities by introducing new traits (genes) from unrelated plant species, microorganisms, or even animal genes into plants, not possible by classical genetics and breeding. Transformation technologies have given rapid progress in crop improvement via genetic engineering and can be applied to forages. These reliable technologies are described below.

A. Transformation Techniques

1. *Agrobacterium*-Mediated Transformation

The soil bacterium, *Agrobacterium tumefaciens*, that causes crown gall in many dicot plant species does so by transferring and integrating tumor genes from the bacterium's Ti (Tumor-inducing) plasmid into the chromosomes of the plant. This transformation system that evolved in nature is very complex and also very efficient. Studies over the years have determined that a portion of the Ti plasmid (the T-DNA), bordered on the right and left by 25 base direct repeats and containing the tumor genes, was inserted into the plant genome.[5] Another important region (the Vir region) of the Ti plasmid was found to contain *vir* genes that are responsible for plant infection and gene transfer. It was also discovered that the tumor and other associated genes could be removed from the T-DNA (disarming it) and could be replaced with genes of interest to be transformed into the plant.[6] Since the Ti plasmid was very large and difficult to manipulate, two types of transformation vectors were developed: cointegrative and the more commonly used binary vectors. These vectors utilize a small *E. coli* plasmid cloning vector where the genes of interest and other associated sequences can be easily assembled using recombinant DNA techniques. The cointegrative vector is integrated into the Ti plasmid by homologous recombination via homologous sequences engineered into the plasmids, following its insertion into the disarmed *Agrobacterium*. With the binary vector, it was found that a small plasmid vector with a wide host range could be independently maintained in both *E. coli* and *Agrobacterium*.[7] This *Agrobacterium*, containing both the Ti and binary plasmids, can efficiently transfer the DNA sequences to the left of the right 25-base border of the binary vector into the plant genome as long as the *vir* function is provided *in trans* by a wild-type or disarmed Ti plasmid. Production of transgenic plants using *Agrobacterium* involves three steps: incubate wounded plant tissues with the *Agrobacterium* strain containing the vector to be transferred, screen for transformed cells, and finally regenerate the transformed cells into plants. Of these two systems, the binary vector has been more widely used because it is easier to manipulate and can be used with a broader range of *Agrobacterium* strains.[7] Examples of *Agrobacterium*-mediated transformation of temperate forage legumes are alfalfa (*Medicago sativa* L.),[8] white clover (*Trifolium repens* L.),[9] and red clover (*Trifolium pratense* L.).[10]

Transgenic plants are now routinely produced from tissue explants such as leaves,[11] stems,[12] suspension cells,[13] and embryos.[14] Although dicots are the natural hosts of *Agrobacterium*, successful transformation has been made with maize (*Zea mays* L.),[15] rice (*Oryza sativa* L.),[16,17] wheat (*Triticum aestivum* L.),[18] barley (*Hordeum vulgare* L.),[19] sorghum (*Sorghum bicolor* Moench.),[20] and creeping bentgrass (*Agrostis palustris* Huds.).[21] Also see the review by Smith and Hood.[22] It is possible that *Agrobacterium*-mediated gene transfer may become a major method of producing transgenic monocots. The *Agrobacterium* method has several advantages over other methods: (1) it transfers relatively large segments of DNA with little rearrangement; and (2) it integrates a single copy or low-copy number of genes into the plant chromosome,[23] which avoids some of the instability and possible gene silencing encountered in high-copy number transformation. Most transgenic plants produced to date were created using the *Agrobacterium* system, however, it is still mainly effective with dicots,[24] and with few exceptions as discussed above, other methods must be used for monocots.

2. Microprojectile Bombardment (Biolistics)

While *Agrobacterium*-mediated transformation has been very successful in many dicot plants, most monocot plants are resistant to this transformation technique. Therefore, other methods have been developed. The most important of these is microprojectile bombardment (the gun), also known as biolistics. However, before microprojectile bombardment was developed, direct gene transfer into protoplasts had been effective in developing transgenic plants in some monocot species. The use of this direct transfer method is limited to a few monocot species and genotypes that can be regenerated from protoplasts.

The development of biolistic systems permitted various plant tissues to be used including pollen, pollen microspore callus, meristems, embryos, leaves, embryogenic callus, and suspension cultures.[25] This approach is equally useful in transforming monocots as dicots.[26] Because it can be used on all plant species, it has been the most widely used transformation method during the past several years.[27]

In biolistics, DNA-loaded gold or other microprojectiles (0.4 to 2 μm in diameter) are accelerated into plant cells by a gun cartridge, explosive helium blasts, or electrical charge.[25] The various systems use devices to scatter the microprojectiles and adjust the projectile velocity and their penetration into the cells upon impact. For stable transformation to occur, the DNA coated on the projectile has to enter the nucleus of the cell and be integrated into the chromosome. Because this is a rare event, a selection system must be incorporated to recover transgenic plants. Microprojectile bombardment has been successfully used to produce fertile transgenic plants in most major crops.[28-30]

3. Other Direct Gene Transfer Methods

Before the biolistics method was developed, transformation of monocots depended mainly on the direct gene transfer methods. These methods include: (1) introduction of foreign DNA into protoplasts by electroporation or chemical treatment, (2) silicone fiber (whisker)-mediated transformation, (3) fusion of DNA-loaded liposomes with protoplasts or tissues, and (4) microinjection. Of these methods, introduction of DNA into protoplasts by electroporation or chemical treatment has proven to be most reproducible and successful for developing transgenic monocot plants.

a. Protoplast transformation

Plant cell walls are generally considered to be a barrier to efficient DNA transfer directly into plant cells.[31] However, foreign DNA can be readily transferred into protoplasts formed by the removal of these cell walls. Isolated protoplasts are suspended in a buffer solution containing the DNA to be transferred; then the protoplast plasma membrane is briefly disrupted either by an electric pulse (electroporation) or by chemicals so that the DNA can enter into the protoplasts through the pores created in the plasma membrane. Some long chain polycations are known to protect and stimulate the uptake of DNA into protoplasts.[32] These chemicals bind with strongly negatively charged nucleic acids and reduce the repulsion between the DNA and the negatively charged plasma membrane. Of these chemicals, polyethylene glycol (PEG) has been the most extensively used. Direct gene transfer into protoplasts by electroporation or PEG treatment is efficient. However, because plant regeneration from protoplasts has proven to be difficult for most species, these methods have largely been replaced by biolistics. Electroporation and PEG treatment have been used to stably introduce foreign genes into several monocots including rice,[33] maize,[34] and tall fescue (*Festuca arundinacea* Schreb).[36,60]

b. Silicon carbide whisker-mediated transformation

The major attraction of this method is its simplicity. It does not require expensive equipment and may be an option for a minimally equipped lab. It involves mixing plant cells in liquid medium with silicon carbide fibers coated with DNA containing the gene(s) of interest along with the necessary regulating elements. This mixture is then vortexed with common lab vortexing equipment, which causes the silicon carbide fibers to penetrate the cells, thus delivering the DNA into the cell.[37,38]

B. Tissue Culture and Plant Regeneration

Successful regeneration of plants from protoplasts, cultured cells, or callus has been the limiting factor for transforming many crop and forage species. However, during the last several years, rapid progress has been made on plant regeneration of grass species. Much of this research was done empirically by trying to identify regenerable plant explants and evaluating an array of hormone combinations, and other media and culture options. Immature seed embryos and inflorescences have proven to be good sources of embryogenic regenerable callus in many monocot species. Culturing protoplasts in agar with nurse cells has proven to be very effective for cell division of monocot protoplasts.[39] Often the ability of cultures to regenerate is genotype specific, that is, only a few genotypes within a species will produce tissue that will regenerate. The ability to regenerate is under genetic control[40] and may be modified through breeding. In both alfalfa[41] and red clover,[42] highly regenerable lines have been developed by several cycles of selecting and intercrossing the most regenerable plants. These lines eased the problems encountered in plant transformation, however, it would be more desirable to have regeneration methodology that is independent of genotype in order to insert gene(s) into most if not all advanced breeding lines.

Several tropical forage legume species have been regenerated. Meijer[43] reported high-frequency regeneration of *Stylosanthes humilis* Kunth using both hypocotyl and leaf-derived explant tissues. Wofford et al.[44] evaluated regeneration in six genotypes representing two species and one subspecies of *Desmodium* using two media protocols. Only one protocol gave satisfactory results, with only one genotype not regenerating, however, regeneration among the lines was variable. That same lab[45] regenerated alyceclover [*Alysicarpus vaginalis* (L.)]. As before with *Desmodium*, six genotypes were evaluated, but with four media protocols. Responses of the genotypes to the culture protocols were described with four of the six genotypes being regenerated. *Centrosema brasilianum* (L.) Benth. cultured from leaf explants was regenerated by Angeloni et al.[46] Other *Centrosema* species would not produce shoots using their methods.

With tropical grasses, Ross et al.[47] reported regenerating buffelgrass (*Cenchrus ciliaris* L.) from embryogenic cell cultured from germinating mature seeds. Multiple green shoots regenerated on 20 to 50% of the embryogenic calli. In addition, a high frequency of transient expression of the gene coding for β-glucuronidase gene (GUS) was observed following particle bombardment gene transfer. Ahn et al.[48] developed a culture method to regenerate common Bermuda grass [*Cynodon dactylon* (L.) Pers.] from immature inflorescences measuring less than 5 mm. Later, regeneration of six improved Bermuda grass cultivars was evaluated using that culture method.[49] Calli could easily be induced with all of the cultivars, however, only three cultivars developed embryogenic calli that could be regenerated and one did so at a very low rate. This work emphasized the genotype specificity of regeneration. Croughan et al.[50] studied somaclonal variation from cultured Bermuda grass, and were able to identify and regenerate several lines with improved resistance to armyworm. Redway and Vasil[51] used the basal portions of young leaves from field-grown elephant grass (*Pennisetum purpureum* Schum.) to initiate white compact embryogenic calli that readily regenerated whether or not they were selected for tolerance to S-(2-aminoethyl)-L-cysteine, a lysine analog. They expected that tolerance (insensitivity) to that analog would make increasing lysine content possible. They obtained tolerant calli and regenerated plants with significantly higher free lsyine content.

Paspalum species have been regenerated from immature inflorescences, basal meristem/leaf, and germinating seedling explants. Marousky and West[52] regenerated Bahia grass (*P. notatum* Fluegge) from callus that developed at the coleoptile base of germinating mature caryopses. Later, a similar method was used to evaluate six Bahia grass cultivars.[53] The portion of embryogenic calli formed, ranged from 12 to 40%. Eight *Paspalum* species and one interspecific hybrid were evaluated for plant regeneration from cultured young unemerged inflorescences.[54] They observed callus proliferation from the glume bases, especially near cut areas, and reported regeneration by a "Relative Shoot Yield Index" (RSYI). This index ranged from 0 to 45 among the species tested, with *P. notatum* having a RSYI of 5.54. Explants containing basal meristem and basal leaf tissue obtained from the ends of stolons were also successfully used to regenerate Bahia grass.[55] Of these three types of explants, the germinating seedlings can be readily obtained at any season without maintaining plants. The basal meristem/leaf explant method requires that plants be maintained and the immature inflorescence is least convenient, being available for only a short time prior to flowering. Akashi and Adachi[56,57] reported regenerating dallis grass from both immature inflorescences and suspension culture-derived protoplasts. Venuto et al.[58] reported on experiments with dallis grass designed to generate somaclonal variation for the improvement of persistence, plant vigor, forage yield and quality, and seed production. Dallis grass is an apomictic species and is not amenable to plant breeding so their best lines are undergoing field testing. In other studies of somaclonal variation, regenerated Pangola digitgrass (*Digitaria decumbens*), a sterile triploid, was evaluated as a means of generating variability for a breeding program.[59] An additional EMS mutagenic treatment was given to long-term tissue culture to increase the variation generated. Variation was monitored by RAPD analyses.

C. Selection Systems and Selectable Markers

Transformation is a rare event with a very low frequency of cells being transformed; therefore, efficient selection strategies are required for the recovery of transformants. In general, transformation constructs contain antibiotic- or herbicide-resistant genes so that transformed cells can grow on the antibiotic- or herbicide-containing medium while the growth of untransformed cells is suppressed. Several selectable markers have been used for plant transformation experiments. These include the neomycin phosphotransferase II gene (*nptII*) and the hygromycin B phosphotransferase gene (*hph*) that provide resistance to the antibiotics kanamycin and hygromycin, respectively. Another selectable marker, the *bar* gene encoding phosphinothricin acetyltransferase provides resistance to herbicides such as bialaphos or glufosinate (vis. commercial formulations Liberty™ or Ignite™). Kanamycin selection using the nptII gene has been extensively used for selecting transformed dicot plants. However, this selection has not been effective for many monocots, since they have naturally high levels of resistance to this antibiotic. Hygromycin B and phosphinothricin have been more efficiently used for selection of transgenic rice,[33] maize,[28] and tall fescue.[36,60]

D. Promoters

In plant genetic engineering, the level of transgene expression generally depends on the strength of promoter that is required to initiate transcription of a transgene, and new trait phenotype largely depends upon that expression level. The cauliflower mosaic virus (CaMV) 35S promoter has been widely used because of its strong promoter activity in many dicot species.[61] However, the activity of this promoter is much weaker in many monocots than in dicots.[62] In monocots, several studies have shown that gene expression can be enhanced by the presence of introns of some monocot genes as in: maize alcohol dehydrogenase gene (*adhl*) introns 1,2,3,6,8, and 9,[63,64] maize shrunken gene (*shl*) intron 1,[65] rice actin gene (*act1*) intron 1,[66] and maize ubiquitin (ubi-1) intron 1.[67] The enhancement of gene expression by the introns appears to be due to efficient splicing and the increased level of mature mRNA.[64] The rice actin (*act1*) promoter and its first intron or the maize ubiquitin (ubi-1) promoter and its first intron appear to be the promoters of choice for many monocot transformations since they are very active in rice, maize, barley, wheat, and tall fescue.[68,69] Convenient gene expression vectors using these regulatory sequences have recently been constructed so that high levels of foreign gene expression can be easily achieved in transgenic monocots. The above-mentioned promoters are constitutive, that is, always turned on. As genetic engineering becomes more advanced, tissue-specific and/or developmentally regulated promoters will be used for certain applications.

E. Transformation of Tropical Forages

Very limited efforts have been made toward developing transgenic tropical forages. For tropical forage legumes, it was demonstrated by Vlachova et al.[70] that *Sesbania rostrata* Bremek and Oberm is susceptible to infection by several *Agrobacterium* strains and therefore subject to transformation by those strains. Manners and Way[71] used *Agrobacterium* and a binary vector system to produce transgenic plants of the tropical forage legume species, *Stylosanthes humilis* Kunth. Transformed cells were selected on kanamycin-containing medium and fertile transgenic plants were produced. Half of the transgenic plants had normal growth, while the rest were stunted. The abnormal growth of some transgenic plants was caused by the cotransfer of T-DNA genes into transformed cells, since *Agrobacterium* used in this experiment was not disarmed. Stable genetic transformation of elephant grass was accomplished by electroporation of protoplasts and biolistics of embryogenic calli.[72] Embryogenic calli derived from both young leaf and immature inflorescence segments were regenerated, however, transgenic plants were not recovered. In addition to the transient gene expression reported in buffelgrass in the plant regeneration section above, Grando et al.[73] reported transient GUS expression in bombarded embryogenic Bahia grass calli. They also bombarded a large number of calli with the soybean (*Glycine max* L.) vegetative storage protein gene, but have not reported transgenic plants. Yuge et al.[74] reported the stable transformation of dallis grass (*Paspalum dilatatum* Poir) and Bahia grass calli, but in this case transgenic plants were not recovered.

F. Traits of Interest for Genetic Engineering Tropical Forages

Some of the traits (genes) being used in genetic engineering of the major crops and temperate forages are of interest for tropical forages. Many of the gene constructs used in transforming those crops can be "borrowed" and used with little or no modification to transform tropical forages. Potentially useful gene constructs for disease and insect resistance include those for overproduction of plant defense-responsive genes that are induced by plant pathogens such as phytoalexins, lignins, hydrolytic enzymes (chitinases, glucanases), and viral coat and movement proteins to mention a few.[75-78] General use of Bt toxin genes in tropical forages may be difficult because the projected profitability may not offset the additional risk of developing resistant insects and the loss of insecticide effectiveness on major crops. Also, because forages have many wild weedy relatives, there will be strong resistance to engineering herbicide resistance into these species due to the fear of developing herbicide-resistant weeds. To dispel these concerns, gene containment mechanisms as ploidy incompatabilities and apomixis in forages may make herbicide resistance escape unlikely and make engineering herbicide resistance feasible, where those mechanisms exist.

Genetic engineering offers unique opportunities for improving the nutritional value of tropical forage grasses, which is often inadequate for satisfactory livestock performance. In my laboratory, we have just completed a project using a maize model transformation system to demonstrate that forage grass digestibility can be improved 5 to 8% by down-regulating the lignin biosynthetic enzyme, O-methyltransferase (OMT).[35] To do that we transformed maize with the antisense version of the sorghum OMT gene, i.e., the OMT gene (cDNA) was assembled in the construct in the reverse order relative to its promoter. We also included the *bar* gene as a selectable marker gene. About 80% of the 350 recovered plants were resistant to 1% glufosinate, indicating they were transgenic. The transgenic origin of selected plants was verified by PCR, northern and Southern analyses. Several of the transgenic plants also expressed the brown midrib phenotype. The OMT activity of the transgenic was significantly reduced with some plants showing up to 60% reduction. Lignin was reduced an average of 16% in the transgenic plants.[35] We are applying that technology to Bahia grass and have bombarded calli under selection.

There is growing interest in isolating genes from forages, especially those that confer stress or disease tolerance. Plants, including forages, growing in extreme conditions are of interest for gene isolation. For example, three sulfate transporter cDNAs (genes) have been isolated from *Stylosanthes hamata* (L.) Taub.. These genes mediate plant uptake of sulfate and may lead to a better understanding of uptake mechanisms, especially from low sulfate soils.[79]

An Australian group has targeted *Stylosanthes* for molecular research. They postulated down-regulating lignin by using either antisene or ribozyme technologies[80] and isolated phenylalanine ammonia-lyase,[81] O-methyltransferase,[82] and coniferyl alcohol dehydrogenase,[83] of the phenylpropanoid pathway. This is a key pathway for producing lignins (key antiquality factors), phytoalexins for disease resistance, and flavonoids for signal molecules. In addition, that group has isolated four peroxidase cDNAs.[83,84] Curtis et al.[85] reported that two *Stylosanthes* peroxidases (Shpx6a and Shpx6b) were rapidly induced following inoculation with a *Colletotrichum* fungal pathogen, wounding, and treatment with the methyl jasmondate. The upstream promoter region of the Shpx6b peroxidase was isolated and incorporated into a vector driving the GUS reporter gene. That promoter was able to function in transgenic tobacco (*Nicotiana tobacum* L.) plants in response to fungal inoculation, wounding, and treatment to methyl jasmondate. The effect of the peroxidase encoding cDNA, Shpx6a, expressed in tobacco and canola (*Brassica napus*) driven by the constitutive 35S CaMV promoter was studied.[86] The peroxidase activity was increased two- to threefold, and a small but statistically significant increase in the level of resistance to fungal attack was observed. This research supports the role of peroxidase in plant defenses and the use of biotechnology in improving disease resistance. Curtis et al.[83] used the above-mentioned cDNAs as probes in a phylogenetic analysis of *Stylosanthes* species.

With today's powerful functional genomics methodology, many such genes may soon be available, with some being isolated from tropical forages. There is also considerable research progress in developing improved legume-rhizobium symbioses in suboptimal conditions.

III. GENETIC AND DNA MARKERS

A. Morphological and Isozyme Markers

Classical morphological genetic markers have been important tools in genetic studies and plant breeding for many years. They are inexpensive and easy to use, however, their utility is seriously limited by the lack of distinguishing morphological features in many species. With the development of high-resolution gel electrophoresis during the 1960s, proteins (mostly seed proteins) and alternate enzyme forms differing in size (isozymes) came into use as molecular markers. These markers complemented and extended the scope of morphological markers. Because proteins and isozyme markers are gene products, they offered, for the first time, direct observation of gene expression. They were used extensively in characterizing germplasm by "fingerprinting" plant genotypes, defining genotypes for plant patents and ensuring varietal security, evaluating genetic diversity, and studying phylogenetic relationships. With seed and other protein markers, crude protein extracts are fractionated using polyacrylamide gel electrophoresis (PAGE), stained with a general protein stain; then the stained gels are photographed for pattern comparisons. With isozymes, the crude protein extract is fractionated as above, then reacted with specific enzyme substrates coupled to various dye indicators to visualize the isozyme pattern. Isozymes are relatively inexpensive to use, requiring only modest equipment and substrate expenditures. Their drawback is that the alleles available are very limited relative to DNA markers and only a very small portion of the genome can be sampled.

B. DNA Markers

During the 1980s and 1990s, DNA marker technologies were developed that offered unlimited loci over the entire genome. Because of the greater number of DNA markers available, they are much more useful than isozymes. DNA markers are actual sites on the chromosome and behave as loci. DNA markers have been used for identifying genotypes,[87] studying genetic diversity,[88] monitoring genetic events, elucidating evolutionary pathways,[89] and facilitating the manipulation of genes in breeding programs as was mentioned for isozymes. More importantly, by using linkage analyses, DNA markers can be mapped (genomic mapping) to linkage groups and chromosome locations. Also, through linkage analyses, they can be used to locate and map single-gene and quantitatively inherited trait loci to chromosomal sites.[90-93] Quantitative trait loci (QTL) constitute the majority of economically important genes, but they cannot be identified and studied individually by classical genetic means. Use of DNA markers in genetic studies of important, complex, quantitatively inherited traits offers the potential of dissecting those traits into single marker-gene-linked components. In addition, the relative importance of marker-linked genes in trait determination and their chromosomal location can be determined, as was demonstrated by Martin et al.[91] when they broke down the quantitatively inherited water-use efficiency trait in tomato (*Lycopersicon esculentum* L.) into three marker-gene components with additive inheritance. Lastly, closely linked DNA markers are now being used to locate and clone (map-based cloning) economically important genes, not accessible by other means. The types of DNA markers are reviewed below.

1. Restriction Fragment Length Polymorphism (RFLP)

"Restriction fragment length polymorphisms" (RFLPs)[87,94] were the first DNA markers to be developed and are still regarded as the most reliable. They are identified by first cleaving genomic DNA with restriction endonucleases into fragments, fractionating the resulting fragments by gel electrophoresis, then identifying specific fragments by hybridization to a labeled DNA probe. Pattern comparisons of labeled fragments can then be used to determine phylogenetic relationships along with the many uses outlined above. RFLP development and utilization is labor intensive, requires hazardous isotopes for labeling, and requires maintaining and distributing probe clone libraries. Tanksley et al.[95] reviewed the use of RFLPs in plant breeding.

2. Random Amplified Polymorphic DNA (RAPD)

The "random amplified polymorphic DNA" (RAPD) marker[96] is much less labor intensive and does not use isotopes, making it a popular DNA marker replacing RFLPs. It uses the polymerase chain reaction (PCR) to amplify specific short fragments of DNA from the genomic DNA template using short random primers. Those fragments are visualized by DNA staining after fractionation by gel electrophoresis. RAPD fragment patterns are used in similar ways to RFLP patterns. The disadvantages of RAPDs relative to RFLPs are that they are less reliable and yield less genetic information since they are inherited in a dominant fashion (3:1), whereas, RFLPs are inherited codominantly (1:2:1), i.e., RAPDs do not distinguish the heterozygote.

3. Microsatellite and AFLP DNA Markers

Microsatellite and "amplified fragment length polymorphism" (AFLP) DNA markers are now the most used DNA marker systems. They do not have the intensive labor requirement of RFLP or the low reliability of RAPD. Both microsatellite and AFLP DNA markers are popular because of the high degree of polymorphism they generate and their ease of use.

Other names for microsatellite markers are "short tandem repeat" (STR), "simple sequence repeats" (SSR), and "variable number tandem repeats" (VNTR). These markers utilize repetitive short sequences, sometimes only two or three nucleotides long, that are abundant and distributed throughout the genome.[97,98] Most often they are visualized by amplification with PCR, but can be hybridized to specific probes.

The AFLP system is capable of generating a great deal of polymorphism with few reactions.[99] It involves restricted genomic DNA fragments that are ligated to proprietary adapters, to which special proprietary primers bind for PCR amplification. These special primers are designed to subdivide the population of restriction fragments into many subsets by the PCR amplification. This complex system comes in kit form (Life Technologies, Rockville, MD) and amplifies 50 to 100 fragments per reaction, which are visualized in the electrophoresis gel. This system is especially useful in creating high-density maps and where genomic variability is low.[100]

4. Other DNA Markers

"DNA amplified fingerprinting" (DAF), which is a variation of RAPD, uses shorter primers and different electrophoresis and staining procedures.[101] The "sequence tagged site" (STS) marker uses the sequence of an RFLP marker to design primers for its PCR amplification,[102] thereby reducing the effort required, eliminating the need to maintain a probe clone library, and eliminating the need of hybridization of Southern blots and associated isotope exposure. The "sequence characterized amplified region" (SCAR) uses the sequence of a RAPD marker to design primers for greatly improved reliability of PCR amplification.[103]

C. Germplasm Characterization and Phylogenetic Studies

Genetic and DNA markers have been of great value in facilitating plant breeding and genetic studies in the world's major crop species. Tropical forages have not commanded a resource commitment as relatively high as the major food and cash crops, therefore, marker research, especially concerning DNA markers, has been limited. Isozyme and DNA marker research on tropical forages is reviewed below.

An isozyme "fingerprinting" system was used to differentiate and catalog Florida's digitgrass (*Digitaria*) germplasm collection and breeding lines.[104] Although distinguishing morphological characteristics were lacking in these grasses, peroxidase and esterase isozymes were found to be highly variable with polymorphism in at least nine peroxidases and seven esterases. That polymorphism, along with several esterase and peroxidase quantitative levels, gave a highly specific isozyme fingerprinting system. Isozymes from leaf tissue, standardized for maturity, were found to be consistent under different field and greenhouse growing conditions.[105] The fingerprinting system was used in the variety development program to identify genotypes and monitor hybridization and was especially useful in eliminating duplicate genotypes from the advanced line testing program. In Australia, Hayward and Hacker[106] used the highly variable esterases as isozyme markers to monitor species hybridization and verify hybrid origin of breeding lines in their *Digitaria* breeding program. They ascertained the mode of inheritance of those esterases to improve their efficiency as markers. Five to seven genes, some multiallelic, were found to control the esterases within the five *Digitaria* species of their study.

The isozyme diversity in Rhodes grass *[Chloris gayana* (Kunth)] and *Cynodon* species was evaluated along with that of the digitgrasses.[105] Both *Chloris* and *Cynodon* accessions were found to be only moderately variable relative to digitgrass. Only three polymorphic peroxidases were found in *Chloris*, whereas the digitgrasses had nine. Isozyme fingerprinting was used by Vermeulen[107] to identify turf type *Cynodons*. In that work crude protein was extracted from newly emerged leaf blades and subjected to six selected enzyme-specific staining systems. The staining systems were selected on the basis of polymorphism and resolving power in the *Cynodons*. Of 21 genotypes tested, 16 were found to be unique; the other five were partitioned into two groups. That fingerprinting system proved useful in identifying genotypes, but was not able to distinguish all genotypes in their collection. The more powerful "DNA amplified fingerprinting" (DAF) method was used to characterize the genetic variation and study the phylogeny of Bermuda grasses.[108,109] Parsimony and cluster analysis were used to group 13 Bermuda grass cultivars according to genetic relatedness and successfully distinguished between closely related cultivars. The DAF technique was demonstrated to be excellent for Bermuda grass genotype identification and other applications, including detecting planting errors and plant mixtures, mislabeled plant materials, and for genotype verification in varietal protection enforcement. This DAF procedure appears to be far superior to the isozyme method in distinguishing *Cynodon* variability.

Genetic markers have demonstrated their usefulness in determining the mode of reproduction in guinea grass (*Panicum maximum* Jacq.) and Bahia grass. Isozymes were used in a search to find sexual plants in *P. maximum*, a species then classified as obligate apomictic.[110] Morphological and isozyme markers detected five segregating single panicle progenies from sexual maternal plants among the 742 screened. Whereas morphological markers were uniform or indicated only two plant types within a 10-plant sexual progeny, esterase isozymes showed four to nine different plant types and peroxidase isozymes, three to six. Apomictic progenies gave uniform patterns for esterase and peroxidase isozymes, and morphological traits. Morphological variation in the progenies of sexual origin was not sufficient to conclusively demonstrate sexuality, but the high variability of the isozyme data was decisive and saved one or more years of testing. In addition, the isozyme variability in the progenies indicated that the parental plants were extremely heterozygous and also established that the species was highly polymorphic. In tetraploid Bahia grass, three RFLP and RAPD markers (using 4 to 6 primers) were used to verify cyto-embryological data on the mode of reproduction.[111] Marker patterns of selfed progenies of three plants identified cytologically as fully sexual, fully apomictic, and intermediate were in close agreement with the cytological data, i.e., progeny from F131 segregated indicating that it was completely sexual, progeny from Q4117 gave a single marker pattern indicating it was highly apomictic, and progeny from Q3664 was intermediate in marker polymorphism.

In phylogenetic studies, Lagudah and Hanna[112] grouped *Pennisetum* species into primary and secondary gene pools based on interspecific hybridization studies of *Pennisetum* species crossed to pearl millet. Then they evaluated that grouping using isozyme markers of six enzymes. As with the *Digitaria*, esterases gave good results with high resolution and identified interspecific differences. Other isozymes were less definitive. Their isozyme data were complex, but generally supported the phylogenetic groupings made by interspecific hybridization. Isozyme diversity of pearl millet was extensively studied by Tostain et al.[113] They placed West African wild and cultivated pearl millets into three groups, wild, early maturing, and late maturing. No new isozyme diversity was found in pearl millets from other African regions or India.[114] The isozyme diversity data led to an evolutionary hypothesis suggesting that multiple domestications of the pearl millet had taken place in the Sahel region of Africa creating early-maturing cultivars that were thought to have been disseminated eastward to India and southward to the Sudanian region.

DNA markers permit using both the chloroplast and mitochondrial genomes in phylogenetic studies. Both genomes are maternally inherited in plants and do not segregate with sexual recombination. Their polymorphism is due to mutational types of events. Clegg et al.[115] found that pearl millet chloroplast DNA (CpDNA) was polymorphic and that CpDNA restriction fragments were suitable for phylogenetic study within that species. Since the chloroplast genome is small, the restriction fragments could be visualized directly by DNA staining in the gel. In other phylogenetic studies of pearl millet and six *Pennisetum* species, Chowdhury and Smith[116] found that mitochondrial DNA (MtDNA) restriction patterns showed little polymorphism among pearl millet accessions, however, the patterns among the seven *Pennisetum* species were highly variable. They compared restriction fragments visualized directly by DNA staining with those visualized by hybridization to a probe (Southern blot). The DNA staining method revealed significant relationships between *P. polystachyon* (L.) Schult. and *P. pedicellatum* Trin. and between elephant grass and

P. squamulatum Fresen. In addition to those relationships, the MtDNA Southern analysis also showed significant relationships of pearl millet to elephant grass and *P. squamulatum*. Relationships determined by Southern analyses were in better agreement with cytological data and that method is preferable for phylogenetic studies in *Pennisetum*.

Shechter et al.,[117] in their investigation of the origin of sorghum, found the four sorghum races, caudatum, durra, guinea, and bicolor, to have close isozyme affinities, while the kafir race was unique. This supported the hypothesis that the four similar races arose from a primitive, early domesticated bicolor type with kafir developing independently. Morden et al.[118] used the patterns of 30 isozyme loci in a broader taxonomic study of sorghum races, tetraploid Columbus grass (*S. almum* Parodi) and Johnson grass [*S. halepense* (L.) Pers.]. They found low levels of variability within and among accessions. Cultivated sorghum contained less variability than *S. bicolor* subsp. *arundinaceum*, its presumed progenitor. Their isozyme data were consistent with the hypothesized origins of *S. almum* and *S. halepense* via autopolyploidy and segmental alloploidy.

Isozymes were used by Chow and Crowder[119] to investigate the origin of a *Desmodium* forage legume selection of unknown species that has unusual silver markings and brown flecks on the leaves. They questioned whether it was an interspecific hybrid of *D. intortum* (Mill.) Urb (having brown flecks) and *D. sandwicense* E. Mey or *D. uncinatum* Jacq. DC. (having silver markings). Isozyme data indicated the unusual plant selection was closely related to *D. intortum* and should be considered a biotype of that species and that the unusual markings could have been a product of gene introgression. Imrie and Blogg[120] used phosphoglucoisomerase and phosphoglucomutase isozymes to study the genetic drift that occurred over time in *Desmodium intortum* cv. Greenleaf in different regions of Australia. Their results showed the presence of *D. sandwicense* alleles in Greenleaf, and that the frequency of these alleles was reduced after being grown in one production area, whereas no changes occurred in another production area. They concluded that natural selection in response to environment caused the reduction in *D. sandwicense* alleles. The genetic variability and mating system of three *Centrosema* species were studied using isozyme and DNA markers.[121] Populations of all three species expressed high polymorphism with both types of markers and within population diversity higher than between population diversity. These species, previously considered predominately as self-pollinators, showed outcrossing rates higher than expected (up to 41%). This higher outcrossing rate should be considered in breeding and germplasm maintenance programs. Genome specificity of 13 sequence-tagged-site (STS) genetic markers was evaluated on 12 *Stylosanthes* species[122] and was found to be low. This suggested that there was a low degree of genomic divergence with the group of *Stylosanthes* species.

RFLP markers were used to fingerprint the University of Florida elephant grass germplasm collection, and two polymorphic probes were able to differentiate 80% of that germplasm. Use of various probes made it possible to detect unwanted mixtures and off types in breeding stocks and advanced lines previously thought to be pure.[123] Genetic diversity of the elephant grass germplasm collection was measured by pairwise comparisons and an index of genetic diversity was calculated. Cultivar Mott and P.I. 300086 were the most diverse. The genetic diversity information is helpful to the plant breeder because a positive relationship has been shown between molecular genetic diversity of parents and grain yield of F_1 hybrids, as demonstrated with maize.[124] The heterozygosity of elephant grass parents was measured as a ratio of segregating fragments divided by the total number of elephant grass fragments in the interspecific progenies. Based on segregation of morphological traits, Mott appears to be much more heterozygous than P.I. 300086; however, segregation of restriction fragments indicated that both had a heterozygosity ratio of near 30%.[125] RFLP markers were used to measure the hybridization rate of crosses made with fresh and frozen stored pollen.[126] Crosses were made without emasculation, so self-fertilization was also possible. Hybrid progeny were identified by transmission of RFLP markers via the pollen. Hybrids obtained using the stored frozen pollen demonstrated that it could be used when fresh pollen is not available.[125] This research facilitated crossing Mott and P.I. 300086, a difficult cross because the flowering periods of the two grasses do not overlap, requiring the use of stored frozen pollen. One hundred progeny pollinated with fresh pollen were screened with over 80% being hybrid.

DNA markers are being used to study relationships within plant pathogen species. Anthracnose is a serious disease of the tropical pasture legume *Stylosanthes* that has restricted its use in South America and Australia. RFLP analyses of the pathogen [*Colletotrichum gloeosporioides* (Penz.) Penz. and Sacc.] were able to differentiate the two forms of the pathogen, Type A and Type B. These analyses facilitated studies that determined Type A infects most *Stylosanthes* species, while Type B infects only *S. guianensis* (Abl.) Sw.[127] These studies verified the existence of two distinct pathogen populations in Australia and assisted in management of the disease. Markers are also being used to monitor disease susceptibility within plant species. Chahal et al.[128] found that peroxidase patterns correlated with downy mildew resistance in pearl millet and suggested that peroxidases may be involved in the resistance mechanism.

D. Genomic and Trait Mapping

Beginning in the 1980s, extensive genomic mapping projects were initiated to construct high density genomic maps using RFLP, RAPD, and other DNA markers for all major crop species. The greater the density of a genomic map, the more useful it is, as the linkages are tighter and the distances between markers or marker-genes are shorter; consequently efforts have been continuous to increase the map density.

Markers are being used to identify and locate genes of economic importance that could not be identified by other means.[129-132] Not only are markers used to locate and map genes of economic importance on the chromosome, markers in high density genomic maps are being used for map-based cloning of genes. Arondel et al.[133] used map-based cloning to clone the omega-3 desaturase gene in *Arabidopsis thaliana* L. This gene regulates fatty acid type. Martin et al.[134] have cloned a tomato protein kinase gene conferring disease resistance.

Of the crops used as tropical forages, high-density genomic maps are being constructed for both sorghum and pearl millet using both RFLP and RAPD markers.[135,136] These and other genomic maps and the probes used in their construction could be useful in tropical forage research. Studies using maize genomic and cDNA probes showed that these probes hybridized strongly to sorghum Southern blots and worked well in mapping the sorghum genome.[137,138] The maize probes also hybridized strongly to sugarcane (*Saccharum officinarum* L.) and foxtail millet [*Setaria italica* (L.) Beauv.], but less strongly to pearl millet and only occasionally to barley.[137] Research using maize probes in sorghum mapping and tomato probes in potato (*Solanum tuberosum* L.) mapping, demonstrated that gene order on the chromosome is generally conserved across species unless the chromosome is modified by rearrangements.[89]

A DNA marker system was developed in elephant grass to assist genetic and breeding studies and to initiate genetic mapping. Lack of genetic information, difficulty in controlling hybridization, and polyploidy established a need for RFLP markers to facilitate genetic study and improvement.[125] Elephant grass did not have inbred or other defined genetic stocks needed for the standard mapping approach using segregating F_2 or backcross progeny. Mapping was further complicated by allotetraploidy, the elephant grass genomic composition being A′A′B B. Delay of setting up F_2 or backcross mapping populations was avoided by using an interspecific hybridization scheme with pearl millet (genomic composition AA) and elephant grass (A′A′B B) to produce the segregating haploid progenies of genomic composition A′B A. In this scheme, the parental genes are distributed to the progeny according to the gametic ratio of 1:1 for single locus segregation. Generally, there was sufficient difference in restriction fragment size between the pearl millet restriction fragments (A genome) and the elephant grass fragments (A′B genomes) so that they did not obscure one another. The pearl millet parent was highly inbred and homozygous so it made a uniform and defined contribution to each individual of a progeny. Thus, the polymorphism observed was due to segregation of the elephant grass genome. Two progenies from highly heterozygous, unrelated elephant grass parents were used. When a probe did not identify a polymorphic elephant grass RFLP in one progeny, it had a good opportunity to do so in the other. This unique interspecific scheme may be a useful approach in other complex genomes to produce suitable segregating progenies for mapping markers.[125]

Marker-plant trait linkage studies were conducted for yield, height, culm weight, tiller number, node number, leaf width, leaf length, leaf angle (uprightness), cool season growth, winter hardiness, leaf mottling, propagation efficiency (vegetative), and pubescence of collar, stem, and sheath on replicated, field-grown plots of the hybrid segregating progenies. Additional marker-linkage analyses were conducted on forage quality traits, i.e., percentage crude protein, phosphorus, fiber, ash, and *in vitro* organic matter digestibility (IVOMD).[139]

In the two progenies, 64 markers were found to be linked to QTL of 26 plant traits. Of these markers, seven marker-linked genes affected the IVOMD trait, seven affected fiber, and seven affected protein content. As these traits are somewhat interrelated, some of the markers linked to genes of one trait overlap other traits, confirming that some QTL are associated with more than one trait. The gene linked to marker A208 showed the largest effect of any marker-linked gene, accounting for 63% of the segregating variation in stem IVOMD in the 'Mott' progeny. Linkage analyses of traits affecting yield show four markers linked to yield, two to height, four to stem weight, five to node number, four to tiller number, three to leaf length, and five to leaf width. In addition, three markers were found linked to genes affecting winter survival and five to propagation efficiency. These results and those reported for other crops show that DNA markers can be effective in facilitating genetic studies and breeding.[139] The usefulness of DNA markers is increased as their density is increased on the map and marker-gene loci are more tightly linked.

The isolation and use of a gene to control apomixis would be very valuable to fix hybrid vigor in crops. Apomixis is common in forage grass species and efforts are in progress to identify and isolate an apomixis gene. In the ongoing program to transfer apomixis from *Pennisetum squamulatum* Fresen to cultivated pearl millet [*P. glaucum* (L.) R. Br.], Ozais-Atkins et al.[140] identified two markers, RFLP UTG197 and RAPD $CO4_{600}$, associated with apomixis in interspecific backcross lines.[141] Later these markers were used to screen 11 apomictic and eight sexual species of *Pennisetum*,[142] where the UTG197 RFLP (converted to an STS marker) was found to be associated with all taxa having apomictic reproduction except those in section *Brevivalvula*. The cloned $CO4_{600}$ RAPD was less specific and was associated with three of the 11 apomictic species. Neither marker hybridized to any sexually reproducing representatives of the genera. Gustine et al.,[143] using bulk segregant analysis,[144] found that the sequence-tagged-site (STS) UGT197-144 (listed as STS UGT197[142]) was closely linked to the apospory gene in buffelgrass (*Pennisetum ciliare* L. Link; synonymous with *Cenchrus ciliaris* L.). In addition, they identified and mapped four other RAPDs linked to the apospory locus. Pessino et al.[145] also used bulked segregant analysis to identify RFLP and RAPD markers co-segregating with apomixis in *Brachiaria*. They used an F_1 of an interspecific cross of a sexual and an apomictic species and used maize cDNA and genomic clones covering the maize genome as probes. They found a clone, umc147, co-segregating with apomixis that led to identifying another linked clone, umc72, in that same area. Of 184 PCR primers, OPC4, (as in *Pennisetum*) was significantly linked to apospory. These studies have developed markers that not only can assist in the management of apomixis, but may also be instrumental in isolating the gene for apomixis to be used in genetic engineering. The *Pennisetum* and *Brachiaria* data demonstrate the conserved gene order not only in the *Pennisetum* genus, but across genera to *Brachiaria* and maize.

IV. HELP FROM MODEL GENOME PROJECTS

Public genome projects of rice and maize, model species for monocots, and *Arabidopsis*, model species for dicots, have extensive databases, genomic maps, cDNA (EST) clone banks, marker primers, and probes. Since these are freely available, they can be very useful in reducing the work and expense of forage biotechnological research. The best source of this information is on the World Wide Web. The rice genome project (RGP) database can be accessed at http://www.staff.or.jp. This project is sponsored by the Japanese government. The maize database (MaizeDB) can be accessed at http://www.agron.missouri.edu/. The Arabidopsis Genome Initiative (AGI) database can be accessed from http://genome-www.stanford.edu/Arabidopsis/agi.html. There are other ways to access this information, but these sites give excellent links to the extensive data and clones available. The nucleotide sequence databases of GenBank and EMBL, as well as the amino acid sequence databases, can be accessed through the Website of the National Center for Biotechnology Information at http://www3.ncbi.nlm. The "expressed sequence tag" database (dbEST) is of special interest as it is a source of genes and is located in GB-EST subdivision of Genbank. The November 26, 1999 release summary in the "NCBl dbEST; database of Expressed Sequence Tags," listed rice as having 47,402 ESTs, maize 43,008 ESTs, and Arabidopsis 45,752 ESTs. These large collections of expressed genes (cDNAs) represent partial sequences of most, if not all, of the genes expressed in the different plant tissues (root, stem, leaf, etc.) of these species. This partial sequence data is the "sequence tag" that is cataloged in the GenBank GB-EST subdivision. This database can be searched and sequences compared with other sequences making it a valuable source of heterologous genes (cDNAs), and the clones can be obtained from the various genome project centers for use as probes or in vector construction for genetic engineering or other uses.

Gene sequences are conserved across the plant kingdom and, generally, the closer the relationship, the higher the level of conservation. Gene sequences within monocots are generally more highly conserved than between monocots and dicots. This conservation of genes makes the use of heterologous probes practical, i.e., probes made from genes of one species can be used in another species. The use of these heterologous probes may reduce the work and expense of isolating genes and using DNA markers. For example, several rice cinnamyl alcohol dehydrogenase-like EST clones were used as a heterologous probe to isolate the sorghum cinnamyl alcohol dehydrogenase (lignin biosynthetic) gene from a sorghum cDNA library (author's unpublished data). One can expect heterologous probes to be more useful if obtained from closely related species.

There is also conservation of gene order on the chromosome across species. When genomic relationships of the major cereals were compared with rice, conservation of gene order was observed, but was restricted to linkage segments. Maize, wheat, and other cereal chromosomes could be described in terms of those segments being like building blocks. These blocks could be assembled into a single stack for the *Triticeae* genome and in different ways for other grass genomes.[146] This is probably also true with many of our undefined forage grass chromosomes.

The *Arabidopsis* dicot model species has the smallest plant genome which, along with its ultra high-density genetic linkage map, makes it the ideal species for map-based cloning. In addition, its genome has been cloned into the large "artifical yeast chromosome" (YAC) and "artificial bacterial chromosome" (BAC) clones and these large clones have been mapped with the genomic markers. An unknown gene can be traced to one of the YAC clones by linkage to markers. That clone can then be obtained for gene recovery. Probably all of the *Arabidopsis* expressed genes have been cloned and sequenced as ESTs with the data and the cDNA clones now available. Chromosome-wise sequencing of the entire genome is underway. The ways in which this data and these clones can be used in forage molecular research mentioned above also apply here.

In conclusion, to maximize research progress in forages with the limited resources at hand, it is important that we use the information, genes, and probes developed in the model and closely related systems. With the development of new, more efficient molecular biological methodology and the extreme competitiveness of private companies in major crop biotechnology, public scientists will probably direct more of their research to forage and minor crops than they have in the past.

ACKNOWLEDGMENT

This chapter is published as Florida Agricultural Experiment Station journal series number N-01927.

REFERENCES

1. Mirasol, F., Bioengineered foods gain wider acceptance in the US, *Chemical Market Reporter*, 256 (10), 5, 1999.
2. Biopharm, Biotech at the end of the century, *BioPharm*, 12 (9), 56, 1999.
3. Lesney, M. S., The garden of Dr. Moreau: Plant biotechnology as a chemical enterprise, *Today's Chemist at Work*, June 28, 1999.
4. Crop Science Society of America, *Molecular and Cellular Technologies for Forage Improvement*, CSSA Publication Number 26, Brummer, E. C., Ed., Crop Science Society of America, Madison, WI, 1998.
5. Wang, Z., Herrera-Estrella, L., Van Montagu, M., and Zambryski, P., Right 25 bp terminus sequence of the nopaline T-DNA is essential for and determines direction of DNA transfer from *Agrobacterium* to the plant genome, *Cell*, 38, 455, 1984.
6. Rubin, R. A., Genetic studies on the role of octopine T-DNA border regions in crown gall tumor formation, *Mol. Gen. Genet.*, 302, 312, 1986.
7. An, G., Ebert, P. R., Mitra, A., and Ha, S. B., Binary vectors, in *Plant Molecular Biology Manual*, Gelyin, S. B., Schilpertoort, R. A., and Verma, D. P. S., Eds., Martinus Nijhoff, 1988, A3-1.
8. Shahin, E. A., Spielmann, A., Sukhapinda, K., Simpson, R. B., and Yashar, M., Transformation of cultivared alfalfa using disarmed *Agrobacterium tumefaciens*, *Crop Sci.*, 26, 1235, 1986.
9. White, D. W. R. and Greenwood, D., Transformation of the forage legume *Trifolium repens* L. using binary *Agrobacterium* vectors, *Plant Mol. Biol.*, 8, 461, 1987.
10. Quesenberry, K. H., Wofford, D. S., Smith, R. L., Krottje, P. A., and Tcacenco, F., Production of red clover transgenic for neomycin phosphotransferance II using *Agrobacterium, Crop Sci.*, 36, 1045, 1996.
11. Horsh, R. B., Fry, J. E., Hoffmann, N. L., Eichhltz, D., Rogers, S. G., and Fraley, R. T., A simple and general method for transferring genes into plants, *Science*, 227, 1229, 1985.
12. Lloyd, A. M., Barnason, A. R., Rogers, S. G., Byrne, M. C., Fraley, R. T., and Horsh, R. B., Transformation of *Arabidopsis thaliana* with *Agrobacterium tumefaciens, Science*, 234, 464, 1986.
13. An, G., High efficiency transformation of cultured tobacco cells, *Plant Physiol.*, 79, 568, 1985.
14. McGranahan, G. H., Leslie, C. A., Uratsu, S. L., Martin, L. A., and Dandekar, A. M., *Agrobacterium*-mediated transformation of Walnut somatic embryos and regeneration of transgenic plants, *Bio/Tech*, 6, 800, 1988.
15. Ishida, Y. J., Saito, H., Ohta, S., Hiei, Y., Komari, T., and Kumashiro, T., High efficiency transformation of maize (*Zea mays* L.) mediated by *Agrobacterium tumefaciens, Nature Biotech.*, 14, 745, 1996.
16. Chan, M. T., Transformation of indica rice mediated by *Agrobacterium tumefaciens, Plant Cell Physiol.*, 33, 577, 1992.
17. Park, S. H., Pinson, S. R. M., and Smith, R. H., T-integration into genomic DNA of rice following *Agrobacterium* inoculation of isolated shoot apices, *Plant Mol. Biol.*, 32, 1135, 1996.
18. Mooney, P. A. and Goodwin, P. B., *Agrobacterium tumefaciens* gene transfer into wheat tissues, *Plant Cell Tissue Organ Cult.*, 25, 209, 1991.
19. Tingay, S., McIory, D., Kalla, R., Fieg, S., Wang, M., and Thornton, S., *Agrobacterium tumefaciens*-mediated barley transformation, *Plant J.*, 11, 1369, 1997
20. Jeoung, J. M., Trick, H. N., Muthukrishnan, S., and Liang, G. H., *Agrobacterium*-mediated transformation in sorghum, in *1999 Annual Meeting Abstracts*, Amer. Soc. Agron., 1999, 157.
21. Yu, T. T., Trick, N. H., Huang, B., and Liang, G. H., *Agrobacterium*-mediated transformation of creeping bentgrass, in *1999 Annual Meeting Abstracts*, Am. Soc. Agron., 1999, 157.
22. Smith, R. H. and Hood, E. E., *Agrobacteriaum tumefaciens* transformation of monocotyledons. Review, *Crop Sci.*, 35, 301, 1995.
23. Jahne, A., Becker, D., and Lorz, H., Genetic engineering of cereal crop plants: a review, *Euphytica*, 85, 34, 1995.

24. Gasser, C. S. and Fraley, R. T., Genetically engineering plants for crop improvement, *Science*, 244, 1293, 1989.
25. Morrish, F., Songstad, D. D., Armstrong, C. L., and Fromm, M., Microprojectile bombardment: A method for the production of transgenic cereal crop plants and functional analysis of genes, in *Transgenic Plants*, Hiatt, A., Ed., Marcel Dekker, New York, 1993, 133.
26. Sanford, J. C., Klein, T. M., Wolf, E. D., and Allen, N., Delivery of substances into cells and tissues using a particle bombardment process, *Particle Science and Technology*, 5, 27, 1987.
27. Christou, P., Strategies for variety-independent genetic transformation of important cereals, legumes and wood species utilizing particle bombardment, *Euphytica*, 85, 13, 1995.
28. Gordon-Kamm, W. J., Spencer, T. M., Mangano, M. L., Adams, T. R., Daines, R. J., Start,, W. G., O'Brien, J. V., Chambers, S. A., Adams, W. R., Jr., Willetts, N. G., Rice, T. B., Mackey, C. J., Krueger, R. W., Kausch, A. P., and Lemaux, P. G., Transformation of maize cells and regeneration of fertile transgenic plants, *Plant Cell*, 2, 603, 1990.
29. Vasil, V., Castillo, A. M., Fromm, M. E., and Vasil, I. K., Herbicide resistant fertile transgenic wheat plants obtained by microprojectile bombardment of regenerable embryogenic callus, *Bio/Tech*, 10, 667, 1992.
30. Lazzeri, P. A. and Shewry, P. R., Biotechnology of cereals, *Biotechnology & Genet. Engineering Reviews*, 13, 79, 1993.
31. Potrykus, I., Gene transfer to plants: assessment of published approaches and results, *Ann. Rev. Plant Physiol. and Plant Mol. Biol.*, 42, 205, 1991.
32. Draper, J. and Scott, R., Gene transfer, in *Plant Genetic Engineering*, Grierson, D., Ed., Chapman and Hall, New York, 1991, 39.
33. Shimamoto, K., Terada, R., Izawa, T., and Fujimoto, H., Fertile rice plants regenerated from transformed protoplasts, *Nature*, 338, 274, 1989.
34. Rhodes, C. A., Pierce, D. A., Mettler, I. J., Mascarenhas, D., and Detmer, I. J., Genetically transformed maize plants from protoplasts, *Science*, 240, 204, 1989.
35. He, X., Down-regulating the expression of a lignin biosynthetic gene by antisense technology to improve forage quality, Ph.D. dissertation, University of Florida, Gainesville, Florida, 1999.
36. Wang, Z., Takamizo, T., Iglesias, V. A., Osusky, M., Nagel, J., Portykus, I., and Sprangenburg, G., Transgenic plants of tall fescue (*Festuca arundinacea* Schreb.) obtained by direct gene transfer to protoplasts, *Bio/Tech.*, 10, 691, 1992.
37. Frame, B. R., Drayton, P. R., Bagnall, S. V., Lewnau, C. J., Bullock, W. P., Wilson, H. M., Donwell, J. M., Thompson, J. A., and Wang, K., Production of fertile transgenic maize plants by silicon carbide whisker-mediated transformation, *Plant J.*, 6, 941, 1994.
38. Thompson, J. A., Drayton, P. R., Frame, B. R., Wang, K., and Dunwell, J. M., Maize transformation utilizing silicon carbide whiskers: a review, *Euphytica*, 85, 75, 1995.
39. Kyozuka, J., Hayashi, Y., and Shimamoto, K., High frequency plant regeneration from rice protoplasts by novel nurse culture methods, *Mol. Gen. Genet.*, 206, 408, 1987.
40. Kielly, G. A. and Bowley, S. R., Genetic control of somatic embryogenesis in alfalfa, *Genome*, 35, 474, 1992.
41. Bingham, E. T., Hurley, L. V., Kaatz, D. M., and Saunders, J. W., Breeding alfalfa which regenerates from callus tissue in culture, *Crop Sci.*, 15, 719, 1975.
42. Quesenberry, K. H. and Smith, R. R., Recurrent selection for plant regeneration from red clover tissue culture, *Crop Sci.*, 33, 585, 1993.
43. Meijer, E. G. M., High frequency plant regeneration from hypocotyl- and leaf-derived tissue cultures of the tropical pasture legume *Stylosanthes humilis*, *Physiologia Plantarum*, 56, 381, 1982.
44. Wofford, D. S., Baltensperger, D. D., and Quesenberry, K. H., *In vitro* culture responses of alyceclover genotypes on four media systems, *Crop Sci.*, 32, 261, 1992.
45. Wofford, D. S., Quesenberry, K. H., and Baltensperger, D. D., Tissue culture regeneration of *Desmodium, Crop Sci.*, 32, 266, 1992.
46. Angeloni, P. N., Rey, H. Y., and Mroginsky, L. A., Regeneration of plants from callus tissue of the pasture legume *Centrosema brasilianum, Plant Cell Reports*, 11, 519, 1992.
47. Ross, A. H., Manners, J. M., and Birch, R. G., Embryonic callus production, plant regeneration, and transient gene expression following particle bombardment in the pasture grass, *Cenchrus ciliaris* (Gramineae), *Aust. J. Bot.*, 43, 193, 1995.

48. Ahn, B. J., Huang, F. H., and King, J. W., Plant regeneration through somatic embryogenesis in common bermudagrass tissue culture, *Crop Sci.*, 25, 1107, 1985.
49. Ahn, B. J., Huang, F. H., and King, J. W., Regeneration of bermudagrass cultivars and evidence of somatic embryogenesis, *Crop Sci.*, 27, 594, 1987.
50. Croughan, S. S., Quisenberry, S. S., and Cuomo, G. J., Bermudagrass somaclone resistance to fall armyworm (Lipidoperta:Noctuidae), *J. Agric. Entomol.*, 14, 73, 1996.
51. Redway, A. R. and Vasil, I. K., Selection of S-(2-Aminoethyl)-L-cysteine tolerance in embryogenic calli and regenerated plants of *Pennisetum purpureum* Schum., *Plant Sci.*, 67, 203, 1990.
52. Marousky, F. J. and West, S. H., Somatic embryogensis from cultured mature caryopses of bahiagrass (*Paspalum notatum* Fluegge), *Plant Cell Tissue Organ Cult.*, 20, 124, 1990.
53. Akashi, R., Hashimoto, A., and Adachi, T., Plant regeneration from seed-derived embryogenic callus and cell suspension cultures of bahiagrass (*Paspalum notatum*), *Plant Sci.*, 90, 73, 1993.
54. Bovo, O. A. and Mroginski, L. A., Tissue culture in *Paspalum (Gramineae)*: Plant regeneration from cultured inflorescenes, *J. Plant Physiol.*, 124, 481, 1986.
55. Shatters, R. G., Wheeler, R. A., and West, S. H., Somatic embryogenesis and plant regeneration from callus cultures of 'Tifton 9' bahiagrass, *Crop Sci.*, 34, 1378, 1994.
56. Akashi, R. and Adachi, R., Somatic embryogenesis and plant regeneration from cultured immature inflorescences of apomictic dallisgrass (*Paspalum dilatatum* Poir.), *Plant Sci.*, 82, 213, 1992.
57. Akashi, R. and Adachi, T., Plant regeneration from suspension culture-derived protoplasts of apomictic dallisgrass (*Paspalum dilatatum* Poir.), *Plant Sci.*, 82, 213, 1992.
58. Venuto, B. C., Croughan, S. S., and Pitman, W. D., Field evaluation of dallisgrass somaclones, in *1998 Ann. Meeting Abst.,* Crop Sci. Soc. Amer., 1998, 67.
59. Chen, C. S., Wange, S. M., and Chen, Y. K., Morphological and RAPD variations of regenerants derived from cell suspension culture of pangolagrass, *Proc. XVIII Intl. Grassland Congress*, 1997, 4-15.
60. Ha, S. B., Wu, F. S., and Thorne, T., Transgenic turf-type tall fescue (*Festuca arundinacea* Schreb.) plants regenerated from protoplasts, *Plant Cell Rep.*, 11, 601, 1992.
61. Benfey, P. N. and Chua, N. H., The cauliflower mosaic virus 35S promoter: combinatorial regulation of transcription in plants, *Science*, 250, 959, 1990.
62. Hauptmann, R. M., Ozais-Atkins, P., Tabaeizadeh, Z., Rogers, S. G., Horsh, R. B., Vasil, I. K., and Fraley, R. T., Transient expression of electroporated DNA in monocotyledonous and dicotyledonous species, *Plant Cell Rep.*, 6, 265, 1987.
63. Callis, J., Fromm, M. E., and Walbot, V., Introns increase gene expression in cultured maize cells, *Genes Dev.*, 1, 1183, 1987.
64. Mascarenhas, D., Mettler, I. J., Pierce, D. A., and Lowe, H. W., Intron-mediated enhancement of heterologous gene expression in maize, *Plant Mol. Biol.*, 15, 913, 1990.
65. Hauptmann, R. M., Asraf, M., Vasil, V., Hannah, L. C., Vasil, I. K., and Ferl, R., Promoter strength comparisons of maize shrunken 1 and alcohol dehydrogenase 1 and 2 promoters in mono- and dicotyledonous species, *Plant Physiol.*, 88, 1063, 1988.
66. McElroy, D., Zhang, W., Cao, J., and Wu, R., Isolation of an efficient actin promoter for use in rice transformation, *Plant Cell*, 2, 163, 1990.
67. Christensen, A. H., Sharrock, R. A., and Quail, P. H., Maize polyubiquitin genes: structure, thermal perturbation of expression and transcript splicing, and promoter activity following transfer to protoplasts by electroporation, *Plant Molecular Biology*, 18, 675, 1992.
68. Christensen, A. H. and Quail, P. H., Ubiquitin promoter-based vectors for high-level expression of selectable and/or screenable marker genes in monocotyledonous plants, *Transgenic Research*, 5, 213, 1996.
69. McElroy, D., Xu, D., Zhang, W., Blowers, A. D., and Wu, R., Utilization of the rice actin 1 gene for foreign gene expression in transgenic cereals, Keystone symposia on crop improvement via biotechnology: an international perspective, *J. Cell. Biochem.*, 16F(S), 209, 1992.
70. Vlachova, M., Metz, B. A., Schell, J., and de Bruijn, F. J., The tropical legume *Sesbania rostrata*: tissue culture, plant regeneration and infection with *Agrobacterium tumefaciens* and *rhizogenes* strains, *Plant Sci.*, 50, 213, 1987.
71. Manners, J. M. and Way, H., Efficient transformation with regeneration of the tropical pasture legume *Stylosanthes humilis* using *Agrobacterium rhizogenes* and a Ti plasmid-binary vector system, *Plant Cell Rep.*, 8, 341, 1989.

72. Wan, C. H., Plant regeneration, cryopreservation, and genetic transformation of Napier grass (*Pennisetum purpureum* Schum.), Ph.D. Dissertation, University of Florida, Gainesville, Florida, 1994.
73. Grando, M. F., Shatters, R., and Franklin, C. I., Tissue culture and gene transformation in bahiagrass, *Congress On In Vitro Biology*, 1998, 1049.
74. Yuge, C., Akashi, R., and Kawamura, O., Improvement of forage quality by means of biotechnology: stable transformation of warm-season grasses by particle bombardment, in *Proc. XVIII Intl. Grassland Cong.*, 1997, 18, 4-49.
75. Broglie, K., Chet, I., Holliday, M., Cressman, R., Biddle, P., Knowlton, S., Mauvais, C. J., and Broglie, R., Transgenic plants with enhanced resistance to the fungal pathogen *Rhizoctonia solani, Science*, 254, 1194, 1991.
76. Johnson, R., Narvaez, J., An, G., and Ryan, C. A., Expression of proteinase inhibitors I and II in transgenic tobacco plants: effects on natural defense against *Manduca sexta* larvae, *Proc. Natl. Acad. Sci. U.S.A.*, 86, 9871, 1990.
77. Gatehouse, J. A., Hilder, V. A., and Gatehouse, A. M. R., Genetic engineering of plants for insect resistance, in *Plant Genetic Engineering*, Grierson, D., Ed., Chapman and Hall, New York, 1990, 105.
78. Legrand, M., Kauffmann, S., Geoffroy, P., and Fritig, B., Biological function of pathogenesis-related proteins: four tobacco pathogenesis-related proteins are chitinases, *Proc. Natl. Acad. Sci. U.S.A.*, 84, 6750, 1987.
79. Smith, F. W., Ealing, P. M., Hawksford, M. J., and Clarkston, D. T., Plant members of a family of sulfate transporters reveal functional subtypes, *Proc. Natl. Acad. Sci. U.S.A.*, 92, 9373, 1995.
80. McIntyre, C. L., Abrahams, S. L., Bettenay, H. M., Sandeman, R. A., Hays, C., Sharp, D., Elliot, A., Manners, J. M., and Watson, J. M., Improving pasture digestibility: low lignin forages, in *Proc. XVII Intl. Grassland Cong.*, 1993, 1081.
81. Manners, J. M., McIntyre, C. L., and Nourse, J. P., Cloning and sequence of a cDNA encoding phenylalanine ammonia-lyase from the tropical forage legume *Stylosanthes humilis, Plant Physiol.*, 108, 1301, 1995.
82. McIntyre, C. L., Rae, A. L., Curtis, M. D., and Manners, J. M., Sequence and expression of a caffeic acid O-methyltransferase cDNA homologue in the tropical forage legume *Stylosanthes humilis, Aust. J. Plant Physiol.*, 22, 471, 1995.
83. Curtis, M. D., Cameron, D. F., and Manners, J. M., Molecular evidence that diploid *Stylosanthes humilis* and diploid *Stylosanthes hamata* are progenitors of allotetraploid *Stylosanthes hamata* cv. Verano, *Genome*, 38, 344, 1995.
84. Curtis, M. D., Nourse, J. P., and Manners, J. M., Nucleotide sequence of a cationic peroxidase gene from the tropical forage legume *Stylosanthes humilis, Plant Physiol.*, 108, 1303, 1995.
85. Curtis, M. D., Rae, A. L., Rusu, R. J., Harrison, S. J., and Manners, J. M., A peroxidase gene promoter induced by phytopathogens and methyl jasmondate in transgenic plants, *Mol. Plant-Microbe Int.*, 10, 326, 1997.
86. Kazan, K., Goulter, K. C., Way, H. M., and Manners, J. M., Expression of a pathogenesis-related peroxidase of *Stylosanthes humilis* in transgenic tobacco and canola and its effect on disease development, *Plant Sci.*, 136, 207, 1998.
87. Soller, M. and Beckmann, J. S., Genetic polymorphism in varietal identification and genetic improvement, *Theor. Appl. Genet.*, 67, 25, 1983.
88. Helentjaris, T., King, G., Slocum, M., Siedenstrang, C., and Wegman, S., Restriction fragment length polymorphisms as probes for plant diversity and their development as tools for applied plant breeding, *Plant Mol. Biol.*, 5, 109, 1985.
89. Bonierbale, M. W., Plaisted, R. L., and Tanksley, S. D., RFLP maps based on a common set of clones reveal modes of chromosomal evolution in potato and tomato, *Genetics*, 120, 1095, 1988.
90. Edwards, D. M., Stuber, C. W., and Wendel, J. F., Molecular-marker-facilitated investigations of quantitative-trait loci in maize. I. Number, genomic distribution and types of gene action, *Genetics*, 116, 1987.
91. Martin, B., Nienhuis, J., King, G., and Schaefer, A., Restriction fragment length polymorphisms associated with water use efficiency in tomato, *Science*, 243, 1725, 1989.
92. Tanksley, S. D. and Hewitt, J., Use of molecular markers in breeding for soluble solids content in tomato — a re-examination, *Theor. Appl. Genet.*, 75, 811, 1988.

93. Nienhuis, J., Helentjaris, T., Slocum, M., Ruggero, B., and Schaefer, A., Restriction fragment length polymorphism analysis of loci associated with insect resistance in tomato, *Crop Sci.*, 27, 797, 1987.
94. Botstein, D., White, R. L., Skolnich, M., and Davis, W. R., Construction of a genetic linkage map of man using restriction fragment length polymorphisms, *Am. J. Hum. Genet.*, 32, 314, 1980.
95. Tanksley, S. D., Young, N. D., Paterson, A. H., and Bonierbale, M. W., RFLP mapping in plant breeding: New tools for an old science, *Bio/Tech.*, 7, 257, 1989.
96. Williams, J. G. K., Kubelik, A. R., Livak, K. L., Rafalski, J. A., and Tingey, S. V., DNA polymorphisms amplified by arbitrary primers are useful as genetic markers, *Nucleic Acids Res.*, 18, 6531, 1990.
97. Gill, P., Jeffries, A. J., and Werrett, D. J., Forensic applications of DNA fingerprints, *Nature*, 318, 577, 1985.
98. Akkaya, M. S., Shoemaker, R. C., Specht, J. E., Bhagwat, A. A., and Cregan, P. B., Integration of simple repeat DNA markers into a soybean linkage map, *Crop Sci.*, 35, 1439, 1995.
99. Vos, P., Hogers, R., Bleeker, M., Reijans, M., van de Lee, T., Hornes, M, Frijters, A., Pot, J., Peleman, J., Kuiper, M., and Zabeau, M., AFLP: a new technique for DNA fingerprinting, *Nucleic Acids Res.*, 23, 4407, 1995.
100. Thomas, C. M., Vos, P., Zabeau, M., Jones, D. A., Norcott, K. A., Chadwick, P., and Jones, D. G., Identification of amplified restriction fragment polymorphism (AFLP) markers tightly linked to tomato Cf-9 gene for resistance to *Cladosporium fulvum, Plant J.*, 8, 785, 1995.
101. Caetano-Anolles, G., Bassam, B. J., and Gresshoff, P. M., DNA amplification fingerprinting using very short arbitrary oligonucleotide primers, *Bio/Tech.*, 9, 553, 1992.
102. Talbert, L. E., Blake, N. K., Chee, W. P., Blake, T. K., and Magyar, G. M., Evaluation of "sequence-tagged-site" PCR products as molecular markers in wheat, *Theor. Appl. Genet.*, 87, 789, 1994.
103. Kesseli, R. V., Paran, I., and Mitchelmore, R. W., Efficient mapping of specifically targeted genomic regions and tagging these regions with reliable PCR-based genetic markers, in *Proc. Symp. Applications of RAPD Technology to Plant Breeding*, Joint Plant Breeding Symposia Series of the CSSA, ASHS and AGA, Minneapolis, Minnesota, 1992.
104. Smith, R. L., Use of Isoenzyme "fingerprints" to identify digitgrass varieties, *Soil and Crop Sci. Soc. Fla. Proc.*, 32, 6, 1972.
105. Smith, R. L. and Schank, S. C., Isoenzymes as biological markers in plant genetics and breeding, *Soil and Crop Sci. Soc. Fla. Proc.*, 29, 196, 1969.
106. Hayward, M. D. and Hacker, J. B., Genetic control of some esterase isozymes of *Digitaria* species, and their utility in the identification of hybrids, *Euphytica*, 29, 347, 1979.
107. Vermeulen, P. H., Beard, J. B., Hussey, M. A., and Green, R. L., Starch gel electrophoresis used for identification of turf-type Cynodon genotypes, *Crop Sci.*, 31, 223, 1991.
108. Caetano-Anolles, G., Callahan, L. M., and Gresshoff, P. M., The origin of bermudagrass (*Cynodon*) off-types inferred by DNA amplification fingerprinting, *Crop Sci.*, 37, 81, 1997.
109. Caetano-Anolles, G., Callahan, L. M., Williams, P. E., Weaver, K. R., and Gresshoff, P. M., DNA amplification fingerprinting analysis of bermudagrass (Cynodon): genetic relationships between species and interspecific crosses, *Theor. Appl. Genet.*, 91, 228, 1995.
110. Smith, R. L., Sexual reproduction in *Panicum maximum* Jacq., *Crop Sci.*, 12, 624, 1972.
111. Ortiz, J. P. A., Pessino, S. C., Leblanc, O., Hayward, M. D., and Quarin, C. L., Genetic fingerprinting for determining the mode of reproduction in *Paspalum notatum*, a subtropical apomictic forage grass, *Theor. Appl. Genet.*, 95, 850, 1997.
112. Lagudah, E. S. and Hanna, W. W., Species relationship in the *Pennisetum* gene pool: enzyme polymorphism, *Theor. Appl. Genet.*, 78, 801, 1989.
113. Tostain, S., Riandey, M. F., and Marchais, L., Enzyme diversity in pearl millet (*Pennisetum glaucum*) 1. West Africa, *Theor. Appl. Genet.*, 74, 188, 1987.
114. Tostain, S. and Marchais, L., Enzyme diversity in pearl millet (*Pennisetum glaucum*) 2. Africa and India, *Theor. Appl. Genet.*, 77, 623, 1989.
115. Clegg, M. T., Rawson, J. R. Y., and Thomas, K., Chloroplast DNA variation in pearl millet and related species, *Genetics*, 106, 449, 1984.
116. Chowdhury, M. K. U. and Smith, R. L., Mitochondrial DNA variation in pearl millet and related species, *Theor. Appl. Genet.*, 76, 25, 1988.
117. Shechter, Y. and de Wet, J. M. J., Comparative electrophoresis and isozyme analysis of seed proteins from cultivated sorghum, *Amer. J. Bot.*, 62, 254, 1975.

118. Morden, C. W., Doebley, J., and Schertz, K. F., Allozyme variation among the spontaneous species of *Sorghum* (Poaceae), *Theor. Appl. Genet.*, 80, 296, 1990.
119. Chow, K. H. and Crowder, L. V., Putative parents of a *Desmodium* selection examined morphologically and by isozyme patterns, *Bot. Gaz.*, 135, 180, 1974.
120. Imrie, B. C. and Blogg, D., Variability of isozyme gene frequency in the tropical pasture legume, 'Greenleaf' desmodium, *Trop. Agric.*, 60, 193, 1983.
121. Penteado, M. I., Garcia, P., and Perez de la Vega, M., Genetic variability and mating system of three species of the genus *Centrosema, J. Hered.*, 87, 124, 1996.
122. Liu, C. J., Musial, J. M., and Smith, F. W., Evidence for low genomic specificity of sequence-tagged-sites in Stylosanthes, *Theor. Appl. Genet.*, 93, 864, 1996.
123. Smith, R. L., Schank, S. C., Seib, J. C., Chowdhury, M. K. U., and Wheeler, R. A., Development and application of RFLP genetic markers in breeding a biomass species, in *Energy from Biomass and Wastes XIV*, Klass, D., Ed., Institute of Gas Technology, Chicago, IL, 1991, 433.
124. Lee, M., Godshalk, E. B., Lamkey, K. R., and Woodman, W. W., Association of restriction fragment length polymorphisms among maize inbreds with agronomic performance of their crosses, *Crop Sci.*, 29, 1067, 1989.
125. Smith, R. L., Chowdhury, M. K. U., and Schank, S. C., Use of restriction fragment length polymorphism (RFLP) markers in genetics and breeding of napiergrass, *Soil and Crop Sci. Soc. Fla. Proc.*, 48, 13, 1989.
126. Hanna, W. W., Long term storage of *Pennisetum glaucum* (L.) R. Br. pollen, *Theor. Appl. Genet.*, 79, 605, 1990.
127. Braithwaite, K. S., Irwin, J. A. G., and Manners, J. M., Restriction fragment length polymorphisms in *Colletotrichum gloeosporioides* infecting *Stylosanthes* spp. in Australia, *Mycol-Res.*, 94, 1129, 1990.
128. Chahal, S. S., Kumar, R., Sidhu, J. S., and Minocha, J. L., Peroxidase isozyme pattern in pearl millet lines resistant and susceptible to downy mildew, *Plant Breeding*, 101, 256, 1988.
129. Klein-Lankhorst, R., Rietveld, P., Machiels, M., Verkerk, R., Weide, R., Gebhardt, C., Koornneef, M., and Zabel, P., RFLP markers linked to the root knot nematode resistance gene Mi in tomato, *Theor. Appl. Genet.*, 81, 661, 1991.
130. Yu, Z. H., Mackill, D. J., Bonman, J. M., and Tanksley, S. D., Tagging genes for blast resistance in rice via linkage to RFLP markers, *Theor. Appl. Genet.*, 81, 471, 1991.
131. Young, N. D., Zamir, D., Ganal, M. W., and Tanksley, S. D., Use of isogenic lines and simultaneous probing to identify DNA markers tightly linked to the Tm-2a gene in tomato, *Genetics*, 120, 579, 1988.
132. Hulbert, S. H. and Mitchelmore, R. W., Linkage analysis of genes for resistance to downy mildew (*Bremia lactucae*) in lettuce (*Lactuca sativa*), *Theor. Appl. Genet.*, 70, 520, 1985.
133. Arondel, V., Lemieux, B., Hwang, I., Gibson, S., Goodman, H. M., and Somerville, C., Map-based cloning of a gene controlling omega-3 fatty acid in *Arabidopsis, Science*, 258, 1353, 1992.
134. Martin, G. B., Brommonschenkel, S. H., Chunwongse, J., Frary, A., Ganal, M. W., Spivey, R., Wu, T., Earle, E. D., and Tanksley, S. D., Map-based cloning of a protein kinase gene conferring disease resistance in tomato, *Science*, 262, 1432, 1993.
135. Chittenden, L. M., Schertz, K. F., Lin, Y. R., Wing, R. A., and Patterson, A. H., A detailed RFLP map of *Sorghum bicolor* X *S. propinquum*, suitable for high-density mapping, suggests ancestral duplication of sorghum chromosomes or chromosomal segments, *Theor. Appl. Genet.* 87, 925, 1994.
136. Liu, J. C., Whitcombe, J. R., Hash, C. T., Busco, C. S., Pitway, T. S., Nash, M., and Gale, M. D., Construction and application of RFLP-based genetic maps in pearl millet, in *Use of Molecular Markers in Sorghum and Pearl Millet for Developing Countries*, Whitcombe, J. R. and Duncan, R. R., Eds., Overseas Development Administration (ODA), United Kingdom, 1993.
137. Hulbert, S. T., Richter, T. E., Axtell, J. D., and Bennetzen, J. L., Genetic mapping and characterization of sorghum and related crops by means of maize probes, *Proc. Natl. Acad. Sci. U.S.A.*, 87, 4251, 1990.
138. Whitkus, R., Doebley, J., and Lee, M., Comparative genome mapping of sorghum and maize, *Genetics*, 132, 1119, 1992.
139. Smith, R. L., Schweder, M. E., Chowdhury, M. K. U., Sieb, J. C., and Schank, S. C., Development and application of RFLP and RAPD DNA markers in the genetic improvement of *Pennisetum* for biomass production, *Biomass and Bioenergy*, 5, 51, 1993.
140. Ozais-Atkins, P., Lubbers, E. L., Hanna, W. W., and McNay, J. W., Transmission of apomictic mode of reproduction in *Pennisetum*: co-inheritance of the trait and molecular markers, *Theor. Appl. Genet.*, 85, 632, 1993.

141. Dujardin, M. and Hanna, W. W., Developing apomictic pearl millet — characterization of a BC_3 plant, *J. Genet. Breed.*, 43, 145, 1989.
142. Lubbers, E. L., Arthur, L., Hanna, W. W., and Ozais-Atkins, P., Molecular markers shared by diverse apomictic *Pennisetum* species, *Theor. Appl. Genet.*, 89, 636, 1994.
143. Gustine, D. L., Sherwood, R. T., and Huff, D. R., Apospory-linked molecular markers in buffelgrass, *Crop Sci.*, 37, 947, 1997.
144. Mitchelmore, R. W., Paran, I., and Kesseli, R. V., Identification of markers linked to disease-resistance genes by bulked segregant analysis: A rapid method to detect markers in specific genomic regions by using segregating populations, *Proc. Natl. Acad. Sci. U.S.A.*, 88, 9828, 1991.
145. Pessino, S. C., Ortiz, J. P. A., Leblanc, O., de Valle, C. B., Evans, C., and Hayward, M. D., Identification of a maize linkage group related to apomixis in *Brachiaria, Theor. Appl. Genet.*, 94, 439, 1997.
146. Moore, G., Devos, K. M., Wang, Z., and Gale, M. D., Grasses, line up and form a circle, *Curr. Biol.*, 5, 737, 1995.

Section III

Use of Tropical Forages

9 Seed Production Technology of Tropical Forages

Sherlie H. West and W.D. Pitman

CONTENTS

I. INTRODUCTION

One of the most important components of tropical* pasture production is the establishment of the stand. Obtaining an adequate supply of high-quality seed is paramount to getting a forage program started. Too often, needed quantities of high-quality seed of the desired cultivar are not available at the appropriate time for planting. Unlike the situation in major grain crops, there are no major international companies with the marketing and delivery infrastructure to supply tropical forage and pasture seeds. Australia is and has been a source of some cultivars of some species. In fact, Australian researchers have generously provided seed of the species adapted to their country to researchers in other countries to be used in suitability trials. In some countries, pasture and forage production is largely based on species and cultivars supplied by Australian researchers and commercial seed sources. An example is Brazil where, before 1970, little or no *Brachiaria* was used in pastures. Twenty-five years later, 70% of the 100 million hectares of improved pasture consisted mostly of four species of *Brachiaria* originating in Australia. During the past few years, the source of this tropical pasture seed planted in Brazil has shifted from imported Australian seed to domestic supplies.[1] In fact, Brazil has recently become the leading producer of tropical forage seeds.[1]

A large industry for tropical pasture seed has not developed because of the limitations and constraints associated with all phases of seed production. A description of these limitations is included to increase the awareness of how they affect local seed production and the necessity of taking precautions to secure a reliable source of seed before attempting any tropical forage program. The major portion of the chapter consists of procedures and guidelines to help minimize the effect of these limitations and constraints during tropical forage seed production. The concluding section presents aspects of the current tropical forage seed industry.

For most of the world, "grass" includes all of that plant material growing close to or a few centimeters above the soil and giving a green mantle to the earth's surface. While in general, much of the text in this chapter will apply to both grasses and legumes, the preponderance of research examples will involve grass as a model plant.

II. PLANT LIMITATIONS

Low seed yield has been a characteristic of tropical forage cultivars.[2] Annual commercial seed production of many species rarely exceeds 200 kg ha^{-1}, and of this, only 25% may germinate.

* In this chapter, the tropical area is defined as the land north and south of the equator where climatic conditions dictate the use of tropical forages species. This area generally lies between 30°N and 30°S latitude.

Therefore, effective annual yields of less than 50 kg ha^{-1} of pure, germinating seed are common.[3] Recent advances in cultivar development, seed production, and harvest technology have resulted in commercial seed yields greater than 1000 kg ha^{-1} being reported.[1,4] These high yields are primarily from large-seeded legumes, although *Brachiaria* spp. and *Andropogon gayanus* have provided high yields.[1,4]

Usually, the beginning of the reproductive stage of forage plants is accompanied by a decrease in forage quality and quantity, i.e., reduced protein content and digestibility and decreased vegetative growth. Consequently, cultivars may be selected for their capacity to produce abundant, high-quality forage rather than large quantities of seed. To keep a plant in its vegetative stage longer, a plant breeder may select for late flowering habit and in doing so, may inadvertently select for poor seed yield as well.

Flower induction in many tropical grass genera is not related to day length.[5] Therefore, emergence of the inflorescence occurs over a long period of growth if not limited by temperature and moisture restraints. In other genera, the day length requirement for floral induction is reached early in the growing cycle, and flowering and seed production may be continuous throughout the summer. In either case, unless frequent harvests occur, much of the seed yield may be lost.

Seed quality, as measured by germination, is poor in most tropical forage species. One of the reasons for this poor quality is that seed maturity varies with position along the panicle or raceme. This characteristic causes a large percentage of the seed to be harvested when it is immature or overmature. The immature seeds would not have developed sufficently to germinate, and the overmature may have deteriorated during periods of high humidity and temperature while awaiting harvest.

Seed shattering reduces yield and also increases harvest cost if the shattered seeds have to be recovered from the soil surface. If a grass has a tendency to shatter, germination may be reduced before the mature seeds are removed. Most tropical grasses belong to the subfamily *Panicoideae* in which the entire spikelet shatters by abscission below the glumes, and ripe seed are not retained on the plant.[5] Generally, anthesis on a single seed head begins at the apex of the panicle or raceme and continues to the base of the inflorescence over a period of two weeks.[6] Seeds can be shattered from the apex of an inflorescence before the base has completed anthesis. Ripe-seed shattering in unfavorable weather or during harvesting operations is a major source of loss in yield of tropical seed crops.[7]

To add to the problem, tiller production occurs throughout the season; thus individual plants have inflorescences at various stages of emergence, anthesis, or ripening.

Dormancy is another undesirable characteristic of tropical grass seed. Hardseededness is a characteristic of many tropical legume seed, producing effects similar to dormancy of grass seed. These are survival features of plants grown in the wild. Most of the cultivars in use have not evolved from decades of breeding and selection programs, so this characteristic remains in their genetic background. Dormancy, which ranges from 10 to 90% in some species, may cause stand establishment to be slow and may necessitate excessive seeding rates, which reduce the supply of seed and increase user costs.

Plant characteristics of most tropical pasture crops make them considerably more expensive to harvest than grain crops. In some species, the seed is tenaciously held in a protective covering, a form of dormancy, until sufficient weathering allows it to be released. In many other species, shattering occurs, requiring the seed to be harvested from the soil. In both cases, expensive and unique equipment is required to collect the seed.

III. CLIMATIC LIMITATIONS

Maximum germination is attained in most types of seed when they reach physiological maturity, a stage of development when the moisture level is still too high for harvesting. Physiological deterioration begins after this stage and is accelerated by high temperatures or high humidity. The combination of these two factors, a condition often found in tropical climates, results in a synergistic effect on physiological deterioration that not only reduces germination, but increases the susceptibility of the seed to invasion by seed pathogens, especially fungi.

The wide variation in climate found in the tropics may be a major factor contributing to the lack of development of infrastructure for production and marketing of tropical forage seed. The types of climate have been described and classified as: (1) Tropical rainy, (2) Tropical humid summer, (3) Wet-and-dry tropical, (4) Tropical dry, (5) Tropical semidesert, (6) Dry winter with long, humid summer, and (7) Permanently humid grassland. The types of climates in the tropics represent a wide range of environmental factors that influence plant adaptation. Consequently, different plant species are required to maximize pasture production. In contrast, major grain crops have a large seed infrastructure, and utilize one genera and species wherever the crops are grown. The demand for seed production for specific tropical forage crops may be regional and, therefore, small. Organizations attempting to provide seed for all various climatic types would require extensive inventories. Furthermore, the unique characteristics of the wide range of plants required for the specific climatic types add costs to seed production. The diverse and expensive equipment that is needed for harvesting and cleaning may be detrimental to the growth of the industry.

IV. MARKET LIMITATIONS

Fluctuations in demand for pasture seed as a result of the inconsistent market price of livestock, seasonal climatic extremes, and overall economic instability discourage the establishment of large inventories of pasture seed. Furthermore, in most cases, the seed requested is for a genus, species, or cultivar that is perennial, making repeat sales irregular or unlikely.

In production of grazing livestock, the inefficient conversion of feed to food dictates that the cost of feed be kept to a minimum in order to make the enterprise profitable. Pasture and forages are low-cost feed only if their cost of production is managed properly. The price that a livestock producer can pay for seed does not encourage private institutions to develop breeding programs. In the absence of proprietary cultivars, there are insufficient profit incentives to warrant financial outlays by large companies to produce or market tropical pasture seed.

While the limitations to seed production of tropical forages are many, they are not insurmountable. There are millions of hectares that could be used for improved forage production, and new cultivars are becoming available that can provide the seed required for planting.

Cost is one of the factors to be considered in the decision to enter seed production as a business. Information must be collected on the potential market for the species and cultivars that are adapted to the area available. Since the equipment for harvesting, drying, cleaning, and processing represent a large part of the initial cost, risk can be minimized by selecting crops that maximize the use of the equipment. Crops with sequential maturity dates and, where possible, crops that mature in opposing seasons should be considered.

V. NEED FOR LOCAL SEED

One of the best and most effective ways to reduce the cost of establishing improved forage crops is to produce the seed for them locally. Growing tropical forage seed where it will be utilized provides important advantages. Local inventories of seed are usually more dependable than shipments from abroad. Desired cultivars will be available at the proper planting time. Labor and the cost of land preparation will not be lost because foreign seed supplies did not arrive or were received too late for the recommended planting date. In the case of foreign suppliers, if the quantity of seed is insufficient to meet the demand, as is usually the case, small orders will not be filled. The cost of local seed will be less because packaging and transportation costs will be minimal.

A seed lot in cross-pollinated crops is a mixture of all genotypes in the population. When such seed is produced in an area with day lengths dissimilar to those of the planting area, genetic shifts can occur. Some seed in the lot may have come from plants that produced few seed due to late flowering in the latitude of utilization, but these plants may produce a larger proportion of the seed when subjected to foreign day length. This seed, when sown in the area of utilization, may produce plants that are not able to adapt to soil and pest conditions for which the cultivar was originally selected. Similarly, self-pollinated forage plants can experience genetic shifts by outcrossing. Cultivars that have been selected or developed to perform well under local pest, soil, and nutritional conditions may lose that adaptation when their seed is produced over a long period in dissimilar areas.

Some cultivars, however, do not produce seed in many areas of the tropics because of day length requirements or other factors. Seed for these cultivars can be produced in other latitudes that foster flowering and viable seed set. Some cultivars may be the result of wide genetic crosses, even interspecific, that are incapable of generating seed. In these and similar cases, vegetative seed pieces must be procured.

Often the cost or unavailability of new cultivars prohibits their immediate establishment in large areas. However, they can be made available by first utilizing a small portion of the total target area for seed production or vegetative material. Plant the area with the highest quality material of the most adaptive cultivar obtainable and increase the seed each year until the demand for seed is met.

VI. SEED PRODUCTION

A. Site Selection

After the decision has been made to produce seed, the most important decision affecting seed production is selecting the proper site. A good site can greatly reduce the costs of production, minimize the risk of failure, and contribute to high yields and high-quality seed. Considerations for site selection include the following.

1. Day Length

Floral induction may precede flowering by days, weeks, or an even longer interval depending on plant species and growing conditions after induction. It is a response to several external stimuli. Many species are induced to flower in response to a suitable day length or photoperiod. As discovered by W. W. Garner and H. A. Allard[8] in the early 1920s, there are three classes of plants that vary in their response to day length. The flowering of long-day plants is induced by long days, that of short-day plants by short days, and day-neutral plants flower in both short- and long-day exposure. There have been some refinements to these three classes related to whether flowering is induced by long days followed by short days or short days followed by long days. Day-length response can be modified by temperature.[9] Latitude of origin can dramatically affect day-length response of accessions of some tropical legume species.[10] Once induced to flower, continuation of such an inductive day length can be required for continued flowering of some legumes.[10] The response to day length in photoperiod-sensitive plants is dramatic and has stimulated research of associated changes in induced plants such as stem elongation, seed germination, and petiole development.[11] Another refinement in the original work is the discovery that the flowering response is determined not by day length, but rather by the length of the dark period. However, in nature, the relationship of day length to night length is direct. Furthermore, most of the research on sensitivity of forage and pasture plants to the length of light or dark is reported in day length not the length of the dark period.

The day-length requirement has been reported for many tropical forage species and even cultivars, but the interpretation of results is difficult. In general, plants in the tropics and subtropics are short-day or day-neutral. There are some exceptions, such as *Paspalum notatum* Flügge (Bahia grass), which is a long-day plant, with its specific flowering response dependent upon the genome of the seed lot. Increases in leaf length and dry weight are also functions of daylength in this grass. Threshold day-length response of this species in a given cultivar population varies from 12.75 to 13.75 hours. Day length considerations of the crop to be grown are important because planting cultivars out of the area for which the cultivar was developed may restrict seed production. Before a commitment is made to produce a large quantity of seed of any species or cultivar at a new location, small, exploratory plantings should be made.

2. Duration of Growing Season

Seed development will be limited unless there is sunlight of high intensity during the entire day over a sufficient number of days for the subject species to reach maturity. Except under certain stress conditions that cause some plants to produce seed prematurely, the reproductive stage of a plant coincides with its peak size and vigor. Obviously, adequate intensity and duration of sunlight are required to attain this maximum energy condition. In some high elevations, and in latitudes distant from the equator with climatic conditions that dictate the use of tropical plants, night temperatures drop to or below 15°C. Such low temperatures markedly reduce the growth of tropical plants[12] and lower seed production if the plants have not matured before they are subjected to them.

3. Rainfall

The requirement for soil moisture varies among tropical species and is somewhat related to the extensiveness of the plant root system. Pollination, seed development, and normal seed yield require adequate moisture levels. In legumes, symbiotic nitrogen fixation is affected by low moisture conditions in the soil. The site selected for seed production should have a supply of water available for irrigation at critical periods unless rainfall is adequate. As a general observation, in areas of less than 1000 mm of annual rainfall, irrigation will be necessary unless the distribution coincides with crop production needs.

4. Climate for Maturation

After physiological maturity, i.e., when seed has reached maximum dry weight, rainfall is not as essential, and warm, sunny days without dew and with low humidity are desirable. A site should not be selected if rainfall, excessive moisture, or cloudy days predominate during the maturation stage of the species to be grown. Damp, cloudy conditions during maturation or later phases of crop development and during field drying often result in seed of low viability due mostly to fungal diseases.

5. Soil Characteristics

Because forage crops vary so much in their response to soil type and requirements for soil fertility, the selection of a suitable site or crop is an important consideration.

a. Fertility

Many soils in the tropics are not fertile and have high or low pH values. Whenever possible, a site should be fertile and have pH values near neutral, so that amendment costs will not readily increase the cost of production. Exceptions occur when a particular plant species is best adapted to low or high soil pH. Infertile soils may require amendments that are not readily available. It is wise to remember that diligence exercised in selecting the proper site will have a great impact on the cost and quality of the seed produced.

b. Texture

The texture of the soil, i.e., the proportions of sand, silt, and clay it contains, is an important consideration. Heavy clay soils may be difficult to prepare for seeding or to cultivate. Such soils tend to be "cold" and wet, and planting often must be delayed until the soil is warm, dry, and workable. Soils composed mostly of sand cultivate easily and drain well, but do not retain moisture or nutrients well enough to meet the needs of the crop. A sandy loam soil, however, has the proper water- and nutrient-holding capacity and permits adequate root development.

c. Organic matter

Wherever possible, soils containing some organic matter should be selected for seed production. Such soils contain nutrients, especially N, and have better structure for aeration and water retention due to aggregation of soil particles.

d. Drainage

Poor drainage can be a limiting factor. Below the soil surface, layers of clay, compacted organic matter, or rock can impede the downward movement of rainwater, cause the soil to become waterlogged, and, in some cases, restrict root development and plant growth. For example, a legume with a deep taproot should not be considered for a wet site with an impervious layer of soil. If economically feasible, soils that are wet because of water-restricting layers can be used for seed production after being mechanically broken up by deep plowing. Alternatively, elevated beds can be mechanically prepared for successful seed production even on wet soils. Beds can be constructed with a "bedder" plow consisting of two counter-angle discs. Bedding provides the proper soil moisture, loose soil, and increases the soil temperature.

6. Isolation Considerations of the Site for Genetic Purity and Seed Quality

Transportation of pollen by insects and wind makes it imperative to isolate fields when cross-pollinated crops are grown for seed. Reports[13] claim pollen can be carried as far as 300 km by the wind. In certified commercial crops, isolation requirements vary from country to country and species to species with a restrictive range of 50 to 1600 m.[14] The genetic purity and quality of the seed to be produced are the most important criteria in the decision of how far to extend the isolation. The seed to be used directly by the farmer for forage establishment should be of good quality. Seed sold commercially should meet and/or exceed certain quality standards.

Another criterion that should be considered is the size of the seed field. A larger field needs less isolation than a smaller field because it generates a more massive pollen bank from which cross-pollination can occur, thus decreasing the probability of pollination from an outside source. The level of pollen produced by the crop, itself, needs to be considered in determining isolation strategies. The best chance any cross-pollinated plant has for pollination is the nearest plant containing compatible pollen. If the seed crop is a heavy producer of pollen and the duration of stigma receptivity is short, the chances of contamination are reduced. Because of the requirement for synchronization of pollen production and stigma receptivity, contamination will most often come from the same species, but of undesirable origin. In crops like *Paspalum notatum* that are induced to flower throughout the season, the chance for contamination continues for several weeks. In grass crops pollinated by wind-borne pollen, the direction, force, and frequency of the prevailing winds should be considered as factors influencing isolation distances.

The overriding factor to consider in determining a safe isolation distance is knowledge about the reproductive characteristics of the seed crop. Some tropical grasses are apomicts. Others like *Digitaria eriantha* Steud. cv. Pangola produce no viable seed and must be propagated vegetatively. In these crops, unlike outcrossing crops, isolation is not as critical.

In addition to genetic purity, the quality of a seed lot takes into consideration the percent of weed seed and inert material, incidence of diseases, viability, physical damage, and size. Precaution should be practiced and guidelines followed to insure high-quality seed. If there is a local seed certification program, its guidelines should be followed even if the seed is not to be certified. However, there are advantages to growing seed that can be certified. The producer can demand a higher price for higher quality seed. The seedsman, who, if he is not the producer, buys seed from the producer and sells it to the user, will feel secure because he can offer a high-quality product and recommend it with confidence to his customers.

The farmer or rancher will receive the ultimate advantage from buying certified seed. The crop will have the advantages promised by the cultivar developer, such as high yield and resistance to diseases and insects, lodging, and drought, and will express other desirable characteristics as well.

Even though isolation can control genetic purity and protect the plants from contaminating pollen, unwanted genetic material may still be present in residue seed from previous crops. A site should not be selected if another cultivar of the same crop was grown there previously or if a previous crop produced seed that would be difficult to separate from that of the current crop in the cleaning procedure. The prevalence of weeds, especially those that are classified as noxious or whose seed is difficult to separate from the crop seed, should be evaluated at site selection. The inert material component of seed can be reduced by careful site selection. Fields containing small pebbles or small rocks should be avoided as well as fields with excessive stubble from previous crops that may harbor pests and diseases. It also may be necessary to know what types of chemicals, especially herbicides, have been used at the selected site.

B. Planting

The principles of husbandry for a forage seed crop are the same as those employed for forage crops in general. The increased unit cost per hectare and potentially greater economic return for high-quality seed warrant additional care and consideration in its production.

1. Date of Planting

Germination of tropical grasses and legumes[3] is typically low, mostly due to dormancy restrictions. Cultivar 'Pensacola' Bahia grass has an optimum temperature for germination of 32 to 34°C[15] and should be planted during the warmest part of the year. Like plantings for forage production, seed production plantings are made when the soil reaches the correct temperature for maximum germination.

Rainfall history at the site needs to be considered in selecting a planting date. The ideal pattern will provide sufficient rain immediately after planting to support germination and seedling growth, followed by a regime of rain that allows maximum growth of the crop until it has reached the late seed-filling stage. Rain should then subside and finally stop, so that the mature seed can desiccate and harvest can occur. If this kind of rainfall pattern can be predicted from climatological records, and if it occurs when the soil temperature is desirable for maximum germination and sunlight is optimum for abundant crop growth, progress will have been made toward finding the correct planting date. While these are ideal guidelines for good yields, high quality seed can be grown under conditions that are less favorable.

2. Seedbed Preparation

Careful attention to seedbed preparation will help to insure a satisfactory stand, which is essential for high seed yields. Debris from previous crops and rocks and other impediments to the planting operation should be removed prior to seedbed preparation or later if the plowing and disking cause these objects to surface. Usually after clearing, the land is deep plowed and disked to obtain an even seedbed that will facilitate husbandry and harvest operations. After, or concomitantly with planting, a roller or cultipacker is used to compact the seedbed to help cover the seed, reduce soil drying, provide a smoother surface, and improve germination and seedling performance.

3. Seeding Rate

The number of plants per unit of land has a direct effect on the yield. Each crop has a density that is optimum for seed yield, and densities above or below this optimum will reduce the yield. For

TABLE 9.1
Germination of Pensacola Bahia Grass (*Paspalum notatum* Flügge) Seed after Various Storage Periods

Date (1986)	Age (d)	Germination (%)
January	154	11.5 e[a]
February	182	12.0 e
March	213	14.8 de
April	243	20.3 cde
May	274	23.5 cd
June	304	25.3 c
July	345	46.8 b
August	371	54.0 a

[a] Means followed by the same letter are not significantly different at the 0.05 level as determined by Duncan's new multiple range test.

Source: West, S. H. and Marousky, F. J., Mechanism of dormancy in Pensacola bahia grass, *Crop Sci.*, 29, 787, 1989.

grasses, general recommendation for stand density is to provide a space of 7.5 cm between plants. If every seed planted produced an established plant, seeding rates could be calculated from the number of seeds in a gram of the crop seed and the desired plant spacing. Performing such a calculation on Pensacola Bahia grass indicates that approximately 2.5 kg ha^{-1} of seed would produce a satisfactory stand. However, the most commonly recommended seeding rate is 15 to 20 kg ha^{-1}. In circumstances where early forage utilization or rapid ground cover is required, even higher seeding rates are used. Similar management decisions are made for other forage species. Wilson and Rumble[16] recommended 190 plants m^{-2}, but Humphreys and Riveros[17] state that seed growers do not use recommended low rates because nonuniform stands may result.

Two of the reasons for recommending high seeding rates are seed dormancy and poor seedling vigor. Dormancy varies with each seed lot, and the range can be from 10 to 90%. Fortunately, dormancy decreases and germination increases with aging of the seed of many grasses as the data in Table 9.1 indicates for Pensacola Bahia grass. There are many dormancy mechanisms in tropical forage crops. The mechanism for Bahia grass is located in the lemma. The lemma is the round side of the seed visible after the glumes are removed. Germination occurs only through a flap that develops in the lemma. Breaking the fibers that hold the flap closed requires aging.[18] Seed dormancy and slow seedling growth result in slow stand establishment so that two growing seasons are typically required for a full stand of Bahia grass to develop.

4. Vegetative Material

Most of the information on planting forages pertains to crops that are established from seed. However, the guidelines for site selection and preparation apply to crop establishment of vegetative material as well. As an example, the following is a recommended procedure for planting Bermuda grass (*Cynodon dactylon* (L.) Pers.) from vegetative cuttings.[19]

Planted properly, the vegetatively propagated hybrid Bermuda grass can become well established in two to three months and provide grazing or hay in the first season. For success:

1. Choose a reasonably well-drained soil and destroy common Bermuda grass, any other types of native grass, and weeds growing at the site.

2. Plant only in moist soil that has received 500 kg ha^{-1} of a complete fertilizer and top-dress it with 100 kg ha^{-1} of N as soon as runners start to develop.
3. On the same day that you plant, disk the soil to destroy germinating grass seed of other species.
4. Plant fresh, well fertilized 15 to 30 cm sprigs with part of the stem buried, and tips extending above ground.
5. Pack the soil well with a heavy roller or by driving the tractor over the planting to establish the necessary capillarity in the soil to keep it moist around the sprigs.
6. Spray immediately with 2 kg ha^{-1} of 2,4-D to control both grass and broadleaf weeds. A second application of the chemical usually will be required within 30 days.
7. Complete steps 3 through 6 on the same day to retain the soil moisture once the soil is disturbed and to control the weeds.

5. Planting Pattern

Broadcasting the seed is faster and easier than row planting. Broadcasted swards are easier to harvest, provide more competition for the weeds, eliminate cultivation costs, and are more compatible with grazing. On the other hand, there are advantages to row planting if the seed is scarce and expensive. According to Humphreys and Riveros,[17] the benefits of row culture are: (1) seeding rate is reduced, enabling a larger area to be sown using a limited quantity of breeder's or basic seed; (2) off-types may be more reliably identified and rogued, and inspection of the crop is facilitated; (3) weed control by inter-row cultivation, spraying with a herbicide, or flaming is facilitated; (4) a more even supply of moisture and nutrients for better plant development is ensured; and (5) at low plant densities, an appropriate row spacing may provide a better light environment for the flowering shoot. Row planting of some viney climbing legumes can be combined with intercropping of support crops, such as corn and cassava or even *Leucaena*, or use of trellis structures to greatly increase seed yields.[20,21]

6. Depth of Planting

Size of the seed to be planted, moisture content of the soil, soil texture, and the anticipated soil moisture regime are factors influencing depth of planting. Small-seeded crops should be shallow planted so that the seed reserves will be adequate to push the new plant to the soil surface. A general recommendation that would only apply under an ideal soil moisture condition, i.e., field capacity, is to plant at a depth three times the smallest diameter of the seed. One cm of depth is another general planting recommendation. If the soil moisture level is low or if the seedbed soil is loose or fluffy, the seed should be planted deeper. Also, even when the soil is moist, if rain is not expected, the seed should be planted deeper. Compacting the soil around the seed with a roller after planting helps to conserve the soil moisture and reduces the need for deep planting.

C. Irrigation

When economically feasible, the best environment for seed production occurs in a dry season when water can be precisely added by irrigation as crop and seed maturation and harvest require. The income due to an increase in yield may be more than the cost of irrigation, and the risk of crop failure certainly is decreased. For example, the yield of seed may be tripled or quadrupled by irrigation. Advantages of irrigation in a dry environment include: (1) having adequate moisture to insure maximum plant growth, which will result in more inflorescences; (2) stimulating flowering by generating stress at the proper time to terminate vegetative growth and stimulate reproduction, and reducing pathogen levels due to excess rainfall; (3) ensuring adequate moisture for crop maturation; and (4) providing the proper dry environment for seed desiccation and harvest. Additional advantages include being able to plant the crop when other factors such as day length and temperature are favorable and to apply fertilizer and pesticides through an irrigation system. As a word of caution, unwise use of irrigation can reduce seed yield. For example, using spray irrigation during flowering and pollen shed may reduce seed set by preventing fertilization.

D. Culture

After planting, both broadleaf and grass weeds will provide severe competition to the crop. If the seed crop is grass, an application of 1/2 kg ha^{-1} active ingredient of 2,4-D will control broadleaf weeds. The first application should be made when the grass crop is 7 to 9 cm tall. Subsequent applications may be necessary, especially if the crop is planted in rows. Cultivation to control weeds should also be practiced if the crop is row planted. Mowing after the crop has reached 7 to 9 cm may give the seed crop an advantage over the weeds and may prevent weedy plants from producing seed. Limited livestock grazing may be possible after the crop is established for the purpose of controlling weeds and for providing feed for the livestock. Careful scheduling is necessary to insure sufficient time for maximum inflorescence production.

Recent advances in herbicide development provide potential new opportunities for weed control in tropical forage seed fields.[4] As determined from an international survey of tropical forage seed producers, weed control is a major problem or limitation remaining inadequately addressed.[1] Unfortunately, only minimal research efforts are now in progress to assess the potentially available solutions.[1]

E. Fertilization

A correct fertilizer program is the key to maximizing seed yield. Seed producers should take advantage of available agencies performing soil analyses and correct any deficiencies that are found. They should also request fertilizer recommendations for the specific forage crop to be grown.

A proper balance of N, P, K, Ca, and S is required for grasses and legumes. When the appropriate *Rhizobium* inoculant is applied to legume seed at planting, a low level of N is required to initiate plant development in infertile soils. In addition to the above nutrients, legume plants have special requirements for the micronutrients B, Cu, Zn, Mn, Mo, and Fe. If the available soil test does not analyze for micronutrients, make an application of them when the crop is planted as insurance against deficiencies.

1. Nitrogen Requirement

Nitrogen is the main nutritional determinant of grass seed yield, and the use of fertilizer N is now an important practice to maximize yield. Loch[22] obtained an increase in *Chloris gayana* seed yield to N fertilizer under all of the conditions in his study. Seed yields of *Cenchrus ciliaris* were greatest with high rates of both N and P.[23] Though conflicting responses to applications of N have been reported, maximum seed yields have been obtained with some consistency from established stands of many tropical grasses by adding approximately 100 kg ha^{-1} of N. Lodging, however, can be a detrimental consequence of higher levels of N, and excessive N fertilizer can maintain seed crop plants in a vegetative state, delaying seed production and reducing seed yield. Plant growth regulators have shown potential to reduce lodging and enhance seed production of lodging-prone tall grasses grown with high rates of fertilizer.[24] Cultivar, species, rainfall amount and distribution, soil fertility, and the ratio of N to P and K are some of the variables that influence seed crop response to N fertilizer.

2. Phosphorus and Potassium Requirements

As in the case of N, there are conflicting reports that show seed increases or little or no increase to added P and K. The seed producer should take into consideration the environmental conditions,

TABLE 9.2
Seed Yield Response of Pensacola Bahia Grass (*Paspalum notatum* Flügge) to Seven N-P-K Treatments at Five Harvest Dates

N-P-K Treatment kg ha^{-1}	Harvest Date (kg ha^{-1})					
	July 12	July 26	Aug 9	Aug 23	Sept 6	Total
0-100-100	178	54	19	16	9	276
100-100-100	269	239	83	30	4	625
200-100-100	523	291	97	48	5	964
400-100-100	287	182	95	46	12	622
100-0-100	407	108	55	15	2	587
100-100-0	311	95	56	25	4	491
100-0-0	195	109	39	35	4	382

Source: West, S. H. and Jank, L., Bahiagrass seed production, in *Proc. International Conference on Livestock in the Tropics,* University of Florida, Gainesville, FL, 1991, 1.

soil fertility, cultivar characteristics, and a range of management factors that can influence response to any fertilizer regimen. An experiment with Pensacola and 'Argentine' Bahia grass that included combinations of N, P, and K and five harvest dates in Pensacola and six in Argentine demonstrated responses to some of the factors that influence seed production.[25]

The fertilizer treatments were applied 83 days before the first harvest. The best combination (Table 9.2) was 200 kg ha^{-1} of N and 100 kg ha^{-1} of P and K. Total seed yield for this treatment was substantially higher than that of the 400 kg ha^{-1} N treatment combined with the same quantities of P and K. Peak seed yield in Pensacola occurred at the first harvest.

The highest Pensacola seed yields were obtained in the earliest harvests with the 200 kg ha^{-1} treatment, while 400 kg ha^{-1} depressed seed yield in the first two harvests and the last. In contrast, the seed yield of Argentine Bahia grass increased with both 200 and 400 kg ha^{-1} levels of N, and peak seed yields were obtained at the fourth harvest.

Considering all treatments, it is clear that all three fertilizer elements are important in Bahia grass seed production and a combination of the three is better than when N is added alone. N was the most important element for increasing seed yield in both cultivars. In the Pensacola Bahia grass experiment, seed yield was increased with the application of 100 kg ha^{-1} or more (Table 9.2). The 200 kg ha^{-1} of N with 100 kg ha^{-1} of P and K gave the highest seed yield in Pensacola. Argentine Bahia grass responded to the two highest N levels (200 and 400 kg ha^{-1}); but, in Pensacola, the 400 kg ha^{-1} level of N increased lodging and vegetative tillers and depressed seed yield.

The influence of fertilizer treatment and date of harvest on seed quality was evaluated by germination and purity tests of Pensacola Bahia grass. Although the values varied slightly with treatment, there was no consistent pattern of response (Table 9.3).

Seed purity of Pensacola, in general, decreased with additions of N. Seed purity decreased linearly with each successive harvest date. Since germination values also decreased generally with successive harvest dates, it is clear that the seed of later harvests had a lower quality. Germination and purity of the Argentine cultivar were not influenced by fertilizer treatment, but purity decreased with each successive harvest date.

F. Seed Harvesting

If the crop has been appropriately nurtured through all of the plant development stages, and environmental limitations and pest intrusions have not been experienced, the harvest of a profitable

TABLE 9.3
Percentage Germination (GERM) and Purity (PUR) of Pensacola Bahia Grass (*Paspalum notatum* Flügge) Seed for Seven N-P-K Treatments and Four Harvest Dates and Means for These Data

N-P-K Treatment kg ha^{-1}	Harvest Date								Means	
	July 12		July 26		August 9		August 23			
	GERM	PUR	GERM	PUR	GERM	PUR	GERM	PUR	GERM	PUR
0-100-100	93	93	92	78	85	72	91	64	90	77
100-100-100	91	93	91	76	81	68	79	56	85	73
200-100-100	91	88	87	76	81	63	78	51	84	69
400-100-100	92	87	90	72	87	56	70	45	85	65
100-0-100	84	88	92	64	88	67	81	56	86	68
100-100-0	80	87	84	68	83	61	81	45	82	65
100-0-0	92	93	88	75	88	62	85	46	88	69

Source: West, S. H. and Jank, L., Bahiagrass seed production, in *Proc. International Conference on Livestock in the Tropics,* University of Florida, Gainesville, FL, 1991, 1.

seed crop can be anticipated. The decisions of when and how to harvest involve, in large part, the characteristics of the crop, the availability of the necessary harvesting equipment, and labor.

1. Time of Harvest

The observation and measurements available to determine the appropriate harvest date are: calendar date, days after flowering, days after a certain level of seed head appears, loss of chlorophyll or color change of seed pod or seed coat, degree of shattering, moisture content, and weight of the seed. Calendar date is the least reliable criterion. Maturity of the crop depends upon a number of components including season, ambient temperature range, amount and distribution of rainfall during the growing season, nutrient availability, incidence of pests during plant development, and management practices employed throughout the seed production process such as days after last grazing, any one of which can vary from year to year.

Days after flowering, days past anthesis, and days after seed heads appear can be used as criteria to determine harvest date after experience has been gained in relating the stages of plant development to yield and seed quality.

In a grass such as *Panicum maximum* Jacq., the amount of seed shattering is an indicator of crop maturity and also economic loss. If the seeds are to be collected from the plant, the harvest should begin before shattering starts; however, if seed are to be harvested from the ground, complete shattering is needed if a once-over harvest is anticipated.

Loss of chlorophyll from the seed pod or seed coat is a visual observation that is used extensively to indicate sufficient maturity for harvesting grass seed. In some legume crops, this point has been reached when 95% of the seed pods have lost all of their chlorophyll.

Moisture content of the seed is possibly the most dependable measurement of maturity. After physiological maturity, which is the stage of the plant when the seed reaches maximum dry weight, the seed begins to lose moisture. In most cases, the seed has reached its peak germination potential at physiological maturity. In less humid locations, some crops, especially legumes, will desiccate until they reach their harvest moisture level or approximate storage moisture level. After physiological maturity, a careful monitoring of the moisture content of the seed, until a constant percentage occurs, is a good method of determining the appropriate time to harvest. Moisture content can be determined by placing a sample (100 grams) of seed from the crop to be harvested in an oven maintained at approximately 45°C. After 12 hours the seeds are weighed and the percentage moisture loss is calculated. Under no circumstance, should the seed producer depend entirely on natural seed desiccation to dry the seed to a storage moisture level. The concept that harvesting should take place as soon as possible after the seed has reached harvest maturity is important. Any delays will increase the opportunity for seed deterioration. In tropical climates with high humidity and high temperatures, which accelerate the deterioration process, the harvest date should be as early as possible.

2. Method of Harvest

Harvesting seed is the process of removing seed from the field for drying, cleaning, and delivery to storage. The method of harvest depends upon the size of the enterprise and the capital invested in equipment. It can range from a hand operation to an operation that is completely mechanical depending upon the volume of seed to be harvested and availability and cost of labor. Government incentives and available labor have resulted in recent commercial seed production of *Brachiaria* spp. from fields often less than 0.5 ha in Thailand, producing high yields per hectare by hand harvest.[1,26]

The all-crop combine is the most common multipurpose equipment used to harvest tropical forage seed. With it, inflorescences can be removed from the plant, the seed threshed from the inflorescences, separated from the straw, and collected in a temporary repository for drying prior to storage. The rate of travel over the ground, speed of the cylinder, and cylinder clearance are all variables to consider when using the combine and can be adjusted to deliver good-quality seed. Careful attention to the efficiency of the harvesting operation and to maintaining the quality of the seed is important.

Often a general purpose farm mower is used, which harvests a larger portion of the plant with a sickle blade. The material is then dried in a windrow or tied in bundles in the field, or in a dryer. After drying, the seed is threshed on a stationary thresher, cleaned, and further dried or placed in storage.

Mechanical harvesting with a beating or rubbing machine is used to remove only mature seed from grass crops. A second or third harvest may follow to harvest seed that was immature at previous harvests. Customized machines of this type have been designed and effectively used by seed producers in Australia for such tropical forages as dallis grass (*Paspalum dilatatum*),[27] carpetgrass (*Axonopus affinis*),[27] buffelgrass (*Cenchrus ciliaris*),[28] and even leucaena (*Leucaena leucocephala*).[29]

Recovery of fallen seeds by hand labor or machines is important in the harvest of *Stylosanthes* spp.,[30] *Chamaecrista rotundifolia*,[30] guinea grass (*Panicum maximum*),[31] green panic (*Panicum maximum* var. *tricholgume*),[31] Siratro (*Macroptilium atropurpureum*),[32] *Brachiaria* spp.,[4,33] *Andropogon gayanus*,[4] and *Calopogonium mucunoides*.[4] Recovery of fallen seed resulted in yields eight times those previously obtained from *Brachiaria* spp. in Brazil.[4] Development of suction harvesters has allowed mechanical recovery of high proportions of seed similar to levels previously possible only with hand labor.

Underground seed production by *Arachis* spp. has required further innovation in mechanized harvest. Two perennial species, *A. pintoi* and *A. kretschmeri*, produce enough seed to justify propagation by seed.[30] Deterioration of pegs makes it necessary to dig the seeds and separate them from the soil. Repeated modifications have led to development of machines to mechanically harvest this seed on a commercial scale.[30,34]

VII. HANDLING TROPICAL FORAGE SEED

Seed drying, cleaning, and storage are critical operations in any seed production business. In the humid and subhumid tropics, where seed with a high moisture content is harvested to prevent field deterioration, methods for drying and storage have to be planned, developed, and used.

A. Drying

Drying must take place immediately after harvest because seed with a high moisture content will deteriorate rapidly in a warm environment. Drying must start soon after the seed is harvested and continue nonstop until a safe moisture level for storage is reached. Seed having a high moisture content cannot be transported over long distances to a dryer because travel time would allow seed damage. If the seed contains more than 17 to 19% moisture, spontaneous heating can occur that will increase the temperature sufficiently to reduce the viability of the seed. Mold will infect seed with a moisture content above 12%, and storage insects can be a problem if moisture content is above 8%.

1. Method of Drying

Selection of a method of, and equipment for, drying involves many factors including cost and availability of equipment, energy supply and cost, crop or crops to be dried, selling price of the seed, size of the business, and climatic conditions at the location. There are many modifications of the following methods and types of drying: natural drying; sun drying; unheated, heated, and dehumidified air drying; drying in storage; drying with desiccants; vacuum drying; and freeze drying. The ideal situation is to have the seed dried naturally in the field to a level of moisture that makes immediate storage possible. This seldom occurs in the tropics. Many crops can be dried in the field by the sun either in windrows or unraked piles. If drying cannot be accomplished by any of these methods, passing air across the seed that is either unheated, or that has heat introduced into the air handling duct, either from an electrical or petroleum energy source is recommended.

2. Principles of Drying

The principles of seed drying are the same wherever seed is produced and, when understood, form the basis for planning drying facilities. Basically, drying is the evaporation of water from the seed coat. There are two phases or steps in the process: the movement of water from the interior of the seed to the seed coat and the evaporation of water from the seed coat. Generally, a lower temperature for drying is used if the seed moisture content is high. The safe maximum temperatures at which seed with varying moisture levels can be heated are shown in Table 9.4.[35] When the ambient temperature is as high as the safe maximum temperature, a dehumidifier may be used to dry the seed.

B. Cleaning and Processing

Seeds are cleaned to: (1) remove contaminants; (2) size and grade them for plantability; (3) upgrade their quality through removal of damaged and deteriorated seeds; (4) apply seed treatment material; and (5) bag them and make them available for distribution, sale, or storage. These tasks must be accomplished effectively and efficiently and with minimal damage to the seed. The equipment available for processing ranges from simple separation tables and boxes to a highly sophisticated

TABLE 9.4
Maximum Temperatures for Drying Seeds Containing Varying Moisture Percentages

Seed Moisture Range	Drying Temperature
Over 18%	29–32°C
10 to 18%	35–38°C
Under 10%	43°C

Source: Delouche, J. C., Matthes, R. K., Dougherty, G. M., and Boyd, A. H., Storage of seed in sub-tropical and tropical regions, *Seed Science and Technology,* 1, 663, 1973.

electronic sorting machine. All of the equipment separates desirable and undesirable material on the basis of differences in physical properties. Seeds that do not differ in certain physical characteristics cannot be separated. Seeds can be separated if they differ in size, length, shape, weight, surface texture, color, affinity for liquids, electrical conductivity, or specific gravity. There is no machine that can separate seed differing in all these characteristics. A cleaning machine is selected on the basis of specific differences.

1. Precleaning and Conditioning

Often, if not always, seeds that have been combined will contain straw, leaves, and residue from previous crops requiring the use of scalping equipment that uses a high volume air stream. Likewise, some crops will have to be passed through a dehuller or shelling equipment; some will have beards, awns, or hooks that can be removed by passing them through a hammer mill or similar equipment.

2. Cleaning

Almost every kind of seed must be cleaned in an air screen cleaner before any other separation can be attempted. Many kinds of seed can be cleaned completely and be ready for use by passing them through this type of cleaner. Consequently, the air screen cleaner is known as basic equipment in seed-cleaning operations. In this unit, screens of different sizes and shapes are used; and moving air separates the seed according to differences in surface area and density. A seed mixture directly from a combine or a precleaning unit flows by gravity from a hopper of the air screen cleaner to the feeder, which meters it into an air stream. Light, chaffy material is blown out and the seed is distributed uniformly over the top screen. A typical air screen operation employs four different sets of screens of varying size.

The next most widely used cleaning equipment is the gravity separator, which separates seed of the same size but of different densities, or seeds of the same density but of different sizes. However, it will not separate a mixture of sizes and densities.

There are specialized cleaning tools available to clean and improve the purity of seeds that are difficult to separate. These separators take advantage of a difference in seed length. Pockets or indentations are used to lift the short seeds from a mixture and reject the long seeds. The velvet roll separator classifies seeds according to a difference in seed coat texture. The spiral separator makes a division of seeds based on their shape or the degree of their ability to roll. The inclined draper separator uses a difference in shape and surface texture to separate seeds on an inclined surface. The horizontal disk separator takes advantage of a difference in shape and surface texture to determine whether the seeds slide or roll when subjected to centrifugal force. The electronic separator divides seeds on the basis of differences in their electrical properties. The magnetic separator takes advantage of the surface texture and stickiness of the seeds to make a separation. The color separator machine separates the seeds on the basis of difference in color or brightness. No matter what equipment is selected, it is important to remember that every mechanical device used in handling seeds is a potential source of injury and contamination. All conveyors must be thoroughly cleaned before they are used to handle a different crop or type of seed, thus the kind of machinery utilized should be easy to clean thoroughly or preferably be of the self-cleaning type.

Seeds are subject to mechanical damage and should be handled as carefully as possible. However, some kinds of seed, for example, seed with the embryonic axis external to the storage material, are more susceptible to injury than others. Certain types of conveyors cause greater damage to the seed, and it is very important to match the appropriate type of conveyor to the type of seed being cleaned. For instance, augers should never be used to convey soybeans or similar crops in which the morphological placement of the embryonic axis increases the vulnerability of the seed to injury.

3. Processing

During processing after the seeds are cleaned, they are size graded, density graded, and treated. The efficiency and effectiveness with which these processes are accomplished very often determine whether the seeds are marketable. The next step in processing is the application of seed coating materials. One of the purposes of coating is to change the shape or size of seed to improve planting operations and this procedure is referred to as pelleting. Another purpose of coatings is to apply pesticides, beneficial bacteria, or materials to change the pH of the soil in the vicinity of the planted seed. This is called treating the seed. Treating and packaging are processing operations, and both must be performed in accordance with recommended procedures.

4. Storage

Seed viability and vigor are at their peak when the seed is at full maturity on the plant in the field. However, the moisture content of the mature seed (at maximum dry weight) is too high to allow it to be harvested. Consequently, seeds are left in the field until they are sufficiently dry to facilitate harvesting without damage. Storage in the field in a humid, tropical location places the seed in an undesirable environment that is not conducive to maintaining its viability and vigor. Field storage represents the first of several phases of seed production in which quality can deteriorate. Therefore, harvest should occur just as soon as the seed are sufficiently dry to prevent damage by the process. Field storage is followed by harvesting, final drying, cleaning, and preparing the seed for storage; and, in all of these operations, further deterioration and mechanical damage can occur. Finally, since the time for planting the new crop is several months away, the seed must be properly stored. The seed should be kept in proper storage in every step of the seed delivery system including handling and storage by the rancher or farmer. The objective of storage is to maintain the existing level of viability and vigor of the seed by reducing or preventing further deterioration.

Often the relative humidity and temperature of the storage area are the most important factors in maintaining the viability and vigor of seeds in tropical environments.[36] While these two factors have a synergistic effect, relative humidity has the greatest influence. High relative humidity promotes physiological deteriorative processes as well as fungal and insect activity on the seeds.

The seed moisture content is in equilibrium with the relative humidity of the storage atmosphere. High relative humidity of the storage atmosphere increases the moisture level in the seed. With a moisture level above 20%, there is active seed respiration, which causes rapid loss of vigor and eventual loss of viability.[36] The infestation, growth, and reproduction of both fungi and insects on stored seed are regulated by the relative humidity of the storage atmosphere.[37] Studies of grain storage have shown that fungal infestations are a major cause of quality loss in stored grain and seed.[37,38] However, storage fungi cannot grow and reproduce on seed in equilibrium with a relative humidity of less than 65 to 70%. Fungal reproduction and insect activity decrease rapidly as the relative humidity of a storage area dips below 50%. Harrington's rule of thumb emphasizes the degree of influence moisture content has on longevity of seed: "The storage life of seed is doubled for each 1% decrease in moisture content of the seed."[37]

Temperature is the next most important factor when storing seed for long periods in tropical environments. The longevity of seed in storage is approximately doubled for each 5.5°C reduction in seed temperature.[37]

It can be generalized that the key to maintaining viability and vigor in seeds is to provide a dry and cool storage environment. For practical and economic considerations, required relative humidity and temperature levels must be determined for the length of storage desired, along with the manner in which these can be provided under the economic and energy constraints.

Many factors need to be considered in the development of storage facilities: the crop to be grown, type of seed to be stored (because of their high oil and protein content, legume seed deteriorates faster than grass seed), length of storage required (more stringent requirements are needed for longer storage), volume and price of seed to be sold (cost effectiveness is important), quality of seed required by customers, quality of seed going into storage, and energy requirements and availability. The importance of these and possibly other factors varies with each specific operation. Assessments of storage time, effect of fungicides, and seed physiology of guinea grass have suggested considerable interaction among these factors.[39] Because of the paucity of published guidelines for this type of business and the uniqueness of each seed production enterprise, a serious and careful investigation of all the other factors affecting storage should be conducted before seed production begins.

Only limited data are available from studies of proper storage methods for forage crop seed; however, extrapolations can be made from studies of storage procedures for other crops.[36] In general, temperature and humidity requirements for storage depend on length of the storage period. For short-term storage of one to nine months, good-quality forage crop seed can be safely stored at combinations of 30°C and 50% relative humidity or 20°C and 60% relative humidity. For intermediate-term storage of up to 18 months, combinations of 30°C and 40% relative humidity, 20°C and 50% relative humidity, or 10°C and 60% relative humidity are effective. Long-term storage, as long as three to five years, can be achieved with 10°C and 45% relative humidity.

a. *Short-term storage*

This type of storage is sometimes used after harvest and during the seed-cleaning and processing operations. Often it consists of covered, slatted bins containing the seeds through which low-humidity air is passed. Heated air can also be passed through the bins. The aeration equipment is not used at night or when the relative humidity is high.

b. *Intermediate-term storage*

After the seeds have been cleaned and put into open bags, they are stored in closed, permanent-type structures. Window-type air conditioners and/or dehumidifiers are used to reduce temperature and humidity.

c. *Long-term storage*

Facilities for this type of storage consist of tightly sealed buildings, usually with a central air-conditioning and dehumidifying system. Construction and energy operating costs are high for these facilities and require a careful analysis of the possible returns from the financial investment. It is important to remember that any system that lowers air temperature without removing moisture will sharply increase relative humidity in a closed system. Adequate planning for dehumidifying the storage area is essential to prevent the seeds from attaining unfavorably high moisture levels.

5. Packaging

Where relative humidity is high, seeds dried below the equilibrium moisture content will gain in moisture if the container is not airtight. Sealed containers should always be used for seeds in shipment to prevent high relative humidity situations from developing during shipping. The moisture content of seeds placed in airtight containers should be no higher than 8%. When a sealed container is subjected to an ambient temperature lower than that existing at the time the container was sealed, the relative humidity inside the container will increase. When possible, the containers should be sealed inside an air-conditioned room.

These guidelines are intended for the small or large producer of seeds and are purposefully general in nature but specific to tropical forage crops. Any bias toward a species or cultivar have been avoided due to the broad spectrum of climates and soils in the tropics and their varying effect on plant performance. Only a limited number of genera, species, and even cultivars of temperate grain and forage crops are required for large areas of agricultural production, however, the list of potential selections of grasses and legumes for the tropical world is extensive and expanding as introduction and breeding programs continue to add new material to the list. A producer of tropical pasture seed needs to be informed regarding the development of new legumes and grasses that would be of benefit to customers in that area.

VIII. THE TROPICAL FORAGE SEED INDUSTRY

A. Current Production

Commercial production of tropical forage seed is widely dispersed with local production often based on supplying local needs.[40] Both financial limitations, such as inadequate foreign exchange, and importation restrictions to control introduction of undesirable organisms are contributing factors.[4] This trend toward local production has resulted in major shifts in areas of seed production. Brazil has recently replaced Australia, the traditional source of most tropical forage seed, as the major producer of such seed.[1] In the early 1990s, as much as 80% of the seed produced in Brazil was *Brachiaria brizantha* cv. Marandu and *Brachiaria decumbens*.[4] Brazil now supplies an international market in tropical America.[31,33] Tropical forage seed is also produced commercially in Africa, Asia, other South American countries, Central America, the Caribbean, and southern North America.[1,40] Brazil, Australia, Kenya, Zimbabwe, and the U.S. are the major exporting countries.[40] Potential for expansion of production has been suggested for Asia and South America.[2,33,41] A notable deficiency in commercial seed availability is that of legume trees, even though superior genotypes have been identified. Seed quality and identity of genotype of such seed are often not known.[29]

Species produced, area of production, technology used, and yields vary greatly.[1] The demand derived from the extensive planting of *Brachiaria* spp.,[33] guinea grass,[31] and *Andropogon gayanus*[42] in Brazil led to large increases in seed production in that country during the past several years. Reduced demand, associated with prolonged local droughts, reduced profitability of beef cattle enterprises, loss of international markets, and environmental concerns, has contributed to decreased volume of tropical forage seed production in Australia.[1,43] Development of technology and other government incentives have contributed to increased numbers of small-scale seed production enterprises in several Asian countries.[26,41,44,45]

B. Seed for Small, Subsistence Farms

Both financial constraints and the small volume of seed involved effectively isolate many tropical subsistence farmers from any otherwise available commercial seed supply system.[46] Lack of availability of locally relevant information and insufficient buffer to allow acceptance of the risk involved are also particularly acute deterrents to use of nontraditional technology. Recent colonization of some tropical areas has removed the buffer of applicable, traditional technology in some instances.[47] Ferguson and Sauma[47] also pointed out that subsistence farmers lack sufficient market or political force to generate the needed technology or seed supply.

While often inadequate, especially for the generation or introduction of new technology, less structured or informal seed supply systems function in many subsistence farm communities.[47] Development of linkages to provide the required support services from research, development, and community sectors has been suggested as a key for more effective seed supply systems.[47] Although lack of a tradition of forage cultivation is a recognized constraint in some situations[48] and institutional inefficiencies can be overwhelming,[49] the recent progress in small-scale tropical forage seed enterprises in Southeast Asia provide useful models.[45] Such widely dispersed sources make seed quality of any appreciable secondary accumulation of seed a substantial concern. These seed quality aspects of small-scale farming are being addressed.[50] Thus, tropical forage seed supplies are being transformed from a limitation for small-scale, subsistence farming to a product of such farms in some situations.

C. Seed Industry Concerns

Hacker and Loch[1] and de Andrade[4] assessed aspects of tropical forage seed production from the perspective of the commercial seed producer. Several key factors surfaced as constraints to further development and even maintenance of the existing industry in some regions. Some of these constraints are largely biological factors, many of which could be overcome with continuing research and extension programs. Lack of market stability and efficient global trade are also substantial economic constraints to the industry. Strategies to reduce risk associated with market instability have been proposed.[40] Phaikaew et al.[41] suggested that common seed quality standards, seed certification, and guidelines for storage and shipping could enhance regional trade in Southeast Asia. The importance of such government functions to the seed industry have been noted where services have been reduced in Australia.[43,51]

Opportunistic harvest of seeds from pastures which are not necessarily managed for seed production provides seeds of variable quality and fluctuating supply.[43] Users of tropical forage seed often seek the lowest price without consideration of quality or seed certification.[40] Even inferior varieties or species may be planted due to differences in price of seed.

As complexity has increased with the increasing number of divergent species available and production in some regions increasing, publicly funded research and extension efforts to support the tropical forage seed industry have recently decreased. Hacker and Loch[1] suggested that this trend of decreased public research funding in support of the tropical forage seed industry is likely to continue. Thus new species, especially those requiring unique seed-production technologies and demonstrating potential for only limited areas of adaptation, will probably not receive sufficient research or extension support for successful commercialization.

The potential for plant variety protection laws to increase investment in tropical forage seed technology could have a positive effect on the industry.[1,4,40] There are also concerns about the effects of such plant variety protection on competition, profitability, and ultimate availability of superior varieties.[52,53] Older public varieties, although superior, may be less profitable to the seed industry than protected, effectively marketed varieties.[1]

As Ferguson and Loch[40] noted, several characteristics of the tropical forage seed industry provide inherent limitations that must be recognized. Demand is derived from the industry supported. The resulting demand changes, often dramatically. Due to high variation within the germplasm available and among environments where this germplasm is grown, some highly productive cultivars may not support adequate seed trade for sustained commercial seed production. The continuing release of new or novel species does not allow extensive transfer of existing seed industry technology, especially for harvest of some new cultivars. Ferguson and Loch[40] have suggested that the combination of inherent constraints and reduced government funding will, in the long term, result in more focused research on key aspects of tropical forage seed production to facilitate commercialization of only widely adapted superior germplasm.

REFERENCES

1. Hacker, J. B. and Loch, D. S., Tropical forage seed production: producers' views and research opportunities, in *Proc. XVIII Int. Grassland Congress*, Vol. 3, XVIII Int. Grassland Congress, Calgary, Alberta, Canada, 1997, 505.
2. Kabirizi, J. M., Pasture seed production in Uganda, *International Herbage Seed Production Research Group Newsletter*, 13, 17, 1990.
3. Boonman, J. G., Experimental studies on seed production of tropical grasses in Kenya, general introduction and analysis of problems, *Neth. J. Agricultural Science*, 19, 23, 1971.
4. Andrade, R. P. de., Tropical pasture seed production: practice, experiences and perspectives, in *Proc. XVIII Int. Grassland Congress*, Vol. 3, XVIII Int. Grassland Congress, Calgary, Alberta, Canada, 1997, 517.

5. Jones, R. J., Seed production, harvesting and storing, in *Tropical Pasture Research Principles and Methods*, Shaw, N. H. and Bryan, W. W., Eds., Bull. 51, CAB Int., London, 1976, 378.
6. Strickland, R. W., Seed production and testing problems in tropical and subtropical pasture species, in *Proc. Int. Seed Test. Assoc.*, 36, 189, 1971.
7. Stoddart, J. L., Seed ripening in grasses, 1. Changes in CHO content, *J. Agric. Sci.*, 62, 67, 1964.
8. Garner, W. W. and Allard, H. A., Further studies in photoperiodism, the response of the plant to relative length of day and night, *J. Agric. Res.*, 23, 871, 1923.
9. Fisher, M. J., Crop growth and development: flowering physiology, in *Forage Seed Production 2. Tropical and Subtropical Species*, Loch, D. S. and Ferguson, J. E., Eds., CABI Publishing, New York, 1999, 81.
10. Gonzalez, R., Gonzalez, J., Clements, R., and Humphreys, L., Flowering and seed production of *Centrosema* spp. in relation to daylength and temperature, in *Proc. XVII Int. Grassland Congress*, Hamilton, New Zealand, 1993, 1682.
11. Borthwick, H. A., History of Phytochrome, in *Phytochrome*, Mitrakoos, K. and Shropshire, W. Jr., Eds., London, Academic Press, 1972, 3.
12. West, S. H., Carbohydrate metabolism and photosynthesis of tropical grasses subjected to low temperatures, in *Proc. of Symposium on Plant Response to Climatic Factors*, UNESCO, Uppsala, Sweden, 1970, 165.
13. Griffiths, D. J., Standards of spatial isolation for seed production of herbage crops, *Herb. Abstr.*, 26, 1956, 205.
14. Humphreys, L. R., *Tropical Pasture Seed Production*, Food and Agriculture Organization of the United Nations, Rome, 1979.
15. Marousky, F. J. and West, S. H., Germination of bahiagrass in response to temperature and scarification, *J. Amer. Soc. Hort. Sci.*, 113, 845, 1988.
16. Wilson, G. P. M. and Rumble, C. J., The effect of seed rate and nitrogen fertilizer on the yield of seed and by-products of Whittet kikuyu grass at Grafton, New South Wales, *Tropical Grasslands*, 9, 53, 1975.
17. Humphreys, L. R. and Riveros, F., *Tropical Pasture Seed Production*, Food and Agricultural Organization of the United Nations, Rome, 1986.
18. West, S. H. and Marousky, F. J., Mechanism of dormancy in Pensacola bahiagrass, *Crop Sci.*, 29, 787, 1989.
19. Burton, G. W., Personal communication, 1992.
20. Altuve, S. M., Perez, D., and Royo Pallares, O., Seed production in *Vigna adenantha*, in *Proc. XVII Int. Grassland Congress*, Hamilton, New Zealand, 1993.
21. Lusembo, P., Sabiiti, E. N., and Ebong, C. E., Seed production of centro under three support systems, unstaked, staked and growing with cassava, in *Proc. XVII Int. Grassland Congress*, Hamilton, New Zealand, 1993.
22. Loch, D. S., *Constraints on Seed Production of Chloris gayana Cultivars*, Ph.D. Thesis, University of Queensland, Australia, 1983.
23. Jimenez-Merino, A., Castrellon-Montelongo, J., and Cadena-Meneses, A., Yield and quality of buffelgrass seed at different levels of nitrogen and phosphorus, in *Proc. XVII Int. Grassland Congress*, Hamilton, New Zealand, 1993.
24. Ramirez, L. and Hacker, J. B., Effect of Paclobutrazol on seed yield in the subtropical grass *Digitaria eriantha*, in *Proc. XVII Int. Grassland Congress*, Hamilton, New Zealand, 1993.
25. West, S. H. and Jank, L., Bahiagrass seed production, in *Proc. International Conference on Livestock in the Tropics*, University of Florida, Gainesville, Florida, 1991, 1.
26. Phaikaew, C., Manidool, C., and Devahuti, P., Ruzi grass (*Brachiaria ruziziensis*) seed production in north-east Thailand, in *Proc. XVII Int. Grassland Congress*, Hamilton, New Zealand, 1993, 1766.
27. Campbell, L. R. V., *Paspalum dilatatum* and *Axonopus affinis* in Australia, in *Forage Seed Production 2. Tropical and Subtropical Species*, Loch, D. S. and Ferguson, J. E., Eds., CABI Publishing, New York, 1999, 325.
28. Loch, D. S., *Cenchrus ciliaris* in Australia, in *Forage Seed Production 2. Tropical and Subtropical Species*, Loch, D. S. and Ferguson, J. E., Eds., CABI Publishing, New York, 1999, 351.
29. Gutteridge, R. C., Stewart, J., Gunasena, H. P. N., Patil, F. B., Mutty, P. K., and Pathak, N. N., Seed Collection and Multiplication, in *Nitrogen Fixing Trees for Fodder Production, A Field Manual*, Roshetko, J. M. and Gutteridge, R. C., Eds., Winrock International, Morrilton, Arkansas, 1996, 61.

30. Andrade, P. R. de and English, B. H., Seed harvesting and drying: legumes, in *Forage Seed Production 2. Tropical and Subtropical Species*, Loch, D. S. and Ferguson, J. E., Eds., CABI Publishing, New York, 1999, 213.
31. Souza, F. H. D., *Panicum maximum* in Brazil, in *Forage Seed Production 2. Tropical and Subtropical Species*, Loch, D. S. and Ferguson, J. E., Eds., CABI Publishing, New York, 1999, 363.
32. English, B. H., *Macroptilium atropurpureum* in Australia, in *Forage Seed Production 2. Tropical and Subtropical Species*, Loch, D. S. and Ferguson, J. E., Eds., CABI Publishing, New York, 1999, 407.
33. Souza, F. H. D., *Brachiaria* spp. in Brazil, in *Forage Seed Production 2. Tropical and Subtropical Species*, Loch, D. S. and Ferguson, J. E., Eds., CABI Publishing, New York, 1999, 371.
34. Ferguson, J. E. and Loch, D. S., *Arachis pintoi* in Australia and Latin America, in *Forage Seed Production 2. Tropical and Subtropical Species*, Loch, D. S. and Ferguson, J. E., Eds., CABI Publishing, New York, 1999, 427.
35. Delouche, J. C., Precepts of Seed Storage, in *Proc. Short Course for Seedsmen*, Vol. 16, Mississippi State University, Mississippi State, Mississippi, 1973, 97.
36. Delouche, J. C., Matthes, R. K., Dougherty, G. M., and Boyd, A. H., Storage of seed in sub-tropical and tropical regions, *Seed Science and Technology*, 1, 663, 1973.
37. Harrington, J. F., Drying, storing and packaging seeds to maintain germination and vigor, in *Proc. Short Course for Seedsmen*, Mississippi State University, Mississippi State, Mississippi, 1959, 89.
38. Christensen, C. M. and Kaufman, H. H., *Grain Storage, The Role of Fungi in Quality Loss*, University of Minnesota Press, Minneapolis, Minnesota, 1969.
39. Oji, O., Madubuike, F. N., and Ezedinma, F. O. C., Effects of storage period, disease infection and treatment with potassium nitrate on viability and germination rate of guinea grass, in *Proc. XVII Int. Grassland Congress*, Hamilton, New Zealand, 1993.
40. Ferguson, J. E. and Loch, D. S., Tropical and subtropical forage seed production: looking back and to future horizons, in *Forage Seed Production 2. Tropical and Subtropical Species*, Loch, D. S. and Ferguson, J. E., Eds., CABI Publishing, New York, 1999, 445.
41. Phaikaew, C., Guodao, L., Abdullah, A., Tuhulele, M., Magboo, E., Bouahom, B., and Stür, W., Tropical forage seed production in Southeast Asia: current status and prospects, in *Proc. XVIII Int. Grassland Congress*, Vol. 2, XVIII Int. Grassland Congress, Calgary, Alberta, Canada, 1997, 7 (Session 28).
42. Ferguson, J. E. and Andrade, R. P. de, *Andropogon gayanus* in Latin America, in *Forage Seed Production 2. Tropical and Subtropical Species*, Loch, D. S. and Ferguson, J. E., Eds., CABI Publishing, New York, 1999, 381.
43. Smith, P., Pastures for prosperity — seeds forum. 6. What we want from the seed industry in the future — a merchant's viewpoint, *Tropical Grasslands*, 30, 88, 1996.
44. Loch, D. S. and Ferguson, J. E., Tropical and subtropical forage seed production: an overview, in *Forage Seed Production 2. Tropical and Subtropical Species*, Loch, D. S. and Ferguson, J. E., Eds., CABI Publishing, New York, 1999, 1.
45. Kowithayakorn, L. and Phaikaew, C., Harvesting and processing techniques of tropical grass and legume seeds for small farm holders, in *Proc. XVII Int. Grassland Congress*, Hamilton, New Zealand, 1993, 1809.
46. Muliokela, S. W. and Griffiths, R., Sustainable, quality herbage seed supplies for SSA smallholders, in *Proc. XVII Int. Grassland Congress*, Hamilton, New Zealand, 1993.
47. Ferguson, J. E. and Sauma, G., Towards more forage seeds for small farmers in Latin America, in *Proc. XVII Int. Grassland Congress*, Hamilton, New Zealand, 1993, 1751.
48. Barnes, P. B. and Alhassan, W. S., Status of forage seed production in Ghana, in *Proc. XVII Int. Grassland Congress*, Hamilton, New Zealand, 1993.
49. Tedonkeng Pamo, E., Problems hindering forage seed production in Cameroon, in *Proc. XVII Int. Grassland Congress*, Hamilton, New Zealand, 1993.
50. Chin, H. F. and Wong, C. C., The importance of tropical pasture seed quality and factors affecting it in the smallholder's farming system, in *Proc. XVII Int. Grassland Congress*, Hamilton, New Zealand, 1993.
51. Quadrio, N., Pastures for prosperity — seeds forum. 5. The future for the seed industry — a grower's perspective, *Tropical Grasslands*, 30, 86, 1996.

52. Rothwell, P., Pastures for prosperity — seeds forum. 1. Proprietary lines and plant breeders' rights, *Tropical Grasslands*, 30, 74, 1996.
53. Rains, J., Pastures for prosperity — seeds forum. 2. The changing face of tropical pasture seed marketing in northern Australia, *Tropical Grasslands*, 30, 77, 1996.

10 Intensive Management of Forage Grasses in the Humid Tropics

José Vicente-Chandler

CONTENTS

I. INTRODUCTION

Vast areas of rolling to steep land in the humid tropics, if developed as well-managed grazing lands, can provide good-quality forage for the production of milk and meat and, at the same time, offer protection against soil erosion.

Most of the experimental work discussed in this chapter was conducted in Puerto Rico under typical humid tropical conditions (25°C mean annual temperature; 1500 to 2000 mm of annual rainfall) utilizing various types of soils such as Ultisols, Inceptisols, and Oxisols. In general, under these conditions, utilization of the techniques described, i.e., high-yielding grasses, proper fertilization, and intensive systems of management, resulted in cut grass yields of over 30,000 kg/ha/yr

0-8493-2318-5/01/$0.00+$.50

of dry forage (150 t/ha green). In pastures, up to five steers or 2.5 milk cows/ha could be carried, which produced, without additional feed, over 1000 kg/yr of gain in weight or over 7500 L/yr of milk, respectively.

In the humid tropics, fertilization is required for sustained high yields since most soils have low fertility, and high-yielding grasses remove large quantities of nutrients that must be continually replaced. Proper pasture management can assure efficient use of the forage produced and repay the necessary investment. Furthermore, pasture land in the tropics is often expensive if costs of the required infrastructure and improvement are considered and, therefore, should be used efficiently.

Intensively managed, highly productive grasslands can increase production per unit severalfold over typical range-type operations, thus reducing the need to bring more rain forest and other ecologically important areas into agricultural use.

Since grains are scarce and expensive in the tropics, and are used mainly for human consumption, forages must play a far greater role in cattle feeding than in temperate regions, where pastures cannot be utilized in the winter and grain for concentrated feed is usually abundant and cheaper.

This chapter is largely a synopsis of studies and experiences related to the eight perennial grasses listed below that are typical of those used throughout the humid tropics.

Clump Grasses	**Sod Grasses**
Guinea grass	Caribgrass
(*Panicum maximum* Jacq.)	(*Eriochloa polystachya* Kunth)
Napier grass or elephant grass	Congograss
(*Pennisetum purpureum* Schumach.)	(*Brachiaria ruziziensis* Germ. & Ever.)
	Pangola digitgrass
	(*Digitaria decumbens* Stent)
	Paragrass
	(*Brachiaria purpurascens* Raddi)
	Signalgrass
	[*Brachiaria brizantha* (A. Rich.) Stapf]
	Star grass
	(*Cynodon nlemfuensis* Vanderyst variety *nlemfuensis*)

Napier grass is mainly used as a cut grass, but can be grazed if carefully managed. Results of studies evaluating dwarf varieties of the grass at the Tropical Agriculture Research Station, USDA-ARS, in Mayaguez, Puerto Rico, suggest that they may be more suitable for grazing, but can be used under cutting management. Pangola digitgrass, star grass, congograss, caribgrass, guinea grass, signalgrass, and paragrass can be harvested either by cutting or grazing.

Paragrass and caribgrass frequently grow together, are similar in their growth habits and environmental requirements, and often crowd out all other grasses on poorly drained soils. Guinea grass is the best grass for drier areas or shallow soils.

In Puerto Rico and elsewhere in the humid tropics, star grass var. *nlemfuensis* (PR PI 2341) has given excellent results when used as pasture. This grass may occasionally have a fairly high hydrocyanic acid (HCN) content, but it is not toxic to cattle as shown by Caro-Costas and Vicente-Chandler.[1,2]

Examples of yields that may be expected from these grasses when intensively managed on deep soils in humid tropical regions are shown in Table 10.1. Yields would be about 20% higher with irrigation and considerably lower on shallow or sandy soils.

Napier grass produces the highest yields of cut forage, but its digestibility is lower than the other grasses due to the higher proportion of silica and the higher moisture content of its stems. With grazing management, yields of all grasses are similar except paragrass, which has a comparatively lower yield.

TABLE 10.1
Productivity that Can Be Expected from Intensively Managed[a] Forage Grasses on Deep Soils in the Humid Tropics without Irrigation Based on Studies Conducted in Puerto Rico[5]

	Cut Every 40–60 Days			Harvested by Grazing Interval			
Grass	Dry Forage Yield[b] kg/ha/yr	Total Digestible Nutrients Consumed by Cattle[c] kg/ha/yr	Carrying Capacity (270-kg Steers)[d] number/ha	Dry Forage Consumed by Cattle kg/ha/yr	Total Digestible Nutrients Consumed by Cattle[c] kg/ha/yr	Carrying Capacity (270-kg Steers)[d] number/ha	Gain in Liveweight kg/ha/yr
Napier	35,200	14,100	10.0	14,000	7,800	5.6	1,200
Congo	32,600	13,000	9.3	13,000	7,000	5.0	1,000
Guinea	30,800	12,300	8.8	14,000	7,800	5.6	1,200
Star	29,000	11,600	8.3	14,000	7,800	5.6	1,200
Pangola	27,200	10,900	7.8	13,000	7,000	5.0	1,000
Para	27,200	10,900	7.8	10,000	5,500	3.9	800
Carib	27,200	10,900	7.8	13,000	7,000	5.0	1,000
Signal	27,200	10,900	7.8	13,000	7,000	5.0	1,000

[a] About 3 t/ha/yr of 15-5-10 fertilizer for grasses havested by cutting and 2 t for grazed grasses.

[b] Multiply by six for napier grass and by five for all other grasses to obtain approximate yield of green forage.

[c] Assuming 20% waste in feeding and 50% digestibility of forage consumed. Total digestible nutrients (TDN) is the most commonly used term for expressing the energy content of forage. It comprises all digestible organic compounds including proteins, fiber, carbohydrates, and fats x 2.25. One kg of TDN = 4.4 megacalories of digestible energy or 3.6 megacalories of metabolizable energy.

[d] One 270-kg animal + 0.5 kg/head daily gain require 3.9 kg/day of TDN, which is equivalent to 1,424 kg/yr of TDN.

Palatability is not a concern since all of these grasses are relished by cattle, and they vary only slightly in composition and digestibility. Of course, such factors as palatability, digestibility, and nutritive value are dependent on cutting or grazing interval, soil fertility, climate, and pasture management practices.

Improved grasses can be considered a permanent investment since stands last indefinitely on appropriate sites, if properly managed. Brush and trees can be removed with heavy equipment or killed with weed killers. A few trees should be left standing to provide shade for the grazing animals. The grass should be planted in furrows about one meter apart. Establishment of the grass can be hastened by applying 15-5-10 fertilizer at the rate of 200 kg/ha one and three months after planting. Weeds should be controlled mechanically or by applications of an appropriate herbicide. Tall-growing weeds can be controlled by repeated mowing and low-growing ones shaded out simply by delaying grazing.

II. CUT GRASSES

A. Fertilizing Cut Grasses

Fertility is the major factor affecting grass yield. Fertilization requirements are high in the humid tropics since the year-round growing weather favors the production of abundant forage, which removes large quantities of nutrients from the soil. For example, well-fertilized grasses harvested by cutting in Puerto Rico removed nearly 330 kg/ha of N, 55 of P, 420 of K, 128 of Ca, and 75 of Mg.[3-5] This is equivalent to the N, P, and K in two metric tons of 16-6-25 fertilizer. Even fertile soils cannot supply these quantities of nutrients.

In addition, heavy rainfall favors the loss of nutrients through leaching and runoff, which account for about one-half and one-quarter of fertilizer loss of N and K, respectively. Also, two of the most prevalent soil orders in tropical areas utilized for pastures, Ultisols and Oxisols, fix large quantities of fertilizer P, making it unavailable or only partially available to the plant.

1. Nitrogen

Intensively managed cut grasses in Puerto Rico can obtain an average of 120 kg/ha/yr of N from typical Ultisols and Inceptisols,[4] but close to 330 kg/ha/yr of N are removed by such grasses. About one half of the N applied as fertilizer can be lost by leaching.

The effect of various levels of N fertilizer on such grasses as Pangola digitgrass, paragrass, napier grass, guinea grass, and star grass harvested by cutting was determined in numerous experiments in Puerto Rico on otherwise well-fertilized Ultisols and Inceptisols.[3,6-11] Yields increased sharply with levels of N up to about 400 kg/ha/yr, and crude protein content (CPC) of the grasses increased with N rates up to the highest level tested. An average of 52% of the N applied was recovered by the grasses. The response to N became greater as harvest interval increased, and the response to N was greatest during seasons of fast growth.[4,8]

The effect of five N sources on napier grass growing on an Ultisol showed that N source did not affect yield[12] and that sulfur-coated urea can reduce leaching of N into the subsoil.[13]

According to these studies, grasses grown on deep soils under humid tropical conditions and harvested by cutting respond to applications of up to 400 kg/ha/yr of N. The N should be applied after each cutting, and the cheapest source used.

2. Potassium

The K-supplying power of typical Ultisols, Oxisols, Inceptisols, and Mollisols for Pangola digitgrass was determined in Puerto Rico over a period of four years.[14,15] During the first year, large and variable quantities of K were released from the four soil orders, but, during the following years,

K release leveled off at about 100 kg/ha/yr, except for the Inceptisols and Mollisols of the semiarid region of the island, which had high reserves of the element.

Although high-yielding tropical grasses such as napier grass and guinea grass take up about 400 kg/ha/yr of K,[16] some of which represents luxury consumption, the actual requirement of these grasses is about 300 kg/ha/yr. Since the soil in Puerto Rico can provide about 100 kg/ha/yr of K and about 25% of that applied as fertilizer is lost, about 300 kg/ha/yr of the mineral should be applied to sustain high yields under such conditions.

In Puerto Rico, yields of four forage grasses increased sharply with K rates of 220 to 440 kg/ha/yr.[17] A K content of 1.0 to 1.5% was associated with high yields, and higher levels indicated luxury consumption. Recovery of fertilizer K in the forage averaged 70%.

It can be concluded that these grasses harvested by cutting can respond to about 300 kg/ha/yr of K.

3. Phosphorus

Most tropical soils are naturally low in P.[18] To sustain high yields, grasses must take up about 55 kg/ha/yr of the mineral.[5] Although losses by leaching are negligible, the P-fixing capacity of the soil and previous fertilization must be taken into consideration when determining the quantity of fertilizer to add.

Grasses did not respond to P fertilization on an Alfisol that had been used for heavily fertilized sugarcane for many years, but responded strongly to 75 kg/ha/yr of P on an Ultisol and an Inceptisol, which had received no previous fertilization.[19] A level of 0.18% P in 60-day-old grass was sufficient for near-optimum growth.

It can be concluded that these grasses harvested by cutting can respond to about 70 kg/ha/yr of P on soils with little previous fertilization, whereas only a small amount may be needed on soils with sufficient P in available forms. When possible, a soil test should be conducted to ensure that the appropriate quantities of P and K are applied.

4. Liming

Even acid soils generally contain sufficient Ca for grasses, but liming may be needed to prevent Al and/or Mn toxicity.

Heavy applications of residually acid N fertilizers can rapidly increase soil acidity.[20,21] Increases in soil acidity are relatively easy to prevent by proper liming, but are difficult to correct once acidity develops deep in the subsoil.

In Puerto Rico, grasses growing on an Ultisol and an Alfisol responded to applications of limestone, and the response was related to soil properties.[21] A Ca content of about 0.4% in the grass indicated sufficiency for optimum growth. High yields were assured when base saturation of the soil was 50% or more, when the soil contained less than 2 meq/100 g of exchangeable Al, or had a pH above 5. About one ton of limestone/ton of fertilizer applied was required to maintain high grass yields. This quantity neutralizes the residual acidity of most commercially available N fertilizers.

5. Financial Aspects of Fertilization

The strong response of tropical grasses to applications of about 400 kg/ha/yr of N, 70 of P, and 300 of K suggests that a 3-1-2 fertilizer ratio (15-5-10, for example) is well suited for such grasses under most conditions, although the ratios and amounts needed may vary in accordance with soil or plant analysis.

The cost of fertilizer and the value of the forage to be fertilized are the main factors determining how much beef can be produced. The following tabulation shows the number of kg of dry forage produced for each kg increment of fertilizer applied to cut grasses in Puerto Rico. The most economical level of fertilizer to use can be calculated from these data and current prices of fertilizer and feed.

Fertilizer Applied t/ha/yr	Dry Forage/Fertilizer kg/kg
1	6.0
2	5.0
3	4.0
4	1.0

The optimum quantity of fertilizer to apply to forages depends on many factors. Less fertilizer is required if manure is applied to the soil. Where land values are expensive, additional forage can usually be produced more profitably by heavier fertilization than by acquiring additional land, especially if the soil has little natural fertility, as usually occurs in the humid tropics. More fertilizer can be used productively when rainfall is adequate and there is little runoff or on deep soils provided that the forage is used efficiently.

Fertilizers can be useful in regulating forage production when the weather is suitable. A heavy application can increase the supply of forage quickly or a reduced quantity can cut back forage production when fewer cattle are being grazed.

It may be profitable to use up to 3 t/ha/yr of 15-5-10 or a similar fertilizer on grasses used for cut feed. The fertilizer should be applied after each cutting along with one ton of limestone for each ton of fertilizer if applied on acid soils.

B. Harvesting Cut Grasses

Transferring forage from the field to the animal with minimum loss is an essential feature of management. For large-scale operations, grass for cutting should be confined to land where harvesting can be mechanized. The grass is usually fed chopped to reduce loss; but, when forage is abundant, it may be fed unchopped so that cattle can reject the less digestible portions.

The effect of cutting interval on yield and composition of numerous tropical grasses has been determined in a number of experiments in Puerto Rico.[4,6,8,9,22-29] Results of experiments with most of these grasses are summarized in Table 10.2. Forage yields increased with length of cutting interval, but the forage became less nutritious with age. Protein, P, Ca, and Mg contents as well as digestibility decreased with length of cutting interval, while the content of lignin (an almost totally indigestible constituent) increased. A compromise needs to be made between the high yields obtained by long harvest intervals and the better-quality forage obtained by frequent cutting intervals.

One approach is based on the findings of McDonald,[30] who determined that the digestibility of grasses decreased by 0.48% per day, starting with 85% digestibility for very young grass.

The following tabulation shows rather close agreement between percent digestibility of star grass at different ages when calculated by the McDonald and Van Soest methods.[30,31]

	Digestibility of Star Grass	
Age of Grass in Days	McDonald Method %	Van Soest Method %
30	70	70
45	63	66
60	56	59
90	42	51

Tessema[32] found that the digestibility of five tropical grasses decreased from an average of over 80% for seven-day-old grass to a little over 60% at 63 days, figures which also agree substantially

TABLE 10.2
Effect of Frequency of Cutting on the Yield and Composition of Six Well-Fertilized[a] Tropical Forage Grasses over a Two-Year Period in the Humid Mountainous Region of Puerto Rico

		kg/ha/yr			Composition of Forage on a Dry-Weight Basis %						
Grass	Interval between Cuttings	Yield of Green Forage	Yield of Dry Forage	Digestible Dry Matter[b]	Dry Matter Content of Forage	Proportion of Leaf Blades	Protein (N × 6.25)	Phosphorus	Calcium	Magnesium	Lignin
Pangola	30	138,000	21,000	14,700	19	38	12.5	0.22	0.43	—	8.1
	45	149,000	25,000	15,000	24	39	9.6	0.22	0.36	—	8.8
	60	161,000	33,000	18,200	28	40	8.0	0.17	0.34	—	9.2
Star	30	85,000	18,000	12,600	22	53	14.6	0.31	0.47	0.29	7.6
	45	98,000	24,000	14,400	24	52	11.1	0.26	0.50	0.23	8.4
	60	116,000	33,000	18,200	28	50	9.7	0.19	0.50	0.21	10.0
	90	120,000	35,000	14,000	29	—	7.7	0.15	0.52	0.27	10.4
Congo	30	94,000	21,000	14,700	22	32	9.8	0.36	0.60	0.19	6.8
	45	117,000	28,000	16,800	24	32	8.7	0.28	0.64	0.23	7.5
	60	126,000	30,000	16,500	24	35	6.4	0.19	0.62	0.28	8.1
	90	164,000	49,000	19,600	30	33	5.1	0.14	0.51	0.26	10.3
Napier	40	155,000	22,000	14,300	14	55	9.9	0.24	0.35	0.30	6.9
	60	236,000	43,000	23,650	17	42	7.9	0.18	0.28	0.19	8.8
	90	252,000	62,000	24,800	25	30	5.4	0.13	0.23	0.19	11.1
Guinea	40	125,000	26,000	16,900	21	63	9.0	0.27	0.88	0.49	8.2
	60	133,000	32,000	17,600	24	53	7.0	0.22	0.78	0.39	9.4
	90	127,000	40,000	16,000	32	36	5.6	0.16	0.63	0.33	11.4
Para	40	91,000	19,000	12,350	21	—	9.2	0.25	0.39	0.27	7.5
	60	111,000	26,000	14,300	23	28	7.2	0.21	0.25	0.20	8.4
	90	115,000	36,000	14,400	31	24	4.8	0.15	0.29	0.15	9.4

[a] About 3 t/ha/yr of 15-5-10.
[b] Assuming a 0.48% decrease in digestibility/day of increase in harvest interval, starting with 85% digestibility for very young grass.[30]

with those obtained by the McDonald and Van Soest methods. Other studies with five or 10 tropical grasses yielded similar results.[23,24,33]

Based on the calculations of McDonald,[30] the grasses generally produced highest yields of digestible dry matter when cut every 40 to 60 days.

The foregoing data suggest that the grasses studied should be harvested about every 40 days during seasons of fast growth and every 60 days during seasons of slow growth. Several factors such as the type of cattle to be grazed, value of the land, and feasibility of supplementing forages with concentrate feeds should be taken into consideration.

In three studies, higher yields of napier grass, star grass, paragrass, congograss, and Pangola digitgrass cut every 60 days were obtained with a low (0 to 7 cm) rather than a high (17 to 25 cm) cutting height.[11,22,34] Another study[28] determined that Pangola digitgrass, tannergrass (*Brachiaria arrecta* Dur. and Schinz), Bermuda grass [*Cynodon dactylon* (L.) Pers. var. *coursii*], and *Digitaria valida* Stent yielded more when cut at a height of 5 rather than 15 cm. It is recommended that tropical forage grasses in a humid tropical environment be cut at about 5 cm above the ground.

III. GRASS PASTURES

A. Grazing Management

Pastures should be started with as few weeds as possible, and undesirable plants should be kept under control. Weeds are a minor problem in pastures that are well fertilized and properly managed so that the grass can successfully compete with them.

Even badly overgrown pastures of Pangola digitgrass, caribgrass, star grass, or similar grasses generally do not require mowing because cattle trample the grasses down. However, where mowing can be easily done, it may be desirable to cut pastures back occasionally, especially overgrown pastures of napier grass or guinea grass having old, unpalatable stems. On the other hand, it can be harmful to mow pastures of such tall-growing grasses too frequently.[5]

Proper grazing management is essential for the efficient utilization of the forage produced and long-term productivity of pastures. Good management consists of obtaining the highest per hectare yield of animal products without significantly reducing production per animal, i.e., producing a maximum net return from pastures, livestock, and fertilizer applied.

High-yielding pastures should generally be grazed in rotation. Using this type of grazing management, more animal products can be generated per hectare, pastures can be grazed to take advantage of the best growth, and grazing can be deferred to "stockpile" forage for seasons of slow growth or to improve the stand. Also, rotationally grazed pastures are grazed more uniformly and efficiently. At least four or five pastures should be available for rotational grazing by each group of cattle.

Intensive "ration" grazing (the grazing in rotation of small paddocks for short periods of time) of heavily fertilized pastures may be warranted on better, more productive land. However, the advantages of such a grazing practice can be dissipated by improper management procedures such as overstocking, especially during critical periods of plant growth.

Frequency and Height of Grazing — Vicente-Chandler et al.,[35] found that almost 15 t/ha/yr of dry forage were produced when Pangola digitgrass was grazed repeatedly to within 15 cm of the ground compared to only 11 t/ha/yr when grazed to a height of 5 cm. The following tabulation shows the effect of grazing interval and height on grass persistence in plots of star grass following one year of grazing.[36]

Treatment	Star Grass Stand %
Grazed to a height of 4 cm	
Every 14 days	28
Every 21 days	49

Treatment	Star Grass Stand %
Every 28 days	70
Grazed to a height of 15–20 cm	
Every 14 days	53
Every 21 days	82
Every 28 days	91

Too frequent or too close grazing adversely affected the stand.

The following results obtained with star grass in another experiment show no significant differences in liveweight gain or total digestible nutrients produced for three grazing intervals, demonstrating that the grass can be grazed at short intervals, if it is grazed no closer than about 20 cm from the ground.[37]

Grazing Interval in Days	Gain in Liveweight in kg/ha/yr	Total Digestible Nutrients (TDN) in kg/ha/yr
14	1,242	7,540
21	1,239	8,060
28	1,110	7,500
Average	1,197	7,700

From the experiments cited, it can be concluded that the grasses studied and others of similar growth habits should be grazed to a height of 15 to 20 cm at 2 to 3 week intervals during periods of fast growth and 3 to 4 week intervals during seasons of slow growth, although such grasses can withstand occasional overgrazing.

B. Fertilizing Grass Pastures

Fertilizer is the most expensive input in intensive pasture management. Fertilizer requirements of grazed pasture grasses differ from those of cut grasses. Only about half as much forage is taken from the land under grazing; hence, far fewer nutrients are taken from it. About 80% of the N, P, and K consumed is excreted in the urine and feces discharged into pastures. However, the fertility of pastures cannot be effectively maintained by grazing animals primarily because of poor distribution of their excreta.[38]

Although there is little accumulation of N in pastures, P and K fertility can be built up from excretion and fertilization, as these nutrients are held more tightly in the soil. Vicente-Chandler et al.[35] found that, in pastures continuously carrying five head/ha and receiving 2000 kg/ha/yr of 14-4-10 fertilizer for 14 years, exchangeable K in the soil increased by 590 kg/ha or 21% of the fertilizer K applied. They also found that about 20 ppm of the applied P accumulated in the soil.

Caro-Costas et al.[39] found that five grasses grown in the humid, mountainous areas and harvested by simulated grazing responded to N applications of up to 200 kg/ha/yr during the winter months and up to 400 kg/ha/yr during the remainder of the year. Observations made of the strips grazed by cattle showed no apparent effect of fertilization on the palatability of the grasses.

Five different N sources had similar effects on yield and composition of Pangola digitgrass grown under humid tropical conditions and harvested by simulated grazing.[40] About 50% of the applied N was recovered in the forage.

C. Effects of Fertilization on Animal Productivity

The response of pastures to fertilization must, in the final analysis, be measured in terms of animal production under actual grazing conditions. Details of procedures followed and results and data obtained in the experiments cited in this section have been described in bulletin form by Vicente-Chandler, et al.[5]

Napier grass pastures on a steep, deep Ultisol in Puerto Rico responded strongly from 675 to 2025 kg/ha/yr to an increase of a 15-5-10 fertilizer (Table 10.3) in terms of increases in carrying capacity, animal weight gain, TDN consumed, dry forage consumed, digestibility, and forage protein content (which increased from 8.1 to 15.9%). Daily gain/head averaged about 0.55 kg.[41] Calculated digestibility of the forage consumed averaged 52%, and 13 kg of dry forage were consumed for each kg of weight gain. The cattle consumed an average of 7.3 kg of dry forage (about 36 kg of green forage)/head daily, which is equivalent to 2.6 kg of dry forage/100 kg of liveweight.

During the following five years, fertilizer rates tested were 1790, 3140, and 4480 kg/ha/yr of 15-5-10, and management was intensified by using the system of "ration" grazing. Cattle grazing the napier grass pastures had weight gains for each increment of fertilization up to 4480 kg/ha/yr, at which level they had an average of 1770 kg/ha/yr of liveweight gain.[31]

Four levels of fertilization, 450 to 2700 kg/ha/yr, had a positive effect on the productivity of Pangola digitgrass pastures on a steep Ultisol.[42] At the highest level, the equivalent of five 270-kg animals/ha were stocked, which averaged 976 kg/ha/yr of liveweight gain.

Star grass pastures on a steep Ultisol responded strongly to three levels of fertilizer up to 3140 kg/ha/yr. At this level under an intensive system of "ration" grazing,[43] they carried the equivalent of 6.8 head of cattle/ha weighing an average of 270 kg and yielded an average yearly weight gain/ha of 1337 kg. Protein content of the ingested forage ranged from 18 to 24% throughout the year and daily gain averaged 0.64 kg/head. Apparent digestibility of the forage ranged from 58 to 72%; lignin content, 3.5 to 5.8%; silica, 0.9 to 2.6%; P, 0.17 to 0.26%; and Ca, 0.31 to 0.47%.

From the above experiments with napier grass, Pangola digitgrass, and star grasses, increasing fertilizer rates from about 500 kg/ha/yr (the minimum required to maintain a good pasture) to 2500 kg/ha/yr produced about 520 additional kg of weight gain and an increase of 5000 kg/ha/yr of TDN.

The most economical level of fertilization can be determined by comparing cost of the fertilizer applied and value of the increase in beef produced.

In terms of milk production, a cow weighing 550 kg requires 4.4 kg of TDN daily for maintenance and 3.6 kg to produce 12 L of milk, or a total of 8 kg of TDN/cow. Thus, every kg of TDN ingested could theoretically be converted to 1.5 L of milk. Therefore, the 5000 kg/ha/yr of additional TDN produced when fertilization is increased by 2000 kg/ha/yr could theoretically produce an additional 7500 L of milk.

It can be concluded that, on deep soils in humid tropical regions and depending on economic factors such as the price of fertilizer, beef, and milk, intensively managed pastures of improved grasses may be economically fertilized with up to 2000 kg/ha/yr of 15-5-10 or similar fertilizer divided into four equal applications. On shallow soils, where periodic droughts can sharply limit forage production, lower application rates should be used. After five years of heavy fertilization, the P and K contents of the soil may increase sufficiently so that only N need be applied during one out of every four or five years. On acid soils, one ton of limestone should be applied for every ton of fertilizer. Plant and soil analyses should be used to periodically assess fertilizer requirements.

IV. ANIMAL PRODUCTION FROM INTENSIVELY MANAGED PASTURES OF DIFFERENT GRASSES

The carrying capacity and yields of beef and milk produced by intensively managed pastures of different grasses have been determined in a number of grazing experiments by Vicente-Chandler et al.[3,4,44] In these experiments, an almost pure stand of each of the grasses tested was maintained in a well-managed pasture, and there were few problems with weeds.

In a four-year grazing experiment on a steep Ultisol, paragrass and molassesgrass (*Melinis minutiflora* Beauv.) gave much lower gains in weight and had lower carrying capacities than Pangola

TABLE 10.3
Effect of Three Fertilizer Levels on the Productivity of a Steep Napier Grass Pasture Grazed by Young Cattle over a Two-Year Period at Orocovis, Puerto Rico

15-5-10 Fertilizer Applied kg/ha/yr	Gain in Liveweight kg/ha/yr	Carrying Capacity (270-kg Steers)[a] number/ha	Dry Forage Consumed by Cattle[b] kg/ha/yr	Total Digestible Nutrients Consumed by Cattle[c] kg/ha/yr	Digestibility of Dry Forage Consumed by Cattle[d] %	Protein Content of Forage Consumed by Cattle %
675	638 b	3.5 b	9,970 b	4,820 b	48	8.1 b
2,025	1,201 a	5.5 a	15,010 a	7,500 a	50	15.9 a
3,375	1,333 a[e]	6.3 a	15,230 a	9,070 a	59	17.6 a

[a] One 270-kg animal + 0.5 kg/head daily gain require 3.9 kg/day of TDN, which is equivalent to 1,424 kg/yr.

[b] Difference in forage harvested from paired strips cut before and after grazing each pasture.

[c] Calculated from body weights, days of grazing, and gains in weight following recommendations of the Joint Committee of the American Dairy Society, American Dairy Science Association, and American Society of Animal Production.[56]

[d] (TDN consumed ÷ dry forage consumed) × 100.

[e] Means within a column followed by the same letter do not differ significantly at the P = 0.05 level.

TABLE 10.4
Productivity over a Four-Year Period of Well-Fertilized Pastures of Five Grasses on Steep Slopes in the Humid, Mountainous Region of Puerto Rico

Grass	Gain in Liveweight kg/ha/yr	Average Daily Gain per Head kg	Total Digestible Nutrients Consumed by Cattle[a] kg/ha/yr	Carrying Capacity (270-kg Steers)[b] number/ha	Minimum Head Carried number/ha
Guinea	1,319 a[c]	0.60	8,941 a	6.5 a	5.0
Napier	1,110 a	0.60	8,140 a	5.8 a	5.0
Pangola	1,124 a	0.60	8,316 a	6.0 a	5.0
Para	781 b	0.50	6,434 b	4.5 b	3.25
Molasses	644 b	0.45	4,838 b	3.5 b	3.25

[a] Calculated from body weights, days of grazing, and gains in weight following recommendations of the Joint Committee of the American Dairy Society, American Dairy Science Association, and American Society of Animal Production.[56]

[b] One 270-kg animal + 0.5 kg/head daily gain requires 3.9 kg/day of TDN, which is equivalent to 1,424 kg/yr.

[c] Means within a column followed by the same letter do not differ significantly at the P = 0.05 level.

digitgrass, guinea grass, or napier grass (Table 10.4). These three grasses produced an average of 8465 kg/ha/yr of TDN, which resulted in an average of 1184 kg/ha/yr of weight gain for the grazing animals and had a carrying capacity of 6.1 head of 270-kg animals/ha with 15,770 kg/ha/yr of forage being consumed by the cattle. Daily gain/head averaged 0.6 kg, which is considered good for young animals fed exclusively on tropical grass pasture. Daily consumption of forage by the cattle averaged 2.7 kg/100 kg of body weight. An average of 7.2 kg of TDN or 13.3 kg of forage was required to produce one kg of gain in weight. Average digestibility of the forage was calculated at 54%. Crude protein content averaged 18.1% and P content averaged 0.22%. Table 10.5 shows the composition of five well-fertilized grasses harvested by simulated grazing (plucking) as affected by season of the year.

As mentioned previously, star grass may occasionally have a high cyanide (HCN-p) content. However, Caro-Costas et al.[11,43] found that the HCN content of star grass dropped off rapidly with age, as shown below.

Age of Star Grass in Days	HCN Content ppm
14	570
28	320
42	105

Laboratory determinations of HCN are not a good index of the possible toxicity of a grass, because they do not identify the precursor glucosides, which may or may not rapidly release HCN. On the other hand, HCN is gradually destroyed in the rumen. Therefore, only feeding trials can determine whether a grass containing this compound is toxic to a grazing animal.

In Puerto Rico, heifers that had been fasting for 36 hours were repeatedly grazed on young, heavily fertilized star grass containing 500 to 550 ppm of HCN-p with no ill effects. Furthermore, thousands of hectares of star grass have been grazed under widely varying conditions in the tropics for many years with no case of cattle poisoning reported and with high production of beef and milk. The possibility of this grass being toxic to horses, sheep, or goats has not been investigated in Puerto Rico.

TABLE 10.5
Dry Weight Composition (%) of Forage Samples Obtained by Simulated Grazing from Well-Fertilized Pastures of Five Grasses in the Humid, Mountainous Region of Puerto Rico, as Affected by Season of the Year[a]

	Pangola				Guinea				Napier				Star			Congo		
Months	Protein	Lignin	Ca	P	Protein	Lignin	Ca	P	Protein	Lignin	Ca	P	Protein	Ca	P	Protein	Ca	P
Jan.–Feb.	18.1	9.39	0.32	0.19	19.8	7.18	0.67	0.19	20.7	7.20	0.30	0.20	23.8	0.51	0.20	22.3	0.61	0.21
March–April	16.4	9.02	0.33	0.20	17.5	7.73	0.67	0.23	18.6	7.90	0.30	0.25	19.9	0.54	0.18	22.8	0.59	0.24
May–June	14.9	7.95	0.35	0.21	16.6	7.78	0.65	0.23	16.8	8.02	0.29	0.25	17.2	0.44	0.19	15.6	0.53	0.21
July–Aug.	16.7	8.93	0.35	0.18	16.6	7.62	0.60	0.20	18.7	7.68	0.29	0.24	17.3	0.41	0.21	13.3	0.45	0.21
Sept.–Oct.	18.3	7.20	0.35	0.25	18.9	8.30	0.53	0.21	20.0	7.73	0.28	0.25	16.3	0.50	0.25	10.3	0.53	0.21
Nov.–Dec.	17.3	9.45	0.35	0.21	19.5	7.16	0.58	0.21	20.9	8.10	0.30	0.25	24.4	0.55	0.27	19.9	0.58	0.23
Average	16.9	8.99	0.35	0.21	18.2	7.63	0.62	0.21	19.3	7.77	0.29	0.24	19.8	0.49	0.22	17.3	0.55	0.22

[a] All values are averages of forage samples taken every 10 days by plucking so as to simulate grazing. Data are from studies by Caro-Costas et al.[44,46]

In several grazing studies involving star grass and other grasses in the humid, mountainous regions of the island, star grass gave superior results.[45,46] In one study, star grass produced an average weight gain of 1510 kg/ha/yr compared to 1060 kg/ha/yr for Pangola digitgrass. In another experiment, highest weight gains, averaging 1426 kg/ha/yr were on star grass pastures, while the production on congograss and Pangola digitgrass pastures averaged 984 kg/ha/yr.

Analysis of forage samples obtained by "plucking" so as to simulate grazing showed that, throughout the year, *in vitro* digestibility of star grass averaged 65% compared to 59.5% for congograss and Pangola digitgrass, and crude protein ranged from 18 to 23%. Digestibility, calculated from dry forage ingested and TDN requirements of the grazing cattle, averaged 54.5% for all grasses.

Higher milk production was also obtained on star grass pastures on a steep Ultisol.[1] For two years, lactating cows were grazed exclusively on intensively managed pastures of star grass or guinea grass, which received 2.2 t/ha/yr of 15-5-10 fertilizer. An average of 7727 L of milk were produced and 729 days/ha/yr were spent grazing star grass compared to 5593 L of milk and 543 days on guinea grass.

Caro-Costas and Soldevila et al. determined the productivity of pastures of guinea grass, tannergrass, Pangola digitgrass, signalgrass, and woolly fingergrass [*Digitaria eriantha* (Steud.) Stapf] growing on an Ultisol.[47,48] During years of low rainfall, productivity of the grasses was similar, with an average gain in weight of about 850 kg/ha/yr for the grazing animals.

Signalgrass was difficult to establish; the fingergrass did not develop a deep root system and was easily uprooted by grazing cattle. Although tannergrass was easily established and competed well with weeds, Soldevila et al.[48] indicated that this grass can be toxic to cattle under certain conditions.

Caro-Costas et al.[49] determined the productivity of irrigated guinea grass, Pangola digitgrass, and napier grass pastures on a level Vertisol in the semiarid region of the island. Napier grass and guinea grass produced similar gains in weight, averaging 1400 kg/ha/yr, while Pangola digitgrass gave an average gain of 1170 kg/ha/yr.

In a two-year experiment,[50] forage production of four grasses was determined in 10 × 10 m enclosures by sampling each plot before and after each grazing period and then calculating the amount consumed by the cattle. Pangola digitgrass, tannergrass, and signalgrass produced an average of 10,421 kg/ha/yr in the experiment compared to 12,173 kg/ha/yr in an adjacent, large-scale grazing experiment.[47]

In a similar experiment with five grasses of two genera,[51] 'Greenalta' limpograss [*Hemarthria altissima* (Poir.) Stapf and E.C. Hubbard] outyielded the other grasses, averaging 15,370 kg/ha/yr of dry forage over a three-year period compared to a combined average of 12,805 kg/ha/yr for star grass, 'Coast-Cross-1' Bermuda grass, and 'Bigalta' limpograss. *Cynodon plectostachyus* (K. Schum.) Pilger (giant star grass) was eliminated from the trials after the second year because of its poor performance.

Star grass produced an average of 12,445 kg/ha/yr of dry forage in this small-plot experiment and 13,820 kg/ha/yr of dry forage in the large-scale experiment. Similar figures for Bigalta limpograss were 12,740 and 13,685 kg/ha/yr.

The results of these trials show that relatively small plots can be used to evaluate the approximate behavior of grasses under actual grazing management and can serve as an intermediate step between small-plot experiments and expensive, large-scale grazing trials.

These studies have proven that intensively managed pastures of grasses such as star grass, guinea grass, napier grass, congograss, and Pangola digitgrass on deep soils in the humid region of Puerto Rico and elsewhere in the humid tropics with similar soils and climate can support on a yearly basis approximately five head/ha of beef cattle with total weight gains of 1000 kg or 2.5 milk cows/ha producing a total of 7500 L of milk with no concentrate feed. This high productivity has been confirmed in many commercial operations on the island.

V. MILK PRODUCTION FROM ALL-GRASS RATIONS

Caro-Costas and Vicente-Chandler determined milk production of Holstein cows fed exclusively on intensively managed, steep pastures of guinea grass, Pangola digitgrass, star grass, and napier grass.[52] With few exceptions, milk production increased during each of five successive lactations. During the fifth lactation, eight of the nine cows each produced more than 4,000 L of milk. Butterfat content of the milk averaged 3.6%. Overall calving interval was 13 months, and the cows maintained an average weight of about 525 kg, which is normal for Holsteins in Puerto Rico.

A private dairy herd of 185 Holstein cows was fed exclusively on intensively managed pasture over a two-year period to determine its performance.[2,53] About 60 ha were used for the milking cows and 15 ha for the dry cows. The pastures were divided into 18 enclosures, which were grazed for one to two days and allowed to rest for three weeks. All pastures were fertilized every three months with 500 kg/ha of 15-5-10 and occasionally limed. The cows grazed on pasture except when being milked so that they could make full use of the abundant forage available. The herd was milked twice daily, at which time a mixture of bone meal and mineralized salt was available to them.

During the two-year experiment, 271 lactations were completed. The first lactations averaged 3340 L and the second, 3826 L. The cows having the lowest production during the first lactation sharply increased their production during the second lactation, those at the second and third production levels maintained their production, but the five cows having the highest production during the first lactation (an average of 5,700 L) produced significantly less milk during the second lactation probably because the pasture material could not provide sufficient nutrients to sustain such high production.

VI. NUTRITIVE VALUE OF GRASSES FOR BEEF AND MILK PRODUCTION

Forages are typically the least expensive source of nutrients for ruminants. The quantity of forage ingested by cattle is probably the most important factor determining its value and largely depends on the availability, palatability, and digestibility of the forage. All of the aforementioned evaluated grasses, if properly managed, are palatable to cattle.

For high levels of production, every effort should be made to have an adequate amount of forage available to cattle at all times. In the various grazing experiments discussed in this chapter, cattle consumed 12 to 13 kg of green forage (2.5 to 2.7 kg of dry forage)/100 kg of liveweight daily, which closely approximates forage intake determined in temperate regions. The quantity of forage ingested was considerably less when cattle were fed older, chopped forage, which is less palatable and digestible.

The tropical grasses tested are similar in digestibility, averaging about 45% for grasses cut every 60 days and 55% for forage consumed by grazing animals.

The nutritive value of grasses is strongly affected by many factors. Protein content increases with N fertilization, and leaves have about twice as much protein, Ca, and P as stems. Digestibility of plant material decreases with age as does protein, Ca, P, and Mg contents, whereas, lignin and dry matter content increase. These changes result from a higher proportion of stems in older forage and from changes in leaf and stem composition as the plant material ages.

Whether grass is grazed or cut and fed markedly affects the nutritive value of forage because grazing animals eat a higher proportion of high-quality leaves and young stems. Data obtained from two adjacent experiments conducted concurrently by Little et al.[54] and Caro-Costas et al.[49] can be used to compare the productivity of grazed and cut napier grass that had been irrigated and well fertilized.

In the experiment with cut napier grass, the daily gain/head averaged 0.4 kg for 18 bulls/ha for a gain in weight/ha/yr of 2,600 kg, whereas bulls on the adjacent pastures of napier grass had a daily average gain/head of 0.6 kg, with 7.3 bulls/ha having a yearly gain of 1,460 kg/ha.

The studies of Carlo et al.[55] also showed a lower daily gain (0.25 kg/head) by young heifers fed on cut grass compared to those on pasture (0.46 kg).

How far forages can go toward meeting the protein and energy requirements of cattle can be determined by comparing their composition with the energy and protein requirements of different classes of cattle as recommended by the Committee on Animal Nutrition of the National Research Council.[56,57]

The 10% protein content required in the feed of most cattle can be met by using grasses cut about every 40 days. Cattle grazing well-fertilized grass pasture consume forage generally containing about 16 to 18% protein, which far exceeds their requirements.

Forages can also meet the dry matter and TDN requirements for normal growth of heifers and steers making a daily gain of 650 g and of breeding beef cows. The consumption in the grazing trials discussed approximates the 2.5 kg of dry forage/100 kg of liveweight required daily by young cattle, but intake of cut grasses of poor quality would be lower and may not satisfy this requirement. The various grazing trials previously discussed show that excellent gains can be made by young cattle grazing properly managed grass systems.

Well-managed grass pastures can also provide all the feed for a 600-kg cow producing 12 L of milk daily with a daily requirement of 7.9 kg of TDN, 1.4 kg of protein, 51 g of Ca, 37 g of P and vitamins. With good-quality pasture plus salt and bone meal, these nutrient quantities can be obtained in 15 kg of dry forage with 53% digestibility and 12% protein. Data on milk production attained by cows fed exclusively on intensively managed tropical pastures corroborate these analyses and show that the first 10 to 12 L of daily milk production can be based exclusively on high-quality grass pastures without any concentrate feed. The energy required for additional milk production, however, must be provided from concentrate feed, which contains more energy/kg of dry matter than forage.

Well-managed grass pastures can, therefore, provide the nutrients required for normal growth of heifers and steers. Dairy cows producing 10 to 12 L of milk daily can also be fed exclusively on well-managed grass pastures. However, higher-yielding milk cows, rapidly fattening cattle, and calves require supplementary feed with a high-energy concentrate such as corn for maximum production levels.

VII. SUPPLEMENTARY FEEDING OF DAIRY COWS AND YOUNG CATTLE ON INTENSIVELY MANAGED PASTURES

The effect of four levels of concentrate feed on milk produced by Holstein cows grazing intensively managed pastures in Puerto Rico was determined by Caro-Costas et al.[58] All 48 cows in the experiment grazed the same well-fertilized pastures.

Milk production, averaging 4757 L/lactation, was not affected by feeding concentrate at rates of 1 kg/2, 4, or 6 L of milk produced daily. However, reducing the concentrate level to 1 kg/8 L of milk sharply reduced production to 3385 L/lactation. Concentrate level did not affect length of lactation or calving interval, which averaged 13.3 months. The most economical level of concentrate feeding (1 kg/6 L of milk) is roughly equivalent to feeding no concentrate for the first 10 L of milk production with 1 kg of concentrate fed for each two additional L produced. This confirms the findings of Caro-Costas and Vicente-Chandler[2,52,53] and McDowell et al.[59] that all-grass rations can be used for the first 10 to 12 L of daily milk production. However, cows on reduced concentrate rations require more forage to meet their nutrient requirements and, therefore, must have continuous access to abundant forage of good quality.

Caro-Costas carried out a three-year experiment with dairy cows comparing two systems of feeding:[60] (1) recommended system: pastures fertilized with 500 kg/ha of 15-5-10 every three months; 2.5 cows maintained/ha and received 0.45 kg of concentrate daily for each L of milk produced in excess of 10 L, but never less than 1 kg of concentrate; and (2) traditional system (used in most dairies in Puerto Rico): pastures fertilized with 300 kg/ha of 15-5-10 every six months, 2.5 cows maintained/ha and received 0.6 kg of concentrate daily for every L of milk produced.

Results obtained with a total of 110 lactations completed for each of the feeding systems showed that, although milk production was slightly less in the recommended system, the cost of concentrate and fertilizer was approximately one-half that of the traditional system.

Yazman et al.[61] determined the effect of four feeding systems at two stocking rates on dairy cows grazing intensively managed pastures. Cows fed exclusively on grass pastures were the lowest yielders, averaging 3450 L/lactation, but also had the lowest cost for feed (pasture and concentrate). When pasture was supplemented with 0.45 kg of concentrate/L of milk produced daily in excess of 10 (recommended system mentioned previously), milk production increased by 1259 L/lactation with the consumption of 814 additional kg of concentrate feed. There was no further significant increase in milk production when the concentrate ration was increased to 0.6 kg/L of daily milk production (traditional system above) or to higher levels.

Mendoza[62] determined that feeding 14% protein concentrate to Holstein heifers grazing intensively managed grass pastures was not economical. In another study,[58] heifers fed only on intensively managed pasture and those on pasture plus a daily ration of 1.6 kg of molasses had similar daily gains and averaged 171 kg of liveweight/head/yr. Those on pasture plus 1.4 kg of corn daily gained an average of 222 kg/yr, but about 510 kg of corn were required to produce this additional gain, which was not economical.

Caro-Costas and Vicente-Chandler[63] found that the cheapest method of feeding heifers was to use pasture exclusively and, if necessary, during periods when forage was scarce, to restrict hours of grazing and feed a supplement of concentrate rather than use "bulky" feed exclusively.

Heifers of the Brahman and Charolais breeds and their crosses were subjected to three different feeding systems to determine effects on weight gain over a period of 360 days.[55,62] In the first experiment, the heifers made similar gains in weight (0.50 kg/head/day) when fed on bulky feed, intensively managed pasture, or pasture for 180 days then bulky feed for 180 days. Heifers fed cut Pangola digitgrass had much lower gains, averaging only 0.26 kg/head/day. Feeding exclusively on pasture was clearly the most economical alternative. In the second experiment of the same duration, heifers fed on pasture alone or on pasture plus 1.4 kg of molasses/head/day made similar gains, whereas those on pasture plus 1.4 kg of corn/head/day made slightly higher gains, but the cost of corn was too great for this practice to be economical.

Giving young bulls 3 kg/day of a urea-molasses supplement in addition to heavily fertilized, chopped napier grass cut every 45 to 60 days and fed *ad libitum* increased the gain/head from 0.40 to 0.68 kg/day compared to another group fed only chopped napier grass.[54]

From these experiments, we can conclude that: (1) cows producing up to about 3,500 L/lactation can be fed all-grass rations from well-managed pastures; (2) more productive cows can derive all the nutrients required daily for their first 10 to 12 L of milk from well-managed pastures, with higher production supplemented at the rate of about 0.5 kg of concentrate/L of milk in excess of 10 L; (3) feeding corn or molasses to young cattle on intensively managed pastures usually is not economical; and (4) on the other hand, supplementary feeding of concentrate to young cattle given poor-quality cut forage is often warranted.

VIII. MAKING FORAGE AVAILABLE THROUGHOUT THE YEAR

Seasonal variations in growth of grasses result in waste during periods of fast growth and shortages at other times. In much of the tropics, season of the year affects the yield of grasses, which is lower during dry periods and during the "winter" months when shorter days, slightly cooler weather, and increased flowering occur.

One solution to this problem is to store excess forage produced during seasons of fast growth as hay or silage. Composition of silage or hay is similar to that of fresh forage although 10 to 20% of the nutrients may be lost even with good management. Haylage and pelleting should also be considered. Whatever method of forage conservation is selected, the costs involved must be carefully considered.

Grasses should have the same type of management for conservation as green chop feeding. All the grasses mentioned in this chapter can be ensiled, and all except napier grass can produce good hay.

Many experiments have been conducted on the production of high-yielding forages for silage or hay on mechanizable, irrigated lands. Caro-Costas found that four sorghum-sudan grass hybrids fertilized with 15-5-10 at the rate of five t/ha/yr produced an average of 39,815 and 54,810 kg/ha/yr of dry forage when harvested every 45 and 60 days, respectively.[64] These grasses are well suited for ensiling or pelleting.

Signalgrass, Pangola digitgrass, and congograss produced similar yields for Sotomayor-Ríos et al.[27-29] averaging higher dry forage production (36,800 kg/ha/yr) than star grass (30,790 kg/ha/yr). They can be used for ensilage, hay, or haylage.

Little et al.[54] found that, on the south coast of Puerto Rico, well-fertilized, irrigated napier grass produced 57,100 kg of dry forage, which is equivalent to 300 t/ha/yr of green forage.

It is usually not economical to use supplementary irrigation in the humid regions of the tropics; but, in the semiarid regions, irrigation greatly increases forage production and is often justified in Puerto Rico.

The problem caused by seasonal variations in forage yield can be solved or ameliorated in other ways. The simplest solution is to sell part of the grazing livestock before the season of slow grass growth, but this is often impractical and may not be profitable.

Cattle may be carried over during periods when forage is scarce by feeding supplemental concentrate. This practice is useful on small farms or where steepness of the land precludes mechanized haying or ensiling operations.

By deferring the grazing of some pastures during the fall, considerable forage of lower quality can be carried over in the field for grazing during the "winter" or drier season. This practice can also help to improve the stand of desirable grasses. Pastures should be fertilized so that abundant forage is carried over into the season of slower growth, and grazing should be less intensive at this time so that the grasses can build up carbohydrate reserves.

During the critical last weeks of the winter or dry season, cattle can be allowed to graze a pasture close to the ground. This generally undesirable practice, if limited to once a year, has no lasting detrimental effect on the pasture, especially since it will likely be undergrazed during the following season of fast forage growth. Young cattle forced to graze this less nutritious forage of stems and older leaves may not gain or may even lose weight, but these losses are usually made up quickly during the following season of fast forage growth. Supplementing pasture with small quantities of concentrate when forage is scarce may prevent weight loss. Lactating cows should receive more concentrate when forced to overgraze pastures or when feeding on cut grass rather than grazing.

Forage yields can be increased during critical months by shifting from a grazing to a cut-feed system. This results in higher yields and better utilization of forage than can be obtained from grazed pasture. Changing from year-round grazing to grazing in the "summer" months and cut-feed systems accompanied by heavier fertilization of pastures in the "winter" can help to equalize forage availability throughout the year as shown by Vicente-Chandler et al.[5]

Utilization of cut forage can be increased by feeding it chopped, but more concentrate may be required to compensate for the higher proportion of lower-quality stems ingested.

Yields during critical periods can also be increased by using a longer interval between cuttings, but this will also result in lower forage quality.

Thus, the problem posed by the uneven growth of grasses throughout the year can be alleviated by storing forage and by careful management.

IX. HOW MUCH TO INTENSIFY FORAGE PRODUCTION

High and economical yields of forage can be attained only through the use of all required practices properly carried out at the right time with due regard given to their interrelationships. It is wasteful, for example, to fertilize pastures heavily when there is little moisture, or to not utilize them fully by grazing animals inefficiently or by using cattle infested with parasites. Conversely, it is counterproductive to use good, healthy cattle and intensive grazing practices on unfertilized, low-yielding pastures of these grasses in the environment and economic conditions existing in Puerto Rico.

The following tabulation shows how the forage yield of napier grass can be increased by combining all desirable practices:

Practice	Dry Forage Produced (kg/ha/yr)
Unfertilized napier grass	9,000
Well-fertilized napier grass cut too frequently	17,000
Well-fertilized napier grass cut every 40 to 60 days leaving too high a stubble	31,000
Well-fertilized napier grass with best harvest interval and cutting height	37,000
Well-managed, irrigated napier grass	50,000

To determine how much to intensify forage production, the main decisions to be made are: (1) whether to use improved or native forages, (2) whether to use a grass-legume combination or grasses fertilized with N, (3) whether to use cut grass or pasture, (4) how much fertilizer to apply and how often, (5) how much to intensify grazing management, and (6) what are the necessary provisions to assure an adequate supply of forage during seasons of slow pasture growth.

Volunteer native grasses are well suited for the unimproved pasture environment and generally tolerate mismanagement better than improved grasses. However, improved forages respond much more readily to intensive management and, with proper care, far outyield native grasses and are more nutritious.[4] Therefore, the first step in intensifying forage production is to establish the most adaptable improved grasses.

Well-fertilized grasses produce much more forage than do grass-legume mixtures, but the latter have a higher protein content. Caro-Costas and Vicente-Chandler found that napier grass fertilized with 300 kg/ha/yr of N produced almost twice as much forage and protein as a tropical kudzu (*Pueraria phaseoloides*)-napier grass mixture.[65]

It is difficult to maintain a desirable proportion of grass and legume in a pasture association. Even with good management, one of the species (often the grass) will predominate after several years, and the pasture must then be replanted or the forage legume resown.

Grass-legume mixtures produce less forage, but require less fertilization than grasses. N can be provided by the legume and the soil. Ca, K, and Mg can be provided by most soils and P by modest applications of superphosphate. However, legumes are more exacting than grasses in their nutrient requirements and many need more lime and often minor element applications when grown in the leached soils (Ultisols and Oxisols) of the humid tropics.

The productivity of grass-legume mixtures is dependent on the capacity of the legume to fix N; and, when N is added to the mixed pasture, it usually results in greater growth of the grass at the expense of the legume. Caro-Costas and Vicente-Chandler[66] found that the application of 220 kg/ha/yr of N increased the production of tropical kudzu-molassesgrass pastures, but resulted in kudzu being crowded out of the association.

Legume longevity can vary, and the survival of a pasture legume can depend not only on animal stocking rate, competitiveness of the companion grass, type and rate of fertilizer applied, and climatic and soil conditions, but also on factors such as reproduction potential (length of period of seed production, viability of seed, and extent of seedling regeneration), rhizome and stolon development, *Rhizobium* requirement and availability, and susceptibility to diseases and pests.

It can be concluded that, where heavy fertilization is economically warranted and intensive management is possible, grass pastures are usually preferable to grass-legume mixtures since they produce more forage and are easier to manage.

As stated previously, animals consume grass of better quality when grazing rather than feeding on cut material. About 40% less grass is harvested from grazed pasture than under cutting management, but its protein content is about twice as high. Forage consumed by grazing animals also has a higher digestibility, and its Ca and P contents are generally higher.

Examples demonstrating that the superior quality of the forage consumed under grazing management is reflected in better performance by the livestock have already been given.

Although the nutritional aspect of the pasture or cut forage decision is important, the ultimate criterion should be the system that produces animal products more economically. In the cut forage vs. pasture experiment with napier grass previously cited, the additional 1,140 kg of gain in weight produced by feeding cut grass must be weighed against the cost of cutting, chopping, and feeding the forage and the cost of the additional fertilizer and greater investment in cattle.

How much fertilizer to apply to forage grasses depends on many factors as discussed previously.

How much to intensify grazing management depends on water accessibility, topography, land values, current market value of livestock products, forage species utilized, and, particularly, quantity and cost of fertilizer applications. Improved, fertilized grass pastures should be grazed in rotation, and "ration grazing" using small enclosures grazed for one to two days followed by three weeks of rest may often be justified.

For the sake of convenience, six increasingly intensive levels of pasture management and forage production can be distinguished under humid or irrigated tropical conditions, although many combinations of these systems are possible.

1. **Unfertilized, native, volunteer pastures.** These are low yielding, and their use should be limited to areas having adverse conditions such as shallow soils or to semiarid regions.
2. **Lightly fertilized pastures of improved grasses** with a carrying capacity of about two 270-kg animals/ha through the year. Rather large, rotational pastures of this type are grazed for 1 to 2 weeks then given 4 to 5 weeks of rest.
3. **Heavily fertilized pastures** carrying five 270-kg animals or 2.5 dairy cows/ha. About 2 t/ha/yr of fertilizer are applied in four applications. Pastures are rotationally grazed for about 1 week, then given 3 to 4 weeks of rest. Pastures are grazed at close to maximum rates the year round through careful management, with occasional supplementary feeding during periods of slow forage growth.
4. **Very heavily fertilized pastures** carrying up to seven 270-kg animals or 3.5 dairy cows/ha. About 3 t/ha/yr of fertilizer are applied in four applications. Cattle are confined to an area containing sufficient forage for only 2 days of grazing, which is then given about 3 weeks of rest. Pastures are grazed at the maximum rate all year by using careful management and by providing supplementary feed to grazing animals during seasons of slow grass growth.
5. **Freshly cut grass.** Up to twelve 270-kg animals or six dairy cows/ha can be maintained. Four t/ha/yr of fertilizer are applied in six applications. Grass is harvested every 40 to 60 days and supplemented with silage or hay during seasons of slow growth. In addition, some concentrate is needed for young cattle and additional concentrate for dairy cows to compensate for the inferior nutritional quality of the cut forage as compared to the grazed grass.

6. **Silage or hay.** Maximum production and use are attained when the grass is fertilized at the rate of 4 t/ha/yr, irrigated as required, cut at 40 to 60 day intervals, and stored as silage or hay to be used as needed. Under such a system, seasonal variations in growth are only a minor problem, and cattle receive forage of uniform quality throughout the year. Some concentrate should be fed to young cattle and the concentrate ration of dairy cows increased above the level provided when the cows graze good pasture.

Intensification of management involves utilization of the less digestible portions of forage often at the expense of the performance of individual animals, and should be pursued only to the point of producing maximum net returns from forages and livestock.

The most desirable level of intensification of forage production must be determined for each farm based on careful study, planning, and experience. The training and experience of the farmer and his/her desire to make maximum use of the available land are of prime importance. Other factors to consider are the current value of the animal products to be marketed and the price of land, fertilizer, concentrate, and other inputs.

X. EFFECTS OF INTENSIVELY MANAGED GRASSLAND ON SOIL CONDITIONS

The effect of well-managed pastures on reducing erosion has been discussed by Vicente-Chandler et al.[5] and Barnett et al.[67] These authors also found that high yields of Pangola digitgrass could be obtained on both subsoil and topsoil of an Ultisol with proper fertilization.

The following tabulation shows that the organic matter, N content, and volume of large pores in an Ultisol formerly having virgin forest decreased markedly after 15 years as unfertilized pasture, but these factors still remained high by most standards.[5]

Factors Determined	Virgin Woodland %	Unfertilized Pasture for 15 Years %
Organic matter content	6.5	3.7
Nitrogen content	0.3	0.2
Pores drained at 0.3 bars	16.0	9.3

On the other hand, the heavy applications of fertilizer and lime required to produce high yields of forage can drastically alter soil conditions and affect yields of subsequent crops. Heavy applications of N from residually acid sources can rapidly build up exchangeable Al and/or Mn in the soil to levels that are harmful to forages, but this problem can be prevented by proper liming. Abruña et al.[68-70] showed that when various crops were planted in former pastures that had been well fertilized and limed, they had yields several times higher than when planted in unlimed pasture land.

Extremely high levels of forage production and the management practices necessary to achieve such high production in the humid tropics have been demonstrated. Despite potential profitability in many specific situations, such as milk production and finishing grass-fed beef cattle, this technology has not been as widely used beyond Puerto Rico as the economic benefits appear to justify. Recent concerns about contributions of agricultural practices to environmental degradation[71] may appear to further discourage the use of such technology. However, it has also recently been emphasized that global efforts to minimize clearing of rain forests, release of carbon dioxide, and concentration of nutrients must provide alternative pathways of economic development and food production for expanding human populations.[72] Management approaches such as more frequent application of smaller quantities of fertilizer, use of less readily leached nutrient sources, and use of unfertilized buffer strips, perhaps of other species, could reduce the danger of excessive nutrient loss from runoff and leaching. Nutrient management must include protection of groundwater from contamination and surface waters from eutrophication. The intensive management of erodable tropical soils as permanent pasture greatly reduces erosion potential from levels typical of cropland. High levels of production from intensification reduce the area required for a particular amount of production and can serve to reserve even greater areas for other uses such as forest, wildlife, or recreation. Thus, high-input tropical forage production, appropriately practiced, can be one component among the alternatives available to develop an agriculture capable of feeding an increasing global population, meeting local economic needs, and maintaining environmental integrity.

REFERENCES

1. Caro-Costas, R. and Vicente-Chandler, J., Comparative productivity of intensively managed Star and Guinea grass pastures in terms of milk production in the humid mountain region of Puerto Rico, *J. Agric. Univ. P. R.*, 63, 436, 1979.
2. Caro-Costas, R. and Vicente-Chandler, J., Feeding Holstein cows exclusively on well managed tropical grass pastures, *J. Agric. Univ. P. R.*, 64, 47, 1980.
3. Vicente-Chandler, J., Abruña, F., Caro-Costas, R., Figarella, J., Silva, S., and Pearson, R. W., *Intensive Grassland Management in the Humid Tropics of Puerto Rico*, Agric. Exp. Stn., Univ. P. R., Bull. 233, 1974.
4. Vicente-Chandler, J., Caro-Costas, R., Pearson, R. W., Abruña, F., Figarella, J., and Silva, S., *The Intensive Management of Tropical Forages in Puerto Rico*, Agric. Exp. Stn., Univ. P. R., Bull. 187, 1964.
5. Vicente-Chandler, J., Caro-Costas, R., Abruña, F., and Silva, S., *Producción y Utilización Intensiva de las Forrajeras en Puerto Rico*, Agric. Exp. Stn., Univ. P. R., Bull. 271, 1984.
6. Vicente-Chandler, J., Figarella, J., and Silva, S., Effects of nitrogen fertilization and frequency of cutting on the yield and composition of Pangola grass in Puerto Rico, *J. Agric. Univ. P. R.*, 45, 37, 1961.
7. Vicente-Chandler, J. and Pearson, R. W., Nitrogen fertilization of hot climate grasses, *Soil Conserv. Mag.*, 26 (12), 269, 1960.
8. Vicente-Chandler, J., Silva, S., and Rodríguez, J., Effect of nitrogen fertilization and frequency of cutting on the yields of: I. Napier grass, II. Guinea grass, and III. Para grass, *J. Agric. Univ. P. R.*, 43, 215, 1959.
9. Vicente-Chandler, J., Silva, S., and Figarella, J., The effect of nitrogen fertilization and frequency of cutting on the yields and composition of three tropical grasses, *Agron. J.*, 51, 202, 1959.
10. Vicente-Chandler, J., Silva, S., and Figarella, J., Effect of frequency of application on response of Guinea to nitrogen fertilization, *J. Agric. Univ. P. R.*, 46, 342, 1962.
11. Caro-Costas, R., Abruña, F., and Figarella, J., Effect of nitrogen rates, harvest interval and cutting height on yield and composition of Star grass in Puerto Rico, *J. Agric. Univ. P. R.*, 56, 267, 1972.
12. Vicente-Chandler, J. and Figarella, J., Effect of five nitrogen sources on yield and composition of Napier grass, *J. Agric. Univ. P. R.*, 46, 102, 1962.
13. Abruña, F., Figarella, J., and Caro-Costas, R., Efficiency of sulfur-coated urea and potassium chloride compounds applied to Star grass growing on an Ultisol under humid tropical conditions, *J. Agric. Univ. P. R.*, 60, 310, 1976.
14. Abruña, F., Potassium supplying power of the major upland Inceptisols of Puerto Rico, *J. Agric. Univ. P. R.*, 64, 91, 1980.
15. Abruña, F., Vicente-Chandler, J., Figarella, J., and Silva, S., Potassium supplying power of the major Ultisols and Oxisols of Puerto Rico, *J. Agric. Univ. P. R.*, 55, 45, 1976.
16. Vicente-Chandler, J., Fertilization of humid tropical grasslands, in *Forage Fertilization*, D. A. Mays, Ed., Am. Soc. of Agron., Madison, WI, 279, 1974.
17. Vicente-Chandler, J., Pearson, R. W., Abruña, F., and Silva S., Potassium fertilization of intensively managed tropical grasses under humid tropical conditions, *Agron. J.*, 54, 450, 1962.
18. McCluny, A. and Quinn, L. R., Sulphur and phosphorus responses of Batatais grass (*Paspalum notatum*), IBEC Res. Inst. Rpt. 18, New York, N.Y., 1968.
19. Figarella, J., Vicente-Chandler, J., Silva, S., and Caro-Costas, R., Effect of phosphorus fertilization on productivity of intensively managed grasses under humid tropical conditions in Puerto Rico, *J. Agric. Univ. P. R.*, 48, 236, 1964.

20. Abruña, F., Pearson, R. W., and Elkins, C., Quantitative evaluation of soil reaction and base status changes resulting from heavy applications of residually acid-forming nitrogen fertilizers, in *Proc. Soil Sci. Soc. Am.*, 22, 539, 1958.
21. Abruña, F., Vicente-Chandler, J., Pearson, R. W., and Figarella, J., Effects of liming on yields and composition of heavily fertilized grasses and on soil condition under humid tropical conditions, in *Proc. Soil Sci. Soc. Am.*, 18, 657, 1964.
22. Vicente-Chandler, J., Silva, S., Abruña, F., and Rodríguez, J. A., Effect of two cutting heights, four harvest intervals and five nitrogen rates on yield and composition of Congo grass under humid tropical conditions, *J. Agric. Univ. P. R.*, 56, 280, 1972.
23. Coward-Lord, J., Arroyo-Aguilú, J. A., and García-Molinari, O., Fibrous carbohydrate fractions and *in vitro* true and apparent digestibility of 10 tropical forage grasses, *J. Agric. Univ. P. R.*, 58, 293, 1974.
24. Arroyo-Aguilú, J. A., Tessema, S., McDowell, R. E., Van Soest, P. J., Ramírez, A., and Randel, P. F., Chemical composition and *in vitro* digestibility of five heavily fertilized tropical grasses in Puerto Rico, *J. Agric. Univ. P. R.*, 59, 186, 1975.
25. Vélez-Santiago, J., Sotomayor-Ríos, A., and Torres-Rivera, S., Effect of three harvest intervals and two fertilizer rates on the yield and HCN content of ten *Cynodon* cultivars, *J. Agric. Univ. P. R.*, 63, 35, 1979.
26. Sotomayor-Ríos, A., Acosta-Matienzo, A., and Vélez-Fortuño, J., Evaluation of seven forage grasses at two cutting stages, *J. Agric. Univ. P. R.*, 57, 173, 1973.
27. Sotomayor-Ríos, A., Juliá, F., and Arroyo-Aguilú, J. A., Effects of harvest intervals on the yield and composition of 10 forage grasses, *J. Agric. Univ. P. R.*, 58, 448, 1974.
28. Sotomayor-Ríos, A., Rodríguez-Garcia, J., and Silva, S., Yield comparison of four forage grasses at two cutting heights and three harvest intervals, *J. Agric. Univ. P. R.*, 58, 26, 1974.
29. Sotomayor-Ríos, A., Vélez-Santiago, J., Torres-Rivera, S., and Silva, S., Effect of three harvest intervals on yield and composition of nineteen forage grasses in the humid mountain region of Puerto Rico, *J. Agric. Univ. P. R.*, 60, 294, 1976.
30. McDonald, H. A., The relationship of the stages of growth to the yield and chemical composition of forage grasses and legumes, unpublished graduate thesis, Cornell Univ., Dept. of Agron., Ithaca, N.Y., 1946.
31. Van Soest, P. J., Wein, R. H., and Moore, L. A., Estimation of the true digestibility of forages by the "in vitro" digestion of cell walls, in *Proc. 10th Intern. Grassl. Conf.*, Helsinki, Finland, 1966, 438.
32. Tessema, S., Nutritional value of some tropical grass species compared to some temperate grass species, unpublished doctoral thesis, Cornell University, Ithaca, N. Y., 1972.
33. Arroyo-Aguilú, J. A. and Coward-Lord, J., Mineral composition of 10 tropical forage grasses in Puerto Rico, *J. Agric. Univ. P. R.*, 58, 426, 1974.
34. Caro-Costas, R. and Vicente-Chandler, J., Effect of two cutting heights on yields of five tropical grasses, *J. Agric. Univ. P. R.*, 45, 46, 1961.
35. Vicente-Chandler, J., Silva, S., Rodríguez, J., and Abruña, F., Effect of two heights and three intervals of grazing on the productivity of a heavily fertilized Pangola grass pasture, *J. Agric. Univ. P. R.*, 56, 110, 1972.
36. Rodríguez, J. and Silva, S., Effect of two heights and three intervals of grazing on stand of a heavily fertilized Star grass pasture, *J. Agric. Univ. P. R.*, 59, 215, 1975.
37. Caro-Costas, R. and Vicente-Chandler, J., Effect of three grazing intervals on carrying capacity and weight gains produced by Star grass pastures, *J. Agric. Univ. P. R.*, 65, 14, 1981.
38. Peterson, R. S., Woodhouse, W. W., and Lucas, H. L., The distribution of excreta by freely grazing cattle and its effect on pasture fertility. II. Effect of returned excreta on the residual concentration of some fertilizer elements, *Agron. J.*, 48, 444, 1956.
39. Caro-Costas, R., Vicente-Chandler, J., and Figarella, J., The yield and composition of five grasses growing in the humid mountains of Puerto Rico as affected by nitrogen fertilization, season, and harvest procedure, *J. Agric. Univ. P. R.*, 44, 107, 1960.
40. Figarella, J., Abruña, F., and Vicente-Chandler, J., Effect of five nitrogen sources applied at four rates to Pangola grass sod under humid tropical conditions, *J. Agric. Univ. P. R.*, 56, 410, 1972.
41. Caro-Costas, R. and Vicente-Chandler, J., Effect of fertilization on carrying capacity and beef produced by Napier grass pastures, *Agron. J.*, 53, 204, 1961.

42. Caro-Costas, R., Vicente-Chandler, J., and Abruña, F., Effect of four levels of fertilization on beef production and carrying capacity of Pangola grass pastures in the humid mountain region of Puerto Rico, *J. Agric. Univ. P. R.*, 56, 219, 1972.
43. Caro-Costas, R., Abruña, F., and Vicente-Chandler, J., Effect of three levels of fertilization on the productivity of Star grass pastures growing on a steep Ultisol in the humid mountain region of Puerto Rico, *J. Agric. Univ. P. R.*, 60, 172, 1976.
44. Caro-Costas, R., Vicente-Chandler, J., and Figarella, J., Productivity of intensively managed pastures of five grasses on steep slopes in the humid mountains of Puerto Rico, *J. Agric. Univ. P. R.*, 40, 99, 1965.
45. Caro-Costas, R., Abruña, F., and Vicente-Chandler, J., Comparison of heavily fertilized Pangola and Star grass pastures in terms of beef production and carrying capacity in the humid mountain region of Puerto Rico, *J. Agric. Univ. P. R.*, 56, 104, 1972.
46. Caro-Costas, R., Vicente-Chandler, J., and Abruña, F., Comparison of heavily fertilized Congo, Star, and Pangola grass pastures in the humid mountain region of Puerto Rico, *J. Agric. Univ. P. R.*, 60, 79, 1976.
47. Caro-Costas, R., Weight gains of heifers fed on five grass pastures intensively managed in the humid hill region of Puerto Rico, *J. Agric. Univ. P. R.*, 64, 43, 1980.
48. Soldevila, M., Green-Ortiz, J. J., Sotomayor-Ríos, A., Arroyo-Aguilú, J. A., and Vélez-Santiago, J., *Alimentación de Novillas Holstein a Base Exclusiva de Pastos Tanner, Pangolas Común y Milanjiana, Guinea y Signal Cultivados Adecuadamente Durante las Cuatro Estaciones del Año*, Agric. Exp. Stn., Univ. P. R., Publ. 127, 1979.
49. Caro-Costas, R., Vicente-Chandler, J., and Burleigh, C., Beef production and carrying capacity of heavily fertilized irrigated Guinea, Napier, and Pangola grass pastures on the semiarid south coast of Puerto Rico, *J. Agric. Univ. P. R.*, 45, 32, 1961.
50. Rodríguez, J., Rivera, E., and Vicente-Chandler, J., Productivity of four intensively managed grasses under grazing management in the humid hill region of Puerto Rico, *J. Agric. Univ. P. R.*, 64, 236, 1980.
51. Rivera, E. and Rodríguez, J., Forage yield of five grasses under intensive grazing management in the humid region of Puerto Rico, *J. Agric. Univ. P. R.*, 64, 259, 1980.
52. Caro-Costas, R. and Vicente-Chandler, J., Milk production by Holstein cows fed only on grass during five consecutive lactations, *J. Agric. Univ. P. R.*, 60, 436, 1979.
53. Caro-Costas, R. and Vicente-Chandler, J., *Producción Comercial de Leche con Vacas Alimentadas Exclusivamente con Buenos Pastos en la Altura Húmeda de Puerto Rico*, Agric. Exp. Stn., Univ. P. R., Publ. 126, 1979.
54. Little, S. N., Vicente-Chandler, J., and Figarella, J., The productivity of irrigated Napier, Guinea, and Pangola grasses as affected by nitrogen fertilization, *Agron. J.*, 51, 111, 1959.
55. Carlo, I., Mendoza, R., and Vélez, J., Evaluation of different feeding systems in beef cattle production, *J. Agric. Univ. P. R.*, 56, 327, 1972.
56. Anonymous, Report on pasture investigation techniques, Joint Committee of the American Dairy Society, American Dairy Science Association, and American Society of Animal Production, *J. Dairy Sci.*, 26, 53, 1943.
57. Anonymous, *Nutrient Requirements of Dairy Cattle* and *Nutrient Requirements of Beef Cattle*, Fifth Revised Edition, Committee on Animal Nutrition of the National Research Council, National Academy of Sciences, Washington, D.C., 1976 and 1978.
58. Caro-Costas, R., Vicente-Chandler, J., and Abruña, F., Effect of four levels of concentrate feeding on milk production by Holstein cows grazing intensively managed tropical grass pastures, *J. Agric. Univ. P. R.*, 56, 104, 1972.
59. McDowell, R. E., Cestero, H., Rivera-Brenes, J. E., Román-Garcia, F., Arroyo-Aguilú, J. A., Berrocal, C. M., Soldevila, M., López-Alberty, J. C., and Metz, S. W., *Tropical Grass Pastures With and Without Supplement for Lactating Cows in Puerto Rico*, Agric. Exp. Stn., Univ. P. R., Bull. 238, 1975.
60. Caro-Costas, R., unpublished data, 1984.
61. Yazman, J. A., McDowell, R. E., Cestero, H., Ròman-Garcia, F., and Arroyo-Aguilú, J. A., Supplementation of dairy cows grazing intensively managed tropical grass pastures at two stocking rates, *J. Agric. Univ. P. R.*, 63, 281, 1979.
62. Mendoza, R., *Comparación de Tres Sistemas de Alimentación en la Crianza de Novillas para Reemplazo*, Agric. Exp. Stn., Univ. P. R., Publ. 115, 1977.

63. Caro-Costas, R. and Vicente-Chandler, J., Effect of supplementary feeding of concentrates and of restricted grazing on Holstein heifers on intensively managed grass pastures, *J. Agric. Univ. P. R.*, 63, 336, 1979.
64. Caro-Costas, R., Effect of harvest interval on yields of four Sudan grass crosses with irrigation on the south coast of Puerto Rico, *J. Agric. Univ. P. R.*, 65, 108, 1981.
65. Caro-Costas, R. and Vicente-Chandler, J., Comparative productivity of merker grass and of a kudzu-merker grass mixture as affected by season and cutting height, *J. Agric. Univ. P. R.*, 40, 144, 1956.
66. Caro-Costas, R. and Vicente-Chandler, J., Effect of liming and fertilization on productivity and species balance of a tropical kudzu-molasses grass pasture under grazing management, *J. Agric. Univ. P. R.*, 47, 231, 1963.
67. Barnett, A. P., Abruña, F., Jackson, W. A., Darling, A. E., and Halloday, J. H., Soil and nutrient losses in runoff with selected cropping treatments in tropical soils, *Agron. J.*, 64, 391, 1972.
68. Abruña, F., Badillo, J., Silva, S., and Vicente-Chandler, J., Crop response to soil acidity factors in Ultisols and Oxisols in Puerto Rico — Soybeans, *J. Agric. Univ. P. R.*, 62, 90, 1978.
69. Abruña, F., Juárez, J., Rodríguez, R., and Vicente-Chandler, J., Effect of soil acidity and liming on yields and composition of sugarcane growing on an Ultisol, *J. Agric. Univ. P. R.*, 52, 85, 1968.
70. Abruña, F., Vicente-Chandler, J., Rodríguez, J., Badillo, J., and Silva, S., Crop response to soil acidity factors in Ultisols and Oxisols in Puerto Rico. V. Sweet potato, *J. Agric. Univ. P. R.*, 63, 250, 1979.
71. Schiere, J. B. and Van Keulen, H., Rethinking high input systems of livestock production: a case study of nitrogen emissions in Dutch dairy farming, *Tropical Grasslands*, 33, 1, 1999.
72. CAST, *Animal Agriculture and Global Food Supply*, Council for Agricultural Science and Technology, Ames, Iowa, 1999.

11 *Pennisetums* and *Sorghums* in an Integrated Feeding System in the Tropics

Wayne W. Hanna and Salvio Torres-Cardona

CONTENTS

I. INTRODUCTION

Sorghum Moench and *Pennisetum* Rich. are two of the most important grass genera for the tropics. Each genus includes an important species used for food, forage, fuel, and building in many parts of the world, while other, lesser-known species in these genera are important forage producers. *Sorghum* and *Pennisetum* species are usually grown in areas where environmental conditions, especially low rainfall, are too harsh to grow maize (*Zea mays* L.).

A. Sorghum

Sorghum bicolor (L.) Moench is the most common sorghum grown throughout the world. It is planted on over 45 million ha around the world and is considered the fifth most important grain crop worldwide, exceeded only by wheat, maize, rice, and barley. Eighty percent of the sorghum-growing area is located in the developing countries, most of which is in the tropics. In these areas, sorghum has the potential to play an important role in reducing hunger either directly through human consumption or indirectly through its use as feed to produce meat, eggs, and milk. The use of hybrids in the tropics is negligible, except in India, Thailand, Zimbabwe, Nicaragua, and

Colombia. Animal use of sorghum grain around the world has more than doubled since the 1960s, while food use has remained about the same.[1]

S. bicolor is also the major sorghum species used for forage. This species is morphologically diverse, ranging from thick-stemmed, low-tillering grain types to thin-stemmed, grassy, multiple-tillering types. Most sorghum used for forage in the tropics is tall (2 to 4 m), of the thick-stemmed landrace type, and is used as a dual purpose grain and forage crop. The most popular improved forage sorghum hybrids are usually from crosses between a 3-dwarf, cytoplasmic-nuclear male sterile (cms) type and Sudan grass (var. *sudanense* Hitchc.). Sudan grass is thin-stemmed, leafy, produces multiple tillers, and imparts these characteristics to the hybrid. Sorghum x Sudan grass hybrids are vigorous with high dry matter yield potential.

Johnsongrass (*S. halepense* (L.) Pers.) is a rhizomatous weed that is well adapted to many areas of the world, including the tropics. It is not usually cultivated, but is probably an important source of forage for livestock of many small farmers. Johnsongrass is thin-stemmed, leafy, produces many tillers, and can be a high-quality forage in the vegetative stage, although its hydrocyanic acid potential (HCN-p) content may be higher in older, more mature material. The reputation of Johnsongrass as a weed comes from its production of rhizomes that are difficult to destroy in unwanted places. Johnsongrass will sometimes cross naturally with *S. bicolor* to produce vigorous, rhizomatous hybrids.

B. *Pennisetum*

The *Pennisetum* genus has over 140 species.[2] Most species are native to Africa and South and Central America, but *Pennisetum* species can be found in almost every country of the tropics and subtropics. Pearl millet, *P. glaucum* (L.) R. Br., grown on over 25 million hectares is the sixth most important grain crop in the world.[3] It is used as a dual purpose annual crop, mainly in the drier areas of the tropics, although mature dry stover with grain removed has low feeding value. Most cultivars are landraces with large diversity in maturity, plant type, and morphological characteristics. Pearl millet hybrids produce high forage yields, but are mainly used in the United States, India, and Australia.

Elephant grass, *P. purpureum* Schumach., is probably the most important forage species in the genus for the tropics. It is perennial, rhizomatous with high dry matter yield potential, and can produce high quality forage if it is managed properly. Interspecific forage hybrids can be readily made between pearl millet and elephant grass that combine the leafiness and quality of the millet with the dry matter yield potential and perennial growth habit of elephant grass. Improved interspecific male-sterile and female-sterile hybrids can be produced commercially by pollinating a cms pearl millet with elephant grass pollen.[4] Superior plants from the interspecific cross such as NB21[3] in India can be vegetatively propagated. Research also has shown that, by doubling the chromosome number of the interspecific hybrids, male and female fertility can be restored and the hybrid propagated by seed.[5]

There are many other *Pennisetum* species that contribute to noncultivated forage production in the tropics. Among these are *P. squamulatum*, *P. flaccidum*, *P. orientale*, *P. polystachion*, and *P. pedicellatum* Trin. (dinanathgrass). Most of these grasses are small-seeded apomicts that combine excellent dry matter yield with good quality and palatability.[6]

C. Characteristics of Genera

Both *Sorghum* and *Pennisetum* species are widely used in the tropics where the moisture and fertility are too low to grow maize. Although both genera tend to make efficient use of soil moisture and fertility, the *Pennisetums* tend to excel in the tropics. One reason for this is that *Pennisetum* roots tolerate lower soil pH and higher Al concentration.[3,7] Ritchey et al.[8] indicated that the potential of sorghum in the tropics is limited by its sensitivity to soil acidity. Maunder[1] stated that high soil acidity or low soil pH limits sorghum production in 75% of Africa. However, sorghums are being selected and developed with tolerance to lower soil pH and higher Al concentrations.[9]

Sorghums have a wider range of adaptability than *Pennisetums*. Most *Pennisetums* do not tolerate waterlogged soils.

One hazard of sorghum or Sudan grass utilization is the potential toxicity of dhurrin, which yields hydrocyanic acid when hydrolized enzymatically in disrupted plant tissues or in the rumen of animals that consume the forage. To alleviate the possible lethal effect of high levels of hydrocyanic acid in the plant, breeders have developed forage sorghums, Sudan grass, and crosses with lower HCN-p. In the tropics, excellent single-cross and three-way forage sorghum hybrids have been identified with low HCN-p and high yielding ability. Hay making also reduces the glucoside content, but cannot eliminate the risk of poisoning in livestock, especially since hay can be ingested at a faster rate than grazed forage.[10] *Pennisetums* do not contain any HCN-p.

A large portion of the tropics has distinct wet and dry seasons with dry periods in the wet season. The drought tolerance, ability to go "semidormant" during stress, rapid response to moisture, and tillering ability to take advantage of ideal growing conditions of *Pennisetums* and *Sorghums* help to make them good crop candidates for tropical growing conditions.

II. AGRONOMIC CHARACTERISTICS

A long and warm growing season favors forage production in the tropics. However, distinct wet and dry seasons can result in an overabundance of forage for part of the year and limited supplies in other parts of the year. Akinola et al.[11] indicated that irrigation could at least double total dry matter and crude protein yields by extending the growing season through the approximately seven dry months per year in Nigeria. French et al.[12] indicated that, although sorghums are used for grazing in Australia, drought makes forage supplies unreliable. The establishment of annuals such as sorghum and pearl millet each year can be risky, especially in areas where moisture supplies are unreliable.

Direct comparisons of forage dry matter yields from different locations in the tropics are difficult to make due to variations in length of growing season (mainly due to moisture) and rainfall amount. Variation in plot size can also significantly affect yields of vigorous tropical grasses such as elephant grass and sorghum x Sudan grass hybrids. Large "border effects" can result from small plots resulting in overestimation of dry matter production under solid stands.

A. Dry Matter Yield and Quality

McDowell[13] stated that a good pasture species in the tropics should (1) have high yields of good quality forage, (2) have good persistence, (3) be easy to propagate, and (4) have the ability to grow with a companion crop. These requirements (except 4) are similar to those of forages for other parts of the world. The ability to grow with a companion crop would tend to make forage production more uniform, and it would improve forage quality of grasses especially if the companion crop were a legume.

Quality is probably the most limiting factor in tropical forages. Higher quality forage can be obtained by increasing *in vitro* dry matter digestibility (IVDMD), raising the crude protein content (CPC), and ensuring adequate levels of essential minerals in the plant material. Cutting regimes that take these factors into account can also help to establish a continuity of nourishing forage. IVDMD and CPC will also fluctuate with the age of the forage, declining as the plants mature or are kept as standover material. Decreasing nondigestible constituents, eliminating or lowering the level of toxic components, and developing disease-resistant genotypes will contribute to obtaining a forage of improved quality. In the tropics, great variation in IVDMD within species exists from region to region due to climate, management, and soil differences (especially available silicon).[13]

The potential of *Sorghum* and *Pennisetum* species for improving forage and animal production in the tropics should be explored to a greater extent. This is especially true for the annuals: *S. bicolor* and *P. glaucum*. Although they will outyield other annual species,[14] the majority of sorghum[1] and pearl millet[15] currently utilized in the tropics consists of landraces grown for grain with the remaining stover (usually of poor quality) used to feed livestock. The landraces probably have undergone some selection by small farmers for grain quality and yield, but none for forage quality. Much of the sorghum and pearl millet is intercropped with legumes in the tropics, which provides some nitrogen for the grass and improves the quality of the forage when fed together.

Research with improved forage sorghum and sorghum x Sudan grass crosses in the tropics show exceptional dry matter yields and demonstrate the high quality (protein and digestibility) potential.[16-20] Data also show that some sorghum cultivars have safe levels of HCN-p.[18,19] For instance, single and three-way forage sorghum hybrids are capable of producing over 25 t ha^{-1} of dry forage with about 10% CPC in 180 days. For the development of the three-way cross, a male sterile F_1 was produced by crossing B Rhodesian (*Sorghum arundinaceum*) onto a cytoplasmic male sterile line. The *S. bicolor* x *S. arundinaceum* x, a desirable Sudan grass line, could be a good mechanism for the development of three-way hybrids with high yielding potentials in the tropics.

Elephant grass has probably made the greatest contribution of these two genera to improved forage production in the tropics. It will outyield other perennial, herbaceous species if cut infrequently.[14,21,22] Tergas and Urrea[23] stated that napier grass consistently produced higher yields of dry matter in Colombia during the rainy and dry seasons than other grasses tested. More effective clonal selections for tropical conditions have been made for this species than for any other species of *Pennisetum*. Elephant grass is a sexual and highly heterozygous species that does not produce true-breeding progeny from seed. The species is also quite heterogenous. A crossed or selfed progeny can produce a wide array of morphological types. Unique high-yielding plants have been propagated vegetatively and easily maintained because of a perennial growth habit.

'Merkeron' was one of the first elephant grass hybrids selected for improved dry matter yield and disease resistance.[24] The cultivar, 'Mott,' has been the most recently released improved elephant grass.[25] It is a dwarf, leafy selection from Merkeron with exceptional high-quality forage.[25-28] The pearl millet x elephant grass interspecific hybrids have also produced high dry matter yields of high protein and highly digestible forage.[28] The vegetative period and forage quality of interspecific hybrids can be extended by crossing elephant grass with short-day-sensitive pearl millet genotypes.[29] Hexaploid interspecific hybrids can be seeded, but hybrids are needed that persist under defoliation.[30]

Effects of harvest frequency on the dry matter yield and quality of tropical forages follow trends similar to those of forages in other areas. As harvests become less frequent, DM yields go up while digestibility and crude protein levels go down.[16,17,25,27,31,32] The objective should be to maximize the yield of digestible forage without sacrificing palatability and quality. Harvest frequency should be about 45 days for most *Pennisetums* and 65 days for most *Sorghums* in the tropics. *Pennisetum* and *Sorghum* species also respond to added soil nutrients. Dry matter yields of elephant grass were increased 35% by adding 276 kg ha^{-1} N.[33]

B. Animal Performance

In Puerto Rico, well-managed elephant grass will produce more animal product than any of the major tropical perennial grasses. Green-chopped elephant grass fertilized with two tons of 15-5-10 per year produced 39,960 kg ha^{-1} dry matter, enough to feed four 272-kg steers. When grazed, this same elephant grass produced enough forage to feed 2.2 272-kg animals, which gained 1232 kg ha^{-1} per year or one 545-kg cow producing 3000 liters of milk per year with no concentrate.[22] Only a small decrease in milk yield from dwarf Mott elephant grass was observed when it was substituted for corn silage. Dwarf elephant grass silage was of high quality with potential for high dry matter intake.[34]

In Australia, cattle intake of millet and sorghum hays decreased with age in 49 to 86 day old hay. A study in which sheep were fed the hay showed that "digestible organic matter intake would decrease to 280 g per day and daily liveweight gain would decline to zero" when organic matter digestibility fell to approximately 60% in hay cut at 68 days.[35] Cattle performed as well on grain sorghum silage of adapted cultivars as maize silage in Kenya. Forage sorghum produced more dry matter than maize and grain sorghums, but required more days on feed to achieve the same carcass weight as the two latter forages.[36]

C. Forage Preservation

Uniform supply of forage is needed throughout the year to supply energy to livestock. However, forage production in much of the tropics is seasonal, unless irrigation is available and affordable, due to dry seasons. Therefore, there is a need for stored forage.

Forages can be stored as silage, stockpiled as stand-over forage or hay, but hay making is unpredictable in the humid tropics.[22] Hay making is especially difficult with sorghum, pearl millet, and elephant grass because of their thick stems, which are difficult to dry. Silage requires less dry weather for preservation (but enough to wilt forage). Brown and Chavalimu[37] showed that hay and silage preserved nutrients with equal efficiency in napier grass. Research in Ghana showed that wilting to get at least 25 to 30% dry matter was necessary for good silage.[38] Intake by sheep increased by over 30% when some ground maize grain was added to the silage. Addition of 1 to 3% urea to 9.6% crude protein elephant grass fodder did not significantly increase digestibility of organic matter by sheep.[39]

Stockpiling as stand-over forage holds potential for meeting forage needs during the dry season. Of three tropical grasses stockpiled in Nigeria for two years, elephant grass gave the highest yields and maintained the highest crude protein.[40]

III. OUTLOOK

The need for *Sorghum* and *Pennisetum* species to produce high quality forages for the tropics will continue to grow in the future. This need can be met in at least four areas: expanded use of new species and derivatives, genetic improvements, better management practices, and forage preservation.

A. New Species or Derivatives

There appear to be a number of species in the *Pennisetum* genus that have desirable characteristics: *P. pedicellatum*,[41-43] *P. polystachyon*, *P. orientale*, *P. flaccidum*, and pearl millet-napier grass — *P. squamulatum* intercrosses.[6] These species and some interspecific crosses combine high yield and high quality. Most are thin-stemmed, leafy, and highly palatable. They also reproduce by apomixis, so superior cultivars can be easily propagated by seed.

B. Genetic Improvement

Significant improvements in forage production could be realized if cultivars and hybrids adapted to the tropics are developed. Duncan and Moss[44] showed that ten tropically adapted sorghum hybrids produced significantly more grain in two harvests than 10 temperately adapted hybrids.

The introduction of short-day sensitivity into tropical cultivars and hybrids would be advantageous. The photoperiodism would lengthen the production of high-quality, leafy forage by keeping the forage plant in the vegetative stage for a longer period of time.

Improvements in forage quality hold the greatest opportunity for improving tropical forages. A number of genetically controlled plant factors for improving forage quality are known to be genetically variable.[45] Forage digestibility (affected by a number of factors) is easily measured and could have a great impact on animal performance in the tropics. The simply-inherited brown-midrib trait, which reduces lignin in corn, sorghum, and pearl millet,[46] has tremendous potential if manipulated correctly in the *Sorghum* and *Pennisetum* genera.

More use of hybrids to take advantage of hybrid vigor is needed in forage crops for the tropics. Hybrids will more than compensate for the higher cost of inputs (seed and fertilizer) by increasing production of high-quality forage.

Hulse indicated a need for improved forage sorghums with whole plant utilization for marginal lands in the tropics.[47] The same need exists for *Pennisetums*.

A significant contribution can be made in developing forages with ability to be stockpiled as stand-over forage during the dry season. In the elephant grass nursery at Tifton, Georgia, significant variation has been observed in a plant's ability to retain leaves during dormancy. Dwarf and semidwarf cultivars retain leaves better than most tall cultivars (Hanna, W., unpublished).

C. Management Practices

Different classes of livestock require different quality forage. Information on the quality factors (digestibility and crude protein) of various local and improved cultivars as affected by soil type, fertility, defoliation, cutting frequency, preservation, etc. needs to be collected so that forages can be produced to meet the needs of the particular livestock class that will utilize them.

D. Forage Preservation

Wet seasons followed by dry seasons in the tropics require that reliable economical systems be developed for preserving forage. Better silage-making methods are needed in the humid tropics and possibly elsewhere. Hay making might be easier if thicker stands were planted to reduce stem size in *Pennisetum* and *Sorghum*. More information is needed on stockpiling stand-over forage. What are the best species and cultivars? What should be the maximum age of forage for stockpiling? How fast does the quality of stockpiled forage deteriorate?

REFERENCES

1. Maunder, A. B., Importance of sorghum on a global scale, in *Proc. Int. Conf. Sorghum Nutritional Quality*, Ejeta, G., Mertz, E. T., Rooney, L., Schaffert, R., and Yohe, J., Eds., Purdue University, W. Lafayette, Indiana, 1990, 8.
2. Brunken, J. N., A systematic study of *Pennisetum* sect. *Pennisetum* (Gramineae). *Amer. J. Bot.* 64, 161, 1977.
3. Rachie, K. O. and Majmudar, J. V., *Pearl Millet*, Pennsylvania State University Press, University Park, PA, 1980, chaps. 4 and 5.
4. Osgood, R. V., Hanna, W. W., and Tew, T. L., Hybrid seed production of pearl millet x napiergrass triploid hybrids. *Crop Sci.*, 37, 998, 1997.
5. Schank, S. C., A seeded type of hybrid hexaploid elephantgrass with potential for livestock production in the tropics and subtropics, in *Proc. Int. Conf. on Livestock in the Tropics*, University of Florida, Gainesville, FL, 1991, A-7.
6. Hanna, W. W., Dujardin, M., and Monson, W. G., Using diverse species to improve quality and yield in the *Pennisetum* genus, in *Proc. XVI Int. Grassl. Congr.*, Nice, France, 1989, 403.
7. Ahlrichs, S. L., Duncan, R. R., Ejeta, G., Hill, P. R., Baligar, V. C., Wright, R. J., and Hanna, W. W., Pearl millet and sorghum tolerance to aluminum in acid soil, in *Plant Soil Interactions at Low pH*, Wright, R. J., Ed., Kluwer Academic Publishers, The Netherlands, 1991, 947.
8. Ritchey, K. D., Golnaga, R., and Sotomayor-Rios, A., Rapid evaluation of juvenile *Sorghum* for tolerance to soil acidity, *J. Plant Nutr.*, 14, 315, 1991.
9. Duncan, R. R., The influence of soil pH on sorghum grain yields, *Commun. Soil Sci. Plant Anal.*, 22, 1605, 1991.

10. Wheeler, S. L. and Mulcahy, C., Consequences for animal production of cyanogenesis in sorghum forage and hay — a review, *Tropical Grasslands* 13, 193, 1989.
11. Akinola, J. O., Addo, A. A., and Olufokunbi, K. C., Improvement of annual forage and seed production in the sub-humid zone of Nigeria through supplement irrigation, *J. Agricultural Science* (Finland), 62, 63, 1990.
12. French, A. V., O'Rourke, P. K., and Cameron, D. G., Beef production from forage crops in the Brigalow region of Central Queensland. 1. Forage sorghum, *Trop. Grassl.*, 22, 79, 1988.
13. McDowell, R. E., *Improvement of Livestock Production in Warm Climates*, W. H. Freeman and Co., San Francisco, CA, 1972, chap. 6.
14. Burton, G. W., Biomass production from herbaceous plants, in *Biomass Energy Development*, Smith, W., Ed., Plenum Publishing Corporation, New York, 1986.
15. Kumar, K. A., Pearl millet: current status and future potential. *Outlook on Agriculture* 18, 46, 1989.
16. Stobbs, T. H., A comparison of zulu sorghum, bulrush millet and white panicum in terms of yield, forage quality and milk production. *Australian J. Exp. Agriculture and Animal Husbandry*, 15, 211, 1975.
17. Mendez-Cruz, A. V., Hepperly, P., Orengo-Santiago, E., and Corchado-Juarbe, N., Performance of two forage sorghum hybrids in southwest Puerto Rico, *J. Agric. Univ. P. R.*, 74, 1, 1990.
18. Sotomayor-Rios, A. and Torres-Cardona, S., Agronomic performance, hydrocyanic acid potential (HCN-p) and heterosis in forage sorghum hybrids, *J. Agric. Univ. P. R.*, 68, 131, 1984.
19. Sotomayor-Rios, A. and Torres-Cardona, S., Agronomic comparison, heterosis and hydrocyanic acid potential (HCN-p) of sudangrass-sorghum and sudangrass-sudangrass hybrids and their parents, *J. Agric. Univ. P. R.*, 67, 143, 1983.
20. Sotomayor-Rios, A. and Telek, L., Forage yield and protein content of milo blanco (*Sorghum bicolor*) and two F_1 hybrids, *J. Agric. Univ. P. R.,* 61, 300, 1977.
21. Hoshino, M., Ono, S., and Sirikiratayanond, N., Dry matter production of tropical grasses and legumes and its seasonal change in Thailand, *J. Japanese Soc. Grassl. Sci.,* 24, 310, 1979.
22. Vincente-Chandler, J., Abruña, F., Caso-Costas, R., Figarella, J., Silva, S., and Pearson, R., Intensive grassland management in the humid tropics of Puerto Rico, University of Puerto Rico Bulletin 233, 164, 1974.
23. Tergas, L. E. and Urrea, G. A., Effect of fertilization on the yield and nutritive value of tropical forages on an ultisol in Colombia, *Tropical Animal Production (Mexico)*, 10, 68, 1985.
24. Burton, G. W., Registration of 'Merkeron' napiergrass, *Crop Sci.*, 29, 1327, 1989.
25. Sollenberger, L. E., Prine, G. M., Ocumpaugh, W. R., Hanna, W. W., Jones, Jr., C. S., Schank, S. C., and Kalmbacher, R. S., 'Mott' dwarf elephantgrass: a high quality forage for the subtropics and tropics, Univ. Fla. Agric. Exp. Stn. Circular S-356, 1988.
26. Sollenberger, L. E., Prine, G. M., Ocumpaugh, W. R., Schank, S. C., Kalmbacher, R. S., and Jones, Jr., C. S., Dwarf elephantgrass: a high quality forage with potential in Florida and the tropics, *Proc. Soil Crop Sci. Soc. Fla* 46, 42, 1987.
27. Boddorff, D. and Ocumpaugh, W. R., Forage quality of pearl millet x napiergrass hybrids and dwarf napiergrass, *Proc. Soil Crop Sci. Soc. Fla.,* 45, 170, 1986.
28. Cuomo, G.J., Blouin, D.C., and Beatty, J.F., Forage potential of dwarf napiergrass and a pearl millet x napiergrass hybrid. *Agron. J.*, 88, 434, 1996.
29. Aken'ova, M.E. and Chheda, H.R., Interspecific hybrids of *Pennisetum typhoides* S & H x *P. purpureum* Schum for forage in the hot humid lowland tropics of West Africa. I. Seedling studies and preliminary evaluation. *Nigerian Agric. J.*, 10, 82, 1993.
30. Spitaleri, R.F., Sollenberger, L.E., Schank, S.C., and Staples, C.R., Defoliation effects on agronomic performance of seeded *Pennisetum* hexaploid hybrids. *Agron. J.,* 86, 695, 1994.
31. Sotomayor-Rios, A., Torres-Cardona, S., Hanna, W., and Quiles-Belen, A., Response of a *Pennisetum* interspecific hybrid and local cultivar Milo Blanco, *Sorghum bicolor* L. Moench, to cutting intervals and locations, in *Proc. XVI Int. Grassl. Congr.*, Nice, France, 1989, 585.
32. Farias, J. M. and Winch, J. E., Effect of planting date and harvest stage upon yield, yield distribution and quality of sorghum sudangrass in northern Mexico, *Tropical Agriculture,* 64, 87, 1987.
33. Mora, B. V. and Rodiles, J. A. F., The effect of application of nitrogen on the forage yield of 4 tropical grasses, *Veterinaria* (Mexico), 20, 265, (in Spanish), 1989.

34. Ruiz, T. M., Sanchez, W. K., Staples, C. R., and Sollenberger, L. E., Comparison of 'Mott' dwarf elephantgrass silage and corn silage for lactating dairy cows, *J. Dairy Sci.,* 25, 533, 1992.
35. Hedges, D. A., Wheeler, J. L., and Muldoon, D. K., Effect of age of millet and sorghum hays on their composition, digestibility and intake by sheep, *Tropical Grasslands,* 23, 203, 1989.
36. van Arkel, H., Creek, M. J., and Squire, H. A., Cold tolerant sorghums: a spectacular forage crop for specific tropical applications, *World Rev. Animal Prod,* 13, 75, 1977.
37. Brown, D. L. and Chevalimu, E., Effects of ensiling or drying on five forage species in western Kenya; *Zea mays* (maize stover), *Pennisetum purpureum* (Pakistan napiergrass), *Pennisetum* sp. (bana grass), *Impomea batala* (sweet potato vines) and *Cajanus cajan* (pigeon pea leaves), *Animal Feed Sci. Tech.,* 13, 1, 1985.
38. Tuah, A. K. and Okyere, O., Preliminary studies on the ensilage of some species of tropical grasses in the Ashanti forest belt of Ghana, *Ghana J. Agric. Sci.,* 7, 81, 1974.
39. Jayasuriya, M. C. N. and Sriskandarajaii, N., Urea as a source of non-protein nitrogen for ruminants. *J. National Sci. Coun. Sci. Lanka* 4, 1, 1976.
40. Omaliko, C. P. E., Stockpiling of three tropical forage grass species, *Agron. J.,* 75, 677, 1983.
41. Tyagi, G. D. and Singh, V., Effect of cutting management and nitrogen fertilization on yield and quality of *Pennisetum pedicellatum* Trin. (Dinanath grass), *Trop. Agric* (*Trinidad*) 63, 121, 1986.
42. Singh, V., Singh, A. K., Verma, S. S., and Joshi, Y. P., Effects of nitrogen fertilization on yield and quality of multicut tropical forages, *Tropical Agriculture* 65, 129, 1988.
43. Mishra, V. A. and Katiyar, D. S., Dinanath grass — a promising fodder for tropical and subtropical regions, *Indian Farming* 40, 17, 1990.
44. Duncan, R. R. and Moss, R. B., Comparative yields of ratoon cropped temperately and tropically adapted grain sorghum hybrids, *Crop Sci.*, 27, 569, 1987.
45. Hanna, W.W., Improving forage quality by breeding. *International Crop Science I,* D.R. Buxton, R. Shibles, R.A. Forsberg, B.L. Blad, K.H. Asay, G.M. Paulsen, and R.F. Wilson, Eds., Crop Science Society of America, Inc., Madison, WI., 1993, chap. 88.
46. Cherney, J. H., Cherney, J. R., Akin, D. E., and Axtell, J. D., Potential of brown midrib, low lignin mutants in improving forage quality, *Adv. Agron.* 46, 157, 1992.
47. Hulse, J. H., Needs, priorities and impact of sorghum nutritional quality research, in *Proc. Int. Conf. Sorghum Nutritional Quality*, Ejeta, G., Mertz, E. T., Rooney, L., Schaffect, R., and Yohe, J., Eds., Purdue University, W. Lafayette, Indiana, 244, 1990.

12 The Effect of Arthropods, Diseases, and Nematodes on Tropical Pastures

Stephen L. Lapointe and Ronald M. Sonoda

CONTENTS

I. INTRODUCTION

Efforts to develop new cultivars and varieties of tropical pasture plants have been accompanied by growing lists of associated pests and diseases. This situation is not surprising because most of these emerging cultivars are being developed from recently domesticated plant species, and their pests and diseases are only of recent concern. The greater productivity of tropical pastures "improved" by these new plants compared to that of native grass communities is unfortunately being gained at the expense of greater vulnerability to pests and pathogens due to genetic uniformity. However, this potential for genetic vulnerability is ameliorated somewhat by the diversity of available grass and legume genotypes that can be deployed spatially or temporally to reduce the risk of catastrophic loss due to a single biotic constraint. Some tropical forage genera and species have severe biotic constraints to their utilization, but it is hoped that the growing number of commercial varieties developed from diverse genera and species of tropical forages can replace these.

Development of "improved" pasture germplasm for tropical systems has traditionally been done through germplasm introduction, i.e., evaluation and selection of naturally occurring genotypes. However, increasing useful genetic diversity through breeding has an expanding role in dealing with biotic constraints such as insects, diseases and nematodes.

Introduced grasses have a relatively longer history of use in tropical pastures than forage legumes. The introduction of African forage grasses such as *Andropogon gayanus* Kunth and species of *Brachiaria* into Latin America has been especially successful, and many more hectares could be utilized for such adaptable grasses. According to Lapointe and Miles,[1] there are over 200 million hectares of well-drained savannas in Brazil, Colombia, and Venezuela on which intensified grazing and crop-pasture systems could be implemented. Introduced accessions of *B. decumbens* Stapf, *B. humidicola* (Rendle) Schweickerdt, *B. dictyoneura* (Figari and De Not.) Stapf, and *B. brizantha* (A. Rich.) Stapf are important in these areas and have been sown throughout Latin America from Mexico to Argentina. Approximately 20 commercial cultivars of *Brachiaria* species have been released in at least 10 Latin American countries in recent years.[2]

Information on the incidence of diseases and their effect on grasses is limited, while insect pest problems of grasses are better defined. The few plant pathologists who work with diseases of tropical pasture plants have focused their attention on those affecting new tropical forage legumes. This is due to the need for a nitrogen-contributing component for low-input pasture systems as well as the apparently greater vulnerability of the legumes to diseases. The use of tropical legume and grass associations in tropical pastures is still a nascent technology. Many fungal pathogens have been reported on tropical forage crops,[3] however, most of these reports have little information on the diseases themselves. The status of information available on tropical pasture plant diseases was evaluated by Lenné and Sonoda in 1990.[4] The 1994 publication *Diseases of Tropical Pasture Plants* edited by Lenné and Trutmann brought together a comprehensive assessment of studies on fungal, bacterial, and nematode problems of tropical forage legumes and grasses.[5] These two reviews indicate intensive studies of only a few fungal diseases of tropical legumes and grasses. Although virus diseases are common, their identity and impact are just now being studied.

While diseases are often transported with introduced germplasm, new insect pest problems usually do not become evident until a genus or species has been grown extensively at its new site. Both *Andropogon gayanus* and *Brachiaria* species have been host to significant insect pests in the neotropics: the former is susceptible to leaf-cutter ants and the latter to a complex of spittlebug species (Cercopidae).[6-10] Although the existing grazing resources in the savannas of Colombia and Venezuela and the cerrados of Brazil can maintain only very low stocking rates (approx. 0.1 head/ha), this native savanna also has few severe insect outbreaks due perhaps to its species complexity and coevolution with native insect herbivores such as leaf-cutter ants and cercopids. However, graziers prefer introduced grasses over native savanna for their increased productivity (weight gain/animal/year and weight gain/ha/year) and their resistance to drought for animal maintenance through the dry season.[1] These same characteristics (increased biomass production and resistance to drought) are probably also related to increased host susceptibility to pests and diseases.

In the past, organisms associated with diseased plants in native grassland areas, if identified, were often recorded with no effort made to determine if the identified organism had caused the disease. Often, even if a disease had a deleterious effect on one or a few components in these native swards, other plants may have buffered the area against severe losses in forage availability. With the development of pastures with limited plant heterogeneity, especially in the case of legumes, a few instances of severe losses to disease have occurred. The most frequently cited example is anthracnose caused by the fungus *Colletotrichum gloeosporioides* (Penz.) Penz. and Sacc. on members of the genus *Stylosanthes*. The annual *S. humilis* HBK, presumed to be an introduction, was first observed in grazing lands in Australia in 1913. Although leaf and stem lesions were reported as early as 1937 on *Stylosanthes*,[11] it was not until the mid-1970s, when relatively large areas of northern Australia had been planted with *S. humilis*, that epidemic losses to this disease occurred. The species is no longer being planted in northern Australia.

In the sections that follow, the major pest and disease problems of tropical pastures are presented along with discussions of the most appropriate approaches for their control. In many cases, however, because of the lack of published material, potential problems can only be mentioned and suggestions made as to where problems may arise as new forages become adopted. The disproportionate emphasis on insect pests in Latin America and diseases in Florida in this chapter reflects the authors' experience in those regions. The final section outlines future areas of research that may be most fruitful in furthering the development of productive, stable, and persistent forage plants for the tropics.

II. ARTHROPOD PESTS OF TROPICAL PASTURES

Several arthropod pests have been recognized as significant constraints to pasture production. The most important pests of pastures in Latin America are native to the Neotropics. In several instances, pests of plants have become more of a problem under "improved" conditions.

Spittlebugs (genera of the Homopteran family Cercopidae) are generally recognized as the most important limitation to the utilization of *Brachiaria* species. Consequently, this pest complex has received the greatest attention from national and international research institutes in the region including major efforts in augmentative biological control and host-plant resistance. While its upright growth habit confers resistance to spittlebugs,[10] *Andropogon gayanus* CIAT 621, valued for its forage quality on poor soils, is extremely susceptible to species of leaf-cutter ants, particularly *Acromyrmex landolti*,[7,12] in the savannas of Colombia and Venezuela and the cerrados of Brazil. Both *Brachiaria* species and *Andropogon gayanus* are examples of introduced forage grasses attacked by pest species that are native to Latin America and have few effective natural enemies that could be used for biocontrol.

In the following discussion, pests of tropical grasses are arranged by arthropod order and pests of tropical forage legumes arranged by host plant genus.

A. Principal Pests of Tropical Pasture Grasses

1. Orthoptera and Isoptera

The black field cricket (*Teleogryllus commodus* Walker) is a pest of pastures in northern New Zealand and Australia. Damage is similar to drought symptoms and has been quantified.[13]

Studies in British Guiana demonstrated that termites reduced the fertility of soil surrounding their mounds by concentrating nutrients in the termitaria, but also increased soil drainage and aeration.[14] Termite activity may also have contributed to the uneven distribution of grassland species in northeastern Australia due to modification of organic matter and plant nutients in the soil.[15]

2. Homoptera and Hemiptera

Aphids are sporadic and localized pests of *Andropogon gayanus*, and probably well controlled by natural enemies. In 1982, the sugarcane yellow aphid, *Sipha flava* Forbes, was reported as a potential pest of *Andropogon gayanus* in the savannas of Colombia,[16] however, no reports since then have confirmed the pest status for this species. In tropical Australia, susceptibility of six grasses to a species of the aphid *Schizaphis* was recorded — *Digitaria decumbens* (pangola grass) suffered more damage compared with *Setaria splendida* Stapf, *Brachiaria decumbens*, *D. pentzii* (L.) Stent, and *Panicum maximum* Jacq.[17]

At least six genera of cercopids, collectively known as pasture spittlebugs, are a major constraint to the productivity and persistance of pastures of *Brachiaria* species, *Panicum maximum*, and other susceptible grasses in humid areas throughout Mexico, Central, and South America (Table 12.1). Most of these Cercopidae have similar habits and cause similar damage to pasture grasses throughout Latin America.[6]

Grasses of the genus *Brachiaria* are infested by at least 13 species of spittlebug, and spittlebug species of the genus *Aeneolamia* are known to have at least nine species of grass host (Table 12.1). This is a conservative estimate of the number of species involved, based only on specimens found in CIAT's insect collection.

Feeding by adult spittlebugs produces foliar chlorosis and necrosis, reduces root biomass, and contributes to weed invasion and gradual pasture degradation.[18] It also has been shown to reduce dry matter production and affect forage quality by lowering crude protein and mineral contents.[9] An important, practical consideration when attempting to manage spittlebugs is that the full effect of damage appears up to three weeks after feeding occurs.[19] Since adult spittlebugs live approximately 10 days, insecticide applications or other control actions based on appearance of damage symptoms are not effective. Pasture spittlebugs are also potential pests of upland rice. Movement of spittlebug adults of *Deois flavopicta* and *Zulia entreriana* from adjoining *Brachiaria* pastures into rice fields was documented.[20] In pastures, spittlebugs have multiple generations per year. In the Brazilian species *Deois flavopicta*, the first generation of adults to emerge in the rainy season produce nondiapause eggs and the proportion of diapause eggs increases until the end of the rainy season.[21] Other studies are beginning to elucidate aspects of spittlebug natural history.[22,23]

A newly described species of burrowing bug, *Atarsocoris brachiariae* Becker (Cydnidae) may be a serious pest of *Brachiaria* in sandy soils. It has been recognized as a problem in the state of Mato Grosso, Brazil;[24] however, little information has been published on this pest.

Greber reported that the delphacid leafhopper *Sogatella kolophon* Kirkaldy transmitted a rhabdovirus of *Digitaria decumbens* and *D. ciliaris* (Retz.) Koeler (summer grass) in Australia.[25] The new host range of the virus included *Brachiaria miliiformis* (J. and C. Presl.) Chase and weed grasses.

3. Lepidoptera

Members of the genera *Spodoptera*, *Mocis*, *Anticarsia*, and *Elasmopalpus* are occasional pests that cause sporadic damage to grasses. Some, such as elasmo, are more severe in Brazil during pasture establishment.

4. Hymenoptera

Robinson and Fowler suggested that leaf-cutting ants are potentially the most damaging pests of Neotropical rangelands.[26] They estimated that the number of colonies/ha of *Acromyrmex landolti fracticornis* in Paraguayan grassland ranges from 1050 to 4400.[27] Lapointe et al.[12] found about 600 colonies/ha of *A. landolti* in infested areas of native savanna in Colombia, but over 5000/ha in a susceptible, introduced grass. Complete loss of swards of *A. gayanus* due to the ant has occurred

TABLE 12.1
Genera and Species of Spittlebug (Cercopidae) in the Insect Collection of CIAT,[1] Country of Origin, and Associated Tropical Forage Grass Hosts in Latin America

Genus and Species	Country of Origin	Associated Tropical Forage Grass Hosts
Aeneolamia flavilatera	Colombia	*Bothriochloa pertusa* (L.) A. Camus
		Brachiaria decumbens Stapf.
		Saccharum sinense
Aeneolamia lepidior	Colombia	*Bothriochloa pertusa*
		Brachiaria decumbens
		Cynodon sp
		Dichanthium aristatum (Poiret) C.E. Hubbard
		Melinis minutiflora Beauv.
		Panicum maximum Jacq.
		Paspalum clandestinum Chiov.
	Costa Rica	Unspecified pasture
Aeneolamia reducta	Colombia	*Bothriochloa pertusa*
		Brachiaria spp.
		Dichanthium aristatum
		Hyparrhenia rufa (Nees) Stapf.
		Melinis minutiflora
		Panicum maximum
		Paspalum clandestinum
	Costa Rica	*Brachiaria dictyoneura* (Figari and De Not.)
Aeneolamia selecta	Brazil	*Brachiaria arrecta*
Aeneolamia varia	Colombia	*Brachiaria* spp.
Deois flavopicta	Brazil	*Brachiaria* spp.
Deois incompleta	Brazil	*Brachiaria* spp.
Deois schach	Brazil	*Brachiaria arrecta*
Mahanarva sp.	Brazil	*Brachiaria* spp.
	Colombia	*Brachiaria* spp.
		Pennisetum clandestinum
	Ecuador	*Brachiaria* spp.
Prosapia fasciata[b]	Costa Rica	Unspecified pasture
Prosapia inferens[b]	Costa Rica	Unspecified pasture
Tomaspis phantastica[b]	Colombia	*Pennisetum clandestinum*
Zulia colombiana	Colombia	*Brachiaria* spp.
		Brachiaria plantaginea
		Digitaria decumbens
		Pennisetum clandestinum
Zulia entreriana	Brazil	*Brachiaria* spp.
Zulia pubescens	Colombia	*Brachiaria* spp.
		Hyparrhenia rufa
		Panicum maximum
		Pennisetum clandestinum
Zulia sp.	Colombia	*Brachiaria brizantha*
		Pennisetum clandestinum

[a] Centro Internacional de Agricultura Tropical (International Center for Tropical Agriculture), Cali, Colombia.
[b] Species identification tentative.

in the eastern plains of Colombia and in northern Minas Gerais, Brazil, (S. L. Lapointe, personal observation).

The role of leaf-cutting ants in the invasion and establishment of woody plant species in pastures was studied by Jonkman,[28] who noted the development of "wood nuclei" on dead colonies of leaf-cutters. Jonkman also studied the change in population density of nests of *Atta vollenweideri* in Paraguayan pastures over a 24-year period using aerial photographs.[29] He estimated nest longevity at 10 years and suggested that changes in land use, particularly regular burning to renovate grasslands, favored increased colony density. Young foliage such as the regrowth induced by burning is preferred by leaf-cutters. Part of the resistance of "improved" pasture grasses such as *Brachiaria* to ant infestation may be related to the fact that pastures containing these grasses retain green leaf through the dry season and are less often burned.

Ants and termites can modify the structure, porosity, chemistry, and other aspects of the soil,[30] resulting in deterioration of pastures. However, inconsistency in species and environments studied and methods used make it difficult to generalize concerning treatments and solutions to control these insects. It should be noted that, while some species deserve pest status, they may also be desirable in pastures by contributing to water infiltration, increasing nutrient levels, and improving soil structure. Lapointe et al.[7] estimated that the leaf-cutter *Acromyrmex landolti* is capable of moving 1.5 m^3 of subsoil to the surface per hectare of infested pasture per year. Grass seeds are low in toxins and can be heavily predated by ants (Myrmicinae and others),[31] a situation usually more common in drier habitats.[32] Little information is available on species composition, biomass, or importance of either seed-eating ants[33] or predatory ants, although it is suspected that the former can affect establishment success of tropical forage grasses and the latter may be important regulators of arthropod herbivores.[34]

B. Pests of Tropical Forage Legumes

1. Pests of *Stylosanthes*

Species of cicadellid leafhoppers may contribute to the decline of *Stylosanthes* in pastures, but their relative importance in the decline has not been measured. In eastern Colombia, at least three species of cicadellids have been found in *Stylosanthes capitata* Vogue. The larvae of the lepidopteran *Stegasta bosquella* attack seed pods of *Stylosanthes*. However, compared with general problems of edaphic adaptation, insect damage may not be a major factor contributing to decline of *Stylosanthes* in these areas.

An interesting characteristic of *Stylosanthes* spp. is their antitick properties, currently being investigated in Mexico for suppression of *Boophilus microplus*.[35]

2. Pests of *Arachis*

CIAT released a series of cultivars of *Arachis pintoi* Krapovickas and Gregory. The cultivar Mani Forrajero, released in 1992, is perhaps the most promising tropical forage legume due to its high forage quality, persistence under grazing, and compatibility with aggressive grasses.[36] The lesser cornstalk borer (*Elasmopalpus lignosellus* (Zeller)) is a major pest of cultivated peanut (*Arachis hypogaea* L.) in the United States. This lepidopteran occurs throughout Central and South America, where it is found on a wide range of host plants.[37] While attacks of *E. lignosellus* have been observed to hinder pasture establishment in Brazil, there are no reports yet of damage to established stands of *A. pintoi*. Wild species of *Arachis* were surveyed for resistance to *E. lignosellus*, but levels of resistance were considered insufficient to justify breeding the plant with cultivated peanuts.[38] Arthropods have been observed on accessions of *A. pintoi* in introductory plots in Brazil and Costa Rica. These included mites (*Tetranychus* sp.), several species of chrysomelids, and lepidopteran larvae feeding on foliage and flowers. To date, however, there is no evidence of these other arthropod pests limiting *A. pintoi* growth and productivity (S. L. Lapointe, personal observation). The cultivar Florigraze of *Arachis glabrata* Benth. was released in Florida in 1981 with no pests reported.[39]

3. Pests of *Centrosema*

There exists a large complex of leaf-eating and leaf-sucking insects associated with *Centrosema* throughout Latin America, but their impact on the persistence of this genus has not been ascertained.[6] Common Centro (*C. pubescens* Benth.), introduced into Australia in 1930, has few pests.[40] The variety 'Belalto' is considered more resistant to mites (*Tetranychus* spp.) than common Centro.[41]

C. Control Options for Tropical Forage Pests

Of the insect-control options, host-plant resistance (HPR) and chemical, cultural, and biological control methods, HPR is the most appropriate for the low-input, extensive, and less valuable pasture systems, especially given the large and diverse amount of grass and legume germplasm potentially adaptable to such tropical systems. Most natural biological controls of pests are usually modified when ecologically complex native savanna communities are replaced by simplified "improved" grass pastures or by a grass-and-legume pasture. However, natural enemies of many pests may persist in these improved swards. For more intensive, higher input systems such as dual-purpose production in more favorable environments, other options, including augmentative biological control and chemical control, become feasible. HPR in such systems may be complemented by the other options.

Investment in pest control in pastures will become more economically justifiable as (1) the value of pastures increases through intensification, association of grasses with legumes, and improvement of soil fertility and sward productivity, and (2) pastures are incorporated into cropping systems. In the latter case, if pests are common to pastures and higher value crops, pasture control of the pest can be more feasible than crop control. One example is control of spittlebug damage to rice in rice/pasture relay cropping and rotations in the savannas of northern South America and the cerrados of Brazil. Since damage to the rice crop is thought to be due mainly to short migrations of spittlebug adults from nearby infested pastures containing, for example, *B. decumbens*, the use of pasture grasses resistant to the spittlebug should eliminate the need for control in rice. Similarly, in Venezuela and Costa Rica, efforts to control spittlebug species that attack both sugarcane and pastures are directed toward control in pastures as well as in cane plantations even though populations may not be high enough to cause damage to the pastures.

1. Chemical Control

Information on the effectiveness of pesticides on the same or similar pests in other crops is often available. Chemical control of arthropod pests in extensive grazing systems is not, however, generally practiced due to economic considerations, effect of toxic residues on grazing animals, and lack of information on correct and timely application procedures. In particular, graziers of extensive systems differ from agriculturalists in that they do not practice intensive management. Insect problems, when they occur, usually are not recognized in time to make effective control decisions. Other barriers to implementation of chemical control are the difficulty of establishing action thresholds and quantifying economic impact of pests in pastures. Costly, long-term trials are necessary, and it is often impossible to fully control other variables. Exclusion of the pest from noninfested control areas is difficult, and animals cannot be grazed on plots sprayed with pesticide. The result is short-term studies of total biomass production without consideration of the effect of infestation on animal performance or long-term effects of repeated or continued infestation over time. As in the case of fungicides, insecticides may be of economic value when pastures are being established and are not yet being grazed.

2. Biological Control

a. *Classical biological control*

In Latin America, the majority of insect pests are native species. However, an example of successful classical biological control of a tropical pasture pest is the introduction of parasitoids for control of Rhodes grass scale, *Antonina graminis*. Thought to have been introduced into Brazil around 1944, *A. graminis* became widely distributed and attacked over 90 species of grass.[42] Rhodes grass scale has been largely controlled through the introduction from Texas of parasitoids, particularly the microhymenopteran *Neodusmetia sangwani* (Rao).[43] Recent surveys of pastures in the state of São Paulo have found *N. sangwani* and other parasitoids in all sampled pastures where *A. graminis* occurred.[44] Today, Rhodes grass scale is no longer considered an important pest of improved pastures in Brazil.

b. *Natural biological control*

Many potential pests of pastures, particularly lepidopteran larvae, are presumably held in check by natural enemies and do not achieve pest status unless the system is perturbed by such means as the application of pesticides. In Australia as in Latin America, most insect pests of forages are indigenous. According to Davidson, the economic impact of pests on pastures in northern Australia is thought to be neglible due to effective natural controls and the inherent tolerance of pasture systems to temporary depletions caused by insect damage.[45] He suggested that the retention of natural habitat in modern farming systems is essential to maintain populations of spiders, insectivorous birds, and parasitic insects as agents of biological control. On the other hand, the major insect pests of forage grasses in Latin America — spittlebugs and leaf-cutter ants — lack effective natural enemies in improved pasture ecosystems. Outbreaks of spittlebugs occur in susceptible *Brachiaria* cultivars even in areas where natural habitat still abounds as in the Amazon Basin and the savannas of Colombia and Venezuela. This supports the observation of few natural enemies, even in natural habitat refugia.

c. *Augmentative biological control*

Tropical pastures should be an ideal environment for augmentative microbial control. However, microbial control of the spittlebug has not yet been successfully implemented. After a period of intense activity and enthusiasm for control of pasture spittlebugs with the fungus *Metarhizium* in Brazil,[46] this method has not been widely adopted, and its use has declined. The failure of *Metarhizium* to provide the level of control anticipated may have been due to poor timing of applications and/or inadequate application methods.[47] First instar nymphs appear to be especially susceptible to the fungus *Metarhizium* during a critical period subsequent to eclosion as they search for a feeding site, but before a protective spittle mass is produced.[48] Since pasture spittlebugs feed at the soil surface and in the root zone, delivery of the fungus through the grass canopy to the soil surface where spittlebug nymphs are found may be essential to achieve control. Another reason for failure to achieve control with *Metarhizium* may be poor selection of isolates. Development of a rearing method for spittlebugs should facilitate studies of pathogenicity and selection of aggressive strains of the fungus.[49]

3. Cultural Control

The cover and litter produced by *Brachiaria* cultivars provide microenvironmental conditions that contribute to spittlebug nymph survival. Some control of the pest can be obtained by intense grazing prior to heavy infestation to reduce biomass, avoid leaf litter accumulation, and subject nymphs to desiccation.[50] Control by grazing must be utilized as a preventative; it cannot be applied as a cure after damage is observed. Using grazing management for control of spittlebug populations requires careful monitoring of the insect's presence and the availability of sufficient livestock at early stages of infestation to maintain sward height. Recent efforts to define the natural history of spittlebug species may lead to improved methods of forage management.[21-23]

Early land preparation has been suggested to control leaf-cutter ant colonies, particularly the relatively small colonies of species such as *Acromyrmex landolti*. Lapointe et al.[7,12] showed that land preparation was effective in reducing colonies of the ant by as much as 90%. However, the ants in the remaining colonies caused significant damage to the highly susceptible grass *A. gayanus* during germination. While more intense soil preparation such as that used for crop establishment may increase control of leaf-cutter colonies, the damage to soil structure may contribute to serious erosion of fragile soils. Excavations of *A. landolti* nests in the Colombian savanna revealed both seasonal variation in colony depth and size in the soil, and an aggregated pattern of colony distribution due to patterns of surface water runoff.[51]

4. Host Plant Resistance

Collection and selection of germplasm from the wild with characteristics that confer resistance to attack by insect pests are ongoing. Breeding efforts are also underway to develop plants with resistance to pests. Studies to determine mechanisms by which pasture plants ward off pests are also being conducted.

Nilakhe studied spittlebug antibiosis, tolerance of and nonpreference for 20 species of forage grasses in the field and selected species in the glasshouse.[52] He demonstrated nonpreference by spittlebugs for *A. gayanus*, while *B. brizantha* (A. Rich.) Stapf cultivar Marandú was antibiotic to and *B. humidicola* tolerant of *Zulia entreriana*. Cultivar Marandú was also confirmed as antibiotic toward *Z. colombiana*,[10] and additional accessions of *Brachiaria* were identified as resistant to *Aeneolamia varia*.[53] Subsequently, the heritability of spittlebug resistance was determined. While more than a single major resistance gene is probably involved, resistance appears not to be complex and enhancement of resistance in sexually reproducing breeding populations should be feasible.[54] Recent improvements in spittlebug rearing and bioassay methods are contributing to rapid screening and selection.[55]

Panicum maximum accessions selected under acidic soil conditions in the eastern plains of Colombia were evaluated for resistance to *A. varia*. Antibiotic accessions were identified, but have not been rigorously tested for resistance to spittlebugs in the field.[56]

Although the relationship between plant secondary compounds and ant preference behavior has been studied in the tropical forests,[57-60] relatively little work has been done on the basis of leaf cutters' preference among tropical forage grasses. *B. decumbens* was the least preferred of several grasses studied, but was the only species of that genus included in the study.[27] *B. humidicola* was found to be highly resistant to the leaf-cutter *Acromyrmex landolti* in the field.[7,12] Subsequently, resistance was described in *B. brizantha*, *B. decumbens*, and *B. humidicola*. The resistance was attributed to plant factors inhibitory to the attine symbiotic fungus.[61]

III. DISEASES AND NEMATODE PESTS OF TROPICAL PASTURES

A. Their Effect on Tropical Forage Plants

Diseases and nematode infestation of tropical forage plants affect animal production by changing or reducing acceptability, quantity or availability, and quality of forage. They can also reduce the persistence of forage plants.

Pathogens can cause seed rot, pre- and postemergence seedling death, leaf spots, defoliation, stem cankers, root rot, systemic problems, and other abnormalities. For example, Lenné and Sonoda found *Rhizopus stolonifer* (Ehr. ex Fr.) Lind. to be a common fungus on stored *Stylosanthes hamata* (L.) Taubert seeds, causing inflorescence blight.[62] When the seeds were moistened, the fungus invaded them and interfered with germination. *Pythium irregulare* Buisman causes pre- and postemergence damping-off of seeded *Stylosanthes guianensis* (Aublet.) Sw.[63] *Colletotrichum gloeosporioides* produced leaf spot, defoliation, stem canker, and shoot dieback in several forage crops and is the most important disease on *Stylosanthes* species, causing all of the above-noted symptoms.[64] Several biotypes of *Rhizoctonia solani* Kuhn are responsible for severe foliar blight in several forage legumes and in some forage grasses. [65,66] An infection by a virus related to the bean common mosaic virus resulted in mosaic-like symptoms and shoot yield losses in the forage legume *Macroptilium atropurpureum* (DC.) Urb.[67] Nematodes generally affect roots, causing systemic syndromes.[68]

B. Evaluating Importance of Forage Diseases and Nematodes

Although the list of pathogens affecting tropical forages is extensive, only a few are considered as having a significant economic effect on tropical pastures. The detrimental effect of a pathogen should be measured in terms of the extent to which it lowers animal production and quality, but most evaluations have focused on its effect on plants because such studies are cheaper, quicker, and more easily done.

If pathogens cause swift, clearly observable losses in forage yield or quality, their short-term effect on developing animals may be easy to assess. Pasture plants are perennial. If epidemics and/or epiphytotics have little effect on their root systems, they will, with management, or in many cases, without man's assistance, recover and again support the grazing animal. However, plants weakened by disease or nematode infestation have little defense against weed invasion. This may lead to eventual degradation of a pasture and loss of forage material.

Small-plot studies on pathogen incidence and severity and plant agronomic characteristics are often done together. These usually consist of replicated small plots of several accessions, and the presence of pathogens is often the result of natural infestation or infection. Although the evaluations are done under circumstances that might not simulate those of a pasture, results obtained from these studies can provide information on the potential severity of individual pathogens. An example of such studies was the gathering of information on rust, gray leaf spot, and *Rhizoctonia* foliar blight of *M. atropurpureum*.[69]

Small-plot experiments have also been conducted to determine the severity of individual pathogens such as rust, *Uromyces appendiculatus* (Pers.) Unger var. 'Crassitunicatus,'[70] virus,[67] and *R. solani* foliar blight[66] on *M. atropurpureum*, anthracnose on *S. hamata*,[71] and root-knot nematode on *Desmodium* species and other legumes.[68] A few evaluations of the effect of grazing cycles on disease incidence and severity have been conducted. For example, Sonoda and Mislevy found that the incidence of *R. solani* foliar blight was drastically reduced from 82% at seven-week grazing intervals to 16% and 0 at 4- and 2-week grazing intervals on the most severely affected cultivars of *Cynodon* species during the wet summer season in central Florida.[66] Those evaluations were, however, conducted in monoculture plots. Interactions in mixed pastures can occur that affect the severity of, or recovery from a pathogen attack. For example, stem infection and death of *S. guianensis* caused by *Sclerotinia homoeocarpa* occurred in mixed swards of *S. guianensis* and *Paspalum notatum* Flugge, but not in monocultures of *S. guianensis* (Sonoda, personal observation). In cases where several pests and diseases interact in monocultures or in mixtures, studies of disease impact should be conducted in the presence and absence of other pests and diseases during several seasons and in different environments.

C. Managing Diseases

Almost all important tropical forage plants are perennial. Once established in a pasture, some pests and diseases will remain on their hosts as long as the hosts are components of the pasture. These diseases may require constant management to keep them at an acceptable level. On the other hand, a swift-acting, severe, but environment-limited disease may have a temporary effect on a component or components of a pasture. Once disease pressure is relaxed, these pastures may rejuvenate on their own or require managerial input.

D. Chemical Control

Because of the generally low return on applications of pesticides on a broadcast basis over a large area, their use to control diseases of pastures is usually of little economic value. However, fungicides may have value as a seed treatment when pastures are being established. Examples of fungicides that have controlled seed and seedling pathogens under experimental conditions are metalaxyl for *Pythium irregulare* on *S. guianensis* seeds,[63] and captafol and several other fungicides for fungi on *S. hamata* seeds.[62]

E. Quarantine

Many of the most important diseases of newly domesticated tropical forage plants can be found in wild stands of these plants in their area of origin. Some diseases appear to have been introduced along with the plants into areas apart from the native range of the plant genus or species. An example is the introduction of anthracnose of *Stylosanthes* into Australia. The *C. gloeosporioides* attacking introduced *Stylosanthes* species in Australia probably accompanied the seeds as the fungus is transmitted by the seed.[72] Its introduction into Florida, again probably on seeds, provides an example of a disease being brought into an area where susceptible plants were indigenous, but the disease was not present. The disease is found in the few pastures of imported *Stylosanthes* species and in pastures containing Florida species planted near *Stylosanthes* infected with the pathogen.[71] The disease is not found in Florida on native *Stylosanthes* stands, which are confined to eastern coastal areas of the Florida peninsula.[73]

Most of the long distance, man-vectored movement of pathogens occurred during the early period of assessment of these plants for forage use. Since that time, more stringent efforts have been made to reduce the movement of pathogens. Not all potentially important pathogens have been transported to areas where a particular host crop is currently being grown. There are many different strains of pathogens, some in the areas of origin of a particular plant, others in areas where man has changed the genetic makeup of plants and pathogens have evolved with these changes. Those involved in forage research should be constantly aware of the potential for transporting pathogens with seeds or other material.

F. Host Resistance

In centers of origin of forage species, there is a diversity of host genotypes with diverse genetic mechanisms to maintain the species in the presence of pests and pathogens. Utilizing resistant cultivars developed from these sources is probably the most economical means of dealing with pathogens. Some effort has been made to introduce genes in *Stylosanthes* that are genes resistant to anthracnose.[74-75] However, resistance to anthracnose has broken down repeatedly because of the presence of a diversity of pathogen genes.[76,77] In a collection of rust isolates from Mexico, all rust isolates caused disease on *M. atropurpureum* cultivar 'Siratro.'[78] Several lines of *M. atropurpureum* selected for resistance to rust in Australia were resistant to all the isolates of rust collected from Mexico.[79]

Much of the current effort to develop resistant varieties for tropical forages involves the screening of agronomically superior accessions from the wild in the presence of natural infestations of pathogens.

G. Cultural Management

Cultural practices can be an economical and effective means of managing diseases in pastures. Weed control and common management practices involving essentials such as fertilizer and water affect disease control. Generally, but not always, when these practices favor the physiological well-being of the host, losses to diseases are minimized. Beneficial cultural practices that can be used effectively under pasture conditions include burning of aboveground portions of fire-tolerant plants to reduce inoculum of aboveground pathogens. Lenné used this method to control anthracnose of *S. capitata* Vogel.[80] Disease incidence and severity can also be reduced by strategic grazing. Grazing is useful in reducing pathogenic inoculum or in interrupting epidemics. Sonoda and Mislevy,[66] as discussed previously, reported significant reductions in the incidence of foliar blight with more frequent grazing of *Cynodon* species.

Lenné reported a reduction in *Meloidogyne javanica* on *Desmodium ovalifolium* Wallr. when the legume was planted in association with several grasses with potential use in tropical pastures.[81] A relatively new concept is the planting of mixtures of cultivars with different levels or sources of resistance to a pathogen to form a composite variety.[82] Tests with mixtures of *Stylosanthes* species to determine their effectiveness against anthracnose have been and are being conducted.[83] There is some indication of benefit from this strategy; however, more studies appear to be needed to ascertain if mixtures will result in less loss of forage material to disease over time.

H. Diseases That Have Had an Impact on Tropical Forage Plants

1. Anthracnose on *Stylosanthes* Species

The genus *Stylosanthes* is considered one of the most important genera of forage legumes for tropical pasture use. As mentioned previously, the main limiting disease of this genus is anthracnose caused by *C. gloeosporioides*.[4] First reported in Brazil in 1937,[11] it was subsequently reported in replicated single-plant plots of various *Stylosanthes* species in Florida in the early 1970s.[64] Soon after, two distinct diseases of the fungus were reported on *Stylosanthes* species in Australia by Irwin, et al.[84] Subsequent infection of *Stylosanthes* species has been reported in most of the areas where they are being used or tested. The pathogen appears to consist of many biotypes.[84,85] As indicated above, anthracnose is present in the few Florida pastures where imported *Stylosanthes* species are planted, but not in the native stands. Most *Stylosanthes* material used in pastures is derived from imported accessions and has a moderate incidence of anthracnose lesions under Florida conditions. No effort has been made to determine the effect of the pathogen on yield of these imported *Stylosanthes*. The native Florida Stylo, *S. hamata*, is not readily affected by *C. gloeosporioides* strains, but has little forage potential. *S. scabra*-like native plants have better potential for pasture use. The incidence of lesions on them appeared to be moderate when the plants were first evaluated; however, dry-matter yields were reduced from 26 to 58% over a two-year period,[72] and efforts to develop these plants for Florida pastures were discontinued.

Townsville Stylo (*S. humilis*) was once an important pasture plant in northern Australia. It was the first species of *Stylosanthes* cultivated on a large scale. The species was estimated to cover over 50,000 ha in northern Australia by 1969,[86] and more hectares were planted until the onset of problems with anthracnose in the middle 1970s. Anthracnose virtually eliminated the use of *S. humilis* in Australian pastures.

There continues to be an intensive effort to devise methods of managing anthracnose on *Stylosanthes* spp. in Australia, South America, and Africa by scientists in Australia, South America, and Belgium as evidenced by the wide array of studies on this disease.[77,78,85,87]

2. Diseases of *Macroptilium atropurpureum*

Cultivar Siratro of *M. atropurpureum* is a viny legume once considered to have great potential for use in the tropics because of its rapid growth and ready acceptance by foraging animals. Siratro, developed from seeds collected in southern Mexico, was introduced into Australia in the 1960s.[88] However, there the cultivar is susceptible to a rust reported to reduce yield of forage and, in Florida, *R. solani* foliar blight and a strain of bean common mosaic virus have been reported to reduce foliage yield.[65,67] The most severe disease of Siratro in Florida is the foliar blight, which is common throughout the summer rainy season. No resistance to *R. solani* has been identified.

Of the *M. atropurpureum* diseases, rust (*U. appendiculatus* var. *crassitunicatus*) has been the most intensively studied. As in the case of bean rust, *U. appendiculatus*, the *M. atropurpureum* rust appears to consist of different genotypes. Isogenic lines of *M. atropurpureum* containing genes from diverse sources for resistance to an Australian isolate of rust reacted differently to seven rust isolates from southern Mexico.[80] The cultivar Siratro was susceptible to 19 isolates of the rust collected in southern Mexico.[79]

3. Nematodes on *Desmodium* Species

Among forage legumes, nematodes have been most troublesome on *Desmodium* species. Cultivar Florida Carpon of *Desmodium heterocarpon* was found to be susceptible to *Meloidogyne* species.[68] Although Florida Carpon has been successfully grown where pastures were initiated in previously uncultivated areas, forage yields of the cultivar were reduced by almost one-half in replicated plantings in nematode-infested pastures. Two of the other five *D. heterocarpon* accessions tested were highly tolerant of the indigenous nematodes.[68] Root-knot nematodes can cause significant damage when Florida Carpon is grown following several plantings of vegetables, especially tomatoes. A rotation of pangola grass (*Digitaria decumbens* Stent) and tomatoes was effective in reducing root-knot nematodes on the vegetable in Florida.[89] However, sward mixtures of pangola grass and Florida Carpon have not been tested.

IV. RESEARCH PRIORITIES

For the present, host plant resistance is the most efficient, practical, and environmentally sound method of controlling plant diseases and arthropod pests of tropical pastures. Elucidation of resistance mechanisms in tropical forage grasses should receive higher priority in breeding efforts of key genera so that selection of progeny can be based on the presence of specific chemical compounds or pathways. In particular, it would be important to understand the interaction, if any, between compounds conferring resistance to leaf-cutters and those determining forage quality. The development of breeding systems for normally apomictic grasses such as *Brachiaria* and *Panicums* raises the possibility of combining qualities previously found in distinct genotypes in a single or several cultivars.

There is a need to develop rapid, economical methods of field screening large numbers of introductions or hybrid progeny. Artificial infestation techniques or other reliable methods for testing responses of plants to pathogens and pests in the field are needed. Screenhouse and glasshouse evaluations are also important, particularly in the case of sporadic pests and pathogens where the possibility of escape is high. When field evaluation with pathogens or pests is difficult and large numbers of entries need testing, it may be advisable to identify accessions with desired agronomic traits first and then screen promising accessions using appropriate bioassay methods in the screenhouse or glasshouse. Although more costly, this approach would avoid the release of highly susceptible cultivars.

Further study is needed to determine whether mixtures of isolines with different sources of resistance to pathogens and pests will allow members of a sward to persist and produce in the presence of a pathogen or pest.

Constant awareness will be required to minimize the chances of planting new pasture plants that can serve as reservoirs for pests and pathogens of major food crops. As discussed above, movement of spittlebugs from *Brachiaria* to rice has been documented.

An important parasite-plant host relationship for pasture production that has not been discussed is the infection of grasses by endophytic fungi. This phenomenon can have multiple effects that vary according to the endophyte/grass association. Effects include increased plant vigor, modified growth habit, insect resistance, and toxicity to grazing animals.[90] Variation in the kinds of toxins produced by endophytes exists both within isolates of endophyte species[91] and among isolates.[92] It now seems possible to select infected pasture grasses for specific mycotoxin production, resulting in cultivars resistant to specific insect pests, but lacking undesirable mammalian toxins.[93-95] Most tropical forages have not been studied for the presence or possible effects of endophyte infection. Members of the Panicoideae are reported to be infected by species of *Balansia*, *Balansiopsis*, or *Myriogenospora atramentosa*.[95] Although no attempt has yet been made to identify endophytes in *Brachiaria*, stained leaves of some cultivars at CIAT have shown the presence of what appear to be endophytic hyphae (S. L. Lapointe and P. Trutmann, personal observation). The possible implications of endophyte infection for feed quality and pest resistance will require more intensive study.

REFERENCES

1. Lapointe, S. L. and Miles, J. W., Germplasm Case Study: *Brachiaria* Species, in *Pastures for the Tropical Lowlands: CIAT's Contribution*, CIAT, Cali, Colombia, 1992, chap. 4.
2. Miles, J. W. and Lapointe, S. L., Regional germplasm evaluation: A portfolio of germplasm options for the major ecosystems of tropical America, in *Pastures for the Tropical Lowlands: CIAT's Contribution*, CIAT, Cali, Colombia, 1992, chap. 3.
3. Lenné, J. M., *A World List of Fungal Diseases of Tropical Pasture Species*, Phytopathological paper No. 31, International Mycological Institute and Centro Internacional de Agricultura Tropical, University Press, Cambridge, 1990.
4. Lenne, J. M. and Sonoda, R. M., Tropical pasture pathology: a pioneering and challenging endeavour, *Plant Disease,* 74, 945, 1990.
5. Lenné, J. M. and Trutmann, P., Eds., *Diseases of Tropical Pasture Plants,* CAB International, Cambridge, U.K., 1994.
6. Lapointe, S. L. and Ferrufino C., A., Plagas que atacan los pastos durante su establecimiento, in *Establecimiento y Renovación de Pasturas*, Lascano, C. and Spain, J., Eds., CIAT, Cali, Colombia, 1991, 81.
7. Lapointe, S. L., García, C. A., and Serrano, M. S., Control of *Acromyrmex landolti* in improved pastures in the Colombian savanna, in *Applied Myrmecology: A World Perspective,* Vander Meer, R. K., Jaffe, K., and Cedeño, A., Eds., Westview Press, Boulder, Colorado, 1990, chap. 47.
8. Lapointe, S. L., Arango, G., and Sotelo, G., A methodology for evaluation of host plant resistance in *Brachiaria* to spittlebug species (Homoptera: Cercopidae), in *Proc. 16th International Grassland Congress*, Nice, France, 1989, 731.
9. Valério, J. R. and Nakano, O., Danos causados pelo adulto da cigarrinha *Zulia entreriana* na produção e qualidade de *Brachiaria decumbens*, *Pesq. Agropec. Bras.*, 23, 447, 1988.
10. Ferrufino, A. and Lapointe, S. L., Host plant resistance in *Brachiaria* grasses to the spittlebug *Zulia colombiana*. *Entomol. Exp. Appl.*, 51, 155, 1989.
11. Anon., Informação sobre algumas plantas forrageiras, Ministerio da Agricultura, Rio de Janeiro, Brasil, 1937.
12. Lapointe, S. L., Serrano, M. S., and Villegas, A., Colonization by and distribution of leaf-cutter ants (*Acromyrmex landolti*, Hymenoptera: Formicidae) in two tropical forage grasses in eastern Colombia. *Fla. Entomologist*, 76, 359, 1993.
13. Blank, R. H., Olson, M. H., and Bell, D. S., Pasture production losses from black field cricket (*Teleogryllus commodus*) attack, *New Zealand J. of Exp. Agriculture,* 13, 375, 1985.
14. Goodland, R. J. A., On termitaria in a savanna ecosystem, *Can. J. Zoology,* 43, 641, 1965.
15. Holt, J. A. and Coventry, R. J., The effects of mound-building termites on some chemical properties of soils in northeastern Australia, in *Proc. 3rd Australasian Conf. Grassl. Invert. Ecol.*, Adelaide, Australia, 1982, 313.
16. Varela, F. A. and Calderón, M., Afido amarillo de la caña de azucar, *Sipha flava*, plaga potencial del pasto Carimagua, *Andropogon gayanus*, en los llanos orientales de Colombia, *Revista Colombiana de Entomologia*, 8, 4, 1982.

17. Teitzel, J. K. and Wilson, R. J., An observation on the reaction of six grasses to aphids and rust in the wet tropics, *Queensland J. of Agric. and Animal Sci.*, 31, 31, 1974.
18. Valério, J. R. and Nakano, O., Dano causado por adultos da cigarrinha *Zulia entreriana* (Berg, 1879) (Homoptera, Cercopidae) na procução de raízes de *Brachiaria decumbens* Stapf, *An. Soc. Entomol. Brasil*, 16, 205, 1987.
19. Valério, J. R. and Nakano, O., Sintomatologia dos danos causados pelo adulto da cigarrinha *Zulia entreriana* (Berg, 1879) (Homoptera, Cercopidae) em *Brachiaria decumbens* Stapf, *An. Soc. Entomol. Brasil*, 21, 95, 1992.
20. Nilakhe, S. S., Ecological observations on spittlebugs with emphasis on their occurrence in rice, *Pesq. Agropec. Bras.*, 20, 407, 1985.
21. Fontes, E. G., Pires, C. S. S., and Sujii, E. R., Mixed risk-spreading strategies and the population dynamics of a Brazilian pasture pest, *Deois flavopicta* (Homoptera: Cercopidae), *J. Econ. Entomol.* 88, 1256, 1995.
22. Peck, D. C., Natural history of the spittlebug *Prosapia* nr. *bicincta* (Homoptera: Cercopidae) in association with dairy pastures of Costa Rica, *Annals of the Entomol. Soc. of America*, 91, 435, 1998.
23. Peck, D. C., Use of alternative food plants exclusively by adult male froghoppers (Homoptera: Cercopidae), *Biotropica,* 30, 639, 1998.
24. Picanco, A., Leite, G. L. D., Mendes, M. C., and Borges, V. E., Attack of *Atarsocoris brachiariae* Becker, a new pest of pasturelands in Mato Grosso, Brazil, *Pesq. Agropecuaria Bras.*, 34, 885, 1999.
25. Greber, R. S., Digitaria striate virus — A Rhabdovirus of grasses transmitted by *Sogatella kolophon* (Kirk.), *Aust. J. Agric. Res.*, 30, 40, 1979.
26. Robinson, S. W. and Fowler, H. G., Foraging and pest potential of Paraguayan grass-cutting ants (*Atta* and *Acromyrmex*) to the cattle industry. *Z. Ang. Ent.*, 93, 42, 1982.
27. Fowler, H. G. and Robinson, S. W., Foraging and grass selection by the grass-cutting ant *Acromyrmex landolti fracticornis* (Forel) (Hymenoptera: Formicidae) in habitats of introduced forage grasses in Paraguay, *Bull. Entomol. Res.*, 67, 659, 1977.
28. Jonkman, J. C. M., Nests of the leaf-cutting ant *Atta vollenweideri* as accelerators of succession in pastures, *Z. Ang. Ent.*, 86, 25, 1978.
29. Jonkman, J. C. M., Population dynamics of leaf-cutting ant nests in a Paraguayan pasture, *Z. Ang. Ent.* 87, 281, 1979.
30. Lobry de Bruyn, L. A. and Conacher, A. J., The role of termites and ants in soil modification: A review, *Aust. J. Soil Res.*, 28, 55, 1990.
31. Janzen, D. H., Seed predation by animals, *Ann. Rev. Ecol. Syst.*, 2, 465, 1971.
32. Carroll, C. R. and Janzen, D. H., Ecology of foraging by ants, *Ann. Rev. Ecol. Syst.*, 4, 231, 1972.
33. Beattie, A. J., The effects of ants on grasslands, in *Grassland Structure and Function: California Annual Grassland*, Huenneke, L. F. and Mooney, H., Eds., Kluwer Academic Pub., Dordrecht, The Netherlands, 1989, 105.
34. Medina, C. A., Lapointe, S. L., and Chacon, P., Fauna de hormigas asociadas con forrajes tropicales y su implicación como depredadoras de huevos y ninfas del salivazo de los pastos *Aeneolamia* spp., *Revista Colombiana de Entomología* 19, 143, 1993.
35. Fernandez Ruvalcaba, M., Cruz Vazquez, C., Solano Vergara, J., and Garcia Vazquez, Z., Anti-tick effects of *Stylosanthes hamata* on plots experimentally infested with *Boophilus microplus* larvae in Morelos, Mexico, *Exp. and Appl. Acarology* 23, 171, 1999.
36. Rincón C., A., Cuesta M., P. A., Pérez B., R., Lascano, C. E., and Ferguson, J., Maní forrajero perenne (*Arachis pintoi* Krapovickas y Gregory): Una alternativa para ganaderos y agricultores, Boletín técnico ICA (Instituto Colombiano Agropecuario) No. 219, 1992.
37. King, A. B. S. and Saunders, J. L., *Las Plagas Invertebradas de Cultivos Anuales Alimenticios en América Central,* Administración de Desarrollo Extranjero (ODA), London, U.K., 1984.
38. Stalker, H. T., Campbell, W. V., and Wynne, J. C., Evaluation of cultivated and wild peanut species for resistance to the lesser cornstalk borer (Lepidoptera: Pyralidae), *J. Econ. Entomol.*, 77, 53, 1984.
39. Prine, G. M., Dunavin, L. S., Moore, J. E., and Roush, R. D., 'Florigraze' rhizoma peanut: A perennial forage legume, Univ. of Fla., Gainesville, FL, Circular S-275, 1981.
40. Teitzel, J. K. and Burt, R. L., *Centrosema pubescens* in Australia, *Trop. Grasslands*, 10, 5, 1976.
41. Grof, B. and Harding, W. A. T., Yield attributes of some species and ecotypes of *Centrosema* in North Queensland, *Queensland J. of Agric. and Animal Sci.*, 27, 237, 1970.

42. Williams, R. N. and Schuster, M. F., Cochonilha dos capins (*Antonina graminis*) no Brasil. I — Distribuição e plantas hospedeiras, *Pesq. Agropec. Bras.*, 5, 215, 1970.
43. Costa, J. M., Williams, R. N., and Schuster, M. F., Cochonilha dos capins (*Antonina graminis*) no Brasil. II — Introdução de *Neodusmetia sangwani*, inimigo natural da cochonilha, *Pesq. Agropec. Bras.*, 5, 334, 1970.
44. Gabriel, D., Levantamento da ocorrência da cochonilha *Antonina graminis* (Maskell, 1897) (Homoptera: Pseudococcidae) e de seus parasitos, em áreas de pastagens do estado de São Paulo. III — Municípios de Martinópolis, Caiabú, Regente Feijó, Indiana, Onda Verde, Bady Bassit e Nova Granada, *Biológico,* 49, 207, 1983.
45. Davidson, R. L., Correcting past mistakes — Loss of habitat for predators and parasites of pasture pests, in *Proc. 3rd Australasian Conf. Grassl. Invert. Ecol.*, Adelaide, Australia, 1981, 199.
46. Pimentel, D., *CRC Handbook of Pest Management in Agriculture*, 2nd Edition, Vol. II, CRC Press Inc., Boca Raton, FL., 1991.
47. Alves, S. B., Risco, S. H., and Almeida, L. C., Influence of photoperiod and temperature on the development and sporulation of *Metarhizium anisopliae* (Metsch.) Sorok, *Z. Ang. Ent.*, 97, 127, 1984.
48. Arango, G. L., Torres, C., and Lapointe, S. L., Patogenicidad de tres cepas de *Metarhizium anisopliae* sobre huevos y ninfas de *Aeneolamia varia* (Fabricius) (Homoptera: Cercopidae), *Revista Colombiana de Entomología*, 20, 43, 1994.
49. Lapointe, S. L., Sotelo, G., and Arango, G., Improved technique rearing spittlebugs (Homoptera: Cercopidae), *J. Econ. Entomol.*, 82, 1764, 1989.
50. Koller, W. W. and Valério, J. R., Efeito da remoção da palha acumulada ao nível do solo sobre a população de cigarrinhas (Homoptera: Cercopidae) em pastagens de *Brachiaria decumbens*, *Anais Soc. Entomológica do Brasil*, 17, 209, 1988.
51. Lapointe, S. L., Serrano, M. S., and Jones, P. G., Microgeographic and vertical distribution of *Acromyrmex landolti* (Hymenoptera: Formicidae) nests in a neotropical savanna, *Environ. Entomol.*, 27, 636, 1998.
52. Nilakhe, S. S., Evaluation of grasses for resistance to spittlebugs, *Pesq. Agropec. Bras.*, 22, 767, 1987.
53. Lapointe, S. L., Serrano, M. S., Arango, G. L., Sotelo, G., and Cordoba, F., Antibiosis to spittlebugs (Homoptera: Cercopidae) in accessions of *Brachiaria* spp., *J. Econ. Entomol.*, 85, 1485, 1992.
54. Miles, J. W., Lapointe, S. L., Escandón, M. L., and Sotelo, G., Inheritance of resistance to spittlebug (Homoptera: Cercopidae) in interspecific *Brachiaria* spp. Hybrids, *J. Econ. Entomol.*, 88, 1477, 1995.
55. Cardona, C., Miles, J., and Sotelo, G., An improved methodology for massive screening of *Brachiaria* spp. genotypes for resistance to *Aeneolamia varia* (Homoptera: Cercopidae), *J. Econ. Entomol.*, 92, 490, 1999.
56. Thomas, D. and Lapointe, S. L., Testing new accessions of Guinea Grass (*Panicum maximum*) for acid soils and resistance to spittlebug (*Aeneolamia reducta*), *Tropical Grasslands*, 23, 232, 1989.
57. Howard, J. J., Leafcutting ant diet selection: The role of nutrients, water, and secondary chemistry, *Ecology*, 68, 503, 1987.
58. Howard, J. J., Leafcutting ant diet selection: Relative influence of leaf chemistry and physical features, *Ecology*, 69, 250, 1988.
59. Howard, J. J., Cazin, Jr., J., and Wiemer, D. F., Toxicity of terpenoid deterrents to the leafcutting ant *Atta cephalotes* and its mutualistic fungus, *J. Chem. Ecology*, 14, 59, 1988.
60. Hubbell, S. P., Weimer, D. F., and Adejare, A., An antifungal terpenoid defends a neotropical tree (*Hymenaea*) against attack by fungus-growing ants (*Atta*), *Oecologia*, 60, 321, 1983.
61. Lapointe, S. L., Serrano, M. S., and Corrales, I. I., Resistance to leafcutter ants (Hymenoptera: Formicidae) and inhibition of their fungal symbiont by tropical forage grasses, *J. Econ. Entomol.*, 89, 757, 1996.
62. Lenné, J. M. and Sonoda, R. M., *Rhizopus stolonifer* (Ehr. ex. Fr.) Lind., a seed-borne fungus of *Stylosanthes hamata* (L.) Taub. in Florida, *Soil and Crop Sci. Soc. Fla. Proc.*, 37, 39, 1978.
63. Gould, A. B. and Sonoda, R. M., Damping-off of *Stylosanthes guianensis* by *Pythium irregulare*, *Trop. Grassl.*, 23, 250, 1990.
64. Sonoda, R. M., Kretschmer, Jr., A. E., and Brolmann, J. B., *Colletotrichum* leaf spot and stem canker of *Stylosanthes* spp. in Florida, *Trop. Agric. (Trinidad)*, 51, 75, 1974.
65. Sonoda, R. M., Reduction in forage yield of Siratro by *Rhizoctonia* foliar blight, *Plant Disease*, 64, 667, 1980.

66. Sonoda, R. M. and Mislevy, P., Foliar blight of *Cynodon* spp. caused by *Rhizoctonia solani* in Florida, *Plant Disease*, 69, 811, 1985.
67. Sonoda, R. M., Effect of a mosaic virus on Siratro and other *Macroptilium atropurpureum* (DC) Urb. accessions, *Soil Crop Sci. Soc. Fla. Proc.*, 36, 147, 1977.
68. Kretschmer, Jr., A. E., Sonoda, R. M., and Snyder, G. H., Resistance of *Desmodium heterocarpon* and other tropical legumes to root knot nematodes, *Trop. Grassl.*, 14, 115, 1980.
69. Sonoda, R. M., Kretschmer, Jr., A. E., and Wilson, T. C., Evaluation of *Macroptilium atropurpureum* (DC) Urb. germplasm for reaction to foliar diseases, *Soil Crop Sci. Soc. Fla. Proc.*, 51, 71, 1992.
70. Jones, R. J., The effect of rust (*Uromyces appendiculatus*) on yield and digestibility of *Macroptilium atropurpureum* cv. Siratro, *Trop. Grassl.*, 16, 130, 1982.
71. Lenné, J. M., and Sonoda, R. M., Effect of anthracnose on yield of the tropical forage legume *Stylosanthes hamata*, *Phytopathology*, 72, 207, 1982.
72. Ellis, M. A., Ferguson, J. E., Grof, B., and Sinclair, J. B., Transmission of *Colletotrichum gloeosporioides* and effect of sulfuric acid scarification on internally-borne fungi in seeds of *Stylosanthes* spp., *Plant Dis. Rept.*, 60, 844, 1976.
73. Sonoda, R. M. and Brolmann, J. B., Reaction of *Stylosanthes hamata* (L.) Taub. indigenous to southeast Florida to *Colletotrichum gloeosporioides* (Penz.) Sacc, *Soil Crop Sci. Soc. Fla. Proc.*, 39, 39, 1979.
74. Chakraborty, S., Pettit, A. N., Boland, R. M., and Cameron, D. F., Field evaluation of quantitative resistance to anthracnose in *Stylosanthes scabra*, *Phytopathology*, 80, 1147, 1990.
75. Cameron, D. F. and Irwin, J. A. G., Use of natural outcrossing to improve the anthracnose resistance of *Stylosanthes guianensis*, *Special Publication — Agron. Soc. Of New Zealand*, 1986, 224.
76. Chakraborty, S., Thomas, M. R., and Ellis, N., A multivariate analysis of pathogenic variation in *Colletotrichum gloeosporioides* infecting the tropical pasture legume, *Stylosanthes scabra*, *Phytopathology*, 86, 283, 1996.
77. Kelemu, S., Badel, J. L., Moreno, C. X., and Miles, J. W., Virulence spectrum of South American isolates of *Colletotrichum gloeosporioides* on selected *Stylosanthes guianensis* genotypes, *Phytopathology*, 80, 1355, 1996.
78. Sonoda, R. M. and Kretschmer Jr., A. E., Reaction of *Macroptilium atropurpureum* cv. Siratro to inoculation with *Uromyces appendiculatus* collected in southern Mexico, *Proc. Soil Crop Science Society of Florida*, 49, 185, 1989.
79. Bray, R. A., Sonoda, R. M., and Kretschmer, Jr., A. E., Pathotype variability of rust caused by *Uromyces appendiculatus* on *Macroptilium atropurpureum*, *Plant Dis.*, 75, 430, 1991.
80. Lenné, J. M., Control of anthracnose in the tropical pasture legume *Stylosanthes capitata* by burning, *Trop. Pest Manag.*, 28, 223, 1982.
81. Lenné, J. M., Controlling *Meloidogyne javanica* on *Desmodium ovalifolium* with grasses, *Plant Dis.*, 65, 870, 1981.
82. Wolfe, M. S. and Barrett, J. A., Population genetics of powdery mildew epidemics, *Ann. NY Acad. Sci.*, 287, 151, 1977.
83. Chakraborty, S., Pettit, A. N., Cameron, D. F., Irwin, J. A. G., and Davis, R. D., Anthracnose development in pure and mixed stands of the pasture legume *Stylosanthes scabra*, *Phytopathology*, 81, 788, 1991.
84. Irwin, J. A. G., Cameron, D. F., and Lenné, J. M., Responses of *Stylosanthes* to anthracnose, in *The Biology and Agronomy of Stylosanthes*, Stace, H. M. and Edye, L. A., Eds., Academic Press, Sydney, Australia, 1984, 295.
85. Munaut, F., Haide, N., Stappen, J. V., and Maraite, H., Genetic relationships among isolates of *Colletotrichum gloeosporioides* from *Stylosanthes* spp. in Africa and Australia using RAPD and ribosomal DNA markers, *Plant Pathology*, 47, 641, 1998.
86. Woods, L. E., A survey of Townsville stylo (Townsville lucerne) pastures established in the Northern Territory up to 1969, *Trop. Grasslands*, 3, 91, 1969.
87. Wilson, P. A. and Chakraborty, S., The virtual plant: a new tool for the study and management of plant diseases, *Crop Protection*, 17, 231, 1998.
88. Hutton, E. M., Siratro: A tropical legume bred from *Phaseolus atropurpureus*, *Aust. J. Exp. Agr. Anim. Husb.*, 2, 117, 1962.

89. Winchester, J. A., Preliminary investigations on the mode of action of Pangolagrass roots in reducing cotton root-knot nematode (*Meloidogyne incognita acrita*) populations, *Soil Crop Sci. Soc. Fla. Proc.*, 20, 178, 1960.
90. Bacon, C. W., Lyons, P. C., Porter, J. K., and Robbins, J. D., Ergot toxicity from endophyte-infected grasses: A review, *Agron. J.*, 78, 106, 1986.
91. Christensen, M. J., Latch, G. C. M., and Tapper, B. A., Variation within isolates of *Acremonium* endophytes from perennial rye-grasses, *Mycological Res.*, 95, 918, 1991.
92. Christensen, M. J. and Latch, G. C. M., Variation among isolates of *Acremonium* endophytes (*A. coenophialum* and possible *A. typhinum*) from tall fescue (*Festuca arundinacea*), *Mycological Res.*, 95, 1123, 1991.
93. Latch, G. C. M. and Tapper, B. A., *Lolium* endophytes — problems and progress, *Proc. Jap. Assn. Mycotoxicology, Supplement 1*, 1988, 220.
94. Siegel, M. R., Latch, G. C. M., Bush, L. P., Fannin, F. F., Rowan, D. D., Tapper, B. A., Bacon, C. W., and Johnson, M. C., Fungal endophyte-infected grasses: alkaloid accumulation and aphid response, *J. Chem. Ecology*, 16, 3301, 1990.
95. Clay, K., Clavicipitaceous endophytes of grasses: their potential as biocontrol agents, *Mycol. Res.*, 92, 1, 1989.

13 Animal Production in Grass-Legume Pastures in the Tropics

Carlos E. Lascano

CONTENTS

I. ABSTRACT

Animal liveweight gain can be augmented with the addition of a legume to a pasture system in the tropics, whether or not there is dry-season stress. However, where this occurs, the gain/unit area in pastures is highly dependent on its effect on the quality and quantity of plant material available to the grazing animal. When there is sufficient rainfall, the milk yield/unit area in nitrogen-fertilized grass pastures can be increased significantly over grass-legume pastures mainly due to higher carrying capacity.

The use of herbaceous and shrub legumes in pure stands (i.e., protein banks) to complement improved grasses can result in increased animal production by partially overcoming nutritional constraints in the dry season. The impact on animal performance is greater, however, using grass-legume pastures than grass pastures complemented with protein banks sown with herbaceous legumes.

Animal production in grass-legume pastures is also affected by grazing management. The effect of stocking rate on animal performance is variable and dependent on changes in botanical composition, which, in turn, are affected by factors such as the palatability and recovery rate of the grazed components. Based on grazing selectivity studies, it would appear that certain grass-legume pastures should be managed with some form of deferred or rotational grazing in order to favor persistence of the more palatable species.

In tropical America, adoption of legume technology by farmers has been slow, and many are not aware of the benefits that such technology could provide. To change this trend, a major emphasis should be placed on on-farm participatory evaluations of grass-legume pastures in order to dem-

0-8493-2318-5/01/$0.00+$.50

onstrate their benefits and on the development of commercial seed supply systems to implement the technology.

II. INTRODUCTION

The value of pasture for a livestock producer depends, to a great extent, on its capacity to produce animal products, which, in turn, is related to the quantity and quality of the forage offered. Tropical pastures generally have high photosynthetic capacity and thus have the potential to produce large quantities of biomass. However, both the quantity and quality of biomass produced can change during the year and over time depending on the length of the dry season and the soil fertility of the location (e.g., soil N). In tropical areas with long dry seasons, liveweight gain and milk yield can be significantly depressed in pastures based on grass alone. Grass pastures sown in areas with low soil fertility degrade over time if no fertilizer is applied. This degradation process is partially reflected in loss of grass productivity and weed invasion, which affect carrying capacity and animal performance.

An alternative to minimize short- and long-term declines in quality and quantity of forage biomass and thus increase livestock production is to utilize legumes in pastures. The rationale for this alternative is that tropical legumes have a higher nutritive value than grasses and, through symbiotic nitrogen fixation, can enhance production and quality of the companion grass and improve soil fertility.

This chapter points out the potential benefits and limitations of animal production on grass-legume pastures compared to those containing grass in pure stand and discusses the effect of grazing management on their productivity.

III. ANIMAL PRODUCTION ON GRASS-LEGUME PASTURES

A. Liveweight Gain

The pioneer work on evaluation and use of tropical grass-legume pastures was done in Australia. As a result, most of the scientific literature on animal liveweight gain in pastures containing tropical legumes comes from there. Table 13.1 summarizes results from some grazing experiments of three or more years conducted in Australia. Annual liveweight gain in grass-legume pastures in dry areas (≤ 1000 mm/yr) ranged from approximately 165 to 205 kg/head and 90 to 180 kg/ha. In areas with higher rainfall (> 1250 mm/yr), gains were similar to those in the drier areas (i.e., 130 to 230 kg/head/yr), but production/unit area was two to three times greater (Table 13.1).

Research on grass-legume pastures in Africa has been mainly conducted in Uganda, Nigeria, and Zimbabwe. Results from this work show that animal production can reach approximately 140 to 155 kg/head/yr, and production/unit area 525 to 650 kg/ha/yr in grass-legume pastures in these countries of Africa (Table 13.2). In tropical America, research on grass-legume pastures has been rather limited. Stobbs[1] reviewed the literature on the effect of tropical grass-legume pastures on milk production and, of 17 references cited, only one came from tropical America. This, in part, reflects the large number of failures that researchers and producers in that area have had with legumes.[2] These failures have been attributed to poor adaptation of commercial legume cultivars, mostly developed in Australia, and to abiotic and biotic factors prevailing in livestock production areas of the American tropics.[3] It follows that in order to exploit the benefits of legumes in animal production in tropical America, it is necessary to have species that are tolerant of grazing and adapted to the stresses prevailing in a given ecosystem.

Even though tropical America is a center of diversity for many important forage legumes such as *Stylosanthes*, *Centrosema*, and *Arachis*, most of the scientific evaluations have been limited to a few species [e.g., *Neonotonia wightii* (Wight & Arn.) Lackey (glycine), *Macroptilium atropurpureum* (DC.) Urb. (siratro), *Pueraria phaseoloides* (Roxb.) Benth. (kudzu), and *Centrosema*

TABLE 13.1
Range in Annual Liveweight Gain (LWG) of Animals Grazing Tropical Grass-Legume Pastures in Australia

Pastures	Rainfall (mm)	Grazing Period (years)	Annual LWG (kg/hd)	(kg/ha)	Reference
Panicum maximum/Macroptilium atropurpureum	700	5	206	103	53
Cenchrus ciliaris/M. atropurpureum	722	5	172	93	54
Urochloa mosambicensis/Stylosanthes hamata	937	3	166	123	55
Setaria anceps/M. atropurpureum	1000	3	163	181	56
P. maximum/Neonotonia wightii	1251	3	232	580	57
Paspalum dilatatum/Centrosema pubescens	1600	7	149	298	58
Phaseolus lathyroides/Trifolium repens					
Brachiaria decumbens/P. maximum Stylosanthes guianensis/M. atropurpureum	1700	3	128	153	59
P. maximum/C. pubescens and *N. wightii*	3200	3	152	565	60

pubescens Benth. (centro)] which are marginally adapted to the edaphic conditions prevailing in major livestock areas of tropical America.[4] This was recognized by the Red Internacional de Evaluación de Pastos Tropicales (RIEPT) and the former Tropical Pastures Program of CIAT.[5] This recognition was translated into a major effort that began in the early 1980s to evaluate and select grass and legume germplasm for important ecosystems such as savannas and humid forests in tropical America.[4] A tangible result of this effort has been the release of legume cultivars such as *Stylosanthes capitata* Vogel cv. Capica for the savannas of Colombia by the Instituto Colombiano Agropecuario (ICA)[6] and *S. guianensis* (Aublet) Sw. (stylo) cv. Pucallpa for the humid forests by a national research organization in Peru.[7] More recently, *Arachis pintoi* Krapov. & D. Gregory (perennial peanut) cv. Maní Forrajero was released as a forage legume in Colombia.[8] This perennial

TABLE 13.2
Range in Annual Liveweight Gain (LWG) of Animals Grazing Tropical Grass-Legume Pastures in Africa and Tropical America

Pasture	Rainfall (mm)	Grazing Period (years)	Annual LWG (kg/hd)	(kg/ha)	Reference
	Africa				
Panicum maximum/Desmodium intortum	1400	4	156	656	48
P. maximum/Stylosanthes guianensis	1400	4	151	636	48
Cynodon plectostachyus/Centrosema pubescens	1800	3	142	526	61
Digitaria decumbens/C. pubescens S. humiliss	1800	3	153	567	61
	Tropical America				
Andropogon gayanus/S. capitata	1570	3	140	160	9
A. gayanus/S. capitata	2500	4	180	285	9
A. gayanus/Centrosema acutifolium	2500	3	155	310	9
Brachiaria decumbens/Pueraria phaseoloides	2500	9	174	296	38
B. dictyoneura/Arachis pintoi	2500	4	183	432	12
B. decumbens/A. pintoi	2900	3	200	600	12
B. brizantha/A. pintoi	4500	3	183	550	12

and stoloniferous forage peanut has high nutritional quality,[9,10] is compatible with aggressive and stoloniferous grasses,[11] and persistent under very heavy grazing.[12]

It was recognized at an early stage of forage legume dissemination that germplasm distribution was not sufficient to promote adoption of grass-legume technology by farmers. Therefore, in RIEPT, considerable effort was made to further pasture establishment,[13] develop alternative methodologies for grazing experiments,[14] and integrate improved grass-legume pastures into production systems.[15] In addition, much emphasis has been given to the development of seed supply systems, which, if not in place, will limit adoption of forage legumes by farmers.[16]

Experimental results of animal production in grass-legume pastures selected by RIEPT for acid, infertile soils are summarized in Table 13.2. In these pastures, mostly evaluated in the Llanos of Colombia, annual liveweight gain has ranged from 135 to 200 kg/head and from 160 to 600 kg/ha. A closer analysis of these results indicates that, in locations with dry-season stress (e.g., three-five months/yr), liveweight gain in legume-containing pastures ranges from 200 to 400 kg/ha; whereas, in areas with no dry-season stress, the gain is on the order of 500 to 600 kg/ha.

In some grazing experiments conducted in Australia, Zimbabwe, Costa Rica, and Colombia, animal liveweight gain was measured on grass-legume and grass-only pastures. Liveweight gain on the mixed pastures was greater by 30% than that on grass pastures (Table 13.3). In a recent review of beef cattle liveweight gain responses to introduction of *Stylosanthes* species in Australia, increases of 30 to 60 kg/head were attributed to the legume component of pastures.[17] However, in some experiments conducted in the Llanos of Colombia, animal weight gain on grass-legume pastures was twice as high as that recorded on grass pastures. In those cases, the larger increments in liveweight gain in the mixed pastures were related to the nutritional quality of the companion grass and to the legume content of the pasture. For example, in pastures of *Brachiaria humidicola* (Rendle) Schweick. (koroniviagrass) (known to be protein deficient[18]), animal production on grass associated with 30% perennial peanut (*A. pintoi*) was two times higher than on grass alone.[12] In contrast, pastures having the same grass associated with 10% of the legume resulted in only 35% more liveweight gain in the grazing animals (Table 13.3).

In general, the potential increases in animal liveweight gain are very high with tropical grass-legume pastures in areas with or without dry-season stress. However, animal liveweight gain/unit

TABLE 13.3
Percentage of Increase in Individual Liveweight Gain (LWG) of Animals Grazing Tropical Grass-Legume Compared to Grass Alone Pastures in Australia, Zimbabwe (Africa), and Colombia

Location	Grass-Legume Pasture	Increase in LWG over Grass Pasture (%)	Reference
	Native Pastures		
Katherine, Australia	Native grass/*Stylosanthes humilis*	78	62
Lansdown, Australia	Native grass/*S. hamata*	24	50
Zimbabwe, Africa	Native grass/*S. guianensis*	36	63
Zimbabwe, Africa	Native pasture/*Macroptilium atropurpureum*	25	64
	Improved Grass Pastures		
Carimagua, Llanos of Colombia	*Brachiaria decumbens/Arachis pintoi*	28	8
	B. humidicola/A. pintoi	213	8
	B. decumbens/Pueraria phaseoloides	43	39
Atlantic Zone of Costa Rica	*B. brizantha/A. pintoi* (low stocking rate)	11	65
	(high stocking rate)	30	65

TABLE 13.4
Range in Stocking Rate (SR) and Milk Yield (MY) of Cows Grazing Tropical Grass and Grass-Legume Pastures

Pasture System	Range in SR (cows/ha)	Range in MY (kg/cow/day)	(kg/ha)	Reference
Grasses with no fertilization; cows of medium production potential.	1.0–1.5	6–8	1000–4000	20, 66, 67, 68
Grasses in association with legumes; cows of medium production potential.	1–2	8–12	3000–6000	66, 69, 70, 71, 72, 73
Grasses with nitrogen fertilization; cows of medium production potential.	3–6	7–9	6000–15,000	19, 74, 75, 76
Grasses with nitrogen fertilization; cows of high production potential.	2–8	10–14	7000–18,000	77, 78, 79

area in these pastures is highly related to dry season stress of the region. The advantage of mixed pastures over grass pastures in terms of animal production is decisive but variable, and dependent on the quality of the companion grass and legume content of the pasture.

B. Milk Production

A strategy pursued by some researchers to alleviate deficiencies in quantity and quality of tropical grass-based pastures has been to supplement grazing material with concentrates, urea/molasses, and agricultural by-products. Supplementation for animals grazing tropical pastures has been the subject of many papers. In most cases, research on supplementation for milk cows has attempted to define responses in animal production in terms of type and level of supplement, and interaction between level of supplementation and forage offered. Results from some of the studies carried out with grazing cows indicate a positive response in milk yield with dietary supplementation over most or all of the lactation (i.e., 0.9 to 1.2 kg of milk/kg of supplement) when cows have medium to high production potential.[19-21] However, in spite of these positive responses, supplementation of milk cows in dual production systems in the tropics is limited and extremely variable due, in many cases, to an unfavorable market price for milk relative to the additional cost of the supplement.[22]

Another alternative to augment milk production is the addition of N fertilizer to tropical grass pastures, which results in increased biomass offered and crude protein concentration in the tissue and thus in higher milk yield/unit area, particularly when cows have high production potential, as indicated in Table 13.4. Tropical pasture N technology has been primarily based on the use of grasses that require relatively fertile soils and that can readily respond to fertilization such as *Cynodon dactylon* (L.) Pers. (Bermuda grass), *C. nlemfuensis* Vanderyst (star grass), *Digitaria decumbens* Stent (pangola grass), and *Panicum maximum* Jacq. (guinea grass). These grasses have been adopted for production in the tropics, but, in general, with little or no N fertilization.[3] As a result, milk yield/unit area in unfertilized pastures containing the grasses is considerably lower than that which could be obtained with relatively high N inputs (Table 13.4).

The use of grass-legume pastures to increase milk yield in the tropics has been an option that has received little attention by researchers in Latin America. Most of the research on milk production in grass-legume pastures has been carried out in Australia. Results from several sources indicate that milk yield/unit area is higher in grass-legume pastures than in grass pastures, but lower than that which can be obtained in N-fertilized pastures (Table 13.4).

Studies carried out in the American tropics have compared short-term milk-yield response in cows grazing grass and grass-legume pastures on acid soils. In Quilichao, Colombia, cows grazing *Andropogon gayanus* Kunth (gambagrass) and *Brachiaria dictyoneura* (Figari & De Not.) Stapf (koroniviagrass), in mixture with *Centrosema acutifolium* Benth., cv. Vichada, and *C. macrocarpum* Benth., produced 20% more milk than those grazing grass alone.[23] A similar increase in milk yield was found at CATIE in Turrialba, Costa Rica, when cows grazed pastures of star grass in mixture with perennial peanut rather than in pure stand.[24]

The potential advantages of using grass-legume pastures for milk production in the tropics are quite evident from the experimental data available. However, a recent study suggests that milk-yield response in grass-legume pastures can be affected by the production potential of the cows. In short-term experiments carried out by CIAT in Colombia, milk-yield increments in grass-legume pastures were lower than those in grass-based pastures for cows having low production potential (an 8% increase) compared to those having medium genetic potential (a 20% increase).[25] These results could have important implications for the utilization of legumes in traditional dual-purpose systems in tropical areas since cows used in such systems usually have low milk-yield potential.[26] However, the overall nutrition of cows that graze grass-legume pastures in dual-purpose systems could be increased by their consumption of the legume. This would affect reproduction positively, which would, in turn, have important economic implications for the farmer.

In general, milk production in legume-grass pastures is higher than in grass pastures without N fertilization, the benefit being mostly in terms of individual daily milk yield. However, milk yield/unit area in N-fertilized grass pastures can be increased significantly over that in unfertilized grass or grass-legume pastures, mainly due to higher carrying capacity.

IV. ANIMAL PRODUCTION ON PASTURES COMPLEMENTED WITH PROTEIN BANKS

A. Native Grass Pastures

There is abundant evidence that losses in animal production during the dry season can be minimized by using nonprotein N or true protein supplementation.[27-29] However, supplementation with purchased protein sources is not always possible due to lack of supply in certain tropical regions or prohibitive cost.[30] An alternative form of protein supplementation is the use of legumes in pure stand, commonly referred to as "protein banks."

The initial evaluation of protein banks in tropical areas was pioneered by Australian researchers in studies carried out in the early and mid-1960s. Results from this work showed that weight loss of steers grazing native pastures in northern Australia could be significantly reduced when animals had frequent access to protein banks of *Stylosanthes humilis* HBK (Townsville stylo).[31]

The idea of a protein bank to complement native grasses was also tested in Nigeria in a location with a six-month dry season. In this case, steers grazing the native grass pasture gained weight during the dry season when they were allowed access to a pure stand of stylo during the night. Animals that grazed the protein bank every two days maintained their weight during the season, whereas animals with access to the legume every four days lost weight. The liveweight gain of animals grazing the legume during the night was comparable to that obtained with supplementation of 800 g/animal/day of cotton seed.[32]

In tropical America, most of the research on protein banks as a complement to native grasses has been restricted to savanna grasslands in the Llanos of Colombia and the Cerrados of Brazil. In the Llanos of Colombia, steers grazing native savanna grasses managed with fire and supplemented with small areas (2000 m^2/head) of kudzu gained 30% more weight during the year than those grazing only the native grass.[33] A positive effect on dry-season animal performance was also observed in the Cerrados of Brazil with heifers grazing pastures complemented by a protein bank of *S. guianensis* var. Pauciflora, cv. Bandeirante.[34]

Subsequent studies carried out in the Llanos of Colombia indicated that steers grazing native pastures and having access to small areas (2000 m^2/head) of mainly grass with a low legume content (i.e., "energy banks") gained 50% more weight than those with access to a protein bank.[35] The higher animal gains with an energy bank are proof that what limits animal production in native savanna pastures of the Colombian Llanos is energy and not protein.[36]

Although, the liveweight gain of steers grazing native pastures can be increased through the use of small areas of legumes in pure stand or protein banks, particularly during the dry season, their use requires managerial and financial inputs and a certain amount of infrastructure (e.g., fencing). These requirements would be difficult to meet in regions with extensive cattle production systems based on native pastures such as tropical American savannas.[37]

B. Improved Grass Pastures

Studies designed to measure liveweight gain or milk production on improved grasses complemented by protein banks are reviewed in this section. In a five-year study in the Llanos of Colombia,[38] liveweight gain was highest for steers grazing fertilized (10 P, 13 K, 10 Mg, and 16 S kg/ha) *Brachiaria decumbens* Stapf (signalgrass) pasture in mixed stand with kudzú compared to the fertilized grass with a protein bank of the legume that comprised 30% of the grazing area or the fertilized grass alone. In that study, animal performance was consistently higher in the grass-legume pasture than in the grass pasture complemented with the legume. Furthermore, while weight gain after five years of grazing showed an upward trend in the grass-legume pasture, the reverse was true in the other two pastures. The legume in the protein bank did not persist after the fifth year of grazing, which was not the case in the association, where, after nine years of grazing, the liveweight gain was 40% greater than in the grass pasture.[39]

In some studies, the potential benefits of protein banks have been assessed with milk cows. For example, in Mexico, milk yield of cows was 9% higher when they were grazing star grass complemented by *Leucaena leucocephala* (Lam.) de Wit (leucaena) than when grazing grass alone.[40] In addition, calves from cows that had access to the protein bank gained 10% more weight than those from cows grazing only the grass pasture; this was attributed to 30% more milk consumption. In Bolivia, Paterson et al.[41] found a 10 to 20% increase in milk yield over normal production during the dry season when cows had access to a protein bank of *Macrotyloma axillare* (E. Meyer) Verdc. (perennial horsegram) cv. Archer and glycine cv. Tinaroo.

More recently in Colombia, Mosquera and Lascano[42] studied the use of protein banks in the dry and wet seasons of the year to complement grass pastures sown in acid soils. Results indicated that crossbred cows in their first month of lactation produced 14% more milk in the dry season when they had access to protein banks sown with *C. acutifolium* and *C. macrocarpum*. However, milk yield of these cows in their fourth month of lactation did not increase in the dry season under this arrangement, nor did the milk yield of Holstein and crossbred cows in the rainy season. Yet, in the same location, the milk yield of Holstein cows increased by 5 to 10% in the rainy season and by 15 to 30% in the dry season when they grazed associations of grasses with these legumes regardless of the stage of lactation.[43]

The use of herbaceous legumes as protein banks to supplement improved tropical forage grasses would seem an attractive option to overcome the problem of legume persistence often found in tropical grass-legume associations. However, the beneficial effect of protein banks on animal production is apparently determined by a number of factors such as the legume species used, the nutritional requirements of the grazing animals, and possibly the quantity and quality of the grass being supplemented. Results of short- and long-term experiments indicate that animal production is higher in grass-legume pastures than in grass pastures complemented with herbaceous legumes in pure stand. This could probably be attributed to more efficient N cycling in the mixed pasture and to a better balance of nutrients (i.e., energy and protein sources) in the forage consumed by the animals in this type of pasture than in a grass pasture with a supplemental protein bank.

The use of legumes such as leucaena that can be maintained as shrubs in protein banks is an alternative that has proven successful in tropical areas with dry seasons. However, commercial cultivars of leucaena (e.g., Cunningham) are restricted to soils with moderate acidity, high calcium content, and low Al saturation. Thus, there is an urgent need to identify shrub-type legume germplasm for the acid soils that are predominant in many cattle-producing areas of the tropics.

V. GRAZING MANAGEMENT OF GRASS-LEGUME PASTURES

A. Stocking Rate and Grazing Method

It is well known that animal production in tropical pastures can be greatly affected by grazing management. Most of the research on grazing management of grass-legume pastures has emphasized the effect of stocking rate on legume persistence and animal performance. Results of such studies have shown that stocking rate can have a variable effect on animal production in these pastures (Table 13.5). In some of the studies cited, increasing the stocking rate resulted in a linear reduction in liveweight gain/animal. However, in other experiments, stocking rate had little or no effect on individual liveweight gain and, in several cases, a higher liveweight gain/head was recorded at the highest stocking rate in the experiment. In some of these experiments, liveweight gain/ha either peaked at an intermediate stocking rate or was highest at the maximum stocking rate.

In most grazing experiments with grass-legume pastures, legume content of the forage offered is monitored over time, and stocking rate is recognized as having a marked effect on botanical composition of the pasture. Roberts[44] reviewed results of some grazing trials involving grass-legume pasture and stocking rates. An important conclusion from this review was that the animal gain/stock-

TABLE 13.5
Effect of Stocking Rate (SR) on Animal Liveweight Gain (LWG) in Some Tropical Grass-Legume Pastures

Pasture	SR (head/ha)	LWG (kg/hd)	LWG (kg/ha)	Reference
Setaria anceps/Macroptilium atropurpureum	1.1	163	181	56
	1.7	136	235	
	2.4	110	256	
	2.9	86	243	
Hyparrhenia rufa/Stylosanthes guianensis	1.6	103	165	80
	2.5	98	246	
	5.0	84	422	
Panicum maximum/Neonotonia wightii	2.5	232	580	57
	4.0	157	629	
	5.0	68	340	
Brachiaria humidicola/Arachis pintoi	2.0	151	302	12
	3.0	130	390	
	4.0	89	356	
B. decumbens or *P. maximum/ S. guianensis/M. atropurpureum*	0.7	122	86	59
	1.2	128	153	
	1.7	124	210	
	1.9	117	221	
Urochloa/Mosambicensis/Stylosanhtes humilis	0.6	124	74	59
	1.25	124	155	
	2.50	146	365	
B. brizantha/A. pintoi	3.0	178	534	65
	6.0	154	937	

ing rate relationship can be affected by botanical composition of the pasture. In grass-legume pastures managed at a low stocking rate, taller species can suppress low-growing species. If the dominant species have a lower feeding value than the species they replace, liveweight gain can drop. In pastures managed at a high stocking rate, the species least tolerant of grazing will not persist. If the remaining species are less palatable and have lower nutritional value than those they suppress, animal weight gain will be less. For example, Partridge[45] found that siratro did not persist in association with *Dichanthium caricosum* (L.) A. Camus (nadi bluegrass) at medium and high stocking rates. The sown legume was replaced in the pasture by *Desmodium heterophyllum* (Willd.) DC. (hetero) and by naturalized legumes of low acceptability at the medium and high stocking rates, respectively. The net result of losing siratro was little change in animal liveweight gain at the middle stocking rate, but a sharp drop in gain at the high stocking rate, when the sown legume was replaced by unpalatable, naturalized legumes.

The experience with tropical grass-legume pastures is that, in most cases, the grass suppresses the legume. However, there are examples where the opposite situation occurs under grazing. For example, cv. Itabela of the stoloniferous legume *Desmodium ovalifolium* Wallick ex Gagnep is not readily consumed by grazing animals. As a consequence, the animals selectively graze-out the grass component regardless of the management system applied.[46] Animal production in pastures dominated by *D. ovalifolium* is low both in terms of individual liveweight gain and production/ha. In contrast, in a grass pasture mixture with perennial peanut, where the legume is in higher proportion than the grass, animal performance has not been affected, mainly due to high intake of the legume.[11]

In most cases, grazing management studies with grass-legume pastures have emphasized legume persistence and animal production as functions of stocking rate under continuous grazing. There are few studies with grass-legume pastures that compare grazing methods at the same stocking rate. In Australia, Grof and Harding[47] found that an association of guinea grass/centro produced 15% more liveweight gain under alternate grazing than under continuous grazing. In Uganda, Stobbs[48] determined that liveweight gain in cattle grazing guinea grass/siratro was similar under continuous and rotational grazing. However, after three years, there was a better grass-legume balance and fewer weeds in pastures under rotational grazing than under continuous grazing. More recently, pastures of koroniviagrass/perennial peanut grown in the Llanos of Colombia produced 20% more liveweight gain under rotational grazing (7 days on/21 days off) than under alternate grazing (7 to 14 days on/7 to 14 days off) when pastures were grazed with three head/ha.[12]

B. Animal Selectivity and Grazing Management

On the basis of the experimental evidence available, it is difficult to decide on the best strategy to manage tropical grass-legume pastures. However, it would seem that long-term productivity of these pastures can be greatly influenced by the relative acceptability of the grasses and legumes being used. It is well documented that the preference by cattle for grasses and legumes can change within the year. Stobbs[49] found that, in a pasture of *Setaria anceps* Massey (nardi grass, cola de perro, setaria) and siratro, legume percentage in the diet was low (2 to 10%) in the spring and the first part of the summer and high (62 to 73%) in the fall. In northern Australia, Gardener[50] found that cattle in native pasture/*Stylosanthes hamata* (L.) Taub. (Caribbean stylo) selected more legume in the dry season than in the rainy season. The same situation was observed in the Llanos of Colombia with cattle grazing associations of the erect gambagrass with kudzu and *S. capitata*.[51] However, these seasonal trends in legume selection by grazing cattle have not been observed in associations of prostrate grasses with stoloniferous legumes such as *D. ovalifolium* and perennial peanut.[23,52] In those pastures, the proportion of legume in the diet relative to the proportion of legume in the forage offered (i.e., selection index) changes with legume species. In pastures with *D. ovalifolium*, cattle select less legume than offered and, as a consequence, the pastures become dominated by the legume.[46] In contrast, cattle previously accustomed to perennial peanut will select slightly more legume than that in the forage offered.[10,52]

From the results on selective grazing, it would seem that, in certain grass-legume pastures, some form of deferred or rotational grazing system could be utilized to favor persistence of the more palatable species (whether grass or legume).

Although there are decided advantages to adding a legume to a grazing system, managing a grass-legume pasture and maintaining the correct balance between the two components can be more of a challenge to farmers than keeping a grass pasture productive.

VI. FUTURE NEEDS

The inclusion of a legume in a tropical pasture system can have a considerable effect not only on the animals being pastured, as emphasized in this review, but also on the grass being grazed and the soil in the pasture. Legumes are high in nitrogen and, through litter and nodule decay, can contribute to soil organic matter and nitrogen reserves for use by the companion grass. Because of their deep root systems, most tropical legumes usually survive dry seasons in better condition than grasses and can offer more nutritious plant material during such times. However, in tropical America, the incorporation of legumes into pasture systems is slow due in part to lack of awareness of their potential benefits and the technology needed to utilize them. Therefore, a major effort should be placed on grass-legume trials in farmers' fields to demonstrate how legumes can increase beef and milk production and to advise on pasture management. On-farm pasture evaluation programs should preferably be integrated with livestock development projects and should include training in pasture establishment and grazing management and development of a seed supply system.

REFERENCES

1. Stobbs, T. H., Milk production per cow and per hectare from tropical pastures, in *Seminario Internacional de Ganadería Tropical*, F.I.R.A., Acapulco, México, 1976, 129.
2. Spain, J. M. and Vilela, L., Perspectivas para pastagens consorciadas na America Latina nos años 90 e futuros, in *Proc. Sociedade Brasileira de Zootecnia, Pastagens*, Campinas, S. P., Brazil, 1990, 87.
3. Toledo, J. M. and Nores, G., Tropical pastures technology for marginal lands of tropical America, *Outlook on Agriculture*, 15, 3, 1986.
4. Toledo, J. M., Pasture development for cattle production in the major ecosystems of the American lowlands, in *Proc. XV Int. Grassl. Cong.*, Kyoto, Japan, 1985, 77.
5. Toledo, J. M., Objetivos y organización de la Red Internacional de Evaluación de Pastos Tropicales, in Manual para la evaluación agronómica, Red Internacional de Evaluación de Pastos Tropicales, Toledo, J. M. Ed., CIAT, Cali, Colombia, 1982, 13.
6. Anon., Capica (*Stylosanthes capitata* Vog.), Instituto Colombiano Agropecuario, *Boletín Técnico No. 103*, Bogotá, Colombia, 1983.
7. Reyes, C., Ordoñez, H., and Pinedo, L., *Stylosanthes guianensis* cv. Pucallpa. Leguminosa forrajera para la Amazonía, Instituto Veterinario de Investigaciones Tropicales y de Altura and Instituto Nacional de Investigación y Promoción Agropecuaria, *Boletín Técnico No. 3*, 1985, 11.
8. Rincón, A., Cuesta, P., Pérez, R., Lascano, C., and Ferguson, J., Maní forrajero perenne (*Arachis pintoi* Krapovickas and Gregory), Instituto Colombiano Agropecuario, Bogotá, Colombia, *Boletín Técnico No. 219*, 1992, 23.
9. Lascano, C., and Thomas, D., Quality of *Andropogon gayanus* and animal productivity, in *Andropogon gayanus* Kunth: *A Grass for Tropical Acid Soils,* Toledo, J. M., Vera, R., Lascano, C., and Lenné, J. M., Eds., CIAT, Cali, Colombia, 1990, 247.
10. Carulla, J. E., Lascano, C., and Ward, J. K., Selectivity of resident and oesophageal-fistulated steers grazing *Arachis pintoi* and *Brachiaria dictyoneura* in the Llanos of Colombia, *Trop. Grassl.*, 25, 317, 1991.
11. Grof, B., Forage attributes of the perennial groundnut *Arachis pintoi* in a tropical savanna environment in Colombia, in *Proc. XVI Int. Grassl. Cong.,* Kyoto, Japan, 1985, 168.

12. Lascano, C., Nutritive value and animal production of forage *Arachis*, in *Biology and Agronomy of Forage Arachis*, Kerridge, P. C. and Hardy, W., Eds., Centro Internacional de Agricultura Tropical (CIAT), Cali, Colombia, 1994, chapter 10.
13. Lascano, C., and Spain, J. M. (Eds.), Establecimiento y renovación de pasturas, in Sexta Reunión del Comité Asesor de la Red Internacional de Evaluación de Pastos Tropicales, Veracruz, México, 1991, 421.
14. Lascano, C., and Pizarro, E. A., (Eds.), Evaluación de pasturas con animales: Alternativas metodológicas, in *Segunda Reunión del Comité Asesor de la Red Internacional de Evaluación de Pastos Tropicales*, Lima, Perú, 1986, 265.
15. Anon., Investigación con pasturas en fincas, in *Memorias Séptima Reunión del Comité Asesor de la Red Internacional de Evaluación de Pastos Tropicales*, Documento de Trabajo No. 124, CIAT, Palmira, 1993, 277.
16. Ferguson, J. and Reyes, C., Semillas: Su multiplicación y su investigación como actividades integradas a la RIEPT, in Investigaciones de Apoyo para la Evaluación de Pasturas. Tercera Reunión del Comité Asesor de la Red Internacional de Evaluación de Pastos Tropicales, Cali, Colombia, 1987, 195.
17. Coates, D. B., Miller, C. P., Hendrickson, R. E., and Jones, R. J., Stability and productivity of *Stylosanthes* pastures in Australia. II. Animal production from *Stylosanthes* pastures, *Trop. Grassl.*, 31, 494, 1997.
18. Hoyos, P., and Lascano, C., Calidad de *Brachiaria humidicola* en pastoreo en un ecosistema de bosque semi-siempre verde estacional, *Pasturas Tropicales*, 7, 3, 1985.
19. Colman, R. L. and Kaiser, A. G., The effect of stocking rate on milk production from Kikuyu grass pastures fertilized with nitrogen, *Aust. J. Exp. Agric. and Anim. Husb.*, 14, 155, 1974.
20. Rees, M. C., Minson, D. J., and Kerr, J. D., Relation of dairy productivity to feed supply in the Gympie district of south-eastern Queensland, *Aust. J. Exp. Agric. and Anim. Husb.* 12, 553, 1972.
21. Moberly, P. K., Maize meal as an energy supplement for the Jersey cow, *East Afr. Agric. and Forestry J.*, 32, 155, 1966.
22. Seré, C., Socioeconomía de la producción de ganado de doble propósito, in Panorama de la Ganadería de Doble Propósito en la América Tropical, Arango, Nieto, L., Charry, A. and Vera, R. R., Eds., Instituto Colombiano Agropecuario, Bogotá, Colombia, 1989, 13.
23. Lascano, C., Avila, P., Quintero, C. I., and Toledo, J. M., Atributos de una pastura de *Brachiaria dictyoneura-Desmodium ovalifolium* y su relación con producción animal, *Pasturas Tropicales*, 13, 10, 1991.
24. González, S., Selectividad y producción de leche en pasturas de Estrella (*Cynodon nlemfuensis*) solo y asociado con las leguminosas forrajeras *Arachis pintoi* CIAT 17434 y *Desmodium ovalifolium* CIAT 350, MS Thesis, Centro Agronómico Tropical de Investigación y Enseñanza (CATIE), Turrialba, Costa Rica, 1992, 142.
25. Lascano, C., and Avila, P., Milk yield of cows with different genetic potential on grass and grass-legume tropical pastures in *Proc. XVII Int. Grassl. Cong.*, New Zealand/Australia, 1993, 2006.
26. Arango, L., La ganadería de doble propósito: Estudio del caso colombiano, in Panorama de la ganadería de doble propósito en la América tropical, Arango, L., Charry, A., and Vera, R., Eds., Instituto Colombiano Agropecuario, Bogotá, Colombia, 1986, 61.
27. Norman, M. J. T., Dry season protein and energy supplements for beef cattle on native pastures at Katherine, *Aust. J. Exp. Agric. Anim. Husb.*, 3, 280, 1963.
28. Holroyd, R. G., Allan, P. J., and O'Rourke, P. D., Effect of pasture type and supplementary feeding on the reproductive performance of cattle in the dry tropics of North Queensland, *Aust. J. Exp. Agric. and Anim. Husb.*, 17, 197, 1977.
29. Taylor, W. J., Holroyd, R. G., and Rebegetz, J. G., The influence of nitrogen based supplements on liveweight, fertility and mortality of heifers grazing dry season native pastures, *Proc. Aust. Soc. Anim. Prod.*, 14, 277, 1982.
30. Paladines, O. and Leal, L. A., Pasture management and productivity in the Llanos Orientales of Colombia, in *Proc. Pasture production in acid soils of the tropics,* Sánchez, P. A. and Tergas, L. E., Eds., Cali, Colombia, 1979, 311.
31. Norman, M. J. T. and Steward, C. A., Complementary grazing of native pasture and standing Townsville lucerne in the dry season at Katherine, *Aust. J. Exp. Agric. Anim. Husb.*, 7, 225, 1967.

32. Haggar, R. J., de Leeuw, P. N., and Agishi, E., The production and management of *Stylosanthes gracilis* at Shika, Nigeria, *J. Agric. Sci. Camb.*, 77, 437, 1971.
33. Tergas, L. E., Paladines, O., Kleinhesterkamp, I., and Velásquez, J., Animal production from native pastures with complementary grazing of *Pueraria phaseoloides*, in the Eastern Plains of Colombia, *Trop. Anim. Prod.*, 8, 187, 1983.
34. Zobby, J. L., Kornelius, E., Saueressig, M. C., and Affin, O. A. D., Protein banks as a complement to native pastures, in *Proc. XVI Int. Grassl. Cong.*, Nice, France, 1989, 1169.
35. Lascano, C., and Plazas, C., Bancos de proteína y energía en sabanas de los Llanos Orientales de Colombia, *Pasturas Tropicales*, 12, 9, 1990.
36. Alvarez, A., and Lascano, C., Valor nutritivo de la sabana biend drenada en los Llanos Orientales de Colombia, *Pasturas Tropicales*, 9, 9, 1987.
37. Vera, R. and Seré, C. Los sistemas de producción pecuaria extensiva del trópico sudamericano: análisis comparativo, in *Sistemas de producción pecuaria extensiva,* Brasil, Colombia y Venezuela, Vera, R. and Seré, C., Eds., 1985, 431.
38. Tergas, L. E., Paladines, O., Kleinhesterkamp, I., and Velásquez, J., Productividad animal de *Brachiaria decumbens* sola y con pastoreo complementario en *Pueraria phaseoloides* en los Llanos Orientales de Colombia, *Prod. Anim. Trop.*, 9, 1, 1984.
39. Lascano, C., and Estrada, J., Long-term productivity of legume-based and pure grass pastures in the Eastern Plains of Colombia, in *Proc. XVI Int. Grassl. Cong.*, Nice, France, 1989, 1179.
40. Saucedo, G., Alvarez, F. J., and Arriagada, A., *Leucaena leucocephala* como fuente protéica para becerros lactantes criados en sistemas de amamantamiento restringido, *Prod. Anim. Trop.*, 5, 40, 1980.
41. Paterson, R. T., Samur, C., and Bresso, O., Efecto de pastoreo complementario de leguminosas reservadas sobre la producción de leche en la estación seca, *Prod. Anim. Trop.*, 6, 135, 1981.
42. Mosquera, P. and Lascano, C., Producción de leche de vacas en pasturas de *Brachiaria decumbens* solo y con acceso controlado a bancos de proteína, *Pasturas Tropicales*, 14, 2, 1992.
43. Lascano, C., and Avila, P., Potencial de producción de leche en pasturas solas y asociadas con leguminosas adaptadas a suelos ácidos, *Pasturas Tropicales*, 13, 2, 1991.
44. Roberts, C. R., Effect of stocking rate on tropical pastures, *Trop. Grassl.*, 14, 225, 1980.
45. Partridge, I. J., Improvement of Nadi blue grass *Dichantium caricosum* pastures on hill land in Fiji with superphosphate and siratro: Effects of stocking rate on beef production and botanical composition (*Macroptilium atropurpureum*), *Trop. Grassl.*, 13, 156, 1979.
46. Toro, N., Productividad animal en pasturas de *Brachiaria humidicola* (CIAT 679) solo y en asociación con *Desmodium ovalifolium* (CIAT 13089) bajo sistema de manejo flexible, MS Thesis, Centro Agronómico Tropical de Investigación y Enseñanza, Turrialba, Costa Rica, 1990, 111.
47. Grof, B., and Harding, W. A. T., Dry matter yields and animal production on the humid tropical coast of North Queensland, *Trop. Grassl.* 4, 85, 1970.
48. Stobbs, T. H., The effect of grazing management upon pasture productivity in Uganda. III. Rotational and continuous grazing, *Trop. Agric.*, 46, 293, 1969.
49. Stobbs, T. H., Seasonal changes in the preference by cattle for *Macroptilium atropurpureum* cv. Siratro, *Trop. Grassl.*, 11, 87, 1977.
50. Gardener, C. J., Diet selection and liveweight performance of steers on *Stylosanthes humilis* with *S. hamata* and *S. subsericea* in the Queensland dry tropics: Effects on pasture composition and cattle liveweight gain. *Aust. J. Agric. Res.*, 31, 205, 1980.
51. Böhnert, E., Lascano, C., and Weniger, J. H., Botanical and chemical composition of the diet selected by fistulated steers under grazing on improved grass-legume pastures in the tropical savannas of Colombia. I. Botanical composition of forage available and selected, *Z. Tierzuchtg. Zuchtgsbiol.*, 102, 385, 1985.
52. Lascano, C. and Thomas, D., Forage quality and animal selection of *Arachis pintoi* in association with tropical grasses in the eastern plains of Colombia, *Grass and Forage Science*, 43, 433, 1988.
53. Clatworthy, J. N., Legumes and summer fattening, *Modern Farming*, 8, 7, 1980.
54. 't Mannetje, L. and Nicholls, D. F., Beef production from pastures on granitic soils, in Commonwealth Scientific and Industrial Research Organization, Division of Tropical Agronomy, *Annual Report* 1974–1975, Brisbane, Australia, 1975, 13.
55. Gillard, P., Edye, L. A., and Hall, R. L., Comparison of *Stylosanthes humilis* with *S. hamata* and *S. subsericea* in the Queensland dry tropics: Effects on pasture composition and cattle liveweight gain, *Aust. J. Agric. Res.*, 31, 205, 1980.

56. Jones, R. J., The relation of animal and pasture production to stocking rate on legume based and nitrogen fertilized subtropical pastures, in *Proc. Aust. Soc. Anim. Prod.*, 1974, 10, 340.
57. Winks, L., Venamore, P. C., James, T. A., O'Grady, P., and O'Rourke, P. K., Performance of steers grazing a tropical grass-legume pasture on the Atherton tableland, *Qld. J. Agric. Anim. Sci.*, 37, 199, 1980.
58. Bryan, W. W., Grazing trials on the Wallum of Northeastern Queensland. I. A comparison of four pastures, *Aust. J. Exp. Agric. Anim. Husb.*, 8, 512, 1968.
59. Winter, W. H., Edye, L. A., and William, W. T., Effect of fertilizer and stocking rate on pasture and beef production from sown pastures in northern Cape York peninsula. II. Beef production and its relation to blood, fecal and pasture measurements, *Aust. J. Exp. Agric. Anim. Husb.*, 17, 187, 1977.
60. Mellor, W., Hibberd, M. J., and Grof, B., Beef cattle liveweight gains from mixed pastures of some guinea grasses and legumes on the wet tropical coast of Queensland, *Qld. J. Agric. Anim. Sci.*, 30, 259, 1973.
61. Oyenuga, V. A., and Olubayo, F. O., Productivity and nutritive value of tropical pastures at Ibadan, in *Proc. X Int. Grassl. Cong.*, Helsinki, Finland, 1966, 962.
62. Norman, M. J. T., The performance of beef cattle on different sequences of Townsville lucerne and native pastures at Katherine, N. T. Australia, *J. Exp. Agric. Animal Husb.*, 8, 21, 1968.
63. Clatworthy, J. N. and Holland, D. G. E., Effects of legume reinforcement of veld on the performance of beef steers, in *Proc. Grassl. Soc. South Afr. Sunnyside*, 1979, 14, 111.
64. Kelly, R. D. and Tiffin, J. W., Effect on animal production of veld reinforcement with Siratro or fine-stem Stylo in a medium rainfall area, in Zimbabwe. Division of Livestock and Pasture, *Annual Report* 1981–82, 1984, 163.
65. Hernandez, M., Argel, P. J., Ibrahim, M. A., and 't Mannetje, L., Pasture production, diet selection and liveweight gains of cattle grazing *Brachiaria brizantha* with or without *Arachis pintoi* at two stocking rates in the Atlantic Zone of Costa Rica, *Trop. Grassl.*, 29, 134, 1995.
66. Colman, R. L., Holder, J. M., and Swain, F. G., Production from dairy cattle on improved pasture in a subtropical cattle environment, in *Proc. X Int. Grassl. Cong.*, Helsinki, Finland, 1966, 499.
67. Holder, J. M., Milk production from tropical pastures, *Trop. Grassl.*, 1, 135, 1967.
68. Blydenstein, J., Louis, S., Toledo, J., and Camargo, A., Productivity of tropical pastures. 2. Guinea grass, *J. Brit. Grassl. Soc.*, 24, 173, 1969.
69. Holder, J. M., Swain, F. G., and Colman, R. L., The use of sod sown vetch (*Vicia sativa*) as a supplement by dairy cows on the far north coast of New South Wales, *Aust. J. Exp. Agric. Anim. Husb.*, 3, 153, 1963.
70. Colman, R. L., Future use of nitrogen fertilizers on pastures and crops in subtropical New South Wales, *J. Aust. Inst. Agric. Sci.*, 36, 224, 1970.
71. Cowan, R. T., Byford, I. J. R., and Stobbs, T. H., Effects of stocking rate and energy supplementation on milk production from tropical grass-legume pasture, *Aust. J. Exp. Agric. Animal Husb.*, 15, 740, 1975.
72. Cowan, R. T., O'Gradu, P., Moss, R. J., and Byford, T. H., Milk and fat yields of Jersey and Friesian cows grazing tropical grass-legume pastures, *Trop. Grassl.*, 8, 177, 1974.
73. Pérez, I. F. and Camejo, R., Producción de leche con gramíneas tropicales y mezclas con leguminosas, Ciencia y Técnica en la Agricultura, *Pastos y Forrajes*, 2, 69, 1979.
74. Millera, M., and Figueroa, J., in *Resumen* Tercer Seminario Internacional Científico y Tecnológico, Estación Exp. Pastos y Forrajes "Indio Hatuey," Cuba, 1977.
75. Arteaga, O., Ocampo, G., and Chongo, R., Respuesta agronómica y producción de leche en pangola con diferentes niveles de fertilización nitrogenada, Ciencia y Técnica en la Agricultura, *Pastos and Forrajes*, 2, 41, 1979.
76. Cubillos, G., Muñoz, H., Ruiz, M., Deaton, O., and Fuentes, G., in Resumen VII. Reunión Asociación Latinoamericana de Producción Animal, Panamá E-11, 1979.
77. Serpa, A., Kibeiro, H., Matta, H., Lucas, E., de Maltos, S. C., and Aronovich, S., Influencia da adubacao nitrogenada e de leguminosas sobre a producao de leite no periodo seco, em pastagens de Capim-Pangola, *Rev. da Soc. Bras. de Zoot.*, 2, 227, 1973.
78. Vicente-Chandler, J., Abruña, F., Caro-Costas, R., Figarella, J., Silva, S., and Pearson, R. W., *Bulletin*, Univ. Puerto Rico, 1974, 233.

79. Caro-Costas, R., and Vicente-Chandler, J., Comparative productivity of intensively managed star and guineagrass pastures in terms of milk production in the humid mountain region of Puerto Rico, *J. Agric. Univ., P. R.*, 63(4), 436, 1979.
80. Stobbs, T. H., The effect of grazing management upon pasture productivity in Uganda. I. Stocking rate, *Trop. Agric.*, 46, 187, 1969.

14 Conservation of Forages in the Tropics and Subtropics

Raymond N. Gallaher and W.D. Pitman

CONTENTS

I. INTRODUCTION

Conservation of forage refers to the preservation of forage plant materials to provide feed for livestock at a time after the primary period of growth of these plants. Humphreys[1] suggested that the purpose of forage conservation is to enhance the efficiency of animal production by using otherwise wasted pasture surpluses during peak growth periods to reduce declines in production, weight loss, and mortality during periods of pasture deficit. There are predictable periods of forage quantity and quality limitations in the tropics and subtropics as well as in other parts of the world. These differential periods are primarily the result of seasonal changes in temperature and rainfall. Because of the need for high quality and quantity of forages, several forage conservation methods have evolved. Three primary forms of conserved forage exist: conserved standing forage, hay, and silage.

Several approaches to the provision of conserved standing forage for grazing by livestock following the plant growing season can be used in the tropics. The simplest approach is to defer grazing, which means to protect the area of conserved forage from grazing by livestock during a portion of the year. The term stockpiling of forage is commonly used in some areas. This term often implies additional input such as application of nitrogen fertilizer late in the growing season or other management to enhance the quantity or quality of the mature standing forage. Particular grasses or legumes selected for provision of primarily protein or energy reserves are often planted on relatively small areas and intensively managed as forage banks. Deferred grazing typically results in forage that requires supplementation, while forage banks are used as a source of supplement.

Hay is harvested forage preserved by drying, usually natural drying in the field following cutting. Recent technology and international trade have contributed to the increased use, especially in temperate regions, of methods to decrease the bulkiness of hay for more economical transport. Compressed bales, cubes, and pellets are forms of hay with reduced bulk. Silage is the product from harvest of actively growing crops at high moisture contents (40 to 80%), which are usually chopped, packed to eliminate air (oxygen), and allowed to ferment anaerobically to low pH.

0-8493-2318-5/01/$0.00+$.50

Well-preserved, appropriately fermented silage will last indefinitely in well-constructed silos as long as pH does not increase and air and water are excluded. The terms haylage and balage have recently been applied to specific types of silage.

In the tropics, forage conservation has not generally been as widely practiced as in temperate regions, especially the use of harvested, stored forage. This has been due to both climate and economics. Some tropical environments provide suitable conditions for growth of pasture plants throughout the year. Other tropical locations are characterized by dry seasons of various length, where standing mature forage can, to some extent, suffice as the only conserved feed. Livestock often decrease in body condition during the dry season, thus providing some of their own stored nutrients. Livestock production systems employed and markets available have sometimes made such approaches the most economical alternative. In general, tropical forages are relatively low quality and do not always produce a stored feed of sufficient value to justify the investment.

Recent worldwide increases in demand for meat and milk products are increasing the profitability of more intensive livestock production, particularly in tropical locations. International markets are providing increased opportunity for previously isolated producers. Provision of high-quality forages will increase in importance as livestock are fed for marketing of finished, high-value meat products and milk producers attempt to supply a year-round demand for fresh dairy products. These economic incentives, combined with available and developing technologies, will provide the driving force for the increased use of stored forage, both hay and silage, in many tropical areas.

While conserved forage including expensively processed hay and silage will be an increasingly useful approach to forage management in the tropics, such practices must be carefully assessed in each individual situation. Whiteman[2] listed several constraints to the use of forage conservation in the tropics. For some classes of livestock, such as mature sheep used for wool production or a livestock breeding herd, forage conservation may often not be economically feasible. The capacity of many classes of livestock to increase in body condition when forage is adequate and use these stored nutrients when forage is inadequate provides considerable buffering for variations in forage supply. The area available for grazing during the growing season must be reduced for production of forage to be conserved. This may increase grazing pressure or require reductions in herd size to unacceptable levels in some situations. The biological phenomenon of compensatory gain, whereby particularly rapid and efficient gains are made with provision of adequate nutrients after a period of weight loss, can reduce the overall benefit from conserved forage. Production responses from alternative uses of the substantial resources required for stored forage must also be assessed. Such alternatives may include planting of improved forage plants, fertilization, irrigation, and supplementation with purchased protein or energy to reduce the need for conserved forage. Some tropical legumes have been particularly useful in such a role when sown into native pastures in the dry tropics. Often grasses are preferentially grazed during the growing season, and the legume component of the pasture is subsequently grazed primarily during the dry season. Some forage-based livestock enterprises may not benefit economically from the use of forage conservation, even though short-term animal production responses may be obtained. Crowder and Chheda[3] listed the additional constraints of inherent inefficiencies in forage conservation systems, low nutritive value of conserved tropical forages, and typically unfavorable weather.

II. CONSERVED STANDING FORAGE

A. Deferred Grazing

The original version of this form of pasture conservation was, perhaps, the traditional nomadic grazing approach to livestock production that developed over hundreds of years in the Middle East and North Africa. Approaches developed in more recent times generally depend upon fences rather than herding, although some small farm enterprises may utilize tethering to control livestock access to forage. A viable alternative to pasture conservation in some situations is to stock pastures leniently so that residual forage remains following the growing season. This approach has been widely practiced by large extensively managed livestock operations in some tropical regions. The residual forage available in the dry season can be especially low in quality since the more palatable forage has already been consumed. With grass-legume mixtures in which relative palatability may not reflect nutritive value, dry season forage can include a large proportion of relatively high-quality legume forage. Either residual forage from leniently stocked pastures or conserved standing forage following grazing deferment may be low in quality and require supplemental feeding for acceptable animal performance. Often, protein supplementation of low-quality grasses may be required for sufficient intake to maintain livestock, even when sufficient amounts of forage are available.

From a review of literature on forage conservation in the Caribbean region, Kalmbacher[4] emphasized the limitation of declining forage quality of conserved standing forage. Tropical grasses decrease in both protein concentration and digestibility with plant maturity. Additional decreases in quality occur, especially when high humidity, frequent dew, or occasional light rains occur during the dry season. Such dry-season moisture can also contribute to rapid deterioration of conserved standing legume forage.

B. Stockpiling

Fertilization with nitrogen late in the growing season can be used to increase both amount and quality of grasses for conserved standing or stockpiled forage. Harvest, either grazing or cutting, of early wet-season production can improve quality but reduce the amount of stockpiled forage. Earlier application of nitrogen can increase the amount of stockpiled forage but result in lower quality. Also, species of grass, and even cultivar within some species, can greatly affect results from stockpiling. Jaraguágrass (*Hyparrhenia rufa*) translocated nitrogen to the roots in the dry season, which reduced its protein response.[5,6] Gains by steers were much greater from Colonial guinea grass (*Panicum maximum*) than from Tanganyika guinea grass on pastures fertilized at the start of the dry season.[7] Fertilization of guinea grass pastures with nitrogen in the dry season in central Brazil resulted in greater responses by steers than did fertilization in the wet season.[8] Animal production and economic responses to investment in stockpiled forage depend upon many factors including the plants and environment. Local experience will generally be required to adequately assess potential of such approaches even with appropriate plants.

C. Forage Banks

Paterson et al.[9] defined a forage bank as an area of pasture "which is not utilized during the wet season when there is plenty of animal feed available, but is saved for use during times of scarcity." Rather than simply deferring grazing, development of a forage bank involves careful selection and establishment of appropriate plant species. These include legumes for protein banks, highly productive grasses for energy banks, and combinations of these two plant types. Forage banks are primarily used to supplement dry season pastures. Forage banks have been used effectively in the Caribbean region, but some attempts have not been successful.[9] Even when initial plantings have been successful, inadequate amounts and inadequate rationing of the forage supply have contributed to failures. For protein banks, an area of 10% of the available pasture has been suggested as an appropriate size.[9] In some situations, only lactating animals or perhaps small livestock such as young goats or rabbits may be effectively fed from forage banks of small area. Due to intensive management requirements, it has been suggested that protein banks may be more appropriate for medium and small farms than for larger, extensively managed enterprises.[10] Such feeding approaches as "cut-and-carry" or tethering to limit access can substantially increase the efficient utilization of forage banks. Paterson et al.[9] suggested that protein banks may be effectively used to supplement the diets of young growing animals, while energy banks may be particularly useful for high-yielding dairy animals. Mixed forage banks of grasses and legumes may be effectively managed for dry-season supplementation of most classes of grazing livestock.

Most productive and locally adapted legumes can be used for protein banks, as long as the leaves are retained on the plant during the dry season. Species that shed their leaves in response to drought stress are not useful for this purpose. Species of low productivity are of limited value as forage banks. Woody legumes such as leucaena (*Leucaena leucocephala*)[9,11] and *Gliricidia sepium*[9] have been particularly useful for protein banks.[9] While leucaena is well adapted to alkaline soils, leaf loss has occurred during the dry season on acid soils.[12] Where adapted, the herbaceous legumes *Neonotonia wightii*,[9] *Macroptilium atropurpureum*,[9] *Pueraria phaseoloides*[9] and *Stylosanthes guianensis*[11] have been successfully used. For energy banks, elephant grass (*Pennisetum purpureum*) has typically been the recommended grass where rainfall is sufficient during the rainy season, although sugarcane (*Saccharum officinarum*) is also an excellent alternative.[9] Annual grasses such as sorghum (*Sorghum bicolor*) hybrids and pearl millet (*Pennisetum americanum*) can also be used. These grasses can produce large amounts of forage, and such energy banks work best where forage is cut, chopped, and carried to the livestock. Banks to provide both protein and energy from grass-legume mixtures must be based on compatible species. These should be selected from among the more productive, locally adapted species that grow successfully in combination and provide the necessary quality characteristics when mature for dry season use. Some varieties of *Hemarthria altissima* and *Digitaria eriantha* produce more leafy, palatable forage at maturity than many upright growing grasses.

Location and management of forage banks must be planned to fit into an overall forage management system. A fertile, well-drained, readily accessible location is preferred. Adapted, productive species that are palatable to the livestock to be fed should be selected. Appropriate management for the selected species, environment, and economic circumstances are important. By selecting the site and species carefully, dependence upon fertilizer and weed control can be minimized. As suggested by Paterson et al.[9] multiple uses can be obtained through careful planning of woody legume plantings for protein banks. Some species can provide shade and living fence posts. Distribution along fence lines allows branches to be cut in place for feeding in the dry season.

The amount of forage production to be expected from forage banks must be estimated from local experience based on the species, soil characteristics, and management imposed. Forage needs also must be estimated in advance for use in planning the size of a forage bank. The daily nutrient requirements of the livestock to be supplemented, the proportion of these nutrients to be supplied by the forage bank, and the length of the supplementation period must all be estimated in advance. In addition, plans for adjustments in the use of the forage bank must be kept in mind as drought, pest infestations, and other unexpected circumstances impact the anticipated forage supply. Where climate and economic conditions limit the feasibility of extensive deferred grazing, hay, silage, and forage banks can be very effective means of dry season forage management. Forage banks are a form of forage management and require substantial planning, oversight, and labor for best results.

III. HAY

When the foliage of herbaceous plants has been cut, dried, harvested, and stored, the product is called hay. Hay can be baled, stacked, chopped, or compressed into pellets. When carefully protected from rain and excess drying, high quality hay will keep without decomposition because it is too dry to sustain rotting organisms. Most hay is field cured to about 20% moisture, but after a few weeks of storage, it will be from 12 to 15% moisture depending upon the ambient humidity.

Even when harvested at the most appropriate stage of growth and development, biological, chemical, and physical processes begin to decrease the quantity and nutritive value of forages at the instant the forage is cut for conservation. Minimization of these adverse effects on the conserved forage product should be an objective of the hay-making process. Degradation processes must be considered carefully in management for the preservation and conservation of forage as a hay crop.

These degradation processes must be minimized, and this can be done by drying the forage from 15 to 20% moisture content. When cut at the optimum time, most hay-crop forages will contain 80 to 85% moisture and must be dried quickly to less than 20% moisture to stop degradation processes by molds and yeasts and reduce the degradation caused by plant enzymes, excess heat, and resulting browning reactions. If the drying process is accomplished correctly on forage cut at the proper stage of growth and development, followed by timely raking, baling, and storage, the resulting hay will be leafy, retain its original color, be soft and pliable, have a good aroma, and will be palatable when fed to animals. The resulting hay will have a good portion of the digestible fiber, protein, and energy conserved for efficient utilization by ruminants.

Forage plants well suited for hay production have the following characteristics: high dry-matter yield; high dry-matter concentration; good, sustainable level of crude protein; high palatability; high digestibility; growth pattern suitable for hay cutting equipment; readily field cured; low lignin content; adequate carotene and vitamin D levels; mineral content sufficient for animal growth and performance; low levels or lack of antinutritional substances; and good response to fertilization. Factors to consider in selecting hay plants for a specific environment include type of soil and its pH, mineral content of soil, amount and pattern of yearly rainfall, and temperature range and yearly fluctuations. Several grasses are suitable for hay production in the tropics and subtropics. Five tropical grasses were evaluated for hay in Puerto Rico from May 1977 to April 1984.[13] These included 'Coastcross 1' Bermuda grass (*Cynodon dactylon*), star grass (*Cynodon nlemfuensis*), 'Callie' Bermuda grass (*Cynodon plectostachyus*), slenderstem digitgrass (*Digitaria pentzii*, now *D. eriantha*), and guinea grass (*Panicum maximum*). These grasses were produced under irrigation, harvested at 35, 45, and 55-day intervals, and were cut 5 cm above the soil surface. From the basis of hay production on a Vertisol soil, guinea grass, star grass, and slenderstem digitgrass were the best grasses, although all five grasses produced hay crops in Puerto Rico. On the basis of yield performance, nutritive value, and crude protein content, the 45-day harvesting interval was the best overall for hay production for all the species studied. Although all five grasses produced good quality hay, guinea grass, star grass, and slenderstem digitgrass harvested at 45-day intervals were recommended as the best alternatives for hay production under conditions of irrigation and high fertilizer application in Vertisol soil on the south coast of Puerto Rico.[13] Further evaluation of the five hay crops stored over four periods from less than 4 months to 12 months showed that all were approximately equal in several quality measurements.[14] For example, all hay crops decreased in digestibility as the harvest interval increased from 35 days to 55 days and decreased steadily the longer the hay was stored from less than 4 months to 12 months. Data reported by these authors suggest that best quality hay would be from short-interval cuttings and short storage periods.[14]

In another study in Puerto Rico, star grass was fed to milking cattle in confinement either as green fodder, chopped in the field daily, or baled hay that averaged 6.21 and 5.59% crude protein on a dry matter basis, respectively.[15] Both forages were harvested at intervals exceeding 55 days. Animal performance was poor in spite of relatively high dry matter intake, indicating inadequate nutritive value of both forages. The authors suggested that higher quality forage, possibly star grass grown under irrigation and cut at earlier maturity, or more liberal concentrate supplementation would be needed for reasonable production with hay from this grass. Perhaps if the forage were cut on 35- to 45-day intervals, as suggested by other research,[13,14] acceptable performance could be obtained. Crowder and Chheda[3] noted that tropical grasses such as *Pennisetum clandestinum*, *Panicum maximum*, *Cenchrus ciliaris*, *Cynodon* spp., and *Digitaria decumbens* (now *D. eriantha*) have potential for use as hay crops. They noted that soil fertility and management were critical. Chambliss et al.[16] added *Chloris gayana* and *Hemarthria altissima* to the list of tropical grasses useful for hay making, also emphasizing management needs. In Brazil, *Panicum maximum* and *Hyparrhenia rufa* were found to be leafier and faster drying than *Andropogon gayanus*, *Brachiaria decumbens*, and *Melinis minutiflora* of a similar age.[17] Drying rate was suggested to be an important criteria for curing hay during the growing season.

Most of the tropical and subtropical areas of the world do not have a ready source of nitrogen (N) fertilizer for growing hay or other forages. Legumes are known to not only produce their own N needs through N_2 fixation in the symbiotic relationship with N-fixing bacteria but also to increase soil N levels[18] and, consequently, the productivity of succeeding crops.[19] Such results with blue lupine (*Lupinus angustifolius*.) followed by sorghum were shown by Gallaher.[20] Tropical forage legumes have been estimated to fix N in the range of 100 to 200 kg N ha^{-1} yr^{-1} when grown in grass-legume mixtures.[21] Several studies[22-24] demonstrated that *Desmodium* spp. when grown alone or in grass-legume mixtures fixed from 47 to 407 kg N ha^{-1} yr^{-1}, and that *Centrosema pubescens* fixed about 270 kg N ha^{-1} yr^{-1} in Hawaii.[22] The leguminous shrub *Leucaena leucocephala* has been shown to fix even higher N levels.[25] Yields of *Desmodium intortum* ranged from 10 to 21 tons ha^{-1} yr^{-1}.[23,26] Yields of *Stylosanthes guianensis*, *Centrosema pubescens*, and *Macroptilium atropurpureum* in mixtures with grass were about 16 tons ha^{-1} yr^{-1} in Malaysia.[27]

Along with the nitrogen enhancement of production, legumes have potential to produce tropical hays of superior crude protein concentrations.[3] However, drying of some tropical legumes for hay can be difficult since leaves may shatter when too dry and stems often do not dry as rapidly as leaves. Thus, not all tropical legumes are suitable for hay making. *Arachis glabrata* and *Alysicarpus vaginalis* have been successfully used as high quality tropical legume hay crops.[16] The residues of some grain legume crops also provide resources of relatively high quality for use as hay in some tropical areas.[28]

Drying or curing of the hay crop is a critical step. The predominant method of curing or drying hay for preservation and storage is sun drying. One of the major problems with this method is unpredictable weather and the high probability of rain damage, resulting in lower quality feed. Collins and Moore[29] suggested that sufficient drying may not occur for safe storage of hay even with extended curing time when humidity exceeds 70 to 75%. In environments with such high humidity during the harvest season, hay is not a viable forage conservation option. Of course, overdrying may be a problem when raking and baling are not done on a timely basis. Leaf shattering can be a major quality reduction factor when hay is left to dry too long. Chambliss et al.[16] recommended that hay should be dried to 12 to 15% moisture before baling in humid environments. The high probability for leaf shatter in legumes requires careful timing in relation to dew and humidity to reduce leaf loss in arid environments. A hay preservative, which consists primarily of propionic acid to suppress molds and other microbes, has allowed hay to be baled at 20 to 25% moisture.

In localized situations to enhance drying of hay and minimize rain damage in the tropics, innovative methods have been used. Stacking hay at high moisture around vents under cover for further drying in the stack has been reported.[1] Use of racks in the field to get the hay off the ground and reduce rain damage has also been employed.[3] Even the occasional use of kilns and smokehouses has been reported in parts of Asia and Africa.[3]

After curing, hay can be stacked as loose hay either under cover or in the field, or it can be mechanically compressed into bales of various sizes and shapes. The traditional rectangular bales of about 25 kg represent a labor-intensive hay system requiring storage facilities for preservation except in dry climates. A manual hay baler for such bales in small-scale operations has been used.[30] More recent development and widespread use of large, cylindrical (typically referred to as round) bales weighing from 270 to 450 kg has led to mechanized handling of hay, especially in temperate regions. Although deterioration can sometimes be substantial, these large bales are often stored in the field without cover. Recent and continuing developments in mechanical aspects of hay cutting, harvesting, and handling promise to enhance the efficiency of hay making and feeding in the tropics.

During storage, both moisture and temperature affect the stability of hay. The optimal temperature range for growth of fungi typically found in stored plant products is 25 to 35°C.[31] Stored hay will absorb moisture in humid environments to the level of the ambient humidity. Thus in the humid tropics, even hay baled dry and stored under cover can be subject to considerable deterioration from mold and microbial deterioration. Restricted air movement in densely packed bales can further contribute to increased temperatures.[31]

After hay has been appropriately stored, effective use of the product depends upon the nutritive value of the hay and the nutrient requirements of the livestock being fed. Wittenberg[31] suggested that a hay having a nutritive profile similar to the nutrient requirements of the livestock to be fed is a high quality hay, provided there are no undesirable characteristics. Such undesirable characteristics include reduced palatability or intake levels, reduced digestibility or availability of nutrients, and adverse effects on animal health. Several chemical analyses can be performed to aid in assessing the quality of hay. These analyses include: crude protein, acid detergent fiber, neutral detergent fiber, *in vitro* digestibility, crude fiber, lignin, and moisture. Analyses for moisture and crude protein can help in balancing the feed ration. Crude fiber, acid detergent fiber, and *in vitro* digestibility provide estimates of digestibility. The cell wall portion of forage is estimated by neutral detergent fiber and is used to estimate intake.

Hay quality can often be judged using physical touch and visual observations. For example, an excellent quality hay is one that has been harvested early, before blossom for legumes[32] and before heading for most tropical grasses on a 28- to 35-day cutting frequency.[13,14,33,34] The hay is very leafy, has a natural green color characteristic of the crop, is soft and pliable, and has little or no trash, weeds, or other foreign matter.[32]

Four quality factors to consider in assessing hay quality are leafiness, color, softness, and foreign matter content. Perhaps the most important consideration in hay quality is stage of growth when harvested. The plants must be young and tender compared to old, tough, stemmy plants. The young plant is more highly digestible. Leafiness is very important for high quality hay, because the leaves will contain the greatest concentration of protein and other nutrients and will be the most highly digestible part of the hay crop. Legume leaves are easily lost if the crop is raked when dry and brittle. Up to one-half of the nutrients of some legume hay crops can be lost through high leaf loss. Quality of hay and standardized means of assessing this quality are taking on greater significance with the ongoing increase in world trade of dried forage products.[31]

Improved hay quality can result from harvesting young, tender forage. However, quality is sometimes sacrificed for quantity by harvesting older crops for hay. Low-quality hay can be improved in feeding value by addition of chemicals such as sodium or calcium hydroxide and urea or ammonia.[35-37] Ammonia treatment has provided substantial benefits. The procedures for treatment of low-quality hay with ammonia require that the hay and ammonia be enclosed in a container for about one month before feeding. The container is usually a sheet of black plastic used to cover the hay with the edges sealed in the ground to enclose the hay stack with an air-tight seal. Caution is suggested when feeding ammoniated hay because of the toxic effects that may occur, especially to young nursing calves. Toxic symptoms include restlessness, impaired vision, loss of balance, sudden stampeding, and running in circles.[35] Although ammonia-treated hay appears to improve forage quality and reduce waste of such hay compared to untreated hay, it has been recommended that ammoniated hay not be fed to lactating cows. It should be reserved for developing heifers, herd bulls, or cull cows.

Dehydrated pellets made from whole-plant corn or Coastal Bermuda grass have been imported by dairymen in Puerto Rico.[38] Production of forage sorghum for dehydration appears practical, particularly under irrigation on the south coast of the island of Puerto Rico.[39] Cows fed hay, pellets, and concentrate equaled those on grazing plus concentrate in milk yield, but cows on grazing supplemented with concentrates and pelleted forages were higher in milk yield.[39] The study showed that good quality pelleted forages may be used effectively in combination with pasture or hay. However, tropical grasses, cut at 45 to 55 days of age and artificially dried, had limitations in nutritive value and palatability for use in conjunction with good quality pastures. Perhaps greater success could be obtained from pelleting tropical grasses if harvested at 28 to 35 days of age when quality would be highest.[13,14,32,33]

IV. SILAGE

When live actively growing crops are harvested (usually chopped) at high moisture (40 to 80%), packed to eliminate O_2, and allowed to partially ferment to low-pH acid conditions for preservation, the end product is a ruminant forage called silage.[32,40-43] Haylage and balage are terms sometimes used for specific types of ensiled forage. Many types of forages can be ensiled. In contrast to the distant past, today most dairymen and increasing numbers of beef cattle producers in temperate regions, especially with cattle in confinement (feedlots), use silage as the major forage or only forage throughout the year. To make good silage, the crop is harvested at the time of greatest feed value and preserved for feeding. Ensiling can only preserve those nutrients already present in crops. It does not improve the feeding value of a crop.

Forage for silage is preserved by anaerobic fermentation. Successful silage fermentation must include all factors necessary for anaerobic fermentation. These are an absence of air, a suitable and adequate source of nutrients, and, once the fermentation has ended, preservation of the ensilage material in a constant state.[44]

Cells of actively growing plants continue to respire for a few hours after material is ensiled. Oxygen is used and replaced with carbon dioxide. Lack of oxygen prevents mold growth. The forage becomes warm; it is moist and very limited in oxygen. These conditions result in rapid growth of bacteria that convert soluble carbohydrates into organic acids, especially lactic acid. Within a few days, the temperature drops and fermentation stops. This "pickled" or fermented material is preserved and will remain in stable condition. However, if air seeps into the preserved mass, mold growth and additional heating will occur. If the initial fermentation does not produce enough acid, other bacteria will grow and produce undesirable fermentation that causes further nutrient loss, undesirable odors, and unpalatable silage.[45] In the ensilage of nontraditional forages, acceptable fermentation may involve production of acetic and propionic acids. Such acetic acid fermentation has resulted in stable silage of tropical forage grasses.[3]

Important plant characteristics for good quality silage include: high dry matter yield; high dry matter concentration; good, sustainable level of crude protein; high palatability; high digestibility; growth pattern suitable for forage cutting equipment; readily field wilted for hay-type crops; low lignin content; adequate carotene and Vitamin D levels; mineral content sufficiently high for animal growth and performance; low levels or lack of antinutritional substances; and good response to fertilization.

Corn (*Zea mays*),[46-55] sorghum,[52,56-59] interplanted corn and sorghum,[60] and other crops that produce high grain yields are the most important crops currently grown for silage. A close estimation of the quality of silage from these crops can be made using several physical and visual characteristics including: grain content or stage of growth, color, odor, and moisture content.[32]

The highest quality part of corn or sorghum silage is the grain. Generally speaking, the higher the grain content, the higher the quality of silage. Corn should be cut for silage in the full dent stage for dent-type corn. For flint or the more tropical corn types, the corn kernels should be carefully monitored, and the crop should be harvested at the soft to hard dough stages because the kernels will not form a dent from which to judge. Corn harvested in the roasting ear (milk) stage usually produces less than one-half the yield of grain as corn cut in the full dent or hard dough stage. It is important to wait for the grain to form before cutting corn or sorghum for silage. Above 35% grain in the silage is considered high.

Three categories of color are considered as follows: desirable color is a bright yellowish-green to slightly brownish color; acceptable color is a dark green or light brown or, if frosted, light yellow color; and undesirable color is deep brown or black indicating excessive heating or putrefaction. Predominantly white or gray color indicates excessive mold development.[32]

Odor falls into three categories as follows: desirable odor is a light, pleasant odor with no indication of putrefaction; acceptable odor includes fruity, yeasty and/or musty odors, which indicate slightly improper fermentation or the silage may have a slightly burnt or sharp vinegar odor; and undesirable odor includes a strong, burnt odor indicating excessive heating or a putrid odor indicating improper fermentation. A very musty odor indicates excessive mold, which will be readily visible throughout the silage.

The final category for estimating silage quality by physical and visual means is moisture. Moisture assessments include: desirable, where no free water is observed when silage is hand squeezed, indicating a well-preserved silage; acceptable, where some moisture can be squeezed from the silage by hand squeezing or the silage will be dry and musty; and undesirable, when the silage is wet, slimy, or soggy and water is easily squeezed from a sample, or the silage is too dry and has a strong burnt odor.

An excellent silage produced from corn, sorghum, or a mixture of the two is one that has greater than 35% grain content, is bright yellowish-green to slightly brownish in color, has a light, pleasant odor with no indications of putrefaction, and has no free water when squeezed in the hand.[32]

Most plant breeding efforts to develop crop varieties suited for silage production have occurred in temperate regions. As silage production increased in warm-temperate and subtropical regions in recent years, developments relevant to silage production in the tropics have been reported. Temperate corn varieties have proven vulnerable to insects and adverse weather in environments similar to those of many tropical regions. Yield losses due to photoperiod effects can be limitations of typical temperate corn hybrids.[48,61,62] Corn genotypes adapted to tropical conditions are available and provide possibilities for overcoming such limitations of temperate corn as adverse photoperiod sensitivity, susceptibility to insect damage, and weather-related problems.[61-67] Although corn has been the premier silage crop, tall-growing grain sorghum hybrids have also been developed for silage production. Sorghum hybrids could have considerable potential for use as silage crops in tropical regions, particularly where corn is not well suited.[68,69] Sorghum hybrids for silage have been selected primarily for plant height, yield, leafiness, grain content, lodging resistance, and general appearance.[57] Pearl millet hybrids also have potential as silage crops in some tropical areas. Low grain levels and high moisture have been limitations for tropical corn, sorghum, and pearl millet. Plant breeding, along with production and utilization research, could readily provide tropical corn and sorghum silage cropping systems for many tropical locations as demand for meat and milk products justifies.

Although not generally providing the yield or forage quality potential of specifically developed annual silage crops, many tropical forages can also be conserved as silage. Crowder and Chheda[3] indicated that *Panicum maximum*, *Pennisetum purpureum*, *Setaria anceps*, and *Tripsacum laxum* have such potential when appropriately managed. Kunkle and Chambliss[70] noted that *Cynodon* spp., *Hemarthria altissima* and *Arachis glabrata* can be harvested for silage. *Vigna unguiculata*, *Indigofera hirsuta*, *Glycine max*, *Lupinus luteus*, *Pennisetum purpureum*, *Digitaria decumbens*, and *Saccharum officinarum* were determined to make acceptable silage in early evaluations.[71,72] Humphreys,[73] however, noted that the bulky nature of most perennial tropical grasses (which makes air exclusion difficult), and insufficient sugar content for effective fermentation are limitations when ensiling these forages. Forage quality of most tropical forages is high enough for production of good silage only in early growth stages when both yield and dry matter concentration are often lower than desired. Both low soluble carbohydrate (sugar) levels and high moisture (low dry matter) concentrations contribute to inadequate fermentation and excessive decomposition during storage.[3] Wilting or partial field drying of *Pennisetum purpureum* in the vegetative stage with added formic acid enhanced ensilage and nutritive value in Brazil.[74] Crowder and Chheda[3] also suggested that low soluble carbohydrate levels, high buffering capacity, woody stems, and soft leaf tissues often result in poor ensilage of tropical legumes.

Several methods or systems of silage storage have been successfully used.[40] These include upright or tower silos and horizontal bunker or trench silos. Plastic-bag silage systems have also been developed in recent years.

The most common upright silos are made of concrete staves. Metal hoops are wrapped around the staves to support the structure. Corn, sorghum, and relatively wet hay-crop silages (55 to 65%

moisture) store well in concrete stave silos. Poured concrete silos are made with a metal form, reinforced with metal rods embedded in the concrete. Oxygen-limiting silos are made of steel panels that have glass fused to the steel to protect it from silage acids. Both the concrete stave silos and poured concrete silos are equipped with top unloaders. The oxygen-limiting silos are equipped with a bottom unloader. Cost estimates for the three types of tower silos of approximately equal capacity in the order of greatest to least expensive is metal oxygen-limiting > poured concrete > concrete stave.[40]

Bunker silos are made by mounding up soil on ground level to form walls or pouring concrete to encase the silage. One end is open for filling and removing the forage. Trench silos are made by digging a long trench in an elevated area of well-drained soil. Earthen walls will work, but concrete walls are also used. In both aboveground bunker silos and trench silos, a concrete floor is desirable to eliminate soggy soil conditions that reduce silage quality and cause difficulties when loading or unloading the silos.

Sealed plastic envelope or plastic bag silos are becoming popular because of the flexibility of such systems. They are well suited for most crops, but especially for hay-crop silages with moisture levels ranging from 50 to 75%. If bags do not tear, quality is good and losses are comparable to tower silos and better than most horizontal silos. The big disadvantage of the bagging system is cost of the machine that blows and packs the forage into the plastic bags. Plastic bags are not reusable, are subject to tears by wild animals, and feeding management requires greater care than with other storage systems to prevent losses.[40]

Various types of structures have been evaluated for use as small silos for small-scale production of silage in the tropics.[3,75] Adequate structural support, as well as exclusion of air and water seepage, appear to be primary considerations in selection of such silo structures. Cost per unit of silage must be assessed for such innovative systems to be economically viable. For small-scale silage production, Paterson et al.[30] suggested that the labor requirement is typically too great without mechanization and investment in machinery is justified only at increased scales of operation.

In order to maximize the preservation of high-quality silage, the silo should be filled as quickly as possible. In most silos, thorough packing to force out the air is essential to prevent spoilage. Since most tropical grasses are coarse and stemmy,[3] adequate consolidation of such silage may require fine chopping and packing. After the air is forced out, water and air should be prevented from reentering the silo. Trench or bunker silos should be completely filled so that after packing and settling the silage forms a rounded or sloping top above the side or ground level permitting water drainage to the sides. The silage should be covered immediately after filling with a plastic covering. A 6 mil black polyethylene plastic cover with about 10 cm of sawdust or some other material on top of the cover is needed to make and store high-quality silage. A fence may be necessary around trench or bunker silos to keep animals from walking over the silage and punching holes in the cover.

Additives have been used in silages for nearly a century, but they remain a constant source of confusion. The purposes of using additives are: to alter fermentation; to decrease losses; and to conveniently introduce materials into the ration.[76] Several classes of additives have been used for silage and specific conditions are necessary for their successful use.[76]

Sterilants such as sodium metabisulfite and antibiotics reduce fermentation. They have worked best with high moisture forage with dry matter in the range of 30%. Generally their effects are erratic, and they are not very reliable. However, they may reduce top spoilage but seldom improve animal production. If residues of sterilants remain in silage, they may contaminate animal products or affect animals.

Mineral acids and organic acids, such as lactic acid or formic acid, have been used for direct acidification. These chemicals aid in rapid reduction in silage pH, which reduces the fermentation required. To work best, the dry matter of the forage should be between 25 and 35%. The organic acids, particularly formic acid, are the most reliable of additives. They are somewhat expensive and corrosive. Formic acid has been shown to repeatedly reduce protein and carbohydrate loss in the silo, and animal performance has been consistently improved.

Lactic acid stimulants include bacterial cultures, enzymes, and limestone. Cultures and enzymes increase the production of lactic acid by either increasing the number of bacteria or improving the medium for their growth. Limestone acts as a buffer and increases total acid production. Forage should have 30% dry matter or less for best results. Cultures and enzymes work best in grasses or legumes. Limestone should be confined to corn or sorghum silage. Lactic acid stimulants are very erratic and should be considered only when an entire silo can be filled with forage of similar moisture content.

Nutrient additives include feedstuffs, urea, N-P-K mixtures, and others. The silage is used as a carrier to get the extra nutrients into the ration. For best results, the forage dry matter should be less than 35% to insure good packing and minimize losses of nutrients in the silage. Reliability of nutrient additives vary with the nutrient. For example, feedstuffs have proven to be excellent additives, but results with urea and N-P-K have been highly variable in subtropical conditions due to difficulty in maintaining proper moisture levels in forage. High moisture forage sorghum harvested at its peak of quality with dry corn grain added to adjust moisture to below 30% while filling the silo, has produced very high-quality silage. Poultry litter added to king grass (*Pennisetum purpureum* X *P. americanum*) at 15% wet basis improved silage characteristics, nutritive value, and animal performance.[77] Where readily available, by-products such as molasses and citrus pulp can be effectively used as nutrient additives and sources of soluble carbohydrates for enhanced ensilage of tropical grasses.[3]

Propionic acid, at the 1% level, has been used to reduce mold and yeast growth. This prevents excessive heating in the silo, and the effect is sustained through feeding the silage. The product works well for corn and sorghum silage.

Haylage is an ensiled feed made from pasture and hay crops. There has been inconsistency in use of terminology regarding ensiled forages and moisture levels. Some references identify all high-moisture (> 50%) ensiled forage as silage, and reserve the use of haylage for low-moisture (near 45%) ensiled forages. Others restrict the use of the term silage to the traditional grain-crop forages that undergo lactic-acid fermentation, and refer to ensiled forage of pasture and hay crops as haylage. Recent use of the term haylage typically refers to ensiled forage of pasture and hay crops at low (45 to 50%) moisture. Any high-quality pasture or hay crop can make good haylage. Of course, the opposite is also true. Any poor quality pasture or hay crop will make only poor-quality haylage. Some of the advantages of hay-crop haylage include: excess growth of pastures can be saved and preserved; in some cool season springs, the first cutting of hay crops can be saved easier as haylage than as hay because of hay curing difficulty; using hay crops for haylage usually saves more feed per hectare than does hay making or grazing; haylage provides stored feed for seasonal dormant periods; and soil erosion is reduced by keeping land in sod hay crops.

Wilting the hay crop results in preservation of greater quantities of feed per hectare. Wilting will also prevent leakage from the silo, which makes a mess around the silo. This leakage contains about 10% dry matter, which is highly digestible. Unwilted, high moisture ensiled hay-crop forage is less palatable to cattle than is wilted haylage. Also more labor is needed to handle the high moisture ensiled forage than wilted haylage.[32]

A 70% moisture content or less is desirable for high-quality haylage. In high-moisture haylage (about 85% moisture for most direct cut and ensiled hay crops), the loss from leakage may be as high as 10 to 15% of the dry matter. On the other hand, haylage at 70% moisture will not have any leakage loss, but as the hay crop for haylage is wilted to moisture contents below 65%, field losses in dry matter typically occur. These losses range from 4% dry matter at 65% moisture to 10% dry matter at 50% moisture.

Good haylage can be made by following appropriate procedures. Cut the hay crop early in its growth cycle (28 to 35 days of age). Quality is higher when the crop is young and early-cut forages are easier to cut and pack in the silo. Wilt the forage to a desirable moisture content of 60 to 70% moisture. Hay conditions speed up the wilting process. Wilted hay handles easily with machinery. A 60 to 70% moisture content hay crop is easy to preserve in the silo. Lower moisture crops require extra management in order to preserve high quality haylage. Air and water must be excluded from the hay crop for good haylage. Chop the hay crop short and clean (about 2 cm setting on the chopper) and distribute it evenly in the silo. Pack trench and bunker silos with heavy tractors. Fill the silo as quickly as possible (one day is ideal). Cover the haylage crop with a black plastic covering and add sawdust or some other material on top of the plastic to hold the cover firmly against the stored haylage. With temperate haylage crops wilted to 60 to 70% moisture, chopped, packed and stored properly, preservatives are not typically needed to make good haylage.[32] In Brazil, acceptable haylage of *Cynodon dactylon* was made without preservatives,[78] but formic acid enhanced ensilage of wilted *Pennisetum purpureum* forage.[74]

High-quality haylage is produced from high-quality hay crops. These crops include most tropical perennial grasses. Much of the criteria used in judging haylage quality is similar to that for judging hay or silage quality. Criteria used in physical and visual determination of haylage quality include stage of growth, color, odor, and moisture content. In haylage, color is an indication of how the crop was treated in the silo and the stage of growth when the forage was harvested. Color will vary from light green to dark green depending on the crop and/or additive that may have been used. A bright color is an indication of best quality haylage. A yellowish-green color is acceptable. It indicates bleaching due to improper fermentation or that excess amounts of chemical preservatives were used, or the chemical preservative was poorly distributed in the haylage. A white or gray color is undesirable and indicates excessive mold due to poor packing and/or air leaks after packing. Dark brown or black indicates excessive heating or putrefaction.

Odor indicates how well the haylage was preserved. The desirable odor is pleasant, indicating a good job of haylage preservation. A slightly burnt smell is acceptable and usually means poor packing and dry haylage. A fruity or yeasty odor indicates slightly improper fermentation, and a musty odor indicates poor packing or air leaks during storage. These indicate acceptable haylage. Undesirable odor of a strong burnt smell means the haylage is dry, was poorly packed, and was subjected to excessively high heat during the ensiling process. A putrid odor accompanied by sliminess indicates excessive air and too much water; this haylage will often have the odor of a wet manure pack.

Haylage is typically preserved near the minimal moisture level for storage of silage. Silage crops should contain from 50 to 70% moisture to preserve well in most storage situations. Only early-cut, young, leafy plants can be wilted to near 50% moisture and preserved as high-quality, low-moisture haylage. Excess moisture, regardless of whether it comes from high-moisture crops, rain, or seepage into trench silos, lowers quality and prevents the best fermentation process. On the other hand, dry haylage will tend to oxidize or "burn up" in storage. Moisture estimates can be obtained by the hand-squeeze method to estimate if the haylage is either desirable, acceptable, or undesirable. Desirable haylage has no free water when squeezed by hand. This indicates a well preserved haylage. For acceptable haylage, some moisture can be squeezed by hand, or the haylage will be somewhat dry and/or musty. Undesirable haylage is wet, slimy, or soggy; water is easily squeezed from the haylage; or it is too dry and has a strong burnt odor. An excellent haylage for ruminant feed will be one that was cut at a young stage of growth and prior to initiation of reproductive growth. It will have a bright color and a pleasant odor and will have no free water when hand squeezed.[32]

Balage is the product of ensiling baled forage. In the evolution of research and technology for making balage, the progression has gone from single-bale bags, multiple-bale bags, and now single-bale stretch-wrap plastic covering.[79] With single-bale bags, bales were individually stored in plastic bags by sealing the open end with twine. Even though the single-bale balage could be stored for up to one year with proper care, it did not become popular because individual bags were expensive and the process required lots of labor. Multiple-bale plastic tubes reduced the cost of producing balage, but this approach was dropped from further research because any undetected holes resulted in all of the bales in the tube being spoiled. Both single-bale bags and multiple-bale tubes had the advantage of not requiring any costly equipment beyond that needed for making round-bale hay.[79]

The round-bale balage system that offers the best choice with present research and technology uses a stretch-wrap plastic to encase the bale.[79-83] Special equipment, a plastic-wrapping machine, is required for the stretch-wrap plastic system. Compared to long-tube storage, the single-bale bag and stretch-wrap systems recapture their greater cost from the use of lower cost plastic and production of higher quality balage.

Successful procedures for making round bale silage depend upon the climate. Allow the morning dew to dry before cutting the forage. Allow sun drying or wilting to occur for 2.5 to 4 hours to drop moisture from about 85% (for most 4 to 5 week old grass crop regrowths) to 50% to 60%. Most newer balers will roll bales of wet forage, which should be no heavier than about 675 kg per bale. Make bales the same day that are uniform with even density for easier manipulation on the plastic wrapping machine. Wrap at least four layers of "stretch wrap" film (50 to 60% stretch); six wraps are even better. Use high-quality "stretch wrap" film that will last at least 6 months. Avoid punctures by using caution in lowering the bales to the ground from the wrapping machine. Use specially designed equipment for lifting and moving bales for final storage so as not to punch holes in the film. If possible, do not stack bales; thus, choose the storage area carefully because holes can be a severe problem from changes in shape of the bales during settling or from birds and wild animals. Holes must be patched quickly. Before feeding, remove the plastic. Generally all forage will be consumed unless spoilage areas have occurred due to air and water leakage. Cattle will usually consume any areas slightly damaged by white mold. It is best for feed to be consumed within 2 to 3 days and no later than 4 days after unwrapping bales due to forage quality damage from heat buildup. It has been suggested that forage producers "make hay when the weather permits and make round bale balage when the weather will not allow timely hay harvest."[81]

Ensiling characteristics of round-bale balage differed with moisture (350 to 400 g dry matter kg^{-1}, wetter; 450 to 500 g dry matter kg^{-1}, drier) contents.[83] Initial fermentation was more rapid in the wetter bales, as shown by a more rapid drop in pH and increase in lactic and acetic acids. The initial lactic acid bacteria counts were higher for the wetter bales through day 9. By day 38, counts had declined with no differences due to moisture. Water soluble carbohydrates were higher in the drier bales. Ammonia N was higher in wetter bales. Nonprotein N did not differ due to moisture level.

Characteristics of round-bale balage (390 g dry matter kg^{-1}) were compared to those of forage chopped and packed under similar conditions in a plastic bag (bag silage).[83] Bag silage had a faster and greater drop in pH and greater increase in lactic and acetic acids. Temperatures were higher in the bag silage than in the balage over the first 30 days. Even though other characteristics differed between the two conservation methods, there were no differences between the two in digestibility by sheep. However, beef calves gained more weight with better feed efficiency from the bag silage. There was a more desirable fermentation pattern in balage made at 350 to 410 g dry matter kg^{-1} than in balage at 460 to 510 g dry matter kg^{-1}. Chopped forage from silage bags had a more desirable fermentation pattern and lower nonprotein N content and resulted in faster animal gains with more efficient feed conversion than balage. Adding soybean meal improved feed efficiency of the balage, probably due to its high nonprotein N content.[83]

V. CONCLUSIONS

Forage conservation systems can be effectively used in many tropical situations. Where the seasonal distribution of either quantity or quality of forage does not meet the nutrient needs of particular classes of livestock, forage conservation options should be assessed. The decision to adopt a particular forage conservation option should not be based simply on positive animal performance responses to such a practice. Economic advantages in the overall livestock production and marketing system should be provided by a viable forage conservation option. The particular approach to forage conservation must consider locally appropriate technology. Availability of land, labor, financial resources, support services, and market opportunities for the products must be considered.

Each forage conservation option includes both benefits and limitations. Simple deferred grazing can be comparatively inexpensive, but requires considerable land area and provides a low quality product. Increasing inputs for greater amounts or quality of stockpiled forage typically results in increased investment of resources for only modest improvement in nutritive value of most mature tropical forages. Further investment of resources to establish superior plant species for either protein or energy banks requires considerable additional management for production and effective utilization of such forages. Such forage banks may be most effectively used in small-scale operations. While small-scale livestock production can benefit from use of stored forage in many situations, the high labor requirement for nonmechanized hay or silage production must be justified by the benefits derived.

Tremendous increases in livestock production in many tropical regions could be achieved through the increased use of mechanized systems of producing and feeding stored forage. Both financial resources and technical support required for such mechanized systems are substantial. Where highly mechanized field crop production has led to development of support industries, advancing technology for both hay and silage production will undoubtedly be introduced. Superior cultivars of hay and silage species adapted to various tropical environments are possible through selection and breeding of existing plant germplasm. Despite the potential provided by this developing technology and an expanding demand, caution is required in most situations. On an individual basis, the large investment required for initial development of these mechanized stored-forage systems could readily exceed the local market capacity to adequately compensate for the increased production potential.

From this perspective, silage systems are particularly vulnerable due to high initial investment required for facilities and equipment. Also, management requirements are high for silage systems to minimize storage losses and spoilage during feedout. Market options for silage not fed where produced are also typically limited due to high transportation cost and susceptibility to spoilage. Compared to hay, however, field loss is typically less, risk of weather damage is reduced, physical labor required is less, a greater variety of crops can be used, and silage is better suited as an ingredient in mixed rations. Haylage stored at reduced moisture ($< 70\%$) provides even further advantages. Nutrient conservation is increased, there is less water to store and handle, intake is greater, and there is less need for preservatives than with high-moisture silage. Disadvantages of this lower moisture product include increased equipment requirements, increased risk of weather damage during wilting, and more difficulty in packing the silo.

Compared to silage, hay is a more concentrated (less water) product allowing transport and marketing options. The extent of mechanization is optional with loose stacks, hand-stacked bales, or completely mechanized handling possible. Field losses are generally higher than for silage and risk of weather damage is greater.

For many small farms in the tropics, use of forage banks could result in an extended period of milk production into the dry season when the price of fresh milk is often greater. In tropical areas where integrated field crop and livestock production enterprises exist, increased use of mechanized systems of stored forage is likely. Where local adaptive research and other support services are provided, dynamic growth of beef and dairy industries can be expected during the next few years due to recent advances in hay and silage technology, especially equipment, and the rapidly developing world market for the products of these industries.

REFERENCES

1. Humphreys, L. R., *Tropical Pastures and Fodder Crops,* Second Edition, Longman, Singapore, 1987.
2. Whiteman, P. C., *Tropical Pasture Science,* Oxford University Press, Oxford, U.K., 1980.
3. Crowder, L. V. and Chheda, H. R., *Tropical Grassland Husbandry,* Longman, London, U.K., 1982.
4. Kalmbacher, R. S., Optimizing seasonal distribution of forage availability, in *Forage-Livestock Research Needs for the Caribbean Basin* (Workshop Proceedings), Moore, J. E., Quesenberry, K. H., and Michaud, M. W., Eds., Caribbean Basin Advisory Group, Gainesville, FL, 1987, 31.
5. Andrade, J. L., Rey, G. E., Ramirez, M. T., Carillo, C. A., and Blue, W. G., Fertilizer responses and dry season changes in jaraguagrass in the Pacific region in Costa Rica, *Tropical Agriculture* (Trinidad), 41, 31, 1964.
6. Blue, W. G. and Tergas, L. E., Dry season deterioration of forage quality in wet-dry tropics, *Soil and Crop Science Society of Florida Proc.*, 29, 224, 1969.
7. Quinn, L. R., Mott, G. O., Bisschoff, W. V. A., and Racha, G. L. da, *Beef Production of Subtropical Grasses*, Bulletin No. 28, IBEC Research Institute, New York, 1963.
8. Quinn, L. R., Mott, G. O., and Bisschoff, W. V. A., *Fertilization of Colonial Guineagrass Pastures and Beef Production with Zebu Steers*, Bulletin No. 24, IBEC Research Institute, New York, 1961.
9. Paterson, R. T., Proverbs, G. A., and Keoghan, J. M., *The Management and Use of Forage Banks*, Caribbean Agricultural Research and Development Institute, St. Augustine, Trinidad, West Indies, 1987.
10. CIAT, Livestock production systems, in *1988 Tropical Pastures Annual Report*, Working Document No. 58, Centro Internacional de Agricultura Tropical, Cali, Colombia, 1989, chap. 17.
11. Milera, M. and Santana, H., Milk production system using *Panicum maximum* cv. Likoni under grazing conditions with *Leucaena leucocephala* proteic bank, in *Proc. XVI Int. Grassland Congress*, The French Grassland Society, Nice, France, 1989, 1161.
12. Zoby, J. L. F., Kornelius, E., Saueressign, M. C., and Affin, O. A. D., Protein bank as a complement to native pasture, in *Proc. XVI Int. Grassland Congress*, The French Grassland Society, Nice, France, 1989, 1169.
13. Mendez-Cruz, A. V., Siberio-Torres, V., Fernandez-van Cleve, J., Fontanet, E., and Ramirez-Oliveras, G., Yield and nutritive value of hay from five tropical grasses at three harvesting intervals, *Journal of Agriculture of the University of Puerto Rico*, 72, 109, 1988.
14. Mendez-Cruz, A. V., Corchado-Juarbe, N., and Siberio-Torres, V., Storage and digestibility, voluntary intake and chemical components of hay of five tropical grasses, *Journal of Agriculture of the University of Puerto Rico*, 72, 531, 1988.
15. Randel, P. F. and Fernandez-van Cleve, J., Confinement feeding of dairy cows based on stargrass as green chopped fodder or hay, *Journal of Agriculture of the University of Puerto Rico*, 72, 231, 1988.
16. Chambliss, C. G., Kunkle, W. E., and Cromwell, R. P., Hay production, in *Florida Forage Handbook*, Chambliss, C. G., Ed., University of Florida, Cooperative Extension Service, Gainesville, FL, 1999, 97.
17. Costa, J. L. and Gomide, J. A., Haymaking from tropical grasses, in *Proc. XVI Int. Grassland Congress*, The French Grassland Society, Nice, France, 1989, 997.
18. National Academy of Science, *Tropical Legumes: Resources for the Future*, National Academy of Science, Washington, D.C., 1979.
19. Singh, K. K. and Awasthi, O. P., Maintenance of soil fertility in the hills with incorporation of legumes in cropping sequences, *Indian Journal of Agricultural Sciences*, 48, 41, 1978.
20. Gallaher, R. N., Value of residues, mulches or sods in cropping systems, *Multicropping Minimum Tillage (MMT)-5*, Florida Cooperative Extension Service, Gainesville, FL, 1980.
21. Date, R. A., Nitrogen, a major limitation in the productivity of natural communities, crops and pastures in the Pacific Area, *Soil Biology and Biochemistry*, 5, 5, 1973.
22. Whitney, A. S., Kanehiro, Y., and Sherman, G. D., Nitrogen relationships of three tropical forage legumes in pure stands and in grass mixtures, *Agronomy Journal*, 59, 47, 1967.
23. Whitney, A. S. and Green, R. E., Legume contributions to yield and composition of *Desmodium* spp. — pangola grass mixtures, *Agronomy Journal*, 61, 741, 1969.
24. Whitney, A. S., Effect of harvesting interval, height of cut, and nitrogen fertilization on the performance of *Desmodium intortum* mixtures in Hawaii, in *Proc. XI Int. Grassland Congress*, University of Queensland Press, St. Lucia, Queensland, Australia, 1970, 631.

25. Brewbaker, J. L., Plucknett, D. L., and Gonzalez, V., *Varietal variation and yield trials of Leucaena leucocephala (Koa Haole) in Hawaii*, Bulletin Number 124, Hawaii Agricultural Experiment Station, Honolulu, HI, 1972.
26. Olsen, F. J. and Moe, P. G., The effect of phosphate and lime on the establishment, productivity, nodulation, and persistence of *Desmodium intortum*, *Medicago sativa* and *Stylosanthes gracilis*, *East African Agricultural and Forestry Journal*, 37, 29, 1971.
27. Ng, T. T. and Wong, T. H., Comparative productivity of two tropical grasses as influenced by fertilizer nitrogen and pasture legumes, *Tropical Grasslands*, 10, 179, 1976.
28. Apori, S. O., Effect of storage on quality of cowpea haulm hay under warm humid conditions, in *Proc. XVIII Int. Grassland Congress*, Vol. 1, XVIII Int. Grassland Congress, Calgary, Alberta, Canada, 1997, 11 (Session 14).
29. Collins, M. and Moore, K. J., Postharvest processing of forages, in *Forages Volume II: The Science of Grassland Agriculture*, Barnes, R. F., Miller, D. A. and Nelson, C. J., Eds., Iowa State University Press, Ames, IA, 1995, 147.
30. Paterson, R. T., Philip, P., and Maynard, P., *A Guide to Improved Pastures for the Drier Areas of the Eastern Caribbean*, Caribbean Agricultural Research and Development Institute, St. John's, Antigua, 1986.
31. Wittenberg, K. M., Microbial and nutritive changes in forage during harvest and storage as hay, in *Proc. XVIII Int. Grassland Congress*, Vol. 3, XVIII Int. Grassland Congress, Calgary, Alberta, Canada, 1997, 265.
32. Burns, J. D., *Silage Production in Tennessee*, Circular 548, University of Tennessee, Cooperative Extension Service, Knoxville, TN, 1967.
33. Mislevy, P., Ed., *Florico Stargrass*, Circular S-361, Florida Agricultural Experiment Station, Gainesville, FL, 1989.
34. Mislevy, P., Ed., *Florona Stargrass*, Circular S-362, Florida Agricultural Experiment Station, Gainesville, FL, 1989.
35. Brown, W. F. and Kunkle, W. E., Improving the feeding value of hay by anhydrous ammonia treatment, Bulletin 888, Florida Agricultural Experiment Station, Gainesville, FL, 1992.
36. Reis, R. A. and Garcia, R., Effect of ammonia level, time of exposure to ammonia and period of storage on chemical composition and digestibility of two tropical grasses hay, in *Proc. XVI Int. Grassland Congress*, The French Grassland Society, Nice, France, 1989, 1001.
37. Rosa, B., Reis, R. A., Resende, K. T., Rodrigues, L. R., and Jobim, C. C., Effect of the ammoniation on *Brachiaria decumbens* hays, in *Proc. XVIII Int. Grassland Congress*, Vol. 1, XVIII Int. Grassland Congress, Calgary, Alberta, Canada, 1997, 25 (Session 14).
38. St. Louis, D. G., Cestero, H., Roman-Garcia, F., Arroyo-Aguilu, J. A., and McSowell, R. E., Value of supplementary feeding with pelleted maize or grass hay for lactating cows grazing fertilized grass pastures in Puerto Rico, *Journal of Agriculture of the University of Puerto Rico*, 63, 390, 1979.
39. St. Louis, D. G., *Evaluation of Grass Hay and Sorghum, Maize, and Soybean Forage for Supplementing Pastures in Puerto Rico*, Ph.D. Dissertation, Cornell University, Ithaca, NY, 1977.
40. Cromwell, R. P., Prevatt, J. W., and Becker, W. J., *Silage Harvesting Equipment and Storage Structures*, Bulletin 240, Florida Cooperative Extension Service, Gainesville, FL, 1989.
41. Harris, J. and Wishart-Smith, L., *Silage Management in Queensland*, Information Series Qi83028, Queensland Department of Primary Industries, Brisbane, Queensland, Australia, 1984.
42. Kunkle, W. E., Cromwell, R. P., Chambliss, C. G., Bates, D. B., and Wasdin, J. G., *Experience with Round Bale Silage*, SS-ANS-12, Department of Animal Science, University of Florida, Gainesville, FL, 1991.
43. Pitt, R. E., *Silage and Hay Preservation*, NRAES-5, Northeast Regional Agricultural Engineering Service, Cornell University, Ithaca, NY, 1990.
44. McCullough, M. E., The production and feeding of silage, Part I: Silage Production, *Feedstuffs*, 43, 32, 1971.
45. Crowley, J. W., Chop at the right moisture level for good silage, *Hoard's Dairyman*, 121, 754, 1976.
46. Cummins, D. G., Quality and yield of corn plants and component parts when harvested for silage at different maturity stages, *Agronomy Journal*, 62, 781, 1970.
47. Cummins, D. G., *Yield and Quality of Corn for Silage Grown under Different Fertilizer Regimes*, Research Bulletin 105, University of Georgia, Athens, GA, 1972.

48. Cummins, D. C., *Influence of Planting Date, Hybrid Maturity and Plant Population and Quality of Irrigated and Non-Irrigated Corn for Silage*, Research Bulletin 178, University of Georgia, Athens, GA, 1975.
49. Cummins, D. G., Production of quality silage, in *Feeds and Feeding Research Day Proc.*, Gallaher, R. N. and Baird, D. M., Eds., University of Georgia, Athens, GA, 1975, 23.
50. Cummins, D. G. and Burns, R. E., Yield and quality of corn silage as influenced by harvest height, *Agronomy Journal*, 61, 468, 1969.
51. Cummins, D. G. and Dobson, J. W., Jr., Corn for silage as influenced by hybrid maturity, row spacing, plant population, and climate, *Agronomy Journal*, 65, 240, 1973.
52. Cummins, D. G. and McCullough, M. E., *A Comparison of Yield and Quality of Corn and Sorghum Silage*, Research Bulletin 67, University of Georgia, Athens, GA, 1969.
53. Cummins, D. G. and McCullough, M. E., Comparison of male sterile and male fertile corn for silage, *Agronomy Journal*, 63, 46, 1971.
54. Cummins, D. G., McCullough, M. E., and Dobson, J. W., Jr., *Evaluation of Corn and Sorghum Hybrids for Silage*, Research Report 72, University of Georgia, Athens, GA, 1970.
55. Jellum, M. D., Cummins, D. G., and Young, C. T., Yield and chemical characteristics of corn (*Zea mays* L.) types, *Agronomy Journal*, 65, 933, 1973.
56. Cummins, D. G., Relationships between tannin content and forage digestibility of sorghum, *Agronomy Journal*, 63, 500, 1971.
57. Cummins, D. G., Methods of evaluation and factors contributing to yield and digestibility of sorghum silage hybrids, in *The 27th Annual Corn and Sorghum Research Conference Proc.*, American Seed Trade Association, Washington, D.C., 1972, 11.
58. Cummins, D. G. and Dobson, J. W., Jr., Digestibility of bloom and bloomless sorghum leaves as determined by a modified *in vitro* technique, *Agronomy Journal*, 64, 682, 1972.
59. Harris, H. B., Cummins, D. G., and Burns, R. E., Tannin content and digestibility of sorghum grain as influenced by bagging, *Agronomy Journal*, 62, 633, 1970.
60. Cummins, D. G., *Interplanting of Corn, Sorghum, and Soybean for Silage*, Research Bulletin 150, University of Georgia, Athens, GA, 1973.
61. Bustillo, J. J. and Gallaher, R. N., Dry matter partitioning in no-tillage tropical corn in Florida, in *1989 Southern Conservation Tillage Conference Proc.*, Teare, I. D., Brown, E., and Trimble, C. A., Eds., University of Florida, Gainesville, FL, 1989, 40 (SB-1).
62. Long, T. A., Overman, D. L., and Gallaher, R. N., Yield and ear leaf nitrogen status in no-tillage second crop temperate and tropical corn, in *1989 Southern Conservation Tillage Conference Proc.*, Teare, I. D., Brown, E., and Trimble, C. A., Eds., University of Florida, Gainesville, FL, 1989, 36 (SB-1).
63. Gallaher, R. N. and Baldwin, J. A., *No-Tillage Corn Results Affected by Hybrids and Pesticides in 1984 at Green Acres Agronomy Farm*, Agronomy Research Report AY-85- 08, University of Florida, Gainesville, FL, 1985.
64. Overman, D. L. and Gallaher, R. N., Growth and partitioning of dry matter between temperate and tropical corn, in *1989 Southern Conservation Tillage Conference Proc.*, Teare, I. D., Brown, E., and Trimble, C. A., Eds., University of Florida, Gainesville, FL, 1989, 38 (SB-1).
65. Wright, D. L., Teare, I. D., and Gallaher, R. N., Corn silage production in North Florida, in *Proc. Southern Regional Tropical Corn Symposium*, Teare, I. D., Ed., Potash and Phosphate Institute, Atlanta, GA, 1992, 15.
66. Gallaher, R. N., McSorley, R., and Biggs, R. H., *Response of Corn Cultivars to Experimental Chemical Spray Treatment*, Agronomy Research Report AY-96-01, University of Florida, Gainesville, FL, 1996.
67. Stanley, R. L., *Corn Performance Trials for 1995*, NFREC Research Report NF-96-02, University of Florida, North Florida Research and Education Center, Quincy, FL, 1996.
68. Arias, F. R. and Gallaher, R. N., Maize-sorghum farming systems in Central American: situational analysis, *Journal of Agronomic Education*, 16, 5, 1987.
69. Gallaher, R. N., Ford, S. A., McSorley, R., and Bennett, J. M., *Corn Forage and Forage Sorghum Double Cropping Yield, Economics, Crop Nutrient Removal and Quality*, Agronomy Research Report AY-91-05, University of Florida, Gainesville, FL, 1991.
70. Kunkle, W. E. and Chambliss, C. G., Silage harvesting, storing, and feeding, in *Florida Forage Handbook*, Chambliss, C. G., Ed., University of Florida, Cooperative Extension Service, Gainesville, FL, 1999, 102.

71. Wing, J. M. and Becker, R. B., *Nutrient Intake of Cows from Silages Made from Typical Florida Forages*, Technical Bulletin 655, University of Florida, Gainesville, FL, 1963.
72. Becker, R. B., Wing, J. M., Arnold, P. T. D., McCall, J. T., and Wilcox, C. J., *Silage Investigations in Florida*, Bulletin 734, University of Florida, Gainesville, FL, 1970.
73. Humphreys, L. R., *Tropical Pasture Utilization*, Cambridge University Press, Cambridge, U.K., 1991.
74. Lavezzo, W., Lavezzo, O. E. N. M., Rossi, C., and Bonassi, I. A., Effects of wilting and formic acid on the chemical composition and nutritive value of elephant grass silage (*Pennisetum purpureum* Schum) purple cultivar, in *Proc. XVI Int. Grassland Congress*, The French Grassland Society, Nice, France, 1989, 965.
75. Asiedu, F. H. K., Alexander, C. A., and Proverbs, G., Ensilage for low resources farmers 1. drum, wirebasket and pit as silos, in *Proc. XVIII Int. Grassland Congress*, Vol. 1, XVIII Int. Grassland Congress, Calgary, Alberta, Canada, 1997, 19 (Session 14).
76. Cummins, D. G., McCullough, M. E., Winsett, I. L., Harter, W. G., and Jarrett, C. J., *Silage Short Course*, University of Georgia, Cooperative Extension Service, Athens, GA, 1972.
77. Asiedu, F. H. K., Alexander, C. A., and Proverbs, G. A., Ensilage for low resources farmers. 2. Poultry litter as cheap additive, in *Proc. XVIII Int. Grassland Congress*, Vol. 1, XVIII Int. Grassland Congress, Calgary, Alberta, Canada, 1997, 17 (Session 14).
78. Andrade, J. B., Ferrari, E., Lavezzo, W., Paulino, V. T., Nogueira, J. R., Braun, G., and Castro, F. B., Dry matter yield and nutritive value of Coast-cross No. 1 preserved as hay, silage, and haylage, in *Proc. XVIII Int. Grassland Congress*, Vol. 1, XVIII Int. Grassland Congress, Calgary, Alberta, Canada, 1997, 3 (Session 14).
79. Cromwell, R. P., Kunkle, W. E., and Chambliss, C. G., *Equipment for Preserving Forage as Round-bale Silage*, Circular 1071, University of Florida, Cooperative Extension Service, Gainesville, FL, 1994.
80. Cromwell, R. P., Kunkle, W. E., Sadler, G. D., and Chambliss, C. G., *The Plastic Wrapper is the Key to Making High Quality Round-Bale Silage*, Circular 1072, University of Florida, Cooperative Extension Service, Gainesville, FL, 1994.
81. Kunkle, W. E. and Bates, D. B., Alternative forage storage-bale silage, in *Proc. Dairy Herd Management Conference*, Spruill, D., Ed., University of Georgia, Athens, GA, 1988.
82. Kunkle, W. E., Cromwell, R. P., Chambliss, C. G., Bates, D. B., and Wasdin, J. G., *Experience with Round Bale Silage*, SS-ANS-12, Department of Animal Science, University of Florida, Gainesville, FL, 1991.
83. Nicholson, J. W. G., McQueen, R. E., Charmley, E., and Bush, R. S., Forage conservation in round bales or silage bags: effect on ensiling characteristics and animal performance, *Canadian Journal of Animal Science*, 71, 1167, 1991.

15 Importance of Symbiotic Nitrogen Fixation in Tropical Forage Legume Production

Eduardo C. Schröder

CONTENTS

I. ABSTRACT

Forage legumes are now prominent components of pastures in many tropical and subtropical areas of the world. Their ability to associate with effective strains of *Rhizobium* bacteria is making an important contribution to sustainable agricultural systems. Through a symbiotic relationship, the bacteria supply nitrogen (N) directly to the legume and indirectly to adjacent grass plants, and add organic matter to the soil upon the decay of their nodules affixed to the roots of the legume. The information regarding nodulation and taxonomy of tropical rhizobia is still incomplete, but, when a new legume is introduced, inoculation with the appropriate *Rhizobium* will ensure that the legume will live up to its potential. There is a wide range of interactions between legume and bacterial strain, and effectiveness of the relationship depends on the specific forage legume's *Rhizobium* requirement, which can vary from being very promiscuous to extremely specific. Acid soils of the tropics require symbiotic systems specially adapted to their conditions. It is now well established that tropical forage legumes can fix as much N as their temperate counterparts. The application of new biotechnology techniques to plants and associated microsymbionts will further increase forage yields and quality.

0-8493-2318-5/01/$0.00+$.50

TABLE 15.1
Groups of Bacteria and Examples of Genera Capable of Fixing Dinitrogen

Physiological Group	Representative Genera
Strict anaerobes	*Clostridium, Desulfovibrio, Methanosarcina*
Facultative	*Bacillus, Erwinia, Klebsiella*
Microaerophile	*Azospirillum, Rhizobium* (S), *Bradyrhizobium* (S), and *Azorhizobium* (S)
Aerobes	*Azotobacter, Beijerinckia, Derxia*
Phototrophic	
Facultative	*Rhodospirillum, Rhodopseudomonas*
Microaerophilic	*Plectonema*
Aerobic	*Anabaena, Nostoc, Synechococcus*

Note: S = Fixes N only in symbiosis.

II. INTRODUCTION

Nitrogen gas (N_2) constitutes almost 80% of the atmosphere, but it cannot be assimilated directly by plants, although mineral N is the most common limiting factor for plant nutrition. Mineralization of soil organic matter supplies all plants some N, but the amount provided by ammonification and nitrification is generally insufficient. To obtain high crop yields, the natural supply is augmented by applications of N fertilizer. However, the prohibitive cost of such fertilizer for forage production and the potential contamination of water sources by fertilizer runoff are compelling reasons for increasing the use of biological nitrogen fixation (BNF) in pasture management.

Nitrogen fixation is the process of converting N_2 gas into ammonia, a biologically available form. Of the thousands of bacteria existing in nature, only a few can transform unavailable N_2 into a reduced form. Bacteria capable of fixing N_2 are listed in Table 15.1. Only through symbiotic associations can enough amounts of N be reduced to be of agricultural value. Plants of the leguminous family have long been known to form associations with members of the genus *Rhizobium*. The main mechanism by which the N is made available to the legume is through the development of bacterial nodules on the roots of the plants. In the nodules, carbohydrates from the plant and N_2 from the air are combined to form organic N compounds. However, not all species of the *Leguminosae* are nodulated. The proportion of nodulated species varies according to their subfamily (Table 15.2). Most members of *Caesalpinioideae*, regarded as the primitive group, are not nodulated.[1] Norris[2] suggested that the most primitive legumes were trees, and that the evolution has been:

$$\text{tree} \rightarrow \text{shrub} \rightarrow \text{woody climber} \rightarrow \text{perennial herb} \rightarrow \text{annual}$$

Rhizobium evolution followed this development and became more specialized and adapted to temperate conditions. Since most tropical legumes originated in poor, acid soils, they show little specialization and nodulate a wide range of species. However, the finding of many fast-growing, highly specific strains in several tropical legumes challenges this theory.

III. THE NODULATION PROCESS

The nodulation process consists of several events that precede active dinitrogen fixation. Before nodules appear on the root, rhizobia in the soil are stimulated to multiply by plant-specific com-

TABLE 15.2
Nodulation Data According to Allen and Allen[1]

Subfamily	Number of Genera	+	+/–	–	Total	Number of Species Estimated
Caesalpinioideae	177	13	13	39	65	2,800
Mimosoideae	66	18	8	5	31	2,900
Papilionoideae	505	241	14	14	269	14,000
TOTAL	748	272	35	58	365	19,700

Note: + means nodulated, – means not nodulated, and +/– conflicting reports.

pounds. The inducer molecules present in the root exudates have been identified as flavonoids, isoflavonoids, and related compounds. The bacteria respond to the plant signal producing a lipo-oligosaccharide molecule (Nod factor), specific to each rhizobial strain.[3] After mutual host-symbiont recognition, specific rhizobial adherence to root hair follows (see Figure 15.1). It is known that rhicadhesin, plant lectins, and cellulose microfibrils are involved. The ability to establish a symbiotic relationship is severely affected in *Rhizobium* mutants lacking the acidic exopolysaccharide (EPS) in all legumes where cell invasion is through infection threads. Exchange of chemical signals initiates root hair curling, root hair infection, and the appearance of root nodules. However, some legumes (*Arachis*, *Stylosanthes*) undergo infection through a more primitive "crack entry" method. Two nodule developmental patterns can occur. One group of plants have a persistent meristem, and

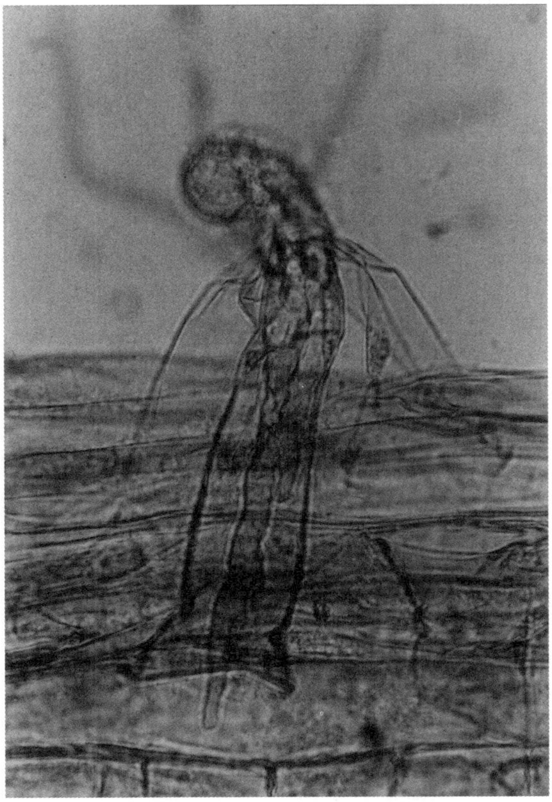

FIGURE 15.1 Infection thread within root hair of *Phaseolus vulgaris*. (Courtesy of Graciela Brelles-Mariño.)

form *indeterminate* nodules (alfalfa, leucaena). In *determinate* nodules, the released bacteria undergo multiple rounds of cell division within plant cells that are also dividing (*Arachis*, *Glycine*).

Within the nodule, bacteria differentiate into enlarged, pleomorphic cells called bacteroids.[4] Bacteroid morphology varies from simple enlargement to very distorted X and Y shaped cells. In addition to this membrane modification, bacteroids produce other proteins, such as cytochromes that function within the low oxygen level tension of the nodule. They are surrounded by a plant-derived membrane, termed the peribacteroid membrane. The compartment is called a symbiosome.

Nitrogen fixation is carried out by a specialized enzyme called nitrogenase, which is synthesized by the bacteria under the low oxygen condition inside the nodule (see "The Central Reaction in N_2 Fixation"). Effective symbiosis is accompanied by the production of leghemoglobin, whose function is to provide enough oxygen for aerobic energy-producing metabolism. The shape, size, and number of nodules are determined by the plant; their duration depends on the strain effectiveness and environmental conditions. Some plant-microbe associations are capable of recycling hydrogen gas released by the nitrogenase reaction (by an uptake hydrogenase enzyme) and form more efficient symbiosis. The nodule "factory" requires large amounts of energy, provided by the plant through photosynthesis. Carbon is transported from the shoot to the roots in the form of sugars. In the nodule, sucrose is transformed into carbon skeletons to accept and transport the N compounds, and to produce ATP and reducing power for nitrogen fixation. Evidence indicates that sugars must be metabolized into dicarboxylic acids (succinate and malate) to be used by bacteroids. The concentration of the ammonia formed by the nitrogenase reaction must be kept low, so it is incorporated into glutamine. Fixed N is exported to the plant as amides (indeterminate nodules) and ureides (determinate growth nodules).

Studies of rhizobial genetics have revealed the nature and location of many bacterial genes involved in symbiosis. The first group of genes discovered were known as Nod proteins, but so many genes are involved that it has been necessary to establish also *nol* and *noe* genes. Many of these genes determine the production or secretion of the Nod factors. The Nod factor alone (natural or synthetic) is sufficient to initiate root-hair deformations and trigger nodule development, but only on specific host legumes. Rhizobial genes known to control host specificity and nodule induction fall into two groups: the *structural nod* genes and the regulatory *nod* genes. The first group are expressed in response to biochemical signals from the plant, usually flavonoid compounds, and a transcriptional activator produced by *nodD* regulatory genes. A group present in all rhizobia (*nodABC*) are called "common" and are involved in the biosynthesis of the basic structure of all Nod factors. Genes determining specificity are called Hsn (Host-specific) and they modify the N-acetylglucosamine core common to all Nod factors.[5]

Although non-nodulating soybean mutants have been known for many years, very few plant genes functioning during nodule development and function have been identified. Those involved in the initial steps of the nodulation process have been designated as ENOD (early nodulins), while genes expressed during nitrogen fixation were named LNOD (late nodulins). Manipulation of plant genes offers the potential to overcome the problem of strain competition.

The Central Reaction in N_2 Fixation:

$$N_2 + 8H^+ + 8e^- + 16MgATP \rightarrow 2NH_3 + H_2 + 16MgADP + 16\ Pi$$

Nitrogen is a very stable molecule, due to its strong triple bond. Therefore, its fixation requires much energy. While biological nitrogen fixation takes place at ambient pressures and temperatures, the industrial process developed by Haber-Bosh requires high temperature and pressure and a metal catalyst.

TABLE 15.3
Species of *Rhizobium* after Bergey's Manual, 8th Edition[7]

Species	Growth Rate	Host Nodulated	Common Name
Rhizobium trifolii	Fast	*Trifolium*	Clover
R. leguminosarum	Fast	*Pisum, Vicia*	Pea
R. phaseoli	Fast	*Phaseolus*	Bean
R. meliloti	Fast	*Medicago*	Alfalfa
R. japonicum	Slow	*Glycine*	Soybean
R. lupini	Slow	*Lupinus*	Lupin
R. spp.	Slow	*Vigna,* others	Cowpea

IV. *RHIZOBIUM* CLASSIFICATION

It was recognized in early studies that leguminous plants have preferences for certain rhizobia and vice versa, but the full implication of this was not understood. Traditionally, *Rhizobium* is one of the two genera of the family *Rhizobiaceae*; *Agrobacterium* causes tumors on many dicots, and rhizobia infect and induce nodule formation on roots of leguminous plants. However, one exception exists; some strains of *Rhizobium* are capable of nodulating the genus *Parasponia*.[6]

Rhizobium species are differentiated by the kind of plants they nodulate. Plants mutually susceptible to nodulation by a particular kind of rhizobia constitute a cross-inoculation group. The bacteria capable of nodulating the plants in a group are considered a *Rhizobium* species.[7] According to this concept, six species were recognized until 1982 (Table 15.3). Strains belonging to a previous established genus, *Photorhizobium*, are now considered *Bradyrhizobium*.

More recently, rhizobial taxonomists have considered speed of growth as a fundamental property and created a new genus, *Bradyrhizobium*, that comprises strains previously classified as *R. japonicum, R. lupini*, and *Rhizobium* spp.[8] This classification can be questioned since growth rate cannot be used to clearly separate strains having intermediate growth rates. The subject has been further complicated by the discovery of fast-growing strains capable of nodulating soybeans,[9] and the discovery of stem-nodulating rhizobia.[10] A molecular approach (multilocus enzyme electrophoresis, MLEE and sequencing ribosomal genes) to taxonomy has further differentiated more genera and species within the *Rhizobiaceae*.[11]

The presence of nodulation genes (*nod*) on a large plasmid (called pSym) is characteristic of *Rhizobium*, while they are located in the chromosome of *Bradyrhizobium*. Other authors have characterized new species, and the presently accepted classification of rhizobia is listed in Table 15.4.

The grouping of leguminous plants on the basis of susceptibility to nodulate and fix N by a particular species of *Rhizobium* is useful to farmers and inoculant manufacturers and will be used here. When specific strains are identified, they can be associated with their parental host plant.

V. AGRICULTURAL INOCULANTS

The addition of bacteria to seeds or soil to enhance productivity can be defined as inoculation. Before bacteria were known, farmers used to transfer and spread soil from a field where legumes were productive to a newly planted field. Soon after Beijerinck[13] isolated rhizobia from nodules, commercial application began. However, early inoculants used agar in a bottle as a carrier, and results were often disappointing. Many forms of inocula are available: the peat-base inoculum, liquid culture, lyophilized rhizobia in talc or calcium carbonate, oil-dried preparations on clay or vermiculite, a frozen concentrate, and the old agar slant. Today, inoculant manufacturers generally use powdered or granular peat as a carrier. The new inoculant technology involves the use of liquid carriers and provides quality, convenience, and planter compatibility.[14]

TABLE 15.4
Present Taxonomic Status of Bacteria Nodulating Legumes

Species	Host Plants
Rhizobium leguminosarum	
Biovar viciae	*Pisum, Vicia, Lens, Lathyrus*
Biovar trifolii	*Trifolium*
Biovar phaseoli	*Phaseolus*
Rhizobium etli	*Phaseolus*
Rhizobium tropici	*Phaseolus, Leucaena*
Rhizobium meliloti	*Medicago, Melilotus, Trigonella*
Rhizobium loti	*Lotus, Anthyllis*
Rhizobium ciceri	*Cicer*
Rhizobium fredii	*Glycine, Vigna*
Rhizobium spp. NGR234	*Macroptilium, Vigna, Glycine*
Rhizobium galegae	*Galega*
Rhizobium huakuii	*Astragalus*
Bradyrhizobium elkanii	*Glycine, Vigna*
Bradyrhizobium japonicum	*Glycine*
Bradyrhizobium spp.	*Arachis, Lupinus, Vigna*
Azorhizobium caulinodans	*Sesbania*

Adapted and updated from Somasegaran and Hoben.[12]

Inoculation methods include both seed-applied or in-furrow. Seeds can be inoculated by sprinkling (moistening the seeds first), making a slurry of the peat with water or adding the inoculant directly to the seed in the drill hopper without any water (the least reliable method). Adverse conditions (acid or dry, hot soils) reduce the number of viable rhizobia in the zone where the young seedling will develop. In such cases, inoculation problems can be solved by coating the seed with finely pulverized calcium carbonate (limestone) or stickers such as a 45% gum arabic solution. Very practical information about inoculation is available in the form of a technical handbook published by FAO.[15]

The use of other seed treatments, such as fungicides, may be toxic to rhizobia. To avoid problems, soil inoculation (granular or liquid) is recommended. All products carry directions on how to use them as well as an expiration date. The quality of inoculants is quite variable, and a program of inoculant quality control (private or governmental) is necessary. In general, inoculants require cool storage, and direct sunlight should be avoided. A short list of some commercial inoculant suppliers is given in Appendix A. Bacterial strains for research can be obtained from several culture collections; Appendix B contains a list of laboratories that carry *Rhizobium* strains. Further information can be obtained from some Websites (Appendix C).

VI. THE NEED FOR INOCULATION

Inoculation is the process of supplying effective nodule-forming bacteria for legumes through the application of high-quality commercial inoculants. There is a commonly held view that tropical legumes are much more promiscuous than their temperate counterparts in that they nodulate freely with a wide range of tropical rhizobia, and that tropical soils are laden with bacteria so that effective nodulation occurs without inoculation. Although it is true that native legumes are usually nodulated by indigenous rhizobia, the introduction of a new species or even cultivars can fail due to the lack of specific rhizobial strains in the soil. Some species and accessions of legume genera previously considered to be promiscuous are now known to require specific strains of *Rhizobium*. Recent work on tropical rhizobia, although inadequate, reveals much greater variation in genetic compatibility and nitrogen fixation than has generally been acknowledged.

A useful grouping of tropical forage legumes according to their *Rhizobium* requirement was postulated by Date.[16] These groups are

1. GROUP PE (**P**romiscuous and **E**ffective). Plants nodulate effectively with a wide range of strains from different genera and species within the group; the symbiotic effectiveness (measured as plant dry weight) is usually 80% or more than the nitrogen control, and only a few strains are not fully effective. This group corresponds most closely to the old cowpea-type group. Some of the genera are: *Arachis, Calopogonium, Cajanus, Canavalia, Clitoria, Crotalaria, Cyamopsis, Desmanthus, Dolichos, Galactica, Gliricidia, Glycine, Indigofera, Lablab, Macroptilium, Macrotyloma, Psophocarpus, Rhynchosia, Stizolobium, Stylosanthes* (several subgroups), *Teramnus, Tephrosia, Vigna,* and *Zornia.*
2. GROUP PI (**P**romiscuous but **I**neffective). These legumes frequently nodulate with a wide range of strains of *Rhizobium*, many of which are ineffective in nitrogen fixation. Genera in this group can be further subdivided: *Rhizobium* for one subgroup does not always form nodules in another, and those strains that do not form nodules have a limited range of effectiveness. Genera of this group are the following: *Adesmia, Aeschynomene, Centrosema* (two subgroups), *Desmodium* (two subgroups), *Psoralea, Sesbania* (two subgroups), and *Stylosanthes.*
3. GROUP S (**S**pecific). These genera and species nodulate effectively with a narrow or restricted range of *Rhizobium* strains that generally originate only from nodules of homologous species. Distinct subgroups can be formed. A large proportion of the members of this group do not nodulate with strains from other genera or species, and many plants nodulate irregularly when the same strain is used on different occasions. Associated forage genera include: *Coronilla, Leucaena, Lotononis, Lotus* (two subgroups), *Lupinus* (two subgroups), *Medicago-Trigonella, Mimosa, Stylosanthes* (two subgroups), and *Trifolium* (African, four subgroups).

Although very limited research on the response of forage legumes to inoculation has been carried out in Puerto Rico, there is evidence to support the need to inoculate alfalfa and leucaena when sown for the first time.

One important characteristic of some tropical *Rhizobium* strains is their ability to grow in acidified media. Date and Halliday[17] were able to separate three groups of strains:

a) Those that grow only at pH 6.8 to 7.0
b) Those that grow only in acidified media (pH 4.5)
c) Those able to grow under both conditions.

Strains capable of growing on a wide range of pH would be more suitable for the manufacture of commercial inoculants.

Inoculation success also depends on the presence of other bacterial species in the soil and, particularly, in the rhizosphere. Studies of the interaction of *Rhizobium* with other rhizosphere bacteria such as *Azospirillum, Bacillus, Erwinia,* and *Pseudomonas* have been initiated.[18] Both detrimental and stimulatory effects have been detected. Several mechanisms like plant hormone synthesis, production of antibiotic substances, P solubilization, and siderophore synthesis have been indicated as responsible for the effects.

VII. AMOUNTS OF NITROGEN FIXED

One of the most important considerations when choosing a forage legume is the amount of N fixed by the plant-bacterial symbiosis. The efficiency of the symbiosis depends on both the bacterial strain and the plant's genetic characteristics. Sometimes, even the variety is a decisive factor.

TABLE 15.5
Estimates of Nitrogen Fixed by Tropical Forage Legumes (in kg.ha^{-1}.yr^{-1})

Species	Average	Range
Acacia sp.	270	
Centrosema sp.	259	126–365
Enterolobium saman	150	
Leucaena leucocephala	277	74–584
Lotononis bainesii	62	
Macroptilium atropurpureum	291	
Mikanea cordata	120	
Pueraria phaseoloides	99	
Sesbania cannabina	542	
Stylosanthes guyanensis	124	34–220

Adapted from Nutman.[20]

The annual nitrogen-fixing capacity of tropical species is within the range of 80 to 300 kg N ha^{-1}, although only a fraction is incorporated into the soil-plant system. Rates of fixation can be estimated by the difference method (Kjeldahl analysis), by using the ^{15}N isotope, or by the acetylene-reduction assay (ARA).[19] Some of the estimates obtained by using these methods are listed in Table 15.5.

Other species fix large amounts of N_2, but estimates are obtained only for a short period. For example, Sanginga et al.,[21] working for a short period of 12 weeks, obtained estimates of up to 167 kg N ha^{-1}. Many environmental factors affect the amount of N fixed under tropical pasture conditions; those more pertinent to agronomists will be dealt with next.

VIII. ENVIRONMENTAL FACTORS AFFECTING SYMBIOSIS

1. **Physical.** Although tropical legumes are adapted to high temperatures, high values affect the symbiosis before the plant or bacteria are affected. In particular, lack of soil humidity severely restricts nodulation, and nodules enter senescence and die. Symbiotic N fixation is also more sensitive than nonfixing plants to other environmental factors that reduce plant growth, such as drought and high temperature. Screening-adapted legume germplasm is a fundamental stage in tropical pasture improvement.
2. **Chemical.** The nodulation process can be adversely affected by alkaline or very acid pH, which contributes to the lack of Ca and augments the toxicity caused by Al and Mn. The effects of major nutrient and microelement excess and deficiencies on symbiosis are discussed below under soil fertility.
3. **Biological.** Several insects tend to feed on mature nodules, reducing the amount of BNF. The interaction with pathogens or deleterious soil microorganisms (including plant virus) can severely affect nodulation. In addition, naturally occurring rhizobia can compete and easily overwhelm the few bacteria added in the inoculum, thus causing ineffective nodulation.

IX. EFFECT OF PASTURE MANAGEMENT ON BIOLOGICAL NITROGEN FIXATION

1. **Defoliation.** Removal of leaves from the legume plants causes nodule senescence and sloughing, particularly if young leaves are removed. The proportion of nodule loss is

determined by the severity of defoliation.[22] Stocking rate of grazing animals can be adjusted to maximize fixation by removal of excess leaves (shading), flowers, or pods. However, careful management is required to maintain enough leaf area. An adequate grass/legume balance increases the amount of N transferred to the soil.

2. **Nitrogen fixation**. The amount of N fixed by legumes is correlated with the total carbon supplied by the top to the nodules. It is regulated by the C/N ratio in the xylem and phloem. Severe defoliation can reduce the amount of N fixed (estimated by the acetylene reduction assay) to almost 0, and 25 to 30 days are needed to recover the original activity.[23]
3. **Reproductive growth**. Maximum amounts of N are fixed during vegetative growth of the plant. During flowering, relative nitrogenase activity is maintained, but the development of pods acts as a competitive sink for assimilate and a pronounced decrease in N fixation occurs.
4. **Soil fertility.** Nodulated legumes require higher supplies of certain essential elements than grasses or non-nodulated legumes. Tropical soils are usually acid and highly leached, and liming and P fertilization are necessary for adequate legume growth. In addition, microelements such as Mo, B, Zn, Mn, and Co can be deficient. One small application (1 to 3 kg ha^{-1}) of the missing element can have a significant effect on BNF. Most legumes form vesicular-arbuscular mycorrhiza (VAM) (see Mycorrhizae below) and this triple symbiosis can improve plant productivity in soils of low fertility by enhancing P uptake. Soil acidity ($pH < 5.0$) and associated toxicities affect the formation and functioning of nodules and mycorrhiza.

For nodulated legumes, P availability is essential. However, there can be significant differences in the P concentration in seeds. Bolland and Paynter[24] suggest that higher concentration may increase root growth of the embryo so that the seedlings start to take nutrients and water from a greater volume of soil earlier than seedlings from seeds containing a low P concentration. Phosphorus has been shown to be required in higher soil concentrations for legumes depending on symbiotic N. All legumes can establish a triple symbiosis (plant, bacteria, and fungi) when efficient vesicular-arbuscular mycorrizae are present.

Mycorrhizae

Mycorrhizae are symbiotic associations between a fungi ("myco") and the roots ("rhizae") of plants. Except for a few species, most plants have mycorrhizae. The fungus is strongly dependent on the plant, but the plant can grow without its partner. Two symbiosis groups exist: ectomycorrhizae and endomycorrhizae. The first group predominates in tree and shrub species and forms a mantle around the root. Endomycorrhizae can grow both outside and inside the root cortex. The most common and important of the several groups that exist is the vesicular-arbuscular mycorriza (VAM). The fungus grows inside the plant cell and forms ovoid bodies (vesicles) and highly branched structures (arbuscules), which appear to be the sites of nutrient transfer. The fungus is a member of the *Endogonaceae*, and the most abundant is the genus *Glomus*.

Mycorrhizal infection can increase the uptake of mineral nutrients, provide heavy metal tolerance, increase plant growth through hormone production, and protect the root against pathogens. In legumes, VAM fungi can improve phosphorus nutrition and, indirectly, increase nodulation and N_2 fixation. This is particularly important in tropical forage legumes, since they are sown in very acid, nutrient-deficient soils, such as the savannas.

X. NITROGEN TRANSFER

In highly extractive agricultural systems such as grain legumes and forages harvested for hay, the amounts of N transfered to the soil are minimal. High productivity of mixed pastures (grasses and legumes) is based on the hypothesis that nonlegumes can utilize the N fixed by the legume component. Several mechanisms of transfer are possible via:

a) Animal urine and dung
b) Decomposition of dead leaves and stems
c) Mineralization of roots and nodules
d) Direct transfer through mycorrhizae
e) N excretion

The amounts of N transferred are affected by factors such as species, soil fertility, and environmental conditions.[25] Rao and Giller[26] quantified the nitrogen transfer of *Leucaena leucocephala* to *Cenchrus ciliaris* using ^{15}N enrichment and determined that biomass production and N accumulation in the grass were enhanced compared to monocultured grass. *Cenchrus* plants growing with detopped plants had a higher percentage of N obtained from *Leucaena* compared to that of non-detopped plants.

The BNF capacity of three pasture legumes (*Centrosema pubescens, Stylosanthes hamata,* and *Pueraria phaseoloides)* grown in monoculture and associated with *Brachiaria humicola* was studied in the acid (pH 4.9) Venezuela savannas.[27] A higher contribution from BNF was observed when the legumes were mixed. However, no clear transfer of N from the legumes to the companion grass was observed.

Land under mature trees (whether for fruit or wood production) offers excellent opportunities for pasture and animal production. For example, in an oil palm plantation in Nigeria, *Panicum* and *Brachiaria* were compared in mixtures with or without tropical kudzu (*Pueraria phaseoloides*). More biomass was produced in plots planted to pastures than in unplanted plots with natural vegetation, both in and outside the plantation. Strategic use may be made of plantation to produce forage reserves for dry season use.[28]

Legume trees can be used in agroforestry as windshields, shade trees, to provide wood, and for browsing by animals. In other systems, pruning of legume trees can be used for fodder or as green manure. Species such as *Acacia, Gliricidia, Leucaena,* and *Prosopis* have been extensively used. Although forage tree legumes have considerable potential as supplements to low-quality grass diets, more data is needed. Further information on legume tree nodulation specificity and effectiveness has been reported by Turk and Keyser (1992).[29] Extensive biodiversity exists in strains, and isolates nodulating and fixing several legume trees in acid soils can be found. The manufacturing of a broad host-range inoculant is possible,[30] however, research funding of the improvement of legume tree symbiosis is minimal.

XI. NITROGEN FIXATION BY GRASSES

Starting in the 1970s, and with the development of the acetylene reduction technique, nitrogenase activity has been detected in the roots of several plants, particularly with tropical grasses (*Digitaria*) and sugarcane. This specific type is now called associative nitrogen fixation and is due to the presence of several diazotrophic bacteria. The most studied genus is *Azospirillum*, of which five species are now recognized (*A. amazonense, A. brasilense, A. halopraeferans, A. irakense,* and *A. lipoferum*). However, several other genera of N_2-fixers (*Acetobacter, Campylobacter, Bacillus, Enterobacter, Erwinia, Herbaspirillum, Klebsiella*, and even *Burholderia*) have been isolated from roots of many plants. Significant BNF contribution of *Acetobacter diazotrophicus* could explain why sugarcane yields in Brazil have not declined after decades of cultivation without added fertilizer.

Many field inoculation experiments have been carried out, but positive crop responses are few and erratic. Even with the use of the ^{15}N isotope technique, the contribution of these microorganisms to plant N nutrition seems limited, and other mechanisms such as plant hormone production and increased mineral nutrient uptake have been postulated to explain this phenomenon, in addition to the N fixed. The practical application of these microorganisms to enhance crop yield has been reviewed,[31] and other aspects of these associations have been reviewed by Boddey and Döbereiner.[32] More research is needed to establish the exact mechanism of plant growth promotion by these bacteria, members of the PGPR (Plant Growth Promoting Rhizobacteria). However, it seems that inoculation with *Azospirillum* can alleviate certain environmental stresses such as drought (osmotic stress) as showed by Sarig et al.[33] High temperature (32°C) enhanced ethylene reduction by *Setaria* inoculated with *Azospirillum*, but plant ontogeny was of fundamental importance to show nitrogen fixation activity.[34] Successful inoculation experiments appear to be those in which researchers have paid special attention to the optimal number of cells of *Azospirillum* in the inoculant, using appropriate inoculation methodology whereby the optimal number of cells remained viable and available to colonize the roots.[35]

The contribution of BNF to tropical pasture grasses has been quantified on four species of *Brachiaria* and 11 ecotypes of *Panicum maximum* in Brazil. Using the ^{15}N technique, *B. humidicola* and *B. decumbens* and several *P. maximum* ecotypes showed up to 7 to 10 kg N $ha^{-1}month^{-1}$ of N fixed during the warm summer months.[36] Inoculation with the specific strain (homologous) is a very important factor in obtaining response among *P. maximum* ecotypes.[37] In Florida, Smith et al.[38] tested 40 genotypes representing five tropical grass genera (3 *Digitaria*, 13 *Panicum*, 8 *Paspalum*, 2 *Cynodon*, and 13 *Cenchrus*). Only two genotypes responded to inoculation: *Digitaria decumbens* and *Panicum maximum*.

A few studies have been reported from the Caribbean region. Taylor[39] conducted a field trial in the Bahamas with pearl millet (*Pennisetum americanum*) and guinea grass (*P. maximum*). Yields were higher in the inoculated treatments, but only pearl millet dry matter yield was significantly higher than the uninoculated control. In Trinidad and Tobago, Collins and Donawa[40] reported that *Digitaria sanguinalis* soil cores showed the highest nitrogenase activity of those tested. Substantially increased rates were obtained when glucose was added, indicating that organic carbon (energy source and oxygen pressure) is a limiting factor. In Cuba, Hernandez and Sarmiento[41] showed that small yield increases were obtained by *Azospirillum* inoculation of *P. maximum* and *B. brizantha*. Inoculation increased root biomass and K content.

XII. FUTURE PROSPECTS

Transfer of the ability to fix nitrogen to plants has been a long-range goal since the beginning of genetic engineering. However, the nitrogen-fixation process itself is not completely understood. Many genes are involved in the process, and the possibility that they would be transferred and operate in the recipient plant remains remote.

Alternatively, present fixing systems can be improved by modification of the plant or the bacteria. Genetic improvement of the symbiotic systems in tropical forage legumes is carried out at three research sites (Queensland, Australia; Cali, Colombia; and Nairobi, Kenya). *Rhizobium meliloti* has been modified by the introduction of an extra *dct* gene and/or modified *nif* expression, but few field experiments show significant beneficial effect.[42]

Alfalfa, a temperate legume adapted to tropical conditions, was the first forage species to benefit from recombinant DNA technology. Researchers at Plant Genetic Systems (PGS, Belgium) developed alfalfa plants resistant to the herbicide glufosinate (Basta).[43] Hill et al.[44] developed virus-resistant, transgenic *Medicago sativa* plants via *Agrobacterium* gene transfer and demonstrated that expression of the coat protein of alfalfa mosaic virus (AMV) confers resistance to infection by the AMV. The introduction of such desirable traits will increase forage yield and quality in tropical legumes.

Molecular methods will be used more and more to study the ecology of BNF bacteria under natural environments. Induced mutations, gene mapping, cloning and sequencing, have been barely employed with tropical rhizobia and legumes. New technologies such as PCR methods and microarrays will enhance our possibilities to improve both symbiotic partners. Significantly, more attention will be given to research the ability of introduced strains to survive in soil and compete with native rhizobia populations. With these new tools, we will make further advances in the management of tropical forage legumes with economic benefits for the farmer.

APPENDICES

APPENDIX A

SOME COMMERCIAL SUPPLIERS OF INOCULANTS

Argentina

Laboratorios ARBO
Padre Ghio 588
6000. Junion (Pcia. Buenos Aires)
Tel.: 54 236 2442 9079
Fax: 54 236 244 7269
E-mail: arbolab@infovia.com.ar

Laboratorio BIAGRO S.A.
Parque Industrial General Las Heras
CC N-4 , 1741 Gral. Las Heras (Pvia. Bs. As.)
Tel.: 54 220 476 1655
Fax: 54 220 476 2170
E-mail: biagro@satlink.com.ar

Nitragin S. A.
Peru 345 4 "C"
1067. Buenos Aires, CF
E-mail: rml@datamrket.com.ar

Calle 10 y 13
Parque industrial Pilar
Unidad Postal 1 (1628)
Pilar, Buenos Aires
www.nitragin.com.ar

Nitrasoil Argentina
Av. Centenario 3359
CP 1878 Quilmes, Buenos Aires
Tel.: 54 11 4278 2670
Fax: 54 11 4278 3386
E-mail: GA_NITRAGIN.ciudad.com.ar

Canada

Agrium Biologicals
402-15 Innovation Blvd
Saskatoon, S.K. S7N 2X8

Mexico

Quimica Lucava, S.A., de C.V.
Camino a Tepalcapa No 224
54900 Tultitlán, Edo. De México,
Tel.: 5884-7144, 5884-6985
Fax: 5884-7145
E-mail: qlucaval@dfl.telmex.net.mx

Uruguay

Calister SA
Carrasco, Montevideo
Tel.: 598-2 600 99 03
E-mail: calister@netgate.com.uy

Enzur S.A.
Azara 3787
CP 11400, Montevideo
Tel.: 598-2 5
Fax: 598-2 5
E-mail: enzur@netgate.com.uy

Lage & Cia., SA
Camino Carrasco 6948
CP 11 500, Montevideo
Tel.: 598-2 600 2714
Fax: 598-2 601 3654
E-mail: lage@adinet.com.uy

United States

AgroForester Tropical Seeds
P.O. Box 428
Holualoa, Hawaii 96725
Tel.: 1-808-324-4427
Fax: 1-808-324-4129
E-mail: seeds@agroforester.com

Celpril
51 Oak Street
Manteca, California 95336
www.celpril.com

Chr. Hansen BioSystems
9015 W. Maple St.,
Milwaukee, Wisconsin 53214
Tel.: 1-888-828-6600

Liphatech, Inc. (ex-Nitragin).
3600 W. Elm Street
P.O. Box 09186
Milwaukee, Wisconsin 53209
www.liphatech.com

Urbana Laboratories
P.O.Box 1393
St. Joseph, Missouri 64502
www.urbana-labs.com

United Kingdom

Micro Bio Ltd.
Unit 2 Centro Boundrt Way
Boundry, Co Maxted Road
Hemel Hempstead
Herts, HP2 7SU

APPENDIX B

LABORATORIES FROM WHICH TO REQUEST SPECIFIC RHIZOBIUM STRAINS (RESEARCH) (Not an exhaustive list)

ATTC. American Type Culture Collection, 12301 Parklawn Dr., Rockville, MD 20852, USA. (For purchase)
E-mail: sales@atcc.org

BNF MIRCENS (UNESCO):

RHIZOBIUM MIRCEN: FEPAGRO/Faculdade de Agronomia, Caixa Postal 776, Porto Alegre, Brazil.

RHIZOBIUM MIRCEN: Department of Soil Science, University of Nairobi, P.O. Box 30197, Nairobi, P.O. Box 30197, Nairobi, Kenya.

RHIZOBIUM MIRCEN: Laboratoire de Microbiologie, ISRA-ORSTOM, B.P. 1386, Dakar, Senegal.

RHIZOBIUM MIRCEN: Soybean and Alfalfa Research Laboratory, BARC-West, USDA-ARS, Bldg-011, HH-19, Beltsville, MD 20705, USA.
E-mail: pberkum@asrr.arsusda.gov

NifTAL MIRCEN: NifTAL Project, 1000 Holumua Rd., Paia, HI 96779, USA.
E-mail: NifTAL@hawaii.edu

Laboratorio de Microbiologia de Suelos e Inoculantes, MGAP, Burgues 3208, CP 11700, Montevideo, Uruguay.
E-mail: lmscilab@adinet.com.uy

WFCC-MIRCEN, World Data Centre for Microorganisms, RIKEN, 2-1 Hirosawa, Wako, Saitama 351-01, Japan.
E-mail: sugawara@viola.riken.go.jo

APPENDIX C

RECOMMENDED WEB SITES

1. American Society of Agronomy: www.agronomy.org
2. American Society for Microbiology: www.asmusa.org
3. Asociación Latino Americana de Rhizobiología (ALAR): www.mscilab@adinet.com.uy
4. Center for Microbial Ecology: www.cme.msu.edu
5. Food and Agriculture Organization: www.fao.org
6. Pasturas de America: www.agro.delmercosur.com/pasturas
7. The Microbiology Learning Center: www.kent/edu/microbiology

REFERENCES

1. Allen, O. N. and Allen, E. K., *The Leguminosae. A Source Book of Characteristics, Uses, and Nodulation,* University of Wisconsin Press, Madison, WI, 1981.
2. Norris, D. O., Legumes and the *Rhizobium* symbiosis, *Empire J. Exptl. Agr.*, 24, 247, 1956.
3. Schultze, M. and Kondorosi, A., The role of Nod signal structures in the determination of host specificity in the *Rhizobium*-legume symbiosis, *World J. Microbiol. & Biotechnol.*, 12, 137, 1996.
4. Oke, V. and Long, S. R., Bacteroid formation in the *Rhizobium*-legume symbiosis, *Current Opinion in Microbiology*, 2, 641, 1999.
5. Downie, J. A., Functions of rhizobial nodulation genes, in *The Rhizobiaceae: Molecular Biology of Model Plant-Associated Bacteria*, Spaink, H. P., Kondorosi, A., and Hooykaas, P. J. J., Eds., Kluwer Academic Publishers, Dordrecht, 1998, chap. 20.
6. Trinick, M. J., Growth of *Parasponia* in agar tube culture and symbiotic effectiveness of isolates from *Parasponia* spp., *New Phytol.*, 85, 37, 1980.
7. Buchanan, R. E. and Gibbons, N., *Bergey's Manual of Determinative Bacteriology*, 8th ed., Williams and Wilkins, Baltimore, MD, 1974.
8. Jordan, D. C., Transfer of *Rhizobium japonicum* Buchanan 1980 to *Bradyrhizobium* gen. nov., a genus of slow-growing, root nodule bacteria from leguminous plants, *Int. J. Syst. Bact.* 32, 136, 1982.
9. Keyser, H. H., Bohlool, B. B., Hu, T. S., and Weber, D. F., Fast growing rhizobia isolated from root nodules of soybean, *Glycine soja, Science*, 215, 1631, 1982.
10. Dreyfus, B. and Dommergues, Y., Non-inhibition de la fixation d'azote combiné chez une legumineuse a nodules caulinaires, *Sesbania rostrata, C. R. Acad. Sc.*, Paris, 291, 767, 1980.
11. van Berkum, P. and Eardly, B. D., Molecular Evolutionary Systematics of the *Rhizobiaceae*, in *The Rhizobiaceae: Molecular Biology of Model Plant-Associated Bacteria*, Spaink, H. P., Kondorosi, A., and Hooykaas, P. J. J., Eds., Kluwer Academic Publishers, Dordrecht, 1998, chap. 1.
12. Somasegaran, P. and Hoben, H. J., *Handbook for Rhizobia. Methods in Legume-Rhizobium Technology.* Springer-Verlag, New York, 1994.
13. Beijerinck, M., Die Bacterien der Papilionaceenknollchen, (The bacteria of root nodules,) *Botanische Zeitung*, 46, 725, 1888.
14. Smith, R. S., New inoculant technology to meet changing legume management, in *Biological Nitrogen Fixation for the 21st Century*, Elmerich, C., Kondorosi, A., and Newton, W. E., Eds., Kluwer Academic Publishers, 1998, 621.
15. FAO, *Technical Handbook on Symbiotic Nitrogen Fixation.* FAO, Rome, 1993.
16. Date R. A., Inoculation of tropical pasture legumes, in *Exploiting the Legume-Rhizobium Symbiosis in Tropical Agriculture*, Vincent, J. M., Whitney, A. S., and Bose, J., Eds., University of Hawaii, College of Tropical Agriculture, Miscellaneous Publication 145. HI, 1977, 293.
17. Date, R. A. and Halliday, J., Selecting *Rhizobium* for acid, infertile soils of the tropics, *Nature*, 277, 62, 1979.
18. Derylo, M. and Skorupska, A., Enhancement of symbiotic nitrogen fixation by vitamin-secreting fluorescent Pseudomonas, *Plant and Soil,* 154, 211, 1993.
19. Bergersen, F.J., *Methods for Evaluating Biological Nitrogen Fixation,* J. Wiley & Sons, New York, 1980.
20. Nutman, P. S., Alternative sources of nitrogen for crops, *Journal of the Royal Agricul. Soc. England*, 137, 86, 1976.
21. Sanginga, N., Ibewiro, B., Houngnandan, P., Vanlauwe, B., Okogun, J.A., Akobundu, I.O., and Versteeg, M., Evaluation of symbiotic properties and nitrogen contribution of mucuna to maize grown in the derived savanna of West Africa, *Plant and Soil,* 179, 119, 1996.
22. Whiteman, P.C., Seasonal changes in growth and nodulation of perennial tropical pasture legumes in the field. II. Effects of controlled defoliation levels on nodulation of *Desmodium intortum* and *Phaseolus atropurpureus, Aust. J. Agric. Res.*, 21, 207, 1970.
23. Othman, W.M.W., Asher, C. J., and Wilson, G. L., ^{14}C-labelled assimilate supply to root nodules and nitrogen fixation of phasey bean plants following defoliation and flower removal, in *Biotechnology of Nitrogen Fixation in the Tropics,* Shamsuddin, Z. H., Othman, W.M.W., Marziah, M., and Sundram, J., Eds., Universiti Pertanian Malaysia, Kuala Lumpur, 1988, 214.
24. Bolland, M. D. A. and Paynter, B. H., Increasing phosphorus concentration in seed of annual pasture legume species increases herbage and seed yield, *Plant and Soil*, 125, 197, 1990.

25. Ta, T. C. and Faris, M.A., Effects of environmental conditions on the fixation and transfer of nitrogen from alfalfa to associated timothy, *Plant and Soil*, 107, 25, 1988.
26. Rao, A.V. and Giller, K. E., Nitrogen fixation and its transfer from *Leucaena* to grass using ^{15}N, *For. Ecol. Manag.*, 61, 221, 1993.
27. Gil, J.L., Guenni, O., and Espinoza, Y., Biological N2-fixation by three tropical forage legumes and its transfer to *Brachiaria humidicola* in mixed swards, *Soil Biology Biochemistry,* 29 (5/6), 999, 1997.
28. Ezenwa, I., Aribsala, O. A., and Aken'ova, M. E., Dry matter yields of *Panicum* and *Brachiaria* with nitrogen fertilization or *Pueraria* in an oil palm plantation, *Tropical Grasslands,* 30, 414, 1996.
29. Turk, D. and Keyser, H. H., Rhizobia that nodulate tree legumes: specificity of the host for nodulation and effectiveness, *Canadian Journal of Microbiology,* 38 (6), 451, 1992.
30. Muok, B.O., Gudu, S.O., and Odee, D. W., A broad-range inoculant for legume trees in acid soils, *Agroforestry Today*, July–September, 12–13, 1998.
31. Okon, Y. and Labandera, C. A., Agronomic applications of *Azospirillum*. An evaluation of 20 years worldwide field inoculation, *Soil Biol. Biochem.*, 26, 1591, 1994.
32. Boddey, R. M. and Döbereiner, J., Nitrogen fixation associated with grasses and cereals: Recent progress and perspectives for the future, *Fertilizer Research*, 42, 241, 1995.
33. Sarig, S., Okon, Y., and Blum, A., Promotion of leaf area development and yield in *Sorghum bicolor* inoculated with *Azospirillum brasilense, Symbiosis,* 9, 235, 1990.
34. Kapulnik, Y., Okon, Y., Kigel, J., Nur, I., and Henis, Y., Effects of temperature, nitrogen-fertilization, and plant-age on nitrogen fixation by *Setaria italica* inoculated with *Azospirillum brasilense* (strain Cd), *Plant Physiology*, 68, 340, 1981.
35. Okon, Y. and Itzigsohn, R., The development of *Azospirillum* as a commercial inoculant for improving crop yields, *Biotech. Adv.*, 13, 415, 1995.
36. Boddey, R.M. and Victoria, R. L., Estimation of biological nitrogen fixation associated with *Brachiaria* and *Paspalum* grasses using ^{15}N labelled organic matter and fertilizer, *Plant and Soil*, 90, 265, 1986.
37. Miranda, C. H. B., Urquiaga, S., and Boddey, R.M., Selection of ecotypes of *Panicum maximum* for associated biological nitrogen fixation using the ^{15}N isotope dilution technique, *Soil Biol. Biochem.*, 22, 657, 1990.
38. Smith, R. L., Bouton, J. H., Schank, S.C., Quesenberry, K. H., Tyler, M. E., Milam, J. R., Gaskins, M. H., and Littell, R. C., Nitrogen fixation in grasses inoculated with *Spirillum lipoferum, Science*, 193, 1003, 1976.
39. Taylor, R. W., Response of two grasses to inoculation with *Azospirillum* spp. in a Bahamian soil, *Trop. Agric.*, 56, 361, 1979.
40. Collins, P. and Donawa, A. L., Nitrogenase activity in soils of Trinidad & Tobago, *Trop. Agric.*, 59, 54, 1982.
41. Hernandez, Y. and Sarmiento, M., Comportamiento de *Chloris gayana* y *Brachiaria brizanta* inoculadas con *Azospirillum, Rev. Cubana Cien. Agríc.*, 29, 115, 1995.
42. Bosworth, A. H., Williams, M. K., Albrecht, K. A., Kwiatkowski, R., Beynon, J., Hawkinson, T. R., Ronson, C. W., Cannon, F., Wacek, T. J., and Triplett, E. W., Alfalfa yield response to inoculation with recombinant strains of *Rhizobium meliloti* with an extra copy of *dct*ABD and/or modified *nif*A expression, *Appl. Environ. Microbiol.*, 60, 3815, 1994.
43. D'Halluin, K., Botterman, J., and De Greef, W., Engineering of herbicide-resistant alfalfa and evaluation under field conditions, *Crop Science*, 30, 866, 1990.
44. Hill, K. K., Jarvis-Eagan, N., Halk, E. L., Krahn, K. J., Liao, L. W., Mathewson, R. S., Merlo, D. J., Nelson, S. E., Rashka, K. E., and Loesch-Fries, L. S., The development of virus-resistant alfalfa, *Medicago sativa* L., *Biotechnology,* 9, 373, 1991.

16 Contribution of Rangelands to Animal Production in the Tropics

José San José

CONTENTS

I. INTRODUCTION

The tropical zone is covered extensively by grazing land ecosystems, where grasses and legume species play a predominant role in livestock production. They cover nearly twenty to thirty percent of the earth's surface[1-4] with trees being the dominant biological form in savannas and grasses in steppes. Savannas are most extensive in Africa, where they make up about 40% of the land area; equivalent less extensive communities are found along the equatorial belt in Australia, Ibero-America, the Pacific islands, and monsoonal and equatorial areas of Asia.[5-8]

In the grazing lands, the major areas dedicated to extensive beef production are the savannas of Ibero-America and Australia, where intensive pressure from recently introduced grazing animals has triggered changes in components and processes of the native ecosystems.[9,10] This chapter will

0-8493-2318-5/01/$0.00+$.50

deal mainly with the Orinoco savannas, Brazilian Cerrados, and Australian grazing lands in order to improve understanding of the biodynamic of animal-pasture-crop systems in these areas.

The well-drained Orinoco savannas cover extensive areas (0.37×10^{12} m^2)[10] of the Orinoco basin as a major unit of homogeneous relief in the northern South America geosyncline with a drained area of 1.1×10^{12} m^2 and a mean annual discharge of 36×10^3 m^3/s.[11] These savannas in the northern region are delimited by the foothills of the coastal Caribbean range (8° 56′ N; 67° 25′ W) and the Orinoco river (7° 46′ N; 64° 25′ W). The western region is located between the latitudes 3 and 6° N to the east of the Andean mountains and between longitudes 71° and 68° W at the western margin of the Orinoco river. The vegetation includes a wide range of floristic and physiognomic types, ranging from herbaceous to woodland communities. Savanna heterogeneity in the Orinoco basin is conditioned by moisture and hydrological features, which act as a major regional determinant, while nutrient levels and surface soil properties provide the subregional determinants.[10,12-17]

The Brazilian Cerrados cover 1.8×10^{12} m^2 and stretch from central Brazil to areas in the southern Mato Grosso, the state of Goias, the state of Tocantins, Mato Grosso do Sul, western Bahia, western Minas Gerais, and the region of Brasilia (Federal District).[18] They also include disjointed areas of Maranhao, northern Piaui, Rondonia, and the state of Sao Paulo. Also, the Cerrados include a wide range of physiognomic types from herbaceous to closed woody formations, with a plant diversity up to 400 vascular species per hectare.[19] The vegetational variation in the Cerrados has been associated with edaphic factors such as soil depth, water table, and soil fertility as well as fire, human disturbance, and hydrological features.[20-31]

The grazing lands of Australia are located in the zone between 11° S and 30° S, forming an arch from the region of Kimberley through the Northern Territory toward Queensland and in a southerly direction down the coast of New South Wales.[32] This land covers 0.45×10^{12} m^2,[33] the vegetational features and woody density of which have been associated with transient variability and gradients in rainfall, soil nutrients, and texture constraints as well as fire impact.[34-37] The savannas encompass a wide range of vegetation systems, which, in the northern region include: the monsoon tallgrass with *Themeda australis* as the dominant understory species, tropical tallgrass with *Heteropogon contortus*, midgrass on clay with *Dichanthium spp.*, tussock grassland with *Astrebla spp.*, and *Acacia* shrubland characterized by *Enneapogon spp.* and *Thyridolepis spp.*[38]

In this chapter, we will analyze the contribution of grazing land systems to annual production in the tropical regions, where traditional land use in grazing land is based on extensive beef production. However, decreasing economic return from surrounding areas as well as the low price of savanna lands have led to a high rate of human migration. Furthermore, in the particular case of the Orinoco savannas, the intensive petroleum activities in a zone covering more than 0.08×10^{12} m^2 have increased opportunities for human use and environmental impact. Therefore, strategies for savanna management as a function of a considered system's components (user motivation, technology, and environment) will be explored by keeping in mind that conventional practices are associated with declining productivity and that unconventional agriculture is not necessarily sustainable.[39] Attempts to ensure long-term productivity and stability in grazing land systems will be discussed.

II. ENVIRONMENTAL CHARACTERIZATION OF MAJOR GRAZING LAND ECOSYSTEMS IN THE TROPICS

A. Climatic Conditions

The Orinoco savannas have a seasonal climate (Figure 16.1) with mean annual precipitation ranging from 1000 to 2400 mm. The vegetative growing season spans April to November, when 95% of the annual precipitation occurs. The mean annual temperature varies from 24 to 27°C. The hottest

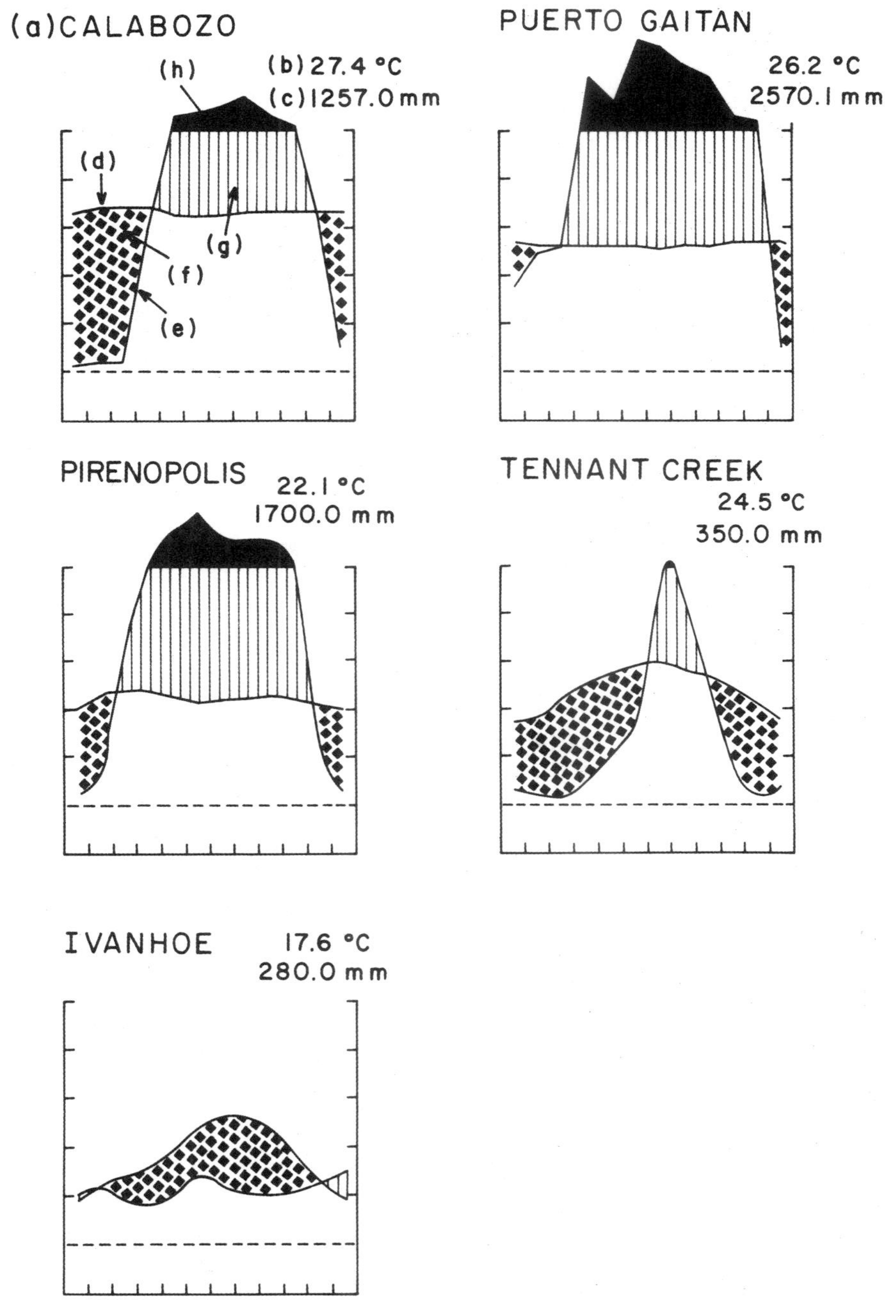

FIGURE 16.1 Climatic diagrams of stations in tropical grazing land ecosystems. Localities: (a) Orinoco Llanos (Calabozo, Venezuela; Puerto Gaitan, Colombia), (b) Cerrados (Pirenopolis, Brasil) and (c) Australian sites (Tennant Creek and Ivanhoe). The symbols and figures on the diagrams have the following meaning. (a) station, (b) mean annual temperature, (c) mean annual precipitation, (d) curve showing monthly mean temperature (scale interval = 10°C), (e) curve showing mean monthly precipitation (scale = 20 mm), (f) drought period (dotted), (g) humid period (vertical hatching), and (h) mean monthly precipitation exceeding 100 mm (scale reduce to 1/10) (black area) according to Walter 1971 (Walter, H., *Ecology of Tropical and Subtropical Vegetation,* Oliver and Boyd, Edinburgh, 1975, 359).

months are March and April and the coldest ones are December and January. The monthly difference between the maximum and minimum temperatures range from 10 to 22°C. The mean annual evaporation throughout the region ranges from 1519 to 2400 mm with maximum monthly evaporation occurring in March.

The Cerrados cover areas with a mean annual precipitation ranging from 800 to 1800 mm.[40] The wet season lasts from five to seven months with erratic dry periods ("veranicos") associated with high evaporation rates in Central Brazil.[41] Mean annual temperatures across the Cerrados range from 20 to 26°C (Figure 16.1). This relatively wide range of mean temperatures is associated with the extensive geographical distribution of the Cerrados, which stretch from 3° N latitude in Roraima State to 23° S in Sao Paulo State.

Australia is the driest continent in the world,[42] with grazing lands mainly distributed in regions having 6 to 8 months of dry seasons and frost periods. The mean annual rainfall throughout the Australian grazing lands ranges from 250 to 1500 mm, with 95% occurring from October to April. Annual and seasonal variability of the rainfall is very high.[43] Thus, comparative climatological studies have shown that annual rainfall variability in the Australian savannas is greater than in other savanna localities.[44] This variability has been associated with the southern oscillation phenomenon,[35] which is a see-saw of atmospheric pressure between the Indian and the South Pacific Oceans with a quasi-periodicity ranging from 2 to 10 years.[45] The mean annual temperatures of the savanna areas range from 20 to 24°C.

B. Edaphic Features

1. Soil Fertility

The soils of the Orinoco savannas and Cerrados present a wide range of morphological, physical, and chemical features.[10,46-50] In well-drained Orinoco savannas, Oxisols, Inceptisols, and Alfisols are present in 0.3×10^{12} m^2 of savannas.[51] In relatively smaller extensions, Ultisols and Entisols are associated with landscape features.[52-54] These soils (Figure 16.2) are acidic, with pH ranging from 4.2 to 4.6, base saturation from 3 to 42%, and available phosphorus from 0.3 to 0.7 meq/100 g. The low organic matter content reflects the high infertility of these soils.

The soils of the Cerrados have acidity characteristics (pH 4.4 to 5.2) similar to those in the Orinoco llanos, with a low cation exchange capacity and high capability for P fixation.[24,25,28,55-58] However, these soils appear to be more highly saturated with Al (0.1 to 1.4 meq/100 g) as compared to those in the Orinoco savannas (0.2 to 0.7 meq/100 g).[49] The high Al concentration in the Cerrados soils is considered a limiting factor for the growth of native [22,24,25,59] and cultivated species, such as cotton and soybeans.[60]

In Australia, the soils have higher contrasting features (Figure 16.2) as compared to those for the Latin American grazing lands. Thus, in northern Australia, there is a predominance of: (1) Oxisols and Alfisols with more fertile features than those in the southeast and subtropical quarters, (2) infertile Entisols and Inceptisols in the northeast region, (3) low-fertility sesquoxidic Oxisols in the northern tropics, and (4) fertile Vertisols in the subinland and inland drier regions.[38,61,62] In most Australian savannas, the soils are low in C, N, P, and S as well as in cation exchangeable basis; however, the Al concentration did not reach toxic levels. One-third of the soils of the northern Australian savannas are infertile and, when the massive earths are included, then the proportion can reach more than half of the soils in the region.[63] The extension of the low fertility soils in western Australia is such that 80 and 95% of the potential areas for pastures and crops, respectively, are located in the eastern part of the region.[35] Furthermore, areas having the highest fertility are frequently dispersed in patches[64] and soil nutrient availability for plant growth is limited in particular situations.[65]

In relation to nutrient dynamics, studies in the Orinoco and Australian savannas have indicated that nutrient accessibility for plant growth is low for all ions.[10,66] During the dry season in the Orinoco savannas, the herbaceous layer conserves 71% of the maximum N accumulated during the

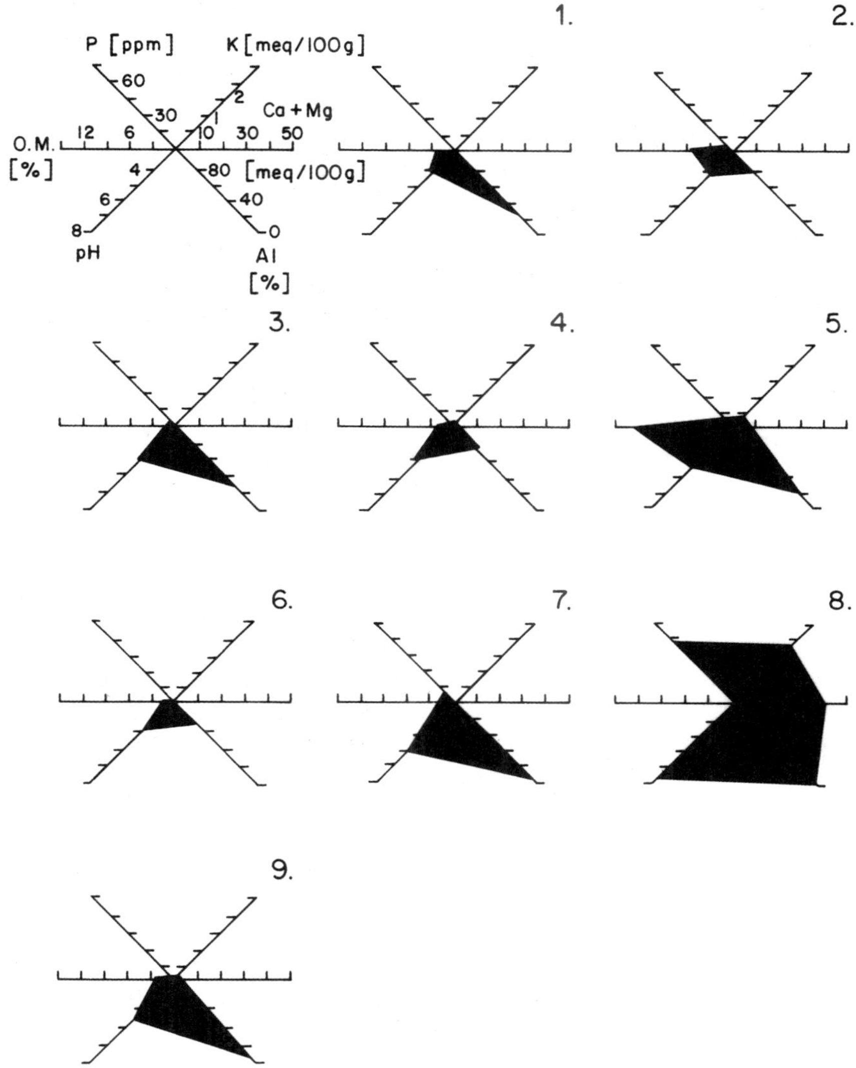

FIGURE 16.2 Polynomial graphs of soil fertility in tropical grazing land ecosystems. Localities: (a) Orinoco Llanos (1 = Calabozo/Oxisol, 2 = Carimagua/Oxisol/Haplustox, and 3 = Santa Maria, Vichada/Ultisol/Aquic Paleustult), (b) Cerrado site (4 = Brasilia/Oxisol/Haplustox) and (c) Australian sites (5 = Malanda/Oxisol/Krasmazen, 6 = McDonnell/Ultisol/yellow earth, 7 = Wenlock/Entisol/siliceus sand, 8 = Hughenden/Vertisol/gray clay, and 9 = Landsdown Sta. /Alfisol/solodic. The area of these graphs is directly proportional to the soil fertility after modifying the system proposed by Alvin and Rosand 1974 (Alvin, P. and Rosand, A., Um novo sistema de presentacao gráfica da fertilidade dos solos para cacau. Cacau Actualidades, 11, 1, 1, 1974). Data taken from references 10, 22, 24, 25, 28, 35, 38, 46–65.

previous wet season.[67] Thus, nutrient redistribution rather than a N loss occurs in the Orinoco savannas. However, N losses from these systems are balanced by the annual nitrogen input to the community from precipitation and biological fixation in legumes and soil crusts.

2. Soil Physical Characteristics

In the Orinoco savannas, soil textural analysis indicates that most soils correspond to the sandy-clay loam group. However, a continuous gradient from clay loam to loamy sand has been evidenced.[68] The effective soil mass, as expressed by the textural proportion of particles less than 2 mm in diameter, ranges from 6 to 100%. In the Australian savannas, the shallow stony and sesquioxidic soils are texturally characterized as having sandy to loamy materials; the cracking clay soils are finely textured with a difference between horizon A, from sandy loam to clay, and horizon B, from clay loam to heavy clays.[38]

From the hydrological point of view, the structure of the Oxisols and Entisols in the Orinoco llanos and Brazilian Cerrados is considered adequate for water movement in the soil profile, whereas water-holding capacity is very low. [10,60,69] Thus, reported data indicate that water availability in the soils of Orinoco savannas (4.9 to 9.1%) [70] and the Cerrados (3.3 to 10.0%)[69] is less than that reported for Alfisols and Vertisols of Australia (8.1 to 11.9%).[38,61] The difference could be attributed to the presence of oxides and hydroxides in the Orinoco and Cerrados soils, which can aggregate the kaolinite to form particles of the size of a grain of sand.

In summary, data on environmental features indicate that in the analyzed savannas, there exists a large spatial and temporal heterogeneity in resources. Therefore, management strategies based on elastic boundaries will be necessary for the sustainable use of these systems.

III. PRODUCTION PROCESSES IN MAJOR GRAZING LAND ECOSYSTEMS IN THE TROPICS

A. Primary Productivity

Long-term studies on the production capacity of the Orinoco savannas encompassing major physiognomic types indicate that maximum aboveground dry mass production of the herbaceous layer ranges from 122 to 688 g/m^2/year[10,68,70-73] and that belowground dry mass shows an accumulation between 142 and 427 g/m^2/year.[74,68] Dry mass decompositions of above and belowground phytomasses were included in these calculations.

The data on the savanna bioproductivity across the Orinoco llanos showed that the variability of the figures is greater than twofold.[68] This wide range is related to environmental gradients, which are related to major regional determinants such as moisture regime and hydrological features.[10]

In the Cerrados, the dry matter production has been evaluated in few localities as compared to extensive heterogeneity in physiognomic features.[30] The reported values for the aboveground production of herbaceous/shrub layers in the campo cerrados (*Pirassununga, S.P.*), campo sujo (Brasilia, D.F.), cerrados *seusu strictu*, (Brasilia, D.F.), and Cerrados *seusu latu* (Brotas — Itirapuaa, S.P.) range from 550 to 780, 97 to 550, 90/202 to 215/327, and 370 to 780 g/m^2/year, respectively.[75-82] In stands that were burned in July and January, the herbaceous/shrub layer is able to accumulate up to 550 and 680 g/m^2/year, respectively.

In the Australian savannas, the annual aboveground phytomass production ranges from 70 to 420 g/m^2/year with the lowest value measured in *Acacia* shrublands, which are dominated by *Enneapogon* and *Thyridolepis spp*[83,84] under 180 to 500 mm annual rainfall. The highest value was harvested in tropical herbaceous savannas, which are dominated by *Heteropogon contortus*, *Themeda australis,* and *Bothriochloa sp.* under 600 to 1400 mm annual rainfall.[85] The annual belowground dry mass accumulation ranges from 181 to 570 g/m^2/year, with the extreme values being measured in an *Acacia* shrubland[83,84] and a monsoon tallgrass savanna dominated by *Themeda australis, Sehima nervosum,* and *Chrysopogon fallax* under a 750 to 1100 mm annual precipitation.[38,86,87]

B. Limiting Ecological Factors on Dry Mass Production and Carrying Capacity of the Analyzed Systems

Calculations of animal-carrying capacity were based on a comsumption of 7.5 kg of dry mass/day/animal unit and an animal unit (A.U.) equal to 450 kg of liveweight.[88] Thus, in the Orinoco savannas, the dry mass accumulated during the wet season (122 to 688 g/m^2/year) could maintain a cattle density ranging from 0.9 to 5.0 A.U./ha. During the dry season, the burned herbaceous vegetation produces less than 100 g/m^2/year and, consequently, the vegetation would maintain 0.7 A.U./ha. In the Cerrados and Australian savannas, the carrying capacities during the wet season range from 0.6 to 5.7 A.U./ha and from 0.5 to 3.1 A.U./ha, respectively. These calculated figures for the analyzed grazing lands do not concur with current requirements for cattle-raising activities. Thus, in the Orinoco llanos, the average grazing area per A.U. is 17 hectares,[89] and particularly in well-managed savannas of the eastern Colombian llanos, from 3 to 5 hectares.[88] In the Cerrados, the actual carrying capacity ranges from 2 to 5 hectares per A.U.[90] For Australian savannas, the actual carrying capacity spans 4 hectares in the productive Queensland areas to 63 to 71 hectares in the Cape York Peninsula and the Northern Territory.[91] The mean annual liveweight gain in native vegetation of Northern Queensland is approximately 91 kg/ha/year, and in Katherine 108 kg/ha/year. McLennan et al.,[92] have reported an average annual liveweight gain of 102 kg per animal for a continuous grazing system in Queensland.

The contrast between calculated carrying capacities and actual values for the analyzed grazing lands appears to be associated partially with their intrinsic capacities for the production of dry mass. Causal factors for this detrimental situation are analyzed below. Climatic and edaphic factors interact to constrain savanna productivity. Seasonal rainfall distribution as well as variation within and among years is a common feature of grazing lands. Typical rainless periods occurring during the wet season of the Cerrados[41,93] and Orinoco llanos[10] can reduce the water supply for vegetative growth. Agroclimatic studies in Australia have proven valuable in determining the effects of large rainfall variability on changes in land use and carrying capacity of the grazing lands. Periods of above-average precipitation are adequate for cropping and sown pastures;[94] however, pastoralism seems to be the most profitable land use over prolonged periods of time because of the adaptation of this system to extreme environmental variability.

The effect of seasonal rainfall and heterogeneity in the Australian savannas is evident when mean animal weight gain is nullified because of unbalanced productivity between the dry and wet seasons.[95] Furthermore, less viable precipitation and low winter temperatures are limiting factors for plant growth in regions with better-quality soils such as Woodstock and Landsdown.[38]

Edaphic constraints are evident in the analyzed grazing lands. In the Orinoco llanos, seasonal net productivity depends on the effective soil mass and, less extensively, in the Cerrados. Thus, the phytomass accumulated by Orinoco savannas growing in deep soils is twice that harvested in shallow soils.[70] Furthermore, low water-holding capacity in the savanna soils leads to a rapid decrease in the amount of water storage at the beginning of the dry season, and consequently the vegetation is affected by water stress. This response is evident in the Orinoco savannas and Cerrados, when the production of dry mass is increased significantly in irrigated stands during the dry season.[96,82]

Dry matter production in Orinoco savannas, the Cerrados, and Australian savannas also depends to a certain extent on the level of soil nutrients. The experimental addition of nutrients (N, P, K) and lime to Orinoco savannas[97] and the Cerrados[82] produces a significant increase in plant phytomass and nutrient concentrations. In the case of the Orinoco savannas, the production increase was due to a higher photosynthetic efficiency and leaf-area development in the dominant grasses as compared to those in the nondominant species.

Nutrient-induced variations in the phytomass of native pastures is also a common threat to animal production in Australian savannas, where nutrient deficiency is a major factor limiting grass productivity on the sesquoxide soils and in tropical tallgrass regions. In this continent, monsoon tallgrass and inland semiarid communities are characterized by nutrient limitations associated with winter drought.[38] Experimental evidence shows that when N and P deficiencies were corrected at Katherine (Northern Territory), the production of native pastures increased up to the potential yield for the season.[98]

An additional factor limiting the carrying capacity of the analyzed grazing lands is the forage quality. In the most productive savannas of the Orinoco llanos, approximately 60% of the accumulated dry mass is wasted as plant mass that is unfit for animal development. A similar experience has been reported for Australian savannas,[99] where pastures are of low quality for animal production, except during the first month of the wet season. Thus, nutrient content in the phytomass rapidly decreases throughout the growing season, lowering the nutritional quality of the forage below that which would meet animal requirements. In the Orinoco savannas, the protein content and *in vitro* digestibility of cellulose in the grasses decrease throughout the wet season from 7.6 and 25.8%, respectively, in May, to 3.7 and 17.6%, respectively, in October. During the same period, the Ca and P contents decrease from 0.12 and 0.10% to 0.8 and 0.07%, respectively.[68,73]

In the tropical monsoon tallgrass savannas of Australia, N and P contents of the leaf phytomass decrease throughout the growing season from 2 and 0.15%, respectively, to less than 1 and 0.08%, respectively.[35]

During the dry season a similar nutrient content decrease occurs following burning of vegetation. Thus, as the season proceeds, the values for protein and digestibility following burning decline from 10.5 and 51.5% to 6.4 and 30.5%, respectively. Consequently, during the dry season, the forage availability and protein content become critical for animal production.

In summary, the limiting nutrient content and deficient amount of forage for animal production are conditioned by biological and environmental factors. Experimental evidence indicates that the grass layer of the Orinoco savannas has a low response capacity to the addition of nutrients and irrigation.[10]

IV. ANIMAL PRODUCTION AND AGRICULTURAL MANAGEMENT IN TROPICAL GRAZING LAND SYSTEMS

A. Traditional Management and System Constraints

To evaluate the current management of tropical grazing lands, we should have an estimate of the upper limit of plant productivity (Table 16.1). Loomis and Williams,[100] Loomis et al.,[101] and Loomis and Connor[102] calculated a theoretical maximum crop productivity of 70 to 100 g/m^2/day on the basis of plant absorption and utilization of solar radiation. However, under experimental research conditions, the maximum crop yield has been recorded at 54 g/m^2/day for a community of *Pennisetum typhoides* growing in Australia,[103] where constraints were reduced to a minimum and the cost of inputs was not taken into account for the analysis of results. However, this figure may be only a partial indicator of the grazing land potential for animal production; hence, the evaluation of land use should be based on integrated features encompassing components of the agricultural systems such as the economic feasibility of land use and the most viable technology.[104] In this sense, optimum yield (Table 16.1) instead of a maximum crop yield must be considered as the basis line. Optimum yield is the highest yield attainable in conditions where rational, technological, and environmental limitations are reduced significantly, and maximum profit is obtained in the short term by using high-input technology. By contrast, the highest levels of yield attained economically in traditional savanna systems is suboptimal because of their lower advantages as compared to agro-potential systems. Exceptions are found in northern Australia, where water and nutrient supplies are adequate for crop yield.[105]

In the traditional system for animal production, economic feasibility is based mainly on manual labor. This situation depends upon factors which are not related to potential land use. Thus, in extremely fragile and stressful environments, land use is limited by scarce resources. In contrast, in less severe environments, yield depends more upon adequate technology and the economic feasibility of land use. Harsh savanna environments with very shallow soils are linked to subsistence activities where economic and structural forces have, in a cyclical manner, contributed to permanent extensive cattle raising. Examples are the San Martin hilly savannas (eastern llanos of Colombia) and the northern savannas of the Venezuelan high central plains. In the Cerrados, there exist extensive areas with quartzitic sandy soils, where soil degradation following agricultural

TABLE 16.1
Input and Production Systems in Tropical Grazing Land Ecosystems

Community Capability for Dry Matter Production	Type of Utilization	Production System: Policy Options	Production System: Components: Technology	Production System: Components: Environment	Condition	Net Return
A. Potential bioproductivity (total ability of the community to accumulate dry matter = 71–100 g/m²/day, with respiration losses equal to 33% of total CO_2 uptake[(1)])					Theoretical conditions	
B. Maximum bioproductivity (= 54 g/m²/day for Pennisetum typhoides planted in Australia for grain and as forage crop[(2)])	Experimental set up				Research-style conditions/eliminating constraints	
C. Optimum bioproductivity	Supra	Comb.[(4)]	Comb.	High	Sustainable production	Maximum profit includ. social costs
		Conv.[(5)]	Conv.	High	Intensive production	Short-term profit
	Sub	Comb.[(6)]	Comb.	Low	Efficient production	Long-term profit
D. Primitive yield[(3)]		Low	Low	High	Temporal production	Marginal
		Labor	Low	Low	Limited production	Marginal
		Labor	Low	Low	Extensive grazing	Marginal

(1) Loomis and Williams[100]
(2) Cooper[103]
(3) Zadock[107]
(4) Comb. = Combines traditional conservation components with modern solutions/sustainable agriculture
(5) Conv. = Conventional agriculture/high input technology
(6) Low input technology

mechanization has resulted in considerable reduction in crop productivity. In northern Australia, the isolated, infertile soils associated with stressful climatic conditions have hindered the spread of agricultural development.[106] Consequently, the yield from the aforementioned systems is considered as "primitive,"[107] and is the result of low resource inputs. The contribution of this agricultural development to the economy is marginal.

In less severe environments, the uncontrolled operation of grazing lands has maintained extensive areas of the Orinoco llanos, the Cerrados, and arid and semiarid zones of western and northern areas of Australia in a subsistence condition. Goedert,[108] Tinker,[109] and Affin and Zinn[110] have analyzed the cause of this situation in the Brazilian Cerrados. There, extensive site-to-site variation with marked differences in resource availability and environmental constraints are factors to be considered. Regrettably, these factors have been scarcely considered. Consequently, economic efforts have failed when local technologies have been extrapolated to a wide range of conditions. Furthermore, imported technologies must be adapted to particular systems and local cultures.

B. Socioeconomic Factors Limiting Land Use

Orinoco savanna people have traditionally obtained their resources mainly from extensive grazing, and, to a lesser degree, from cropping and wood extraction. Pastoralism took place on the basis of sedentary herding societies, and nomadic pastoralism is customary for the flooded savanna people to cope with seasonal changes in forage and land availability. In these practices, the savanna society used labor as the major input for land management and they did not invest effort and knowledge in the development of technologies. This attitude has prevailed in the savanna functioning for a long time, and local populations have been able to survive using this low output system.

The labor as a major input in these systems is extremely inefficient for the management of savannas with relatively higher potential for development. As a consequence of the sparse investments in organization, capital, and knowledge, the traditional savanna society has failed to make appropriate use of the opportunities for land development and human wealth. One of the present drawbacks in savanna use is the lack of organization to take appropriate and timely advantage of the heterogeneous savanna landscape. The lack of planning to efficiently extract the resources and the uneven allocation of land rights and technological development are affecting savanna use. Furthermore, the patterns of resource distribution and property rights have not been adequately assembled for the sustainable use of the savannas. Thus, in Orinoco savannas, land property is concentrated in extensive lots owned by a few people as has been pointed out by Silva and Moreno[111] using the data from the population census. Furthermore, savannas are predominantly held in unregulated private tenure.[112] This situation is also common in Australian and African pastoral communities, where land is in the hands of a society segment.[113] Land access to the savanna people has been proposed by agrarian reforms, but its implementation has been limited due to inadequate financial support and resource distribution unrelated to socioeconomic driving forces. Population pressure on resources using traditional practices to obtain adequate output has had a countereffect on economic and social development. Holmes[112] has pointed out various factors contributing to failure of agrarian reform in Latin American countries. They are related to high land prices, inadequate forms of land distribution, ineffective administration, and political instability of reforms. As a consequence, sustainability of these systems over time has not been achieved. Furthermore, the actual allocation of land rights to the savanna people did not reinforce the maintenance of land tenure forms since they could be opportunistically transferred to other people. The original aims in practice have been frequently diverted from local interests and Orinoco savannas are still characterized by extreme inequality in land distribution and ownership.

Organization of labor and land is also modulated by governmental agencies through subsidies and other financial supports, but the maintenance of the productive capacity of savanna resources is very precarious because the relation between costs and prices is very distorted. The land tenure regime characterized by extensive, unregulated private estates maintains the monopoly of labor by the limited demand. Labor replacement for other inputs from the irrational use of subsidized loans has also increased adverse socioeconomic and environmental effects on savanna development.

Land use conflicts in the Orinoco savannas exist among pastoralists, agriculturists, and oil extractors. As a consequence, oil exploitation expansion has resulted in encroachment of savanna areas, removing vital land resources from pastoral and agricultural production. Even though oil companies pay compensation to savanna people for environmental damages, this activity has irreversible consequences for future land uses.

Savanna people are relatively marginalized in the formulation of trade and land management policies. Resource degradation in nearby areas, low land price, and the development opportunities due to oil industry activities are forcing people to migrate in the savannas. However, the new land use forms in marginal lands have decreased the potential use of savannas to support high production. These systems tend to lead to economically inefficient and unequal outcomes. To overcome these constraints, the elaboration of land use strategies requires a search for sustainable forms of development based on land capability, cultural features, legal framework on resources and land tenure, and policy designs to control the economic driving forces.

Similarly, attention has to be paid to international trading arrangements because, irrespective of material influences that drive savanna use, international policies affect the benefits received by people from products.[114] On the other hand, international trading agreements should overcome the tendency for resource degradation and consequently favor sustainable forms of savanna use.

C. Contrasting the Feasibility of High and Low Input Systems for the Management of Tropical Grazing Lands

Due to demographic pressure and the relative potential of particular grazing lands for expanding agricultural frontiers, high input technologies have encouraged the extension of desirable crops and forage toward native grazing lands. Thus, cattle grazing management has been intensified in less limiting environments and farming extended toward the boundaries of plant adaptation. Experimental findings support the theory that Oxisols and Ultisols in Latin America can be cultivated extensively for pastures and high-yielding crops.[115-117] In the 1980s and 1990s, new programs evolved for intensive use of land with the introduction of forage plants and grain crops. High yields have been obtained in savannas with crops such as pineapples (22 t/ha), sorghum (3.5 t/ha), and peanuts (2.5 t/ha).[118,119] More than twenty-five percent of the total grain production of Brazil is centered in the Cerrados region.[120] *Brachiaria* and *Andropogon* have been widely used forage grasses in Oxisols and Ultisols, mainly due to their high quality and productivity as forage for animal feeding, especially during the dry season.[121,122] In Carimagua (Colombia), the cultivated plains with *Brachiaria decumbens, B. humidicola,* and *Andropogon gayanus* are supplied annually with P, K, Mg, and S, but the required addition of N is not economically feasible.[123,124] In the high rainfall areas of Queensland, mixed pastures of *Panicum maximum, Chloris gayana,* and *B. decumbens* in association with *Stylosanthes guianensis* and *Macroptilium atropurpureum* have been managed intensively by using superphosphates at rates of 300 kg/ha every four years.[125] The aforementioned results were obtained by using intensive agricultural practices with possibly serious socioeconomic and environmental consequences. In general, the implementation of conventional agricultural methods on Oxisols and Ultisols implies high costs, which could be above the returns associated with beef production.[126]

High input technology has been responsible for a remarkable increase in agricultural productivity around the world, but socioeconomic factors and short-term depletion of natural resources (soil, water, and genetic diversity) are factors preventing the widespread use of the present technology for the development of grazing lands. Therefore, the adoption of technological strategies to reduce limitations in grazing systems should be based on calibrated demands from energy sources, especially in those systems having extremely infertile soils and scarce resources where inadequate use of fertilizers and other agricultural supplies produce erosion and threaten water purity. By contrast, adoption of low-input technology for the management of grazing land systems as proposed by Sánchez and Salinas[117,127] is one of the most desirable options for maintaining yield and reducing environmental disturbance. This low-input approach is based on attaining 60% of the maximum yield by means of: (1) introduction of cultivars that are tolerant of acidic soils and high Al and Mn concentrations, (2) minimization of costs, and (3) use of the favorable properties of acidic and infertile soils. Sánchez and Salinas[117,127] have proposed the aforementioned technology for the management of 24% (0.17×10^{12} m^2) of the Latin American savannas, which are characterized by Oxisols and Ultisols with less than 8% slopes and without physical limitations. For soils with slopes from 8 to 30% (0.19×10^{12} m^2), they have developed low-input management technologies utilizing legume-based pastures. A complementary strategy for land management was proposed by Sedek-Leon,[128] who considers agricultural systems less demanding of energy requirements.

Applying low-input technology for the management of Oxisols and Ultisols has been related mainly to reduction of soil chemical constraints rather than to integral edaphic characteristics involving both chemical and physical attributes. Thus, references state frequently that low soil nutrient content and high levels of Al and Mn are the major limiting factors for plant growth in the Latin American savannas.[28,129] Therefore, the development of low-input technology is based mainly on the management of soil mineral limitations. On the other hand, the role of physical limitations in the land use of tropical systems has been scarcely recognized.[10,130] Thus, ecological analysis of the Orinoco savannas and the Cerrados and Australian savannas which relied on climatic, edaphic, and biotic factors indicated that soil physical and chemical gradients account for the compositional variation across the regions.[10,30,37] In the Orinoco savannas, soil physical gradient involving bulk density and water availability accounts for more than 37% of the compositional variation across the basin. In the case of Cerrados soils, low water-holding capacity is a key factor for rain-feeding crops, as rainless periods shorter than a week may result in water stress conditions for cropping in 60% of Central Brazil.[41,56,131] A complementary factor is the effect of agricultural practices on soil physical conditions, which have a tendency to degrade after soils are compacted by heavy agricultural machinery and overgrazing.[129] Consequently, the soil-water balance is affected adversely and plant root systems hindered. For example, over 15 years of cultivation in Oxisols, Moura et al.[132] have found that the infiltration rate decreased from 830 to 120 mm/ha, and Ive et al.[133] reported overland flow above 37% in heavily utilized pastures. In summary, degradation of grazing land has led to desertification and reduced efficiency in conversion of agricultural inputs into meat, milk, and fiber.[134]

D. Exploring Feasibilities for Sustainable Development in Grazing Land Systems

Most advances in management have been obtained by an expansion of the agricultural frontier instead of being related to increases in productivity of native grasslands. This situation is related partially to the fact that the cost of grazing lands is relatively low compared to that for other agricultural inputs.[113]

Sustainable pasture production requires that the vegetation and soil resources be maintained despite user stress. The effects of reduction in animal production due to environmental fluctuations, grazing pressure, forage defoliation, and replacement by woody and unpalatable species have been analyzed during long-term experiments.[135-137] Deforestation, sheet and gully erosion, sedimentation of water sources, fertilizer and pesticide contamination, species replacement, and reduction in biodiversity are occurring at quite a fast pace in the grazing lands of the analyzed systems and elsewhere in the world.[3,138] In such areas, environmental degradation is becoming a social and economic issue. Therefore, the adoption of sustainable and equitable practices is an urgent need for the management of grazing lands, especially because of their low ability to withstand or regenerate from overuse or improper use.[138,139] In systems with high potential for agricultural productivity, sustainable agricultural practices are more productive and often more profitable than the conventional ones.[38,39] The economic feasibility of technologies for the management of grazing land systems and the sustainability of animal production are commensurate with the regenerative maintenance of the savannas.

In grazing lands, there are large-scale attempts to implement sustainable agricultural practices with the participation of local populations in the processes of planning, decision making, and implementation. Activity regarding sustainable practices should be directed toward: (1) increasing agricultural and labor productivity, (2) introducing ecological rationality, and (3) coordinating agricultural, environmental, and economic policies.[140] They should be concerned with balancing traditional uses and agroecological strategies for satisfying the demand for food. This approach takes into account efficiency in water and nutrient uses as well as socioeconomic driving forces, ecological limitations, and policy options.

Tergas et al.,[124] and CIAT[122] have proposed a strategy for the sustainable management of the Colombian high plains, which is based on raising cattle in native savannas using traditional burning practices, combined with technological innovations (i.e., improved species, protein banks, and association of grasses and legumes). This strategy is able to increase animal production by 30%, mainly due to forage availability during the dry season. Similar results have been achieved by Restrepo[141] by using native burned savannas that have been cultivated over 20 to 30% of their surfaces with *Brachiaria decumbens*. In Australia, the use of fire and the control of animal movement have led to sustainable yields and conservation.[138] Furthermore, in northern Australian savannas, system management for sustainability has been developed[138] based on rotational burning and nutrient supplementation for livestock. Results indicate that animal liveweight gain was more than twice as much as that obtained by commercial management. Therefore, the adoption of this system across extensive areas seems to be possible because of a high profit level and minimal changes involved as well as low investment and reduced operational costs. However, even though the burned savannas are apparently stable in terms of their capability for dry matter production and species regrowth under different carrying capacities,[142] the dry matter accumulated in the system has to be carefully managed in order to avoid carbon losses. The effect of using fire on dry matter production, forage quality, and patch grazing has been analyzed experimentally on the Orinoco llanos[71] and the Australian monsoonal tallgrass systems.[143] These results indicate that this tool should be evaluated in a comparative manner over prolonged periods of environmental fluctuations. Also, the extent to which resource-poor food enterprises can be changed into year-round and self-sufficient ones could be assesed in this way. Therefore, burning as a tool for obtaining a tree/grass balance[144-147] has to be evaluated, particularly in combination with agronomical practices.

One of the major challenges for the sustainable management of tropical grazing lands is to use land capability effectively all year long, especially in zones with strong seasonal climatic changes where it is necessary to supplement grazing material to avoid annual net losses of animal liveweight. Thus, overstocking usually results in low animal production in northeastern Australia, where a large difference occurs between wet and dry season carrying capacities.[99] Solutions such as haymaking and the use of forage crops have been adopted widely in regions where costs are relatively low. These practices reduce cattle mortality, but production is modest when the digestibility and overall quality of the forage supplement are low.[95] On the other hand, the use of leguminous trees and browse shrubs can provide high nutrient fodders and seeds for animal feeding during the drought periods of the grazing lands.[148-150] Thus, in the Brazilian Cerrados, 60% of the seasonal feed requirements for cattle grown are supplied with the aforementioned feedstuffs. Another approach to increase carrying capacity during the dry season is interplanting trees and shrubs with grasses. This cropping system is able to simultaneously provide animal shelter, and protect the soil from erosion and water losses. *Acacia* and *Desmodium* are among the most promising species in dry savannas, where pastures grow seasonally and nitrogen fixation via their nodules can improve soil fertility and physical properties.[149] Also, certain grasses and legumes can be used as food sources (e.g., Tamarind and carob). Leguminous trees with edible leaves and pods can be used as nurse plants as well as to support climbing crops. *Acacia albida* provides protein-rich pods at the end of the dry season.

Sorghum, when planted beneath, yields twice as much as that grown in African savannas.[149] Seedheads, leaves, and stems of a perennial grass such as *Thinopyrum intermediun* (intermediate wheatgrass) can all be utilized for animal feeding. This is also true for *Vigna aconctifolia* (moth bean) and other edible bean-producing legumes, whose foliage provides forage for livestock.[148,149] Persistent legumes could be a valid management option because they have much higher dry-season nutritive value than grasses. Furthermore, they furnish more than 2% nitrogen in the foliage of the Australian savannas at the end of the wet season. Research for adapting species to savanna conditions has been done in northern Australia. *Stylosanthes hamata* cv. Verano is an example of a Caribbean legume that has been adapted to the wet regions of Australian savannas.[95] However, legumes are present in a low proportion and rarely constitute more than five present of the understory phytomass.[38] Furthermore, the legume-grass association is difficult to maintain for long periods. A key factor is its upkeep in soil fertility management.[151] In northern Australia savannas, the proportion of legumes (*Stylosanthes humilis* and *Leucaena leucocephala*) is maintained by adjusting the stocking rate. Nevertheless, the amount of N fixed by the legumes and its availability for livestock must be evaluated, due to the high level of nitrogen volatilization under field conditions. [67,152]

The diversification of species and products could be an option for enabling savannas and agricultural fields to be used year round as agro-silvo-pastoral systems. Savanna and agricultural residues can be utilized as bioenergetic sources. A beneficial practice is the use of living mulch[148-150] for conserving soil moisture and reducing leaching and weeds, as well as extending the growing season. The result is enhanced yield using a cheaper solution. Maize planted through peanuts (*Arachis prostata* and *A. pintoi*) and rice grown with *Desmodium triflorum* are examples of living mulch.

Rational options for integrated savanna systems have been based on the selection and adaptation of "ideal-types" to acidic and infertile soils. These systems are geared to solve the local problems of food supply without environmental degradation.

A native grass species of the Orinoco savannas with a high forage potential is *Paspalum coryphaeum*.[153] Experimental data indicate that animal consumption of this species was significantly greater (7.7 kg/day) than that for pastures dominated by *Axonopus purpusii* (5.9 kg/day). Tropical legumes such as species of *Leucaena* and *Desmodium,* which are high-protein browse shrubs, can be planted for green manure. Another option for increasing farm production could be the ecological integration of crops in grazing lands. Thus, cassava (*Manihot esculenta*), a root crop highly tolerant of acidic and oligothropic soils, is an energy source for animal feeding during the dry season. Sweet cassava is able to produce up to 10 t/ha/year of tuberous roots in native savanna soils,[154] where other crops are not able to grow beyond their vegetative phases. This production potential of cassava is associated with its high rate of nutrient extraction in infertile soils.[154] Another adaptable crop is the cowpea (*Vigna unguiculata*), which is one of the most tolerant of acidic soils with high aluminum saturation.[115,155] However, the optional production of cowpea requires an application of 140 kg/ha of P.[156]

Data have been accumulated to emphasize the need for the development of sustainable strategies that take into account not only soil nutritional aspects, but also water supply. This situation is particularly critical in the utilization of arid and semiarid regions of Australia (<350 mm/year), where annual fluctuations in rainfall and temperature are climatic constraints, influencing a year-to-year variability in forage availability.[136] Under these conditions, cattle management is controlled by water supply in 75% of the territory. In northern Australia, rainfall variability affects management policies regarding the establishment of sown pastures, conventional land use, and vegetation resilience.[92,157] Consequently, strategies have been formulated to avoid pasture degradation by adjusting the stocking rate.[158] The effect of seasonal rainfall variation in tropical grazing land is also evident in the Cerrados and Orinoco llanos.[41] Thus, even though corn (*Zea mays*) is cultivated extensively during the wet season of the Orinoco llanos, rainfall distribution is inadequate to maintain cell turgor, especially during grain filling, thereby reducing crop yield.[159] Therefore, complementary water supply is apparently necessary for even the transient use of the grazing lands. However, irrigation of Oxisols and Ultisols is considered unprofitable for some crops and forages.[115] To overcome this problem, technological strategies for the management of adverse soil physical features have to be developed, especially under climatic conditions where the annual evaporation rate is above the amount of annual rainfall. However, there are many gaps in our understanding of the effects of soil physical features on the productivity of grazing lands. Consequently, research for comprehensive and integrated management of the plant/soil/atmosphere continuum is an urgent need in order to reach sustainable management. Furthermore, intensive knowledge is imperative to maintain grazing lands, since there are examples of savannas being withdrawn from production owing to soil degradation. This particular situation has been overcome by selecting suitable genotypes and agricultural practices relating to crop density and geometry. Weed control, as well as monitoring water requirements as a function of crop development,[160] is also necessary. Water-holding capacity and cation exchange of the soil have been improved by engineered moisture barriers such as asphalt layers and emulsions with strong acidic groups, which are low-cost inputs.[161] Polymers, for both aggregation and stabilization of soil conditions, are also beneficial.[162] Slags from mining operations and liming are plausible tools to improve soil physical conditions.[163] Currently, extensive areas in the Cerrados and specific localities in the Orinoco llanos are cultivated by using low-tillage and cropping systems. This is an adequate means for increasing soil water availability during the cropping cycle and the size of soil aggregates, as well as to control erosion.[127] Using low-tillage systems, soil aggregates can range from 9.5 to 5.6 mm. In soils managed with conventional agricultural systems, they vary from 5.6 to 0.5 mm. Furthermore, soil erosion during the growing season can be decreased from 13.6 to 4.3 kg/ha.

Long-term economical solutions have been proposed to improve soil water retention capacity and to avoid soil degradation problems in the management of grazing lands. These solutions are based on cultivation systems of perennial grasses and legumes where costs are held down by using low-resource planting systems. Thus, legumes and intercropping systems are employed as inexpensive practices, which provide food as well as ground covers. An example is corn and beans (e.g., brown tepary) as well as corn and the shrub *Desmodium discolor*; grasses and *Acacia spp.* also grow well in association. Other examples are several *Desmodium* and *Acacia spp.* that are suitable for pasture production in tropical and subtropical areas of variable rainfall.

Forecasting options and optimizing pasture conditions have also been explored by using simulation models based on resources, edaphic conditions, and climatic variability.[35,164]

In summary, strategies utilizing the sustainable management approach to help people working with infertile systems should be included in developmental programs to realize the potential of grazing lands. At the same time, these programs have to incorporate options from local populations, with special emphasis on the spatial and temporal heterogeneity of the grazing resources, and the urgent need for adapting economically plausible technologies. Insight into savanna management for sustainability has been attained through the interactions of different areas of scientific expertise, which, for example, have elucidated the basic functional relationship between the components of the grazing systems. As a consequence, decisions have to be made on both biological and socio-economic grounds. The need for more research in practical areas to increase resource capacity and to reinforce the efforts being made to develop sustainable strategies is imperative.

ACKNOWLEDGMENTS

This work was conducted as part of the "Savanna Bioproductivity MAB (UNESCO) Project" of the Instituto Venezolano de Investigaciones Científicas (IVIC), and partially sponsored by the National Research Council of Venezuela (CONICIT) and "Man and the Biosphere Programme" (MAB/UNESCO). We appreciate the expert technical assistance of R. Bracho, M.Sc. (IVIC), N. Nikonova, M.Sc. (IVIC), and Carmen Buendía, M.Sc. (IVIC).

REFERENCES

1. Lieth, H. F., Ed., *Patterns of Primary Productivity in the Biosphere*. Hutchinson Ross. Stroudberg. PA, 1978.
2. Pemadasa, P. S., Tropical grasslands of Sri Lanka and India, in Werner, P. A., Ed., *Australian Perspectives and Intercontinental Comparisons*. Blackwell Scientific Publications, London, 1991, 51.
3. Werner, P. A., Savanna ecology and management. *Special Edition of the Journal of Biogeography*. Blackwell Scientific, London, 1991, 221.
4. Scholes, R. J. and Hall, D., The carbon budget of tropical savannas, woodlands and grasslands, in Breymeyer, A. I., Hall, D., Melillo, J. M., and Agren, G. I., Eds., *Global Change: Effects on Coniferous Forest and Grasslands SCOPE*. John Wiley & Sons, New York, 1996, 69.
5. Whyte, R. O., Introduction, in *Tropical Grazing Lands*, Dr. Junk Publishers, The Hague, 1974, chap. 1.
6. Whittaker, R. N., *Communities and Ecosystems*. 2nd ed., MacMillan Publishing Co., New York, 1975, 385.
7. Koechlin, J. R. and Legris, P., Introduction, in *Tropical Grazing Land Ecosystems. A State of Knowledge Report*, UNESCO/UNPE/FAG. Natural Resources Research XVI, Paris, 1979, 17.
8. Scholes, R. S. and Walker, B. N., *An African Savanna, Synthesis of the Nylsvley Study*. Cambridge University Press, 1993.
9. Hadley, M., Comparative aspects of land use and resource management in savanna environments, in *Ecology and Management of the World's Savannas*, Tothill, J. C. and Mott, J. J., Eds., Australian Academy for Science, Canberra, 1985, chap. 13.
10. San José, J. J., Montes, R. and Mazorra, M., The nature of savanna heterogeneity in the Orinoco basin, *Global Ecology and Biogeography Letters*, 7, 441, 1998.
11. Meade, R. H., Nordin, C. F. J., Pérez-Hernández, D., Mejía, B., and Pérez Godoy, J. M., Sediment and water discharge in Río Orinoco, Venezuela and Colombia, in *Proceeding in the Second International Symposium in River Sedimentation, Water Resources and Electric Power Press*, Beijing, 1983.
12. Meyer, J., Notes on vegetation of the Venezuelan Llanos, *Journal of Ecology*, 21, 235, 1983.
13. Pittier, H., La mesa de Guanipa, in *Trabajos Escogidos, Ministerio de Agricultura y Cria*, Caracas, 1942, 197.
14. Beard, J., The savanna vegetation of northern tropical America, *Ecology*, 23, 194, 1953.
15. Beard, J., The classification of tropical American vegetation types, *Ecology*, 36, 89, 1957.
16. Cuatrecasas, J., Aspectos de la vegetación natural de Colombia, *Revista de la Academia Colombiana de Ciencias Exactas Física y Naturales*, 10, 221, 1958.
17. Blydestein, J., Tropical savanna vegetation of the Llanos of Colombia, *Ecology*, 48, 1, 1967.
18. Eiten, G., *Vegetacao do Cerrado, in Cerrado: Caracterizacao,* Ocupacao e Perspectivas, Pinto, M. N., Ed., Editora Universidade de Brasilia (UnB), Brasilia 1994, 9.
19. Silberbauer-Gottsberger, I. and Eiten, G., Fitossociologia de um hectare de Cerrado, *Brasil Florestal*, 54, 55, 1983.
20. Alvin, P. and Araújo, W., O solo como factor ecológico no desenvolvimento da vegetacao no Centro-Oeste do Brasil, *Turrialba*, 2(4), 153, 1952.
21. Alvin, P., Teoria sobre formacao dos campos Cerrados, *Revista Brasileira de Geografía*, 16, 4, 496, 1954.
22. Arens, K., O Cerrado como vegetacao oligotrofica, *Revista Brasileira da Faculdade de Ciencias e Letras Universidade de Sao Paulo*, 224, 15, 59, 1958.
23. Askew, G., Moffatt, D., Montgomery, R. and Searl, P., Soil and soil moisture as factors influencing the distribution of the vegetation formations of the Cerrado Roncador, Mato Grosso, in *III Symposio Sobre o Cerrado*, Blucker, E., Ed., Universidad de Sao Paulo, Sao Paulo, 1971, 150.
24. Goodland, R., Oligotrofismo e aluminio no Cerrado, in *III Symposio Sobre o Cerrado*, Blucker, E., Ed., Universidade de Sao Paulo, 1971, 44.
25. Lópes, A. and Cox, F., A survey of the fertility status of surface soils under Cerrado vegetation in Brazil, *Soil Science Society of America Journal*, 41, 742, 1977.
26. Ratter, M., Observations on forest of some mesotrophic soils in Central Brazil, *Revista Brasileira de Botanica*, 1, 47, 1978.
27. Araujo, G. and Haridasan, M., A comparison of the nutritional status of two forests on dystrophic and mesotrophic soils in the Cerrado region Brazil, *Communications in Soil Science and Plant Analysis*, 19, 1091, 1988.

28. Haridasan, M., Solos do Distrito Federal, in *Cerrado: Caracterizacao, Ocupacao e Perspectivas,* Pinto, M. N., Ed., Universidade de Brasilia, Brasilia, 1994, chap. 10.
29. Ratter, J., Transition between cerrado and forest vegetation in Brazil, in *Nature and Dynamic-Savanna Boundaries,* Furley, P. A., Proctor, J. and Ratter, J. A., Eds., Chapman & Hall, London, 1992, chap. 20.
30. Ratter, J. and Ribeiro, J., Biodiversity of the flora of the cerrado, in *VIII Simposio sobre o Cerrado, Anais, Proceedings,* R. Carvalho Pereira, and Bhering Nasser, L. C., Eds., EMBRAPA/CPAC, Brazil, 1996, 3.
31. Ribeiro, J. F. and Walter, B. M. T., Fitofisionomias do bioma cerrado, in *Cerrado: Ambiente e Flora,* Sano, S. M. and Almeida, S. P., Eds., EMBRAPA/CPAC, Brazil, 1998, chap. II.
32. Evans, J., Interpretación de los resultados de investigaciones sobre manejo de praderas tropicales, in *Producción de Pastos en Suelos Acidos de los Trópicos,* Tergas, L. and Sánchez, P., Eds., CIAT, Cali, Colombia, 1988, 291.
33. FAO, *The Ecological Management of Arid and Semi-Arid Range Lands in Africa and the Near East,* Rome, 1974, 1.
34. Scanlan, J. C. and Burrows, W. H., Woody overstory impact on herbaceous understory in *Eucaliptus spp.* community in Central Queensland, *Journal of Ecology,* 15, 191, 1990.
35. McKeon, G., Days, K., Howden, S., Note, J., Orr, D., Scattini, W., and Weston, E., Northern Australian savannas: management for pastoral production, in *Savanna Ecology Management,* Werner, P., Ed., Blackwell Scientific Publications, Oxford, 1991, 11.
36. Andersen, A. N. Fire ecology and management, in *Landscape and Vegetation Ecology of the Kakadu Region, Northern Australia,* Finlayson, C. M. and von Oertzen, I., Eds., Klimer Academic, Netherlands, 1996, 179.
37. Williams, R. J., Duff, G. A., Bowman, D. M., and Cook, G. D., Variation in the composition and structure of tropical savannas as a function of rainfall and soil texture along a large-scale climatic gradient in the Northern Territory, *Australian Journal of Biogeography,* 23, 747, 1996.
38. Mott, J. J., Williams, J., Andrew, M. H., and Gillison, A. N., Australian savanna ecosystems, in *Ecology and Management of the World's Savanna,* Tothill, J. C. and Mott, J. J., Eds., Australian Academy of Science, Australia, 1985, 56.
39. Reganold, R., Papendick, O. K., and Parr, F., Sustainable agriculture, *Scientific American,* 262, 6, 112, 1990.
40. Camargo, M., Alfonsi, R., Pinto, H., and Chiaran, J., Zoneamento da aptidao climática para culturas comerciais em áreas de Cerrado, in *IV Simpósio Sobre o Cerrado, Brasilia 1976. Cerrado: Base para Utilizacao Agropecuária,* São Paulo, Itatiaia, 1977, 89.
41. Cochrane, T., Porras, J., and Hernao, M. R., The relative tendency of the Cerrados to be affected by veranicos, in *VI Simposio Sobre o Cerrado. Savannas: Alimentos e Energia,* EMBRAPA-CPAC, Planaltina, 1988, 229.
42. Brown, J., Australia's surface water resources. Water Consultants Report No. 1, Dep. of Resources and Energy, Canberra, 1983, 1.
43. Nix, H. A., Climate of tropical savannas. Ecosystems of the world 13, in *Tropical Savannas,* Bourliere F., Ed., Elsevier, Amsterdam, 1983, 37.
44. McNahon, J., Hydrological characteristics of selected rivers of the World, *Technical Documents in Hydrology,* UNESCO, Paris, 1982, 1.
45. Allan, R. J., El Niño Southern Oscillation influences in the Australian region, *Progress in Physical Geography,* 12, 313, 1988.
46. Camargo, M. and Falesi, I., Soils of the Central plateau and transamazonian highway of Brazil, in Soil Management in Tropical America, Bornemisza, E. and Alvarado, A., Eds., CIAT, Cali, Colombia, 1975, 25.
47. Malagon, D., Characterization and Genesis of Selected Soils in the Eastern Plain of Colombia, Ph.D. Thesis, University of Nebraska, Lincoln Nebraska, 1973.
48. Malagon, D., *Modelos y caracterización de la génesis de suelos en dos ecosistemas tropicales (sabana y selva pluvial) de Colombia,* Venezuela, CIDIAT, 1977, 1.
49. San José, J. J., and García-Miragaya, J., Contenido de nutrientes en el suelo y en la fitomasa de comunidades de la sabana de trachypogon, Calabozo, Venezuela, *Boletín de la Sociedad Venezolana de Ciencias Naturales,* 136, 114, 1979.
50. San José, J. J. and Montes, R., Regional interpretation of environmental gradients which influence Trachypogon savannas in the Orinoco Llanos, *Vegetation,* 95, 25, 1991.

51. Cochrane, T., Sánchez, L., Azevedo, L., Porras, J., and Garver, C., *Land in Tropical America,* Vol 1, CIAT/EMBRAPA-CPAC, Planaltina-Brasilia, 1985, 1.
52. Comerma, J., La 7º aproximación y los suelos venezolanos, *Agronomía Tropical,* 21, 365, 1971.
53. IGAC (Instituto Geográfico Agustín Codazzi), *Estudio General de Suelos de los Municipios de El Cálvario, Guaranal, Acacias, Villavicencio, Restrepo y Cumaral (Departamento del Meta),* Publicaciones IGAC, Bogotá, 1980, 1.
54. IGAC (Instituto Geográfico Agustín Codazzi), *Estudio General de Suelos de la Comisaría del Vichada,* Publicaciones IGAC, Bogotá, 1983, 1.
55. Kerr, J. C. and Resende M., Recursos edáficos dos Cerrados: ocorrência e potencial, in *VIII Simposio sobre o Cerrado.* Pereira, R. C. and Nasser, L. C. B., Eds., EMBRAPA/CPAC, Planaltina, D. F., Brasil, 1996, 15.
56. Ranzani, G., Solos do Cerrado, in *Simposio Sobre o Cerrado,* Universidade de Sao Paulo, Sao Paulo, 1963, 51.
57. Ranzani, G., Solos do Cerrado no Brasil, in *III Simposio Sobre Cerrado,* Blücher, E., Ed., Universidade de Sao Paulo, Sao Paulo, 1971, 26.
58. Silva, J. E. and Resck, D. V. S., Materia organica do solo, in *Biologia dos Solos dos Cerrados,* Vargas, M. A. T. and Hungria, M., Eds., EMBRAPA, Planaltina, D. F. Brasil, 1997, 465.
59. Malavolta, E. and Kliemann, H., *Desordens nutricionais no Cerrado,* Piracicaba, POTAFOS, 136, 1985, 1.
60. Souza, D. M. and Ritchey, K. D., Adubacao para o establecimento de pastagens consorcidas nos solos do Cerrados, in *VI Simposio Sobre o Cerrado. Savannas: Alimento e Energia,* EMBRAPA-CPAC, Planaltina, Brasil, 1988, 15.
61. Sánchez, P. and Isbell, R., Comparación entre los suelos de los trópicos de América Latina y Australia, in *Producción de Pastos en Suelos Acidos de los Trópicos,* Tergas, L. and Sánchez, P., Eds., CIAT, Cali, Colombia, 1978, 29.
62. Isbell, R. F., Kimberley-Arnhem-Cape York (III) Division of Soils, CSIRO. *Soils: An Australian Viewpoint.* CSIRO/Academic Press, 1983, 189.
63. Tothill, J., Nix, H., Stanton, J., and Russel, M., Land use and productive potential of Australian savanna lands, in *Ecology and Management of the World's Savannas,* Tothill, J. and Mott, J., Eds., The Australian Academy of Science, Canberra, 1985, 125.
64. Nix, H. A., The environment of Terra Australia, in *Ecological Biogeography of Australia,* Keast, A., Ed., Dr. Junk Publishers, The Hague, 1981, 103.
65. Bishop, H., The response to nitrogen and phosphorus fertilizer to native pasture in the Balbiriniland, system in north-west Queensland, *Tropical Grassland,* 11, 257, 1977.
66. Hutton, J., Soluble ions in rainwater collected near Alice Spring and their relation to locally derived atmospheric dust, *Transactions of Royal Society of South Australia,* 107, 138, 1983.
67. San José, J., Montes R., and Nikonova-Crespo, N., Carbon dioxide and ammonia exchange in the *Trachypogon* savannas of the Orinoco Llanos. *Annals of Botany,* 68, 321, 1991.
68. San José, J. J. and Montes, R., An assessment of regional productivity: the Trachypogon savannas at the Orinoco Llanos, *Nature and Resources,* 25, 1, 25, 1989.
69. Cavedon, A. and Sommer, S., *Levantamento Semi-detalhado do Solos.* Fundacao Zoobotanica do Distrito Federal, Jardín Botánico de Brasilia, 1990, 1.
70. San José, J. J., Montes, R., García-Miragaya, J., and Orihuela, B., Bioproduction of Trachypogon savannas in a latitudinal cross-section of the Orinoco Llanos, *Acta Oecologica,* 6, 1, 25, 1985.
71. Blydestein, J., La sabana de Trachypogon del Alto Llano, *Boletín de la Sociedad Venezolana de Ciencias Naturales,* 23, 139, 1962.
72. San José, J. J. and Medina, E., Effect of fire in organic matter production and water balance in a tropical savanna, in *Tropical Ecological System,* Golley, F. and Medina, E., Eds., Springer-Verlag, New York, 1975, 251.
73. Paladines, O. and Leal, J., Manejo y producción de las praderas en los Llanos Orientales de Colombia, in *Producción de Pastos en Suelos Acidos de los Trópicos,* Tergas, L. and Sánchez, P., Eds., CIAT, Cali, Colombia, 1978, 331.
74. San José, J. J., Berrade, F., and Ramirez, J., Seasonal changes of growth, mortality and disappearance of belowground soil biomass in the Trachypogon savanna grass, *Acta Oecologica,* 17, 4, 347, 1982.

75. Souza, M. H. A. O., Alguns aspectos ecologicos do vegetacao na região perimetral do represa do Lobo (Brotas-Itirapina, S. P.) S. Paulo. Tese de Doutorado, Universidade de São Paulo, São Paulo, Brasil, 1977, 369.
76. Cavalcanti, L. H., Efeito das cinzas resultantes da queima sobre a produtividade do estrato herbáceo-subarbustivo do Cerrado de Emas, Tese de Doutorado, Universidade de Sao Paulo, Instituto de Biociencias, São Paulo, 1978.
77. Cesar, H., Efeito da quiema e do corte a vegetacao de um campo sujo na fazenda Agua Limpia, D. F., Dissertacao de Mestrado, Departamento de Biología Vegetal, Universidade de Brasilia, Brasilia, 1980.
78. Meirelles, L., Producao primária e suas relacoes com os fatores ambientais em pastagem artificial e campo sujo de Cerrado queimado e natural. Dissertacao de Mestrado, Dep. de Biología Vegetal, Universidade de Brasilia, Brasilia, 1981.
79. Coutinho, L., Devuono, Y., and Louza, J., Aspectos ecológicos do fogo no Cerrado. IV época da queimada e a productividade primaria liquida do estrato herbáceo subarbustivo, *Revista Brasileira de Botanica,* 5, 37, 1982.
80. Coutinho, L. M., Influencia del fuego en el Cerrado del Brazil, *Boletín de la Sociedad Venezolana de Ciencias Naturales,* 145, 61, 1988.
81. Batmanian, G., Efeito do fogo sobre a producao primária e a acumulacao de nutrientes do estrato rasteiro de um Cerrado. Dissertacao de Mestrado, Dep. Biología Vegetal, Universidade de Brasilia, Brasilia, 1983.
82. Villela, D., Resposta do estrato rasteiro de um Cerrado a irrigacao e á colagem. Dissertacao de Mestrado, Dep. de Biología Vegetal, Universidade de Brasilia, Brasilia, 1990.
83. Christie, E., Ecosystem processes in semi-arid grasslands. I. Primary production and water use of two communities possessing different photosynthetic pathways, *Australian Journal of Agricultural Research,* 29, 775, 1978.
84. Friedel, M., Studies of Central Australia semidesert rangelands. I. Range condition and the biomass dynamics of the herbage layer and litter, *Australian Journal of Botany,* 29, 219, 1981.
85. Holroyd, R., Allan, P., and O'Rourke, P., Effect of pasture type and supplementary feeding on the reproductive performance of cattle in the dry tropics of north Queensland, *Australian Journal of Experimental Agriculture and Animal Husbandry,* 17, 197, 1977.
86. Norman, M., Response of native pasture to nitrogen and phosphorus fertilities at Katherine, N. T., *Australian Journal of Experimental Agriculture and Animal Husbandry,* 2, 27, 1962.
87. Norman, M., Pattern of dry matter and nutrient content changes in native pastures at Katherine, N. T., *Australian Journal of Experimental Agriculture and Animal Husbandry,* 3, 119, 1963.
88. Mullenax, C., Adecuación y manejo de las sabanas naturales en la altillanura de los Llanos Orientales de Colombia, *Carta Agraria,* 278, 2, 1982.
89. Gonzáles-Jimenez, E., Escobar, A., and Parra, R., Productividad primaria, secundaria, sistemas de producción actuales y potenciales para las sabanas de Venezuela, *Boletín de la Sociedad Venezolana de Ciencias Naturales,* 139, 305, 1981.
90. Kornelius, E., Saueressig, M., and Goedert, W., Establecimiento y manejo de praderas en los Cerrados del Brasil, in *Producción de Pastos en Suelos Acidos de los Trópicos,* Tegas, L. and Sáchez, P., Eds., CIAT, Cali, Colombia, 1979, 159.
91. Skerman, P. and Roverpos, F., Types of utilization, in *Tropical Grazing Land Ecosystems. A State of Knowledge Report,* UNESCO/UNEP/FAG Natural Resources Research XVI, Paris, 1979, 366.
92. McLennan, S., Hendricksen, R. B., Beale, I., Winks, L., Miller, C., and Quirk, M., Nutritive value of native pastures in Queensland, in *Native Pastures in Queensland: The Resources and Their Management,* Burrows, W. H., Scanlan, J., and Rutherford, J., Eds., Queensland Dep. of Primary Industries Information Series, Queensland Goverment Printer, Brisbane, 1988, 125.
93. Oliveira, P. E., Fenologia e biologia reprodutiva das espécies de Cerrado, in *Cerrado: Ambiente e Flora,* Sano, S. M. and Almeida, S. P., Eds, EMBRAPA/CPAC, Brazil, 1998, 169.
94. Lloyd, P., *Perspectives in Productivity,* Queensland Dep. of Primary Industries, Brisbane, 1980.
95. Stubbs, W. and Arbuckle, J., These beef cattle had lucerne hay as winter feed supplement, *Queensland Agricultural Journal,* 88, 8, 449, 1962.
96. San José, J. J. and Medina, E., Organic matter production in the Trachypogon savanna at Calabozo, Venezuela, *Tropical Ecology,* 17, 2, 113, 1976.

97. García-Miragaya, J., San José, J. J., and Hernández, J., Effect of added nitrogen, phosphorus and potassium in the aboveground biomass production of Trachypogon savanna grasses, *Tropical Ecology,* 24, 1, 33, 1983.
98. McCown, R. and Williams, J., The water environment and implication for productivity, in *Australian Perspectives and Intercontinental Comparisons,* Werner, P., Ed., Blackwell Scientific Publication, London, 1991, 169.
99. Burrows, W., Carter, J., Scanlan, L., and Anderson, E., Management of savannas for livestock production in north-east Australia: contrast across the tree-grass continuum, in *Savanna Ecology and Management. Australian Perspectives and Intercontinental Comparisons,* Werner, P., Ed., Blackwell Scientific Publication, London, 1991, 159.
100. Loomis, R. and Williams, W., Maximum crop productivity: an estimate, *Crop Science,* 3, 67, 1963.
101. Loomis, R., Williams, W., and Hall, A. E., Agricultural productivity, *Annual Review of Plant Physiology,* 22, 431, 1971.
102. Loomis, R. S. and D. J. Connor, *Crop Ecology. Productivity and Management in Agricultural Systems,* Cambridge University Press, New York, 1992.
103. Cooper, J., Potential production and energy convection in temperate and tropical grasses, *Herb Abstr.,* 40, 1, 15, 1970.
104. Arias, L., Castillo, J., Gracía, R., Gómez, A., Mirelles, M., Rosello, M., and Salazar, L., *Metodología empleada por el FONAIAP para el estudio de los sistemas de producción agropecuaria,* Publicaciones de Investigaciones Generales, FONAIAP/CENIAP, Caracas-Venezuela, 1981.
105. McCown, R., Jones, R., and Hammer, G., Agriculture in Australia's seasonally-dry tropics and subtropics. Climatic and Soil Constraints, in *ACIAR Proceedings Eastern Africa ACIAR Consultation on Agricultural Research Nairobi, Kenya,* National Council for Science and Technology, Kenya and ACIAR, Canberra, 1984.
106. Davison, B., *The Northern Myth: Limits to Agricultural and Pastoral Development in Tropical Australia,* University Press, Melbourne, 1972.
107. Zadoks, J., Yields, losses and cost of crop production. Three views with reference to wheat growing in The Netherlands, in *Crop Loss Assessment,* University of Minnesota, St. Paul, Minnesota, 1980.
108. Goedert, W., Management of the Cerrado soils of Brazil, *Journal of Soil Science,* 34, 405, 1983.
109. Tinker, P., Low input agriculture — has it any logical justification?, in *VI Simposium Sobre o Cerrado.* Savana: Alimento e Energia, EMBRAPA-CPAC, Planaltina, Brasil, 1988, 183.
110. Affin, O. A. D. and Zinn, Y. L., Sustentabilidade dos sistemas nos cerrados, in *VIII Simpósio sobre o Cerrado,* Pereira, R. C. and Nasser, L. C. B., Eds., EMBRAPA-CEPAC, Planatltina, Brazil, 1996, 28.
111. Silva, J. F. and Moreno, A., Land use in Venezuela, in *The World's Savanna,* Young, M. and Solbrig, O., Eds., Man and the Biosphere Series 12, UNESCO and The Parthenon Publishing Group, París-New York, 1993, 239.
112. Holmes, J. H., Land tenure in private and mixed-property regimes in *The World's Savanna,* Young, M. and Solbrig, O., Eds., Man and the Biosphere Series 12, UNESCO and The Parthenon Publishing Group, París-New York, 1993, 67.
113. Scoones, I., Toulmin, C., and Lane, C., Land tenure for pastoral communities, in *The World's Savannas,* Young, M. D. and Solbrig, O. T., Eds., Man and the Biosphere Series, Vol. 12 published by UNESCO, Paris and The Parthenon Publishing Group, U.K., 1993, 49.
114. Young, M. and Solbrig, O., Providing an environmentally sustainable economically profitable and sociably equitable future for the world's savannas, in *The World's Savanna,* Young, M. and Solbrig, O., Eds., Man and the Biosphere Series 12, UNESCO and The Parthenon Publishing Group, Paris-New York, 1993, 321.
115. Sánchez, P., Advances in the management of oxisols and ultisols in tropical South America, in *Proceedings of the International Seminar on Soil Environment and Fertility Management in Intensive Agriculture,* Society of the Science Soil and Manure, Tokyo-Japon, 1977, 535.
116. Vicente-Chandler, J., Abruña, F., Caro-Costas, R., Figuerela, J., Silva, J., and Pearson, R., Intensive grassland management in the humid tropics of Puerto Rico, *University of Puerto Rico Agricultural Experimental Station Bulletin,* 233, 1, 1974.
117. Sánchez, P. and Salinas, J., Low input technology for managing oxisols and ultisols in tropical America, *Advances in Agronomy,* 34, 279, 1981.

118. Sánchez, C., *La fertilización como práctica fundamental en un sistema integral de manejo de los suelos de los Llanos Orientales de Monagas,* Universidad de Oriente, Jusepin, Venezuela, 1975.
119. Michelena, V. A., Transferencia de masa y energía en una comunidad con metabolismo CAM: *Ananas comosus* L. Merr. cultivada en los Llanos Orientales de Venezuela, Trabajo de Grado, Ph. Sc. en Biología, Mención Ecología, CEA-IVIC, Caracas, 1992.
120. Goedert, W., Prefacio, in *VI Simposio Sobre o Cerrado. Savanas: Alimento e Energia,* EMBRAPA-CPAC, Planaltina, Brasil, 1988, 7.
121. Sotomayor-Rios, A., Rodríguez-García, J., and Velez-Santiago, J., Effect of three harvest intervals on the yield and protein content of ten *Brachiaria, Journal of Agriculture of University of Puerto Rico,* 65, 2, 147, 1981.
122. CIAT (Centro Internacional de Agricultura Tropical), Informe Anual del Programa de Pastos Tropicales, CIAT, Cali, Colombia, 1992.
123. Tergas, L., Paladines, O., and Kleinheisterkamp, I., Productividad anual y manejo de pasturas de *Brachiaria decumbens* Staff en los Lanos Orientales de Colombia, *Producción Animal Tropical,* 7, 260, 1982.
124. Tergas, L., Producción animal potencial y manejo de praderas naturales y cultivadas en los Llanos Orientales de Colombia, in *La Capacidad Bioproductiva de las Sabanas,* San José, J. J. and Montes, R., Eds., Publicaciones CIET/UNESCO, Caracas, 1987, 350.
125. Teitzel, J., Formulación de programas de fertilización para la costa húmeda tropical de Australia, in *Producción de Pastos en Suelos Acidos de los Trópicos,* Tergas, L. and Sánchez, P., Eds., CIAT Cali, Colombia, 1979, 103.
126. Espinoza, W. and de Oliveira, A. J., Eficiencia técnica de uso de fertilizantes nos trópicos do Brasil, in *VI Simposio Sobre o Cerrado. Savanas: Alimento e Energia,* EMBRAPA-CPAC, Planaltina, Brasil, 1988, 131.
127. Sánchez, O. and Salinas, J., Suelos Acidos. *Estrategias para su Manejo con Bajos Insumos en América Tropical,* Sociedad Colombiana de Suelos, Bogotá, 1983, 1.
128. Sedek-Leon, J., *La Agricultura como Problema Energético. Elementos para el Análisis de una Nueva Agricultura,* Editorial Metropolis C. A., Venezuela, 1983.
129. Ritchey, K., Silva J., and Costa, V., Calcium deficiency in clay B horizons of savanna oxisols, *Soil Science,* 133, 6, 378, 1982.
130. Lal, R., Modification of soil fertility characteristics by management of soil physical properties, in *Soil Physical Properties and Crop Production in the Tropics,* Lal, R. and Greenland, D., Eds., John Wiley & Son, New York, 1979, 397.
131. Lopes, A., Available water, phosphorus fixation and zinc levels in Brazilian Cerrado soils in relation to their physical, chemical and mineralogical properties, Ph.D. Thesis, North Carolina State University, Raleigh, 1977.
132. Moura, W., Boul, S., and Kamprath, E., Studies of a latosol roso (Entrustox) in Brazil: phosphorus reactions, *Experientiae*, *Brazil*, 13, 235, 1972.
133. Ive, J., Rose, C., Wall, B., and Torsell, B., Estimation and simulation of sheet run-off, *Australian Journal of Soil Research,* 14, 129, 1976.
134. Pearson, C. and Ison, R., *Agronomy of Grassland Systems,* Cambridge University Press, London, 1987.
135. Scattini, W., A Model for Beef Cattle Production from Rangeland and Sown Pasture in South-eastern Queensland Australia. Ph.D. Thesis, University of California, Berkeley, 1973.
136. McKeon, G., Howden, S., Silburn, D., Carter, J., Clewett, J., Hannmer, G., Johnston, P., Lloyd, P., Mott, J., Walker, B., Weston, E., and Willcocks., J., The effect of climate change on crop and pastoral production in Queensland, in *Greenhouse: Planning for Climate Changes,* Pearman, G., Ed., CSIRO, Australia, 1988, 546.
137. San José, J. J. and Fariñas, M., Temporal changes in the structure of a Trachypogon savanna protected for 25 years, *Acta Oecológica,* 129, 2, 237, 1991.
138. Winter, W., Australia's northern savannas: a time for change in management philosophy, in *Savanna Ecology and Management. Australian Perspectives and Intercontinental Comparison,* Blackwell Scientific Publications, London, 1991, 181.
139. Haridasan, M., Impactos de procesos ecologicos, in *Alternativas de Desenvolvimento do Cerrados,* Dias, B. F. S., Ed., FUNATURA, Brasilia, D. F. 1992, 27.

140. Altieri, M., Sustainable agriculture development in Latin America: exploring the possibilities, *Agriculture Ecosystems and Environments,* 39, 1, 1992.
141. Restrepo, J., Alternativas tecnológicas para el uso sostenible de las sabanas colombianas, in *Taller sobre Desarrollo de las Sabanas,* PROCITROPICOS/FONAIAP/OEA, Boa Vista, Brazil, 1999.
142. CIAT (Centro Internacional de Agricultura Tropical), Tropical Pastures Program, Annual Report 1983, Cali, Colombia, 1984.
143. Hodgkinson, K., Harrington, G., Griffin, G., Noble, J., and Young, M., Management of Vegetation with Fire, in *Management of Australia's Rangelands,* Harrington, G., Wilson, A., and Young, M., Eds., CSIRO, Melbourne, 1984, 141.
144. Mott, J., Patch grazing and degradation in native pastures of the tropical savannas in northern Australia, in *Grazing-Land Research at the Plant Animal Interface,* Horn, F., Hodgson, J., Mott, J., and Brougham, R., Eds., Winrock International, Morrilton, Arkansas, 1987, 153.
145. San José, J. J. and Montes, R., Fire effect on the coexistence of trees and grasses in savannas and the resulting outcome on organic matter budget, INTERCIENCIA, 22, 6, 289, 1997.
146. Scholes, R. J. and Archer, S. R., Tree-grass interactions in savannas, *Annual Review of Ecology and Systematics,* 28, 517, 1997.
147. Jackson, J. and Ash, J., Tree-grass relationship in open eucalypt woodlands of northeastern Australia: influence of trees on pasture productivity, forage quality and species distribution, *Agroforestry Systems,* 40, 159, 1998.
148. NAS, *Underexploited Tropical Plants with Promising Economic Value.* National Academy of Sciences, Washington, D.C., 1975, 187.
149. NAS, *Tropical Legumes: Resources for the Future.* National Academy of Sciences, Washington, D.C., 1979, 331.
150. Hernández Bermejo, J. E. and Leon, J., Cultivos Marginados, Colección FAO Producción y Protección Vegetal, No. 26, FAO, Roma, 1992, 338.
151. Spain, J., Pasture establishment and management on the Llanos Orientales of Colombia, in *Pasture Production in Acid Soils of the Tropics,* Sánchez, P. and Tergas, L., Eds., CIAT, Cali, Colombia, 1979, 15.
152. Vallis, I., Peake, D., Jones, R., and McCown, M., The cycling of nitrogen from cattle urine applied during the dry season, in *CSIRO Division of Tropical Crops and Pastures, Annual Report,* CSIRO, Melbourne, 1982.
153. Viera, J., Mejoramiento genético del *Paspalum coryphaeum*, Trim. *Boletín de la Sociedad Venezolana de Ciencias Naturales,* 142, 151, 1984.
154. San José, J. J. and Mayobre, F., Quantitative growth relationship of cassava (*Manihot esculenta* Crantz): crop development in a savanna wet season, *Annals of Botany,* 50, 309, 1982.
155. Malave, V., Análisis comparativo del crecimiento de tres cultivares de frijol (*Vigna unguiculata* L. Walp) en la sabana del Estado Monagas, Venezuela en el período de lluvias de norte, 1991–1992, Trabajo de Grado. MSc. en Agricultura Tropical, Universidad de Oriente, Maturin, 1995, 110.
156. López-Hernández, D., Coronel, I., and San José, J. J., Ensayos de fertilización controlada mediante el uso de la isoterma de absorción de fósforo en suelos de Calabozo (Edo. Guarico), *Boletín de la Sociedad Venezolana de Ciencias Naturales,* 142, 205, 1984.
157. Eyles, A. and Cameron, D., The contribution of science to Australian tropical agriculture. 3. Tropical pasture research, *Journal of the Australian Institute of Agricultural Science,* 51, 17, 1985.
158. Orr, D., Factors affecting the vegetation dynamics of *Astrebla* grassland. Ph.D. Thesis, University of Queensland, Brisbane, 1986.
159. Smith, A. and San José, J. J., Productividad del maíz (*Zea mayz* L.) en las condiciones climáticas de los Llanos Altos Centrales de Venezuela. I. Crecimiento de los híbridos Obregon y FM. 6 sembrados en la temporada de lluvias, *Agronomia Tropical,* 29, 2, 127, 1984.
160. Leslie, J., Soil crop husbandry: recent trends in production research, in *ACIAR Proceedings Eastern Africa-ACIAR Consultation on Agricultural Research, Nairobi, Kenya,* National Council for Science and Technology Kenya and ACIAR: Canberra, 1984, 97.
161. Erickson, A., Improving the water properties and sand soil, in *Optimizing the Soil Physical Environment Toward Greater Crop Yield,* Hillel, D., Ed., Academic Press, New York, 1971, 35.
162. De Boodt, M., Improvement of soil structure by chemical means, in *Optimizing the Soil Physical Environment Toward Greater Crop Yield,* Hilled, D., Ed., Academic Press, New York, 1971, 43.

163. Issa, O., Sobre el control de acidez en tres suelos minerales venezolanos: el mecanismo de acción sobre el aluminio cambiable y el fósforo en solución, Trabajo de Ascenso, Universidad Simón Bolívar, Caracas, 1993, 113.
164. Ludwing, J., Clewett, J., and Foran, B., Meeting the needs of decision support system users, in *Decision Support System for the Management of Grazing Lands,* Vol 11, Stuth, J. and Lyons, B., Eds., Man and the Biosphere Series, UNESCO, Paris, 1993, 209.

Section IV

Global Review of the Latest Research on Tropical Forages

17 Current Developments from Tropical Forage Research in Australia

Raymond J. Jones

CONTENTS

I. INTRODUCTION

Australia's tropical and subtropical areas extend from 10° S to about 30° S. As a developed country, Australia is unique in having such a large area (~50% or >4M km^2) within the tropics. It is perhaps

0-8493-2318-5/01/$0.00+$.50

not surprising then that Australia has taken a major role in the investigation and development of tropical pasture species to improve the productivity of its grazing lands.

In this chapter, the characteristics of the widely diverse tropical and subtropical environments within Australia are briefly described (Section II). This sets the scene for the developments that have occurred in the identification of suitable pasture species, their evaluation, and their subsequent release for use by farmers and graziers (Section III). Section IV discusses the development of legume and grass cultivars and their combination in mixtures for grazing purposes and defines the objectives for improvement in new cultivars. In Section V, additional uses of the grasses and legumes are considered, and the final section (VI) considers the use of new biotechnology approaches to the production of new cultivars.

II. DESCRIPTION OF TROPICAL AND SUBTROPICAL AUSTRALIA

A. Geographic Description

Australia, with a land area of 8.5 M km^2 is the smallest continent, but the second largest island in the world.[1] It is characterized by its flatness, with a low mean altitude of 330 m and with only 2% of its land surface above 1000 m elevation compared with values of 780 m and 27% for North America. It is essentially a land of low plateau, and is the only continent without present volcanic activity, though the presence of eroded volcanic plugs attest to past activity. It is also characterized by its dryness, being the driest inhabited continent in the world, with 65% of its land area classed as desert.[1] Of the total land area, 37% lies north of the Tropic of Capricorn. If the subtropics are included (to 30° S), then the combined tropics and subtropics occupy about 50% of the land mass.

1. Landforms

Like Australia as a whole, tropical Australia is dominated by lowlands. The arid plains of central Australia rise gently to the east toward the commonly called Great Dividing Range, which is not really a range, but rather elevated flat country that ends abruptly in the Great Escarpment. This runs south from northeast Queensland to Victoria at a distance from the coast of from 10 km to 150 km resulting in a generally narrow coastal plain with relatively short seaward-flowing streams. Viewed from the coast, the escarpment sometimes gives the impression of a mountain range running from north to south with highest elevation in the northern sector near Innisfail and Cairns and in the southern tropical sector in northeastern New South Wales.

Less than one third of the continent is drained directly seaward from a relatively narrow subcoastal belt that supports most of the perennial rivers. The remaining drainage system is complex.

Many of the western and southern-flowing streams are ephemeral and flow into Lake Eyre (7 m below sea level), which only fills once or twice each century. The Murray Darling system has a length of 5250 km. Starting in humid southern Queensland and flowing inland, it discharges into the sea in southern Australia after passing through a huge tract of arid and semiarid country. In the northwest, the Hamersley and Kimberley Ranges and the Arnhem Land Plateau are associated with rivers, Fitzroy, Ord, and Daly, which occupy the major structural basins and discharge into the Timor Sea.

2. Soils

The soils of northern Australia are generally old and highly weathered, but are not very acid. In the east, the soil pattern is complex due to a number of factors involving parent material, past history, and variation in rainfall.[2] The summary of the main soils given here is based on Hubble[3] and Isbell.[2]

Shallow stony soils occur mainly in elevated areas on steep and dissected landscapes in north Queensland and in the northern portion of the Northern Territory and the Kimberley region in the northeastern portion of Western Australia. These areas are generally not suitable for the development of improved pastures.

Cracking clay soils occur widely in subhumid and semiarid western Queensland and in the northern part of the Northern Territory. The shrinking and swelling characteristics of the soils with high clay content often result in an uneven microrelief known as gilgai, which is common in subhumid eastern Queensland. Some of these soils, originally supporting leguminous forests (*Acacia harpophylla*), are deep and fertile with high levels of both N and P. These are now used for cropping and for sown grass pastures.

Sesquioxidic soils occur widely in southern, central, and northern Queensland and are scattered throughout the Northern Territory, but are not well represented in the tropical areas of western Australia. The red and yellow earths in this soil group are generally deep with sandy to loamy surface textures and a gradual increase in clay content with depth. Although they have generally good physical features, they have low organic matter and N levels and extremely low levels of soil P.

Sodic soils with marked texture-contrast (sand/loamy surface soils and heavy clay subsoils), are represented in eastern coastal and subcoastal areas by the widespread solodic and solodized-solonetz soils. These have adverse physical features resulting in waterlogging in the wet season through poor penetration of water to the subsoils and low availability of water in the dry season. The soils have a low nutrient status and are often high in exchangeable sodium levels at depth (6 to 30%). Other sodic soils occur in desert areas. Those in southwest Queensland have a surface stone mantle forming a desert pavement and are unsuited climatically as well as physically for improved pastures.

Other soils of importance include the podzolic soils in the more humid areas of eastern Queensland and parts of the Northern Territory. These are of low to moderate nutrient status.

Krasnozems and xanthozems are restricted to small areas of the high rainfall zone of eastern Queensland. These are deep, well structured, highly acidic soils with high P fixing capacity. They may also be low in P, K, and molybdenum.

Soils of higher nutrient content include the non-calcic brown soils and the euchrozems, which occur in eastern Queensland, in drier areas of the Northern Territory, and in the Kimberleys of western Australia. These soils have good physical features.

Alluvial soils are widespread and of importance locally, though they do not occupy large areas. Their generally good fertility, water-holding capacity, and structure enable them to be used for cropping or for development of improved pastures.

3. Climate

Over the wide latitudinal range from the tip of Cape York (10° 40′ S) to Grafton in New South Wales (29° 40′ S), a range of tropical/subtropical climates is experienced.[4] Mean monthly maximum temperatures attain or exceed 30°C during the summer months over most of the zone, with temperatures highest in the north and the interior. In winter, low night temperatures result in large diurnal variations of about 18°C in the northern areas and about 11°C to 15°C south of the tropic where winter mean monthly minimum temperatures fall below 7°C over most of the subtropics.[5] In Figure 17.1, the 500 and 1000 mm mean annual isohyets are shown, and in Figure 17.2 climatic data for the ten centers shown on the map. These range from the northwest to the southeast, and illustrate the temperature and monthly rainfall distribution over the zone.

Solar shortwave radiation rarely falls below 300 cal cm^2/day in winter and can exceed 600 cal cm^2/day in summer in inland northern areas. Here, high values result in an annual pan evaporation range from 1300 mm to 3000 mm (Figure 17.2). Rarely is radiation a major limiting factor to pasture growth.

Radiation frosts occur in most subtropical locations, and they can occur north of the tropic in inland areas. They are particularly heavy in southern inland areas resulting in loss of pasture quality and death of some species.

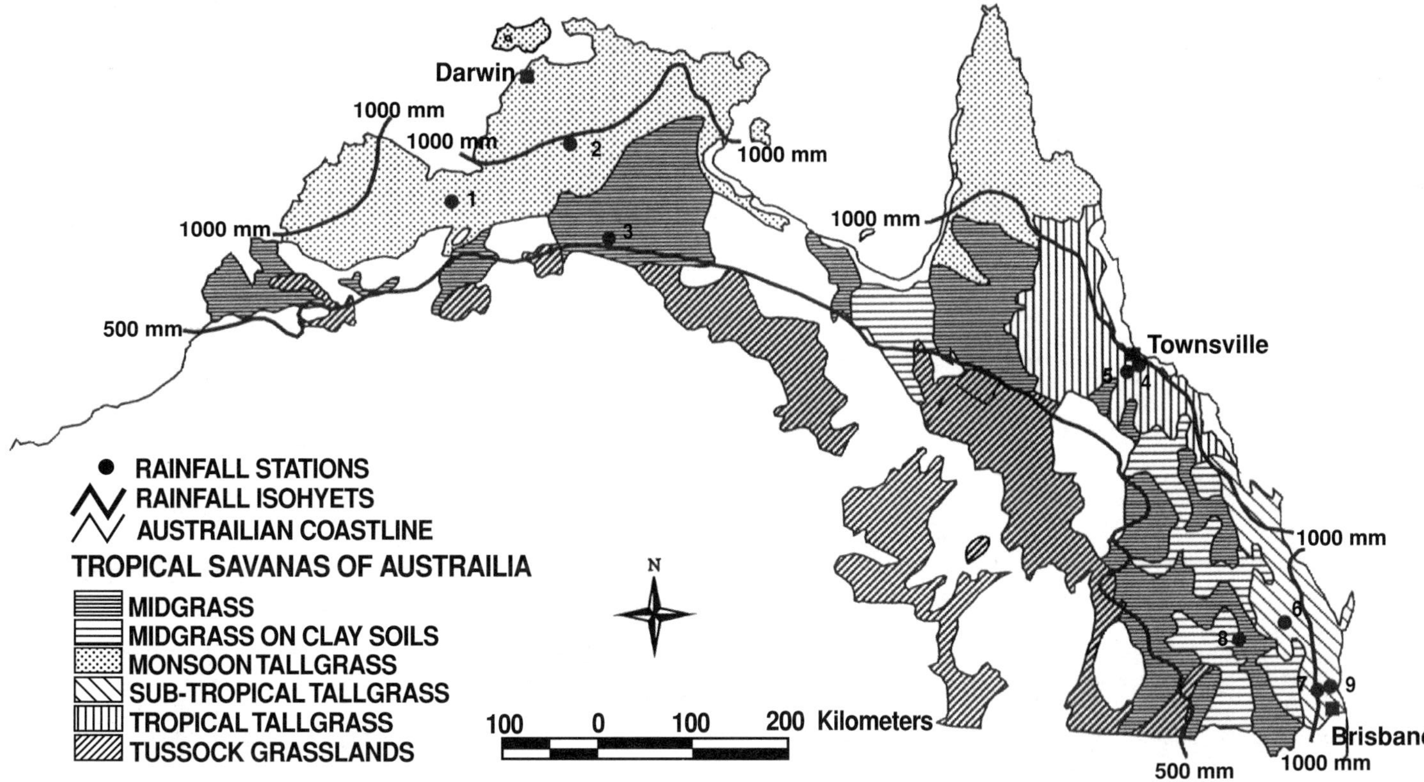

FIGURE 17.1 Main vegetation types in the Australian Savanna Zone with the 500 mm and 1000 mm mean annual rainfall isohyets superimposed. Also marked are 9 of the 10 sites in Figure 17.2. (From Mott et al.[138] With permission of Australian Academy of Science.)

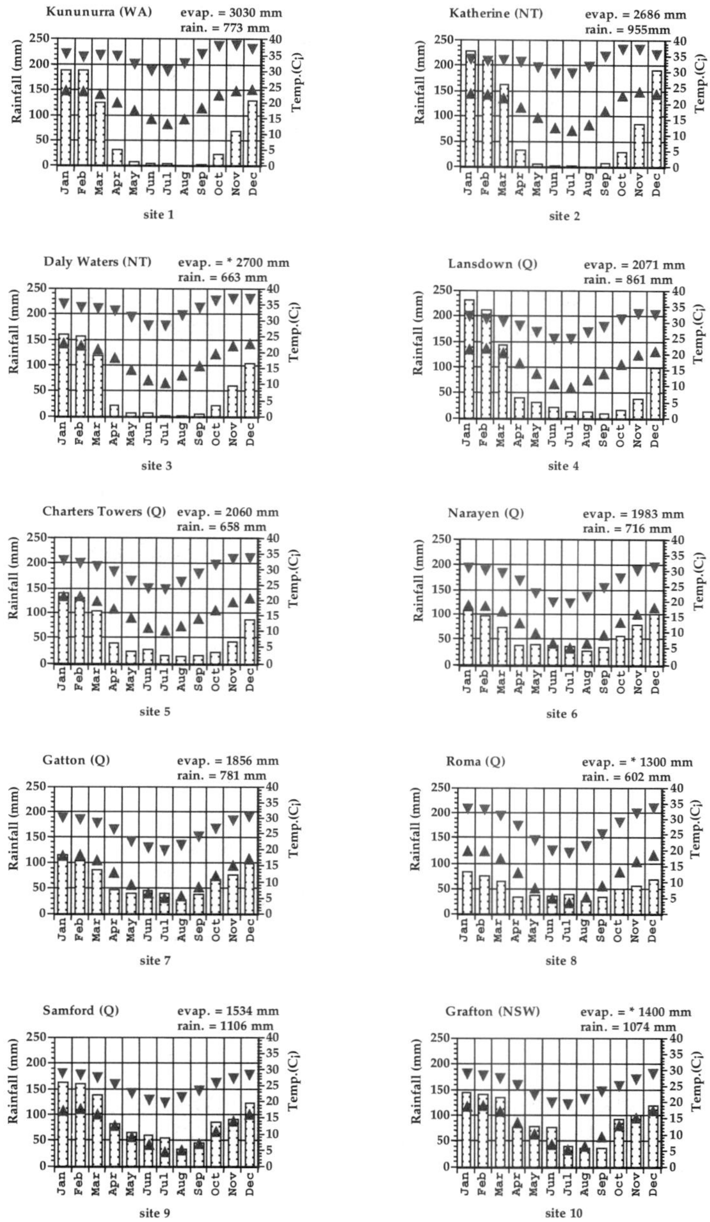

FIGURE 17.2 Climatic information for ten sites from the tropical north to the subtropical southeast of Australia: ▼ mean monthly maximum; ▲ mean monthly minimum temperature (°C). Mean monthly rainfall (mm) is represented by the vertical bars. Sites 1 through 9 are identified on Figure 17.1.

Rainfall is notably variable within and between years, particularly in central and southern areas. Apart from a narrow coastal area in northeast Queensland and the north of Cape York, which receives 1500 to 3000 mm of mean annual rainfall (MAR), the MAR over northern Australia ranges from 1500 mm to 300 mm, with rainfall decreasing with distance from the coast. The rainfall distribution is of mainly summer incidence throughout, although the amount of rain falling outside the summer months increases at southern sites. For example, Katherine in the Northern Territory (NT) has virtually no winter rain, whereas Roma, Samford, and Gatton in southeast Queensland have a significant winter rainfall component (Figure 17.2).

Rain in the north is generally associated with cyclonic activity. The earliest monsoon rains occur in October in the extreme north, moving south to reach their most southerly extent by January. They usually last until March/April, with January and February being the wettest months. Failure of these summer rains is the major cause of droughts. The intensity of the summer rains can result in high runoff rates, reducing rainfall effectiveness, particularly on shallow soils on sloping land.

The effectiveness of the rain that falls is related to its distribution over time, the water storage capacity of the soil, and the amount of rain that falls. Using the water-holding capacity of the soil, the rainfall, the evapotranspiration, and the temperature, the period of the year when pastures grow can be calculated. Within the area where sown pastures are an option, this period varies from about 30 weeks at the tip of Cape York and the wet tropical coast to 15 weeks in the drier inland sites receiving about 500 mm annual rainfall.[5] Rainfall is clearly the most important climatic variable controlling growth of pastures in tropical Australia.

4. Vegetation

The vegetation over much of the region is an open woodland, usually dominated by *Eucalyptus* species, although Acacia-softwood scrubs or natural grasslands often occur on more fertile clay soils (Figure 17.1).[6,7] The herbaceous understory in the eucalypt woodlands is essentially gramineous. The major C4 genera are *Themeda*, *Heteropogon*, *Aristida*, *Bothriochloa*, *Dichanthium*, *Chrysopogon*, and *Sorghum*. There are few prominent native herbaceous legumes that contribute to the diet of cattle. The formation and maintenance of these savannas has been strongly influenced by the burning of the dry understory, which has regularly occurred both before and after white settlement through planned or unplanned fires.

Despite the generally low soil fertility and fluctuating rainfall, a viable beef industry has been developed over the past 150 years. These native pasture communities have several advantages: they are clearly adapted to the climatic and soil constraints described above; they are composed of many species and so are less affected by attacks from pests and diseases than are monospecific swards; they are tolerant of fire; and they have persisted (though with some changes in composition) for more than 100 years under cattle grazing. The extent of these native pastures, some 400 M ha, provides a most valuable resource. Both in total area, and when expressed as a percentage of the total land area, Australia is unique as far as its native pastures are concerned. Queensland alone has 151 M ha or 87% of its total area.[8]

In the past, the maintenance of a viable beef industry on the native pastures has been dependent upon the use of large land area per property with low stocking rates and low inputs. In the north, the aim has been to produce lean beef from 4- to 5-year-old steers and cull cows for the export market — mainly the U.S. More recently, a growing live export trade to Asia is demanding younger steers for use in feedlots overseas.

The combination of generally low fertility soils, a prolonged dry season, and predominantly C4 grasses results in pastures that have low nutritional quality for much of the year, with a trend for nutritional and climatic stresses to increase from south to north to northwest.[9,10]

Total dry-matter yields, even in the absence of grazing, are generally low (1.4 to 4 t/ha). A relatively short period of high-quality leafy forage following the start of the rainy season in November-December is followed by a period of moderate quality through the wet season, and a period of low quality through a prolonged dry period, when grasses mature. During this period, even the most nutritious leaf component of the perennial grasses usually has N concentrations of less than 1%, a P concentration of only 0.05%, and a dry matter digestibility of 40 to 50%.[10] Values for older leaf and stem material would be even lower, and so the pastures are unable to meet the maintenance requirements of cattle. Fluctuations in animal gain from year to year are closely related to the amount and distribution of the annual rainfall,[5] to soil characteristics (depth and fertility), and to stocking rate. Annual steer gains ranging from 7 to 193 kg/steer have been measured from native pastures in Queensland.[11] Those in the Northern Territory and in the tropical area of west Australia would be at the lower end of this range.

Breeding performance, as expressed by calving rate, also reflects the generally low nutritive value of the pastures grazed. Data for Queensland, published in 1974, show low calving rates of from 40% in the far northern Peninsular and Gulf region to 67% in the subtropical southeast of the state.[11] It would be expected that current calving percentages would be higher than this as a result of management strategies adopted since 1974. However, the results from 1974 reflect the general low quality of the native pastures from which most of the beef in northern Australia is produced.

In addition to the problems of low yield and poor nutritional quality of the herbage, many areas have poor natural water supplies by way of permanent streams. Unless dams or bores are constructed, effective utilization of the resource is very limited, except for short periods during the wet season.

5. Areas Suited to Pasture Development

The areas of the tropics and subtropics where some improvement in productivity is possible from the sowing of forage species are restricted to an arc of country running from the northwest of western Australia along the north and down the east coast of Queensland to Grafton on the New South Wales coast. This arc extends inland from the coast to approximately the 500 mm mean annual rainfall isohyet, a distance that varies from about 300 km to 600 km (Figure 17.1).

The area so delineated is similar to that deemed suitable for improved pastures by Davies and Eyles.[12] Of the gross area of 164 M ha, some 110 M ha was regarded as suitable for improved pastures (including crops in some areas).

In Queensland, where most tropical pasture development has occurred, the estimates of the area suited to improved pasture development have ranged from 52 to 58 M ha; but if potential crop land is removed, then the area is reduced to 41 M ha.[13] Revised estimates to exclude areas remote from infrastructure, where costs would be very high or where technical difficulties are foreseen, define the easily attainable sown pasture potential at 22 M ha, of which 10.4 M ha are on infertile earths or infertile duplex soil.[13] Similar estimates for the remainder of tropical and subtropical Australia are not available, but are unlikely to exceed 4 M ha.

III. PLANT INTRODUCTION AND BREEDING PROGRAMS FOR FORAGES

A. Research and Selection of Ecotypes

Sown pastures in the tropics and subtropics of Australia are based on plants introduced (by accident or planned) from overseas. In general, the tropical native grasses and legumes have not shown much promise for sown pastures, which are expected to withstand heavier stocking pressure from introduced ruminants than that experienced in the wild by marsupials. None of the native Australian grasses have been developed as cultivars elsewhere in the world. This may perhaps be due to lack of adequate collections being tested, although extensive collections of the Australian *Bothriochloa-Dichanthium* complex evaluated in the United States of America, proved not to be useful in comparison with introductions from Asia.[14]

One native Australian *Glycine* species has shown promise as a ground cover and pasture plant in South Africa, and *G. latifolia* is currently showing promise as a frost-tolerant, low-growing legume for cracking clay soils in the Australian subtropics,[15] but, in general, there is a paucity of productive, persistent, palatable, and grazing-tolerant native legumes in the Australian tropics.

It is not surprising, therefore, that plant introduction has played a vital role in the development of sown tropical pasture plants in Australia.

The history (up to 1984) of plant improvement has been well documented by Eyles and Cammeron,[16] and includes introduction and breeding of legumes and grasses, involving mainly the

Australian Commonwealth research body (CSIRO) and specifically the Tropical Pastures Division, now called the Division of Tropical Crops and Pastures (DTCP) and the Queensland State Department of Agriculture, now called the Department of Primary Industries (DPI).[16]

Preliminary work, which commenced in the 1920–1930s, gave promising results with introduced species along the moist tropical and subtropical coastal areas and on the more fertile soils in southeast Queensland. In the tropics, the legumes *Stylosanthes guianensis* (Stylo), *Centrosema pubescens* (Centro), *Calopogonium mucunoides* (Calopo), and *Pueraria phaseoloides* (Puero), species that had been used as cover crops in plantation agriculture, and the grasses *Melinis minutiflora* (molassesgrass), *Panicum maximum* (guinea grass), and *Brachiaria mutica* (paragrass) grew well and showed promise for cattle pastures. In the subtropical southeast of the state, the grasses *Paspalum dilatatum* (Paspalum), *Chloris gayana* (Rhodes grass), *Cenchrus ciliaris* (buffelgrass), *Pennisetum clandestinum* (Kikuyu grass), *Sorghum sudanense* (Sudan grass), and lucerne showed promise.

From 1950 to 1975, there was a rapid growth in the number of researchers involved in pasture improvement associated with the government policy of increasing agricultural production.[16] Plant introduction and pasture plant evaluation increased during this period and there were 12 overseas collecting visits involving 12 different collectors compared with one prior to 1950. Major areas for the collections were South and Central America and southern and eastern Africa. During this period, 55 cultivars were released to the industry for use in the tropics and subtropics and of these, only *Macroptilium atropurpureum* cv. Siratro, Krish sorghum, and *Setaria sphacelata* cv. Narok were bred lines.

Since 1975 to the present (2000) there has been a decline in the number of scientists involved in pasture research, though plant collection and plant breeding have continued at a lower level. From 1975, there have been major overseas collections of pasture plants in Central and South America, Africa, and Southeast Asia. Increasingly, the collections have become more specific in nature with better-defined objectives to overcome deficiencies that had become apparent in existing cultivars. Forty-eight cultivars have been released, 12 of which resulted from breeding programs in Australia as opposed to selection from introduced ecotypes or introduced cultivars.

Hand in hand with the collection of legumes has been the collection and evaluation of *Rhizobium* strains to ensure that effective symbioses occurred under field conditions and that these effective strains were developed commercially for all the recommended forage legume cultivars.[16]

B. Development of New Cultivars (Methodology)

Basic to the development of new cultivars is the assembly of as wide a range as possible of genetic material of the species concerned. Early work was often based on a few introductions of species found useful elsewhere. Later evaluation has been based on a much broader range of genetic material (natural ecotypes) often of wide latitudinal or altitudinal range and from different soils. This material was obtained by correspondence with interested parties overseas, through planned major collecting expeditions, usually in conjunction with scientists in overseas countries, and through smaller collections made by scientists who were overseas for other purposes.

To get the most benefit from collecting missions, a knowledge of the geography, geology, climate, and flora of the proposed collecting sites was essential. This information enabled more focused missions to be organized. Useful material has not necessarily come from homoclimatic areas overseas; for example, many valuable tropical grasses have been introduced from east Africa, despite the fact that we do not have comparable climatic conditions in the tropics and subtropics of Australia. However, collecting from homoclimatic areas is a valuable starting point.

The ideal sampling strategy has been to collect 50 to 100 individuals per site, to sample as many sites as possible, and to ensure that the sample sites represent as broad a range of environments as possible.[17] However, it is rarely possible to achieve all of these objectives and the maximum number of sites sampled has been more important than getting a large number of plants per site.[18]

The Australian experience in evaluating tropical forage plants has been documented in more detail elsewhere,[19-21] and will only be briefly dealt with here.

When quarantine procedures have been met, the collection is grown in the glasshouse or as spaced plants in the field to describe the material in terms of morphology, flowering behavior, and dry matter yield. The use of relevant controls is important so that the degree of variation between accessions can be assessed. For example, it is important to know if the variation is reflected in material exceeding that of the control cultivars.

If the collections are large, then some classification procedure may need to be imposed on the data with a view to grouping the accessions.[21-23] Such grouping aids in the selection of material for subsequent evaluation, since it is impractical to subject all the introductions to larger scale field evaluation. Seed multiplication is required for further stages of testing and, hence, this may delay field evaluation on a larger scale. In some instances the plants "selected themselves" in the field, e.g., they may have shown resistance to disease or insect attack or have survived frost damage.

Multisite field testing, using well-documented experimental sites, is an integral part of the evaluation program for introduced material.[20] The sites used would have been evaluated for nutrient deficiencies, and these would have been corrected by appropriate fertilization. In the past, these rates would have been high. Recently only low rates of fertilizer (10 kg P/ha) or no fertilizer would be applied to the plots. Much of the early testing was done collaboratively by the research groups and facilitated by funding from the Meat Research Corporation (MRC), now The Meat and Livestock Authority (MLA), whose funds are 50% from industry and 50% from matching government funds. From these experiments, the site x genotype interactions can be assessed with a view to selecting those with wide ecological adaptations and identifying those with perhaps narrower ecological adaptability, but with high performance at specific sites, e.g., frost tolerance or ability to establish readily on black cracking clays.

Legumes are usually evaluated with an adapted associate grass in sward experiments and exposed to grazing as early as possible in the evaluation program to assess persistence and palatability. Unpalatable legumes can make competitive weeds! Grasses are generally evaluated with an adapted legume or a mixture of adapted legumes. Compatibility with legumes, as well as dry matter production, is an essential characteristic. For coastal pastures, where N fertilization may be economic, the grasses are evaluated in pure stands at one or more rates of N fertilizer. In each situation persistence is required, but difficult or impossible to measure in short term (2 to 3 yr) experiments. Imposition of cutting frequency treatments at this stage may identify accessions more tolerant of heavy defoliation and therefore more persistent under practical conditions.

Larger sward experiments, usually at several sites, compare the performance of selected material with control cultivars. The swards are subjected to intermittent heavy grazing, but sampled for yield and botanical composition just before grazing or more usually at the end of the growing season. Plots may be mown after grazing to a standard height to prevent uneaten residues biasing subsequent samplings, and notes recorded of the relative acceptance by the animals of the various treatments.

Further evaluation of selected lines (which now may number only 2 to 4) occur in large plots or small paddocks where the lines are grazed individually at a commonly used stocking rate for the area, or possibly at two stocking rates. Some animal production measurements may be made, especially in favored areas where high stocking rates enable small paddocks to be used. The experiment may run for a longer period of time and the best species may then be considered for commercialization.

In more recent times, the evaluation procedures have been less rigorous than described above with less involvement of grazing studies. Such "short cuts" are questionable, though justified on the basis that the reaction to grazing of many of the species is known. Although animal production data are the ultimate assessment of the value of the forage species, such information is rarely collected prior to release of the material to the industry. Such data are often measured after release and so help to further promote the cultivar concerned and to obtain information on the relation between pasture characteristics and animal gain. Data associated with animal production can be taken at any stage in the evaluation process, but only when some reduction in the numbers has occurred. Such measurements include mineral composition, *in vitro* digestibility, and levels of known toxins.

One factor often overlooked in the past was the ability to produce commercial seed yields. Unless seed (or vegetative material) can be produced readily and cheaply, the cultivar may have a limited market. Examples in Australia are the low seed yield of *Setaria sphacelata* cv. Narok[24] and the difficulty of seed harvesting with *Desmodium heterophyllum*.[25]

In Queensland, the provision of new cultivars to the grazing industry is now through a formal mechanism via the Queensland Herbage Plant Liaison Committee (QHPLC) established in 1964 from its predecessor, the Queensland Pasture Liaison Committee (QPLC).[26] This committee is comprised of representatives from the CSIRO Division of Tropical Crops and Pastures (3), the Queensland Department of Primary Industries (QDPI) (3) (equivalent of a State Department of Agriculture), the Agriculture Department of the University of Queensland (2), the Seed Industry Association of Australia (SIAA) (2), the Queensland Seed Producers Association (QSPA) (1), the Subtropical Seed Growers Association (STSGA) (1), the Northern Territory Department of Primary Industries and Fisheries (NTDPIF) (1), the New South Wales Department of Agriculture (NSWDA) (1), and the Queensland Department of Environment and Heritage (1). This broadly based committee assesses the evidence presented by the sponsor or sponsors (usually the CSIRO, DTCP, or the DPI) of any new material and, on the basis of this, proceeds to release the material, request more information, or refuse release. Until recent times they facilitated release of new material by establishing Seed Increase Committees to organize production of seed to the stage where supplies are adequate for commercial purposes.[26] The general policy has been to release material when evidence of merit is available without waiting for animal production data. The introduction of Plant Variety Rights (PVR) has encouraged the commonwealth and state governments to recoup some of the costs involved in developing new cultivars, and activity by private companies has also increased, but currently release of new cultivars is through the HPLC.[27] However, since most cultivars are now promoted under PVR, seed increase is the responsibility of the private companies.

Committees similar to the QHPLC are responsible for the release of subtropical and tropical pasture plants in New South Wales (NSWHPLC) and the Northern Territory (NTHPLC), though the majority of the tropical cultivars have been released through the QHPLC.

In Queensland, the production of seed of the most promising materials for experimental evaluation is handled by specialist seed-production groups in the DPI, situated in an environment suitable for seed production on the Atherton Tableland in north Queensland and at Gympie in southeast Queensland. This has broken the bottleneck of producing sufficient seed for a wider evaluation of promising material.

In addition to releasing cultivars for use, the Herbage Plant Liaison Committees encourage the registration of all new cultivars in the *Register of Australian Herbage Plant Cultivars*. This register, published first in 1967, has been updated periodically to handle the new releases. Descriptions of the plants to be registered are submitted to the registrar, usually after having been vetted by the relevant state HPLC. The registrar then circulates copies of the detailed description to other HPLCs for advice and recommendation and may seek additional advice before deciding to publish the information in the register. The registrar then accepts the description for registration on behalf of the Agricultural Council. The originator or breeder of the cultivar supplies a small sample of authentic seed for dry cold storage and reference.

The current register (3rd edition) containing descriptions of 43 temperate grasses, 123 temperate legumes, 67 tropical grasses, and 35 tropical legumes, was published in 1990.[28]

C. Economically Important Cultivars Released

From 1961 to the end of 1999, 121 tropical cultivars have been released through the HPLCs in Queensland, New South Wales, and the Northern Territory.[29] Of these, 58 were legumes and 63 were grasses. Seven grasses and nine legumes resulted from plant breeding work. Over 80% of the releases were from direct introductions (see Tables 17.1, 17.2, and 17.3).

At the time of release, it is implied that the released cultivar will have economic importance to the industry. Otherwise there would be little point in commercial release. Seed growers would not be interested, and so the cultivar would fail to become commercialized. Having said this, the reality is that, of the cultivars released for tropical Australia, relatively few are traded on a large scale. Table 17.4 gives a list of the important species traded.

Diseases and insect pests have caused the demise of some cultivars, e.g., all the *Stylosanthes humilis* cultivars, many of the *S. guianensis* cultivars, and *S. scabra* cv. Fitzroy proved to be susceptible to strains of the fungus anthracnose (*Colletotrichum gleosporoides*). Others had problems of seed production, e.g., Narok setaria and *D. heterophyllum* cv. Johnstone, while others were found to have limited adaptation such as *Neonotonia wightii* cv. Tinaroo.

Among the legumes, susceptibility to heavy grazing or the need to maintain heavy fertilizer application for persistence led to a lack of favor with graziers. In general, the twining tropical legumes fall into this category, a situation which, for *Macroptilium atropurpureum* cv. Siratro, was compounded by an introduced leaf rust (*Uromyces appendiculatus*), which reduced yield and seed production.

In recent times, the advent of the psyllid (*Heteropsylla cubana*) has reduced yields of *Leucaena leucocephala* in the more humid coastal and subcoastal zones of Queensland and New South Wales with a reduction in areas sown.

Some cultivars become popular after many years of low seed sales. A classic example is *Sorghum* sp. hybrid cv. Silk, released in 1977, which only attained high seed sales in the late 1980s and 1990s when sales of seed exceeded 300 t/annum. Its ease of establishment on clay soils and the increased use of short-term pastures in grain-cropping areas contributed to its belated popularity.

The most successful and the most economically important legume released has been *Stylosanthes*; particularly *S. scabra* cv. Seca and *S. hamata* cv. Verano. It is now estimated that 90 to 100,000 ha are sown annually. Two independent benefit/cost analyses have been undertaken using different economic models on the research involved in the development of these cultivars in relation to the benefits to industry.[30,31] Results are summarized below.

Returns to the Stylo Research

	Study 1[30]	Study 2[31]
Present value of benefits ($M)	190	142.7
Present value of research investment ($M)	28	30.4
Benefit: Cost ratio	7	4.7
Internal rate of return (%)	25	16.3

These represent very healthy returns on the research investment, returns which are expected to increase as the stylo technology gains in popularity.

In addition to the seed sales listed in Table 17.4, some species are also grown on farms, and seed is traded by farm-to-farm sales, e.g., *Cenchrus ciliaris*. There is often a time lag between release and the establishment of large areas of any pasture plant, but the time lag is often shortened if beef prices are high. Species with adaptation to large subhumid or semiarid ecological areas will clearly have more potential impact than those adapted to less extensive higher rainfall zones. *Stylosanthes scabra* cv. Seca, which is widely adapted from the subtropics to the tropics in rainfall

TABLE 17.1
Tropical Forage Cultivars Released by the Queensland Pasture Liaison Committee and by the Queensland, New South Wales and Northern Territory Herbage Plant Liaison Committees 1961 to 1970

Species	Common Name	Cultivar Name	Year of Release	Releasing Authority[a]
Cenchrus ciliaris	buffel grass	Nunbank	1961	Queensland
Lotononis bainesii	lotononis	Miles	1962	Queensland
Cenchrus ciliaris	buffel grass	Tarewinnabar	1962	Queensland
Cenchrus ciliaris	buffel grass	Lawes	1962	Queensland
Cenchrus ciliaris	buffel grass	Boorara	1962	Queensland
Neonotonia wightii	glycine	Cooper	1962	Queensland
Panicum coloratum var. makarikariense	makarikari	Burnett	1962	Queensland
Panicum coloratum var. makarikariense	makarikari	Pollock	1962	Queensland
Neonotonia wightii	glycine	Tinaroo	1962	Queensland
Leucaena leucocephala	leucaena	Peru	1962	Queensland
Paspalum commersonii	scrobic	Paltridge	1962	Queensland
Digitaria decumbens	pangola grass	—	1962	Queensland
Desmodium uncinatum	desmodium	Silverleaf	1962	Queensland
Chloris gayana	Rhodes grass	Callide	1963	Queensland
Pennisetum purpureum	elephant grass	Capricorn	1963	Queensland
Desmodium intortum	desmodium	Greenleaf	1963	Queensland
Paspalum plicatulum	plicatulum	Rodd's Bay	1963	Queensland
Paspalum plicatulum var. glabrum	plicatulum	Hartley	1963	Queensland
Setaria sphacelata	setaria	Nandi	1963	Queensland
Chloris gayana	Rhodes grass	Samford	1963	Queensland
Panicum maximum	guinea grass	Coloniao	1963	Queensland
Panicum maximum	panic	Gatton	1964	Queensland
Vigna luteola	vigna	Dalrymple	1963	Queensland
Macrotyloma uniflorum	biflorus	Leichhardt	1965	Queensland
Brachiaria ruziziensis	ruzi grass	Kennedy	1965	Queensland
Sorghum sp hybrid	forage sorghum	Krish	1965	Queensland
Stylosanthes guianensis	finestem stylo	Oxley	1965	Queensland
Macrotyloma axillare	axillaris	Archer	1966	Queensland
Setaria sphacelata	setaria	Kazungula	1967	New South Wales
Lablab purpureus	lablab bean	Rongai	1967	New South Wales
Brachiaria decumbens	signal grass	Basilisk	1967	Queensland
Neonotonia wightii	glycine	Clarence	1967	New South Wales
Stylosanthes humilis	Townsville stylo	Lawson	1968	Queensland
Stylosanthes humilis	Townsville stylo	Gordon	1968	Queensland
Pennisetum americanum	bulrush millet	Ingrid Pearl	1968	N. Territory
Setaria sphacelata	setaria	Narok	1969	Queensland
Stylosanthes humilis	Townsville stylo	Paterson	1969	Queensland
Pennisetum clandestinum	Kikuyu grass	Whittet	1970	New South Wales

[a] Queensland, Queensland Pasture Liaison Committee (until 1963) and Queensland Herbage Plant Liaison Committee (from 1964); New South Wales, New South Wales Herbage Plant Liaison Committee; N. Territory, Northern Territory Herbage Plant Liaison Committee.

Source: From Walker et al., *Tropical Grasslands,* Vol. 31, p. 268, 1997.

TABLE 17.2
Tropical Forage Cultivars Released by the Queensland Pasture Liaison Committee and by the Queensland, New South Wales and Northern Territory Herbage Plant Liaison Committees 1971 to 1988

Species	Common Name	Cultivar Name	Year of Release	Releasing Authority[a]
Stylosanthes guianensis	stylo	Endeavour	1971	Queensland
Pennisetum clandestinum	Kikuyu grass	Breakwell	1971	New South Wales
Stylosanthes guianensis	stylo	Cook	1971	Queensland
Kummerowia striata	Japanese lespedeza	Kaloe	1971	New South Wales
Centrosema pubescens	centro	Belalto	1971	Queensland
Desmodium heterophyllum	hetero	Johnstone	1971	Queensland
Aeschynomene falcata	jointvetch	Bargoo	1973	New South Wales
Trifolium semipilosum	Kenya white clover	Safari	1973	Queensland
Stylosanthes hamata	Caribbean stylo	Verano	1973	Queensland
Lablab purpureus	lablab bean	Highworth	1973	Queensland
Urochloa mosambicensis	sabi grass	Nixon	1973	N. Territory
Paspalum plicatulum	plicatulum	Bryan	1974	Queensland
Panicum maximum	guinea grass	Makueni	1974	Queensland
Neonotonia wightii	glycine	Malawi	1975	Queensland
Panicum maximum	guinea grass	Riversdale	1975	Queensland
Stylosanthes scabra	shrubby stylo	Seca	1976	Queensland
Sorghum sp. hybrid	forage sorghum	Silk	1977	Queensland
Leucaena leucocephala	leucaena	Cunningham	1977	Queensland
Bothriochloa insculpta	creeping bluegrass	Hatch	1978	Queensland
Calopogonium mucunoides	calopo	Tortilla	1978	N. Territory
Andropogon gayanus	gamba grass	Kent	1978	N. Territory
Stylosanthes guianensis	stylo	Graham	1978	Queensland
Stylosanthes scabra	shrubby stylo	Fitzroy	1979	Queensland
Setaria incrassata	purple pigeon grass	Inverell	1980	Queensland
Brachiaria humidicola	koronivia grass	Tully	1980	Queensland
Medicago sativa	lucerne	Trifecta	1982	Queensland
Lolium multiflorum	Italian ryegrass	Midmar	1982	Queensland
Cassia rotundiflora	roundleaf cassia	Wynn	1983	Queensland
Medicago sativa	lucerne	Sequel	1983	Queensland
Aeschynomene americana	jointvetch	Glenn	1983	Queensland
Centrosema pascuorum	centurion	Cavalcade	1984	N. Territory
Setaria sphacelata	setaria	Solander	1984	Queensland
Vigna parkeri	creeping vigna	Shaw	1984	Queensland
Centrosema pascuorum	centurion	Bundey	1986	N. Territory
Digitaria smutsii	digit	Premier	1986	Queensland
Ornithopus compressus	seradella	Madeira	1987	Queensland
Arachis pintoi	Pinto peanut	Amarillo	1987	Queensland
Medicago scutellata	snail medic	Kelson	1987	Queensland
Echinochloa polystachya	aleman	Amity	1987	Queensland
Hymenachne amplexicaulis	hymenachne	Olive	1987	Queensland
Stylosanthes hamata	Caribbean stylo	Amiga	1988	Queensland

[a] Queensland, Queensland Pasture Liaison Committee (until 1963) and Queensland Herbage Plant Liaison Committee (from 1964); New South Wales, New South Wales Herbage Plant Liaison Committee; N. Territory, Northern Territory Herbage Plant Liaison Committee.

TABLE 17.3
Tropical Forage Cultivars Released by the Queensland, New South Wales and Northern Territory Herbage Plant Liaison Committees 1989 to 1999

Species	Common Name	Cultivar Name	Year of Release	Releasing Authority[a]
Bothriochloa insculpta	creeping bluegrass	Bisset	1989	Queensland
Setaria sphacelata	setaria	Splenda	1989	Queensland
Digitaria eriantha	digit grass	Apollo	1989	Queensland
Medicago sativa	lucerne	Quadrella	1989	Queensland
Macroptilium gracile	llano macro	Maldonado	1990	N. Territory
Clitoria ternatea	butterfly pea	Milgarra	1990	Queensland
Stylosanthes scabra	shrubby stylo	Siran	1990	Queensland
Digitaria milanjiana	finger grass	Jarra	1991	Queensland
Desmanthus virgatus	desmanthus	Marc	1991	Queensland
Desmanthus virgatus	desmanthus	Bayamo	1991	Queensland
Desmanthus virgatis	desmanthus	Uman	1991	Queensland
Bothriochloa pertusa	indian bluegrass	Dawson	1991	Queensland
Bothriochloa pertusa	indian bluegrass	Medway	1991	Queensland
Aeschynomene americana	American jointvetch	Lee	1991	Queensland
Chloris gayana	Rhodes grass	Nemkat	1992	Queensland
Macroptilium atropurpureum	atro	Aztec	1993	Queensland
Vetivera zizanioides	vetiver	Monto	1993	Queensland
Echinochloa frumentacea	jap millet	Indus	1993	Queensland
Panicum laxum		Shadegro	1993	Queensland
Chloris gayana	Rhodes grass	Finecut	1993	Queensland
Chloris gayana	Rhodes grass	Topcut	1993	Queensland
Cenchrus ciliaris	buffelgrass	Bella	1994	Queensland
Cenchrus ciliaris	buffelgrass	Viva	1994	Queensland
Bothriochloa bladhii spp. glabr	forest bluegrass	Swann	1994	Queensland
Dichanthium aristatum	bluestem	Floren	1994	Queensland
Leucaena leucocephala	leucaena	Tarramba	1994	Queensland
Aeschynomene villosa	villose jointvetch	Kretschmer	1995	Queensland
Aeschynomene villosa	villose jointvetch	Reid	1995	Queensland
Arachis glabrata	rhizoma peanut	Prine	1995	Queensland
Centrosema pubescens	centro	Cardillo	1995	Queensland
Digitaria milanjiana	fingergrass	Strickland	1995	Queensland
Digitaria milanjiana	fingergrass	Arnhem	1996	N. Territory
Centrosema brasilianum	centro	Oolloo	1996	N. Territory
Stylosanthes seabrana	catinga stylo	Primar	1996	Queensland
Stylosanthes seabrana	catinga stylo	Unica	1996	Queensland
Medicago sativa	lucerne	Sequel HR	1996	Queensland
Pennisetum glaucum	pearl millet	Siromill	1996	Queensland
Medicago sativa	lucerne	Hallmark	1998	Queensland
Lablab purpureus	lablab	Endurance	1999	Queensland
Paspalum atratum	atrapaspalum	Hi gane	1999	Queensland

[a] Queensland, Queensland Pasture Liaison Committee (until 1963) and Queensland Herbage Plant Liaison Committee (from 1964); New South Wales, New South Wales Herbage Plant Liaison Committee; N. Territory, Northern Territory Herbage Plant Liaison Committee.

TABLE 17.4
Tropical Pasture Seed Produced in Queensland (tons in 1995)

Grasses			Legumes		
Forage	Seed Produced (tons)	No. of Cultivars	Forage	Seed Produced (tons)	No. of Cultivars
Silk sorghum	400	1	Shrubby stylo	165	2
Buffel	400	3	Caribbean stylo	85	2
Rhodes	350	3	Jointvetch	50	2
Panics	190	4	Glycine	30	2
Brachiaria	130	2	Siratro	20	2
Setaria	80	3	Cassia	10	1
Purple pigeon	50	1	Leucaena	10	3
Creeping bluegrass	30	1	Desmanthus	7	3
Urochloa	25	1	Arachis	3	1
Paspalum	20	1			
Others	24	8	Others	12	18
Total	1699	28	Total	392	36

From Walker et al.[139] With permission from the Tropical Grassland Society.

zones from 500 to 1200 mm, probably has the greatest potential as a legume in tropical pastures. Seed production now surpasses that of Verano (Table 17.1).

IV. FORAGE COMPONENTS

A. Legumes

The overriding nitrogen deficiency for the growth of pastures and of livestock in tropical Australia has focused attention on the need for effective legumes. The search for adapted legumes for northern Australia has been a qualified success.[16,32,33] Attempts have been made to reproduce the subclover story of southern Australia in the tropics with some degree of success. The successful legumes do not resemble subclover in growth form and most are perennials. Furthermore, there is a wider range of genera used in pastures in the tropics. The Register of Herbage Plant Cultivars lists 6 temperate/Mediterranean genera and 13 tropical genera, although there are 119 temperate legume cultivars (103 in *Trifolium* and *Medicago*) and only 38 tropical cultivars (11 in *Stylosanthes*) that have been released for use in sown pastures in Australia.[28]

The most important genus of tropical legumes currently under evaluation is *Stylosanthes* with about 40 species.[34] The value of *S. humilis* was first recognized in Australia in 1914. It was a chance introduction from South America, which spread naturally in northern Australia and was actively sown in pastures up until the 1970s, when it was decimated by the introduced fungus that causes anthracnose.[35]

Species of the genus are found over a wide latitudinal range from 41° N to 30° S, mainly in the Americas, but also in Africa and Asia.[36] Members of the genus occur in a variety of forms from prostrate herbaceous annuals to taller woody perennials and are adapted to soils that range from very acid to alkaline with the majority adapted to acid soil.[36]

The annual *S. humilis* showed great promise as a self-regenerating annual species, easy to establish, liked by cattle, especially in the dry season, and a good seed producer. However, its success was short lived after the advent of anthracnose. One introduction, cv. Kohn Kaen, released in Thailand, is anthracnose resistant, but not adapted to Australian conditions. Anthracnose also attacked a number of *S. guianensis* cultivars and accessions, so that this species is now of minor importance in Australia. Four species are of current importance and interest.

S. guianensis var. *intermedia* (fine stem stylo) represented by cv. Oxley, has shown little variation between introductions from the subtropical areas of South America. It is well adapted to granitic soils with coarse sandy surfaces. It is climatically well adapted to the subtropics and survives frost and heavy grazing, possibly due to the low crown buds. Attempts have been made to introduce, by crossing, the higher yielding ability, adaptation to a wider range of soils, and ease of nodulation and establishment of the robust forms of *S. guianensis* var. *guianensis* to produce a more widely adapted subtropical stylo. However, research has terminated because of inability to combine the desired traits in a stable, free-seeding cultivar (D.F. Cameron, personal communication).

S. hamata (Caribbean stylo) is a short-lived perennial. Tetraploid and diploid forms of this legume occur. The tetraploids are represented by cv. Verano and the more recently released cultivar Amiga. They are more productive on acid or neutral soils, nodulate more freely with native strains of root nodule bacteria (RNB), and as a result are more persistent under pasture conditions than are the diploids. Variation in susceptibility to anthracnose and in many other plant characteristics exist.[21] The diploids are more specific in their *Rhizobium* affinities[37] and better adapted to alkaline clay soils, but no cultivars have been released.

Current emphasis is in the selection of tetraploid lines with better anthracnose resistance, greater persistence, and wider climatic adaptation, particularly tolerance to cool subtropical conditions, than Verano and Amiga. The material in this program was specifically collected from higher altitude areas in Venezuela, but so far no material suitable for subtropical conditions has emerged.[38]

S. scabra (shrubby stylo) is represented by cv. Seca and is the most widely adapted tropical legume in Australia. The commercial material available differs from the material first released as cv. Seca in being earlier flowering and leafier. Some outcrossing, probably with the anthracnose susceptible but more agronomically desirable cv. Fitzroy, may have occurred in early commercial multiplication. This has enabled the cultivar to be successful in a wider range of environments from the subtropics to the tropics. Although slow to establish, it is eventually productive and may dominate pastures under light grazing in low P situations or in the absence of grazing-tolerant grasses at higher grazing pressures. Its anthracnose resistance, ability to grow on soils of low P status (4 to 6 ppm available P by bicarbonate extraction), to tolerate heavy grazing, to regenerate from seed, and to survive droughts have led to its increasing popularity. Although many other accessions of *S. scabra* have been evaluated since its release, Seca is still among the highest producers across a range of environments, including some of the lighter clay soils, which are not alkaline. Like other *S. scabra* and *S. viscosa* accessions, secretions from glands on the leaf and stems are able to kill the larvae of the cattle tick (*Boophilus microplus*) when they ascend the plants to await a passing host.[39]

The cultivar Siran, released in 1991, was bred specifically to have broad multigenic anthracnose resistance and good agronomic characteristics. It is leafier than cv. Seca, but may not be quite as drought tolerant. Current opportunities for improvement in *S. scabra* include selection for early flowering, frost tolerance, disease resistance, higher seed yield, and better seedling vigor to give higher first-season yields and adaptation to marginal environments.[38]

S. seabrano combines the seedling vigor, early flowering, and prolific seeding ability of *S. hamata* with the perenniality and height of growth of *S. scabra*. It is better adapted to the more fertile clay soils and has greater cold and frost tolerance.[38,40] However, it is more specific in its *Rhizobium* requirements than the current *Stylosanthes* cultivars. Solving the *Rhizobium* problem has enabled its release. Two cultivars are available; cv. Primar is better adapted to subtropical areas with early frosts and a short growing season, and cv. Unica is adapted to more northern areas with a longer growing season. It has the potential to become the stylo of the subtropics and of the more fertile clay soils.

The only other species of *Stylosanthes* being researched is *S. macrocephala*. Its anthracnose resistance and free-seeding habit are important characteristics. Although mainly adapted to very acid soils of pH < 5, some material grows well at up to pH 6. In northern Australia, accessions show large differences in yield and flowering time, but there are also problems with effective nodulation under field conditions. Again, further evaluation is dependent upon effective strains of *Rhizobium* being developed (R.J. Williams, personal communication).

Macroptilium atropurpureum cv. Siratro was the first bred tropical pasture legume and was released in 1960.[41] The breeding program continued to improve yield and stoloniferous development while retaining the other good qualities of Siratro such as seedling vigor and broad adaptation. These aims were not realized for several reasons, including the susceptibility of all the advanced lines under test to leaf rust (*Uromyces appendiculatus* var *crassitunicatus*). This appeared in Australia in 1978 and caused yield reductions of some 30%, a reduction in leaf digestibility and N%,[42] and a reduced seed yield of 20 to 30% (English and Hopkinson, personal communication). Only one of 100 introductions of *Macroptilium atropurpureum* was totally resistant to the disease.[43] Since that time, introductions from Mexico and Central America showed resistance, though agronomically they were not as attractive as Siratro. A successful breeding program incorporating resistant genes from lines of diverse origin into Siratro culminated in a multiline cultivar almost identical to Siratro, but with rust resistance.[44] This material was released as cv. Aztec in 1993 and should replace Siratro in the future since, under challenge from rust, it yields 50% more than Siratro (R.A. Bray, personal communication). The admirable qualities of Siratro as a readily established broadly adapted legume in subtropical and tropical Australia are offset by its intolerance of heavy grazing,[45] lack of adaptation to dry environments (<750 to 850 mm MAR), and higher P requirements for growth than the stylos. Its high palatability, relative to stylos, can also be seen as a factor not conducive to long-term persistence under heavy grazing. However, with good management it has been shown to increase cattle liveweight gains, associated grass yield, and N content, and is still used in mixtures with other legumes.[46]

Incorporation of lower palatability (possibly from some of the Mexican lines), reduction in internode length, increased production of hard seed, and better drought tolerance could improve this useful legume. Currently there is no research geared toward these objectives.

Aeschynomene americana is adapted to the wetter coastal belt of northern Australia. Two cultivars have been released. The first, cv. Glenn, released in 1983, is a self-regenerating annual with good seedling vigor enabling it to reestablish in grazed pastures. Its ease of establishment, tolerance of low fertility, waterlogging, and heavy grazing, together with its acceptance by stock, make it a major contributor to pasture development in coastal areas of tropical Queensland.[47] Seed sales are now over 50 t/yr making it the third most popular legume sown in tropical pastures after *S. scabra* cv. Seca and *S. hamata* cv.Verano.[47] The second cultivar, Lee, released in 1991, has higher DM yield, a longer growing season, and better persistence when grown with competitive grasses than cv. Glenn.[47]

The subtropical *Ae. falcata* cv. Bargoo is a highly persistent perennial adapted to a wide range of soils in the 700 to 1600 mm rainfall zone and tolerant of heavy grazing, although not very high yielding. Despite its desirable features, it has never been widely used due to the difficulties of obtaining commercially rewarding seed yields.[47]

The 316 accessions of *Aeschynomene* introduced to Australia, comprising 29 species, have been classified and assessed for their potential usefulness in Australia.[48] Of these *Ae. elegans*, *Ae. villosa*, *Ae. abyssinica*, *Ae. brasiliana*, *Ae. brevifolia*, and *Ae. histrix* are being evaluated for their suitability to cooler and drier environments than those suited to the released cultivars of *Ae. americana*. *Ae. filosa*, *Ae. flutans*, *Ae. parviflora*, and *Ae. sensitiva* are adapted to swampy areas and may be useful for ponded pastures.[47]

Chamaechrista (*Cassia*) *rotundifolia* cv. Wynn is a free-seeding semierect and short-lived summer growing perennial legume tolerant of heavy grazing. It prefers the lighter textured surface soils in the 700 to 1200 mm rainfall zone.[28,49] Its ease of establishment and ability to spread readily in grazed pastures can lead to legume dominance under heavy grazing if associated with grasses of low vigor. This ability to colonize rapidly may be associated with low palatability relative to associated grasses, though it is eaten in amounts (20 to 30%) that contribute to improved animal performance.[50]

Higher yielding introductions have been identified that may be better suited to more northern environments, though these have not yet been released as cultivars. Selection of lines with greater hardseededness may enable better persistence in areas with marginal rainfall.[49]

Desmanthus includes18 species, although only *D. virgatus* has been widely evaluated in northern Australia. Its specific attribute of worth is its adaptation to clay soils, which have always posed problems for legume persistence. *D. virgatus* is also adapted to dry environments in both North and South America and is palatable to livestock and tolerant of grazing.[51] In Queensland, statewide evaluation of over 40 accessions has resulted in the selection of three cultivars adapted to subhumid and semiarid environments.[52] These are:

cv. Marc	(CPI 78373 ex Argentina) — early flowering, low growing (50 cm) type from a 650 mm rainfall zone.
cv. Bayamo	(CPI 82285 ex Cuba) — mid-season flowering, tall (150 cm) type from a 1200 mm rainfall zone.
cv. Uman	(CPI 92803 ex Mexico) — late flowering decumbent type (80 cm tall) from a 1250 mm rainfall zone.

All are effectively nodulated with rhizobium strain CB3126 and produce good machine-harvested seed yields.[52]

It is too early to assess specific adaptability of these cultivars and what commercial impact *D. virgatus* will make when seed supplies become available.

Indigofera is another genus that includes species adapted to clay soils. Although the genus is known to contain the hepatotoxin, indospicine,[53] not all *Indigofera* species contain the toxin.[54] Among these is *Indigofera schimperi* from Africa. None has been released as a cultivar, but CPI 52621 has been widely tested on clay soils with a view to release. It is a small perennial shrub native to eastern and southern Africa, where it occurs on heavy clay soils. Tolerance to drought, high soil pH, and some salinity are other agronomic characteristics, though low palatability may be a problem. In Queensland, it has persisted on alkaline/saline clay soils and dulplex soils in dry southwest Queensland,[55] and on black earths of the livestock/cropping zone of subtropical Queensland.[15] It should be a valuable addition to the small number of legume cultivars adapted to clay soils.

Centrosema pubescens, a vigorous perennial trailing tropical legume, has been used in the wet tropical coastal areas for many years. The commercially available material combines well with tufted grasses such as *Panicum maximum* in grazed pastures, but is intolerant of low fertility and highly acidic soils. It is also attacked by leaf diseases, particularly in the cool season. *Centrosema scheidianium* (formerly *pubescens*) cv. Belalto has a less twining growth habit, roots well at the nodes, and grows more actively in the cool season than does common centro. In addition, it is less susceptible to *Cercospora* leaf spot and red spider mite (*Tetranychus* spp.) and has persisted well under grazing. Lack of commercially available seed has limited its use in pastures of the wet tropics.

The annual *Centrosema pascuorum* is adapted to the monsoon climate of northern Australia with a reliable wet season rainfall and a clear cut dry season. It is a vigorous twining or scrambling species able to compete well with associated grasses and to survive long periods of waterlogging as well as periods of drought. The bred cultivar Cavalcade is early flowering, nematode-resistant, and adapted to the drier monsoonal regions with 700 to 1500 mm annual rainfall. For the wetter areas of 1300 to 1500 mm, cv. Bundey is later flowering, higher yielding, and better adapted to waterlogged conditions. Both legumes provide good quality feed into the dry season.

Arachis species demonstrating forage potential share the valuable attribute of tolerance of defoliation. They have potential for the moist coastal and subcoastal areas from the subtropics to the tropics. *A. pintoi* cv. Amarillo was released in 1987 and *A. glabrata* cv. Prine in 1995. Both have persisted well in a range of environments.[56] Other promising species are *A. prostrata* and *A. repens*. All have the problem of difficulty in economically producing propagating material, whether seed or vegetative material. This problem needs to be overcome before its potential as a valuable legume associate in grazed pastures can be realized.[57]

There are other successful herbaceous legumes that are commercially used over limited areas. These include the cold tolerant, prostrate *Lotononis bainesii* cv. Bryan in coastal and subcoastal subtropical Queensland; *Vigna parkeri* cv. Shaw, a grazing tolerant, fine-stemmed twining legume adapted to the wetter subtropical areas, and *Neonotonia wightii* cv. Tinaroo, a vigorous late flowering twining legume adapted to deep fertile soils on the Atherton Tablelands in north Queensland. *Clitoria ternatea* cv. Milgarra is being increasingly used in central and southeast Queensland as a ley legume for clay soils. Although not adapted to heavy continuous grazing, it performs well in a ley system and is of high nutritional quality.

Leucaena leucocephala is the only shrub legume to have been grown commercially to any extent. It is estimated to be sown on 35,000 ha in northern Australia.[58] Most sowings have been with two cultivars, Peru and Cunningham, which are more branched and suitable for grazing. They have similar performance in most environments and are renowned for their high beef production potential.[59] Unfortunately, both are susceptible to the psyllid insect, which arrived in Australia in 1986.[60] Through the 1980s, plantings increased sharply, but new sowings mainly occur in the drier areas of central Queensland[61] and in the Ord Irrigation Area in the northwestern portion of western Australia, where humidity is lower than in coastal areas and the psyllid attacks are less frequent or absent. *L. leucocephala* cv. K636, now named cv. Tarramba, recovers faster from psyllid attack than existing cultivars and has good seedling vigor.[62,64] It is, however, taller growing and less branched than cv. Cunningham and may require more frequent slashing to keep it within grazing height. It has given similar production to cv. Cunningham in central and northern Queensland and in the northwestern portion of Western Australia, but higher yields in the more psyllid prone southeast Queensland.[63] Detailed agronomic evaluation of the *Leucaena* genus has identified psyllid tolerant material in *L. pallida*, *L. trichandra*, *L. collinsii*, and *L. diversifolia*. The F_1 hybrid of *L. pallida* (K376) X *L. leucocephala* (K636) (known as KX2) had both high psyllid tolerance and high yield in a range of environments,[64] however, it may be unsuited to extensive as opposed to intensive use in Australia.

The higher tannin levels in some of the psyllid tolerant species compared with those in the *L. leucocephala* cultivars, may be associated with lower nutritive value.[65] The wider germplasm now available may enable breeders to develop cultivars with better cool tolerance for subtropical areas, improved pest tolerance, and also good nutritive value.[66]

Other shrub legumes being evaluated in Queensland include *Calliandra calothyrsus*, which is not attacked by psyllids, grows on more acid soils than *L. leucocephala*, and is more tolerant of cool conditions.[67] Its potential for beef production is not considered to be good in north Queensland.[68] Collections of the genus made by the Oxford Forestry Institute are also being assessed for variation in yield, leafiness, and feed quality.[67]

Gliricidia, with four species, has only one species, *G. sepium*, of agronomic significance.[62] Although variation in yield and forage quality occurs between different accessions tested in tropical Australia,[62] the low palatability to cattle and the leaf fall at the onset of cool conditions in the dry season are serious limitations to the development of suitable cultivars.

Sesbania includes an estimated 70 species, however, only *S. sesban* has much promise as a perennial pasture plant.[62,69] Its ability to tolerate waterlogging, soil salinity, and alkalinity are valuable attributes. Variation in yield and other agronomic attributes has been measured on introductions from Africa, and one of these, selected following animal production measurements, was released in 1993 as cv. Mount Cotton.[69] The cultivar is suitable for low lying wet areas in the subtropics. It establishes rapidly, is liked by cattle, and is highly digestible. It is, however, subject to stem breakage during grazing. As such, it is expected to survive only 2 to 3 years under grazing.[69]

B. Grasses

Far more seed of tropical grasses than of tropical legumes is currently sown in Australia (Table 17.4), with 70% of the sown pasture area sown solely to grasses.[13] Large areas are sown to grass on the fertile Brigalow soils, which once carried forests of *Acacia harpophylla* (Brigalow) and covered about 9 m ha.[70] These and other clay soils, clay loams, and alluvial soils in Queensland and northern New South Wales have been used for cropping and pastures. Their high initial fertility enabled crops and pastures to be grown without fertilization for decades. In the pasture phase, however, decrease in production over time is a feature associated with a decline in available soil N.[71] Plowing and resowing reestablishes another cycle as the N is released for further pasture production.[72]

A number of grasses are available for pasture development; only the more important ones are described below. More detailed descriptions are found elsewhere.[28,73,74]

Cenchrus ciliaris (buffelgrass) is the most drought tolerant and widely sown tropical grass species. Nine cultivars have been released with Biloela and Gayndah the most widely used. American buffel and West Australia buffel are regarded as being better adapted to more acid soils and to low rainfall, respectively.[28] They are not sown in more productive environments. Biloela and Molopo are vigorous cultivars better adapted to clay soils, though less compatible with legumes than the shorter cultivars like Gayndah.[28] Molopo also exhibits more rhizomatous development and has better cold tolerance than other cultivars.

Two new cultivars, Viva and Bella, have been selected for winter/spring growth and good overall yield and were released in 1994.[75]

Panicum maximum var *trichoglume* (green panic) is represented by one cultivar, Petrie green panic, the origin of which is not clear.[28] It is widely adapted to soils of medium to high fertility with rainfall above 600 mm. Although less frost tolerant than Rhodes grass, it is more drought tolerant and produces more early growth in spring.[28] It is not well adapted to wetter areas near the coast, where *P. maximum* cv. Gatton has more vigor and persistence.[28]

Urochloa mosambicensis cv. Nixon is the only *Urochloa* cultivar in use. Although not as drought tolerant as *Cenchrus ciliaris* cultivars, it is adapted to the drier tropical areas, but has poor cold and frost tolerance. It is intolerant of waterlogging and very acid soils, responds rapidly to small falls of rain, and is very palatable to cattle. Compatibility with legumes is good. This may be due to a combination of relatively high palatability and lower growth form compared to some other tropical grasses. Nixon is adapted to a wide range of soils, especially those with higher soil P levels (>6 to 8 ppm bicarbonate extractable P). It is very early flowering and free seeding, but remains palatable throughout in north Queensland.

Chloris gayana (Rhodes grass) is well adapted to a wide range of soils in the subtropics and the wetter parts of the seasonally dry tropics. It is not as drought resistant as *Cenchrus* or *Urochloa* or *P. maximum* var *trichoglume* cv. Petrie though it tends to be better adapted to lower fertility conditions, waterlogging, and saline soils. Good stoloniferous development enables it to give rapid ground cover and to tolerate heavy grazing when not stressed by drought. Seven cultivars have been released in Queensland. They differ in flowering time, leafiness, and acceptability by cattle. The diploid cultivars Pioneer and Katambora are finer leaved, early flowering, and more drought tolerant, though less palatable than the tetraploids Callide and Samford. Stoloniferous development is least in cv. Pioneer, which is the most frost tolerant of the cultivars. The newer cultivars, Topcut and Finecut, are derived from Pioneer and Katambora, respectively, and are higher yielding, more leafy, and have finer stems (D.S. Loch, unpublished data). The cv. Nemkat, also derived from Katambora type material, has specific resistance to the root knot nematode (*Meloidogyne javanica*), and is therefore valuable in rotation with tobacco to reduce the incidence of nematode infestation.

Bothriochloa pertusa (Indian bluegrass), introduced from India and Africa, is a variable species in terms of its morphological and agronomic characteristics. It is adapted to clay soils and to a wide range of other well-drained soils.[76,77] The Bowen strain, naturalized in northeast Queensland, is early flowering, free seeding, grazing tolerant, and moderately drought resistant. It is one of the few grasses that can establish when oversown into grazed native pasture without mechanical disturbance. It is not very responsive to P fertilization relative to other grasses such as *U. mosambicensis*[78] and can grow on P deficient soils. Two cultivars have been released; cv. Medway is much later flowering than the naturalized Bowen strain, more productive, and more resistant to leaf rust. The cv. Dawson is also later flowering but more prostrate. It forms a denser sward with suitability for lawns and recreational areas as well as for use in pastures.

Palatability of these cultivars may be lower than some of the native grass species and also other sown species due to the strong odor from volatiles in the leaves. In experimental pastures, poor compatibility of the Bowen strain with Verano stylo at high stocking rates has resulted in poor steer gains compared with *U. mosambicensis* cv. Nixon/Verano pastures.[77]

Setaria sphacelata var *sericea* is a variable, tufted species of mainly African origin with a range of ploidy levels. In Australia they are mainly grown in the subtropics, near coastal areas receiving about 1000 mm annual rainfall, or in elevated tropical areas. They are not well adapted to drought, but tolerate waterlogging and heavy grazing and combine well with legumes. The cv. Nandi has been in use for many years. It has lower yield and frost tolerance than the other cultivars, Kazungula, Narok, and Solander. The latter has higher seed yields than Narok but has similar frost tolerance.

Setaria sphacelata var *splendida*, a very palatable robust form better adapted to the lowland tropical areas, has been mainly propagated vegetatively from tillers due to its very low seed production. However, cv. Splenda, bred from hybridization of *S. sphacelata* var *sericea* and *S. sphacelata* var *splendida* has similar morphology and adaptation, but can be readily established from seed.

Pennisetum clandestinum (Kikuyu grass) is a strongly perennial, stoloniferous, and rhizomatous grass adapted to the wetter and more fertile subtropics and elevated tropics. It withstands heavy grazing and is associated with dairying in New South Wales and Queensland. It establishes from vegetative propagation and also from seed. Technology for commercial seed production has been devised in Australia. Cultivars have been developed from naturally occurring ecotypes and differ in growth form, seed production, and disease resistance. The cv. Whittet is more robust and erect than cv. Breakwell, which is densely tillered and prostrate. The finer leaved cv. Crofts is more cold tolerant, and cv. Noonan is more tolerant of "kikuyu yellows" disease and has higher seed production.

Brachiaria decumbens (signalgrass) is a decumbent, stoloniferous, and rhizomatous apomictic perennial, which is well adapted to the humid tropical areas but does not tolerate flooding. It is readily established from seed and forms a dense ground cover that is tolerant of heavy grazing. Compatibility with legumes can be a problem, especially in fertile situations. Stoloniferous legumes like *Desmodium heterophyllum* are more compatible than the twining legumes. The only cultivar used in Australia is cv. Basilisk derived from seed introduced from Uganda.

Brachiaria humidicola (koroniviagrass) is an apomictic, prostrate, densely stoloniferous, and rhizomatous perennial adapted to low fertility and poorly drained situations in the wet tropics. However, it responds to fertilization and under better drained conditions can tolerate very heavy grazing. Its feed quality is not high, though the grass is palatable. Compatibility with most legumes is poor, especially in fertile situations. It is used mainly in intensive nitrogen-fertilized pastures in coastal north Queensland. There is only one cultivar, Tully.

Digitaria milanjiana (finger grass) is a variable species both morphologically (erect to prostrate stoloniferous forms) and in its range of adaptation to the wet and dry tropics and subtropics. It is generally a palatable species to stock, and the rhizomatous and stoloniferous forms are tolerant of heavy grazing. Problems of low seed production have been a feature of some introductions. There is only one cultivar in Australia, cv. Jarra. It is drought resistant, grazing tolerant, and persists well with legumes under grazing. It also produces commercially acceptable seed yields. Although adapted to a range of soil types, it does not tolerate waterlogging. It is resistant to the root burrowing nematode (*Radopholus similis*) and has proven useful in breaking the cycle of this pest in banana plantations in north Queensland.

Digitaria eriantha (*D. smutsii*) (common finger grass) is a variable species, with tufted and densely stoloniferous forms, generally palatable and of good nutritive value. Pangola grass (formerly *D. decumbens*) is sterile and needs to be vegetatively propagated. It has been used in northern Australia, but succumbed to a range of pests and diseases. It is still used in some areas, but the cost of establishment is often prohibitive.

Seeding, tufted forms are represented by cv. Premier and Apollo. They are adapted to subtropical areas with 700 to 1000 mm MAR and grow best on sandy to sandy loam soils. Early spring growth, the ability to spread by seed, and acceptability to stock are important attributes. Compatibility with legumes may be a problem, especially in more fertile situations.

Andropogon gayanus (gambagrass) is a highly productive, tall, erect, coarse, tussock grass that is very palatable to cattle when young. Only one cultivar, Kent is used in Australia. It is adapted to a range of soils in the seasonally dry tropics with seasonal rainfall of 750 to 1500 mm MAR. It can grow on soils of low P status and remains green in the dry season. Although compatible with a range of legume species, its tall growth habit (up to 4 m) can produce management problems.

C. Grass/Legume Associations

1. Problems

The C4 grasses, with their higher photosynthetic capacity and potentially higher productivity, can result in compatibility problems when grown with C3 legumes.[79] Management of such mixtures may be more complicated than the management of N-fertilized grass pastures, especially on more fertile sites and with legumes that may be only marginally adapted edaphically and climatically. However, three factors may prevent the anticipated shift to grass dominance.

First, the C4 grasses do not express their potential because nitrogen (N) is a major limiting factor on most soils in Australia, except the fertile clays of the Brigalow region in central and southern Queensland just after cultivation and pasture establishment. On low fertility soils, where N is a major limiting factor, effectively nodulated legumes may have a competitive advantage over grasses.

Second, the selective grazing of domestic stock can greatly modify the grass/legume competition in grazed swards. Following the break of season with spring/summer rains, a flush of C4 grasses could provide powerful competition to reestablishing legumes from seed or regrowing perennial plants. Fortunately, the grass is favored by cattle relative to many legumes at this time of year, thus grazing controls the flush of grass growth and enables the legume to grow unhindered by cattle grazing. When the grasses mature in autumn and the legumes are in the flowering stage, the legumes become more palatable and are preferred.[80,81] Their selection at this time of year, however, is no major penalty, provided seed can be set, as the grasses may be unable to exploit the advantage from lower grazing pressure because of low available soil N, low moisture availability, low temperatures, or a combination of these factors.

Third, it has been established that under low soil P availability, legumes such as *S. hamata* are more competitive and efficient at utilizing soil P and diluting it for growth than are the grasses.[78] In grazed pastures this can result in legume dominance. Conversely, if P fertilizer is applied to these stylo/grass pastures, sown grasses tend to respond more than the legumes resulting in grass dominance.[78,82] By manipulating stocking rate and fertilizer P rates, the legume content of these stylo-sown grass pastures may be kept within a desired range.

In subtropical pastures, the lack of persistence of legumes in mixture with grasses has been a major problem with some legumes, notably the twining legumes, but also with stylo. With frosty wet winters, the situation may be exacerbated. *Vigna parkeri*, although a twining legume, does tolerate heavy grazing in its area of adaptation (moist, subtropical areas) through its ability to form fine prostrate stems near ground level.[83] Both *Chamaechrista rotundifolia* and *Arachis pintoi* can withstand heavy grazing. Their tolerance to heavy grazing may be due in part to relatively low palatability as well as growth form.

In the seasonally dry tropics of northern Australia, a "reverse situation" has occurred where, with increasing grazing pressure, the native perennial grasses such as *Themeda triandra*, *Heteropogon contortus*, *Sehima nervosum*, and *Chrysopogon fallax* have declined or been eliminated and the oversown pastures have become stylo dominant.[84] Introduced grasses to withstand the higher stocking pressures have been sought to address this legume dominance.[77] As far as animal production is concerned, there is little or no penalty in terms of animal production (provided Na and P deficiencies are overcome by supplements) from pure legume swards. However, they are usually unstable ecologically and prone to weed invasion. An associated grass is needed to provide stability and to prevent soil erosion from early storm rains.

With taller shrub legumes such as *Leucaena leucocephala*, competition from tropical grasses after the establishment phase is not a problem. However, such legumes can create management problems if they grow tall and require mechanical slashing. Free-seeding shrub legumes, if not properly managed, can also form thickets. These impose severe problems with grazing animals. Notable examples are impenetrable *L. leucocephala* thickets in Hawaii and in some of the Pacific Islands.[85] Under good commercial management, excess height and reseeding is not usually a problem, especially in the subtropics where top growth may be frosted.

Like temperate and Mediterranean legumes, tropical legumes can acidify the soils in which they grow. It was anticipated that in association with perennial tropical grasses the rates of acidification in tropical Australia would be much lower than rates recorded with annual legume-based pastures in southern Australia. However, comparable rates have been measured on stylo-based and on leucaena-based pastures.[86-88] The acidification rates are more rapid on weakly buffered light textured soils and where the legume is dominant and plant material removed as hay or for seed production. Management should be aimed at maintaining a good perennial grass associate to reduce acidification rates. However, minimizing acidification rates on extensive pasture areas is an ongoing challenge to researchers.

2. Advantages of Grass/Legume Associations

In areas where it is uneconomic to use nitrogen fertilizer (generally in areas not used for dairying, and in areas below 850 to 1000 mm MAR), a legume is necessary to maintain pasture productivity and to improve pasture quality due to the higher N and digestibility of the C3 legume compared with C4 grasses.[89] The main advantages compared with grass-only pastures are higher gains/animal and higher carrying capacity for fattening cattle and in higher carrying capacity, higher calving percentage, and higher weaning weights in cow/calf systems.

In general, good grass/legume pastures can increase steer gains by 30 to 50% and carrying capacity by three- to tenfold compared with native pastures.[46,90-95] These increased gains per steer are mainly measured in the autumn/winter period.

Fewer experiments have been conducted on sown pastures with breeding cattle but, compared with native pastures, conception rates, weaning weights, and cow liveweights have improved.[96-99] With the higher stocking rates possible on the improved pastures, weaning weight/ha can be increased fivefold.[99]

In addition to the benefits to grazing livestock, grass/legume pastures may be used to restore the depleted soil fertility over 1.2 m ha in subtropical Queensland resulting from exploitive grain growing,[100] to increase soil carbon (C) and N, and also to improve soil structure.[100,101] These benefits can then be reflected in improved crop yields and reduced soil loss, when grown in a rotation with crops. These effects have been well documented in southern Australia with subterranean clover based pastures, which have provided the major input of N for following wheat crops.[102] In tropical Australia, pasture/crop rotations have not been as successful as in southern Australia. First, because much of the cropping has occurred on fertile clay soils without the need for inputs of fertilizer N, and, second, because successful perennial tropical legumes for such soils have been difficult to find. The decline over time in N content of these soils has focused attention on the need for improving soil N to maintain or increase wheat protein content.

In the subtropics, the combination of temperate legumes (*Medicago* spp.) with tropical grasses can be used in rotations, resulting in improved yields and quality of subsequent grain crops related to improvements in soil N. Leys containing lucerne also increase crop yield and quality. However, they deplete soil water more than does continuous wheat, and in the year following the ley phase, grain yields are reduced in low rainfall environments.[101] The annual medics do not deplete soil water more than continuous wheat and so grain yields are not adversely affected.[101]

Benefits following long leys (three and a half years) of lucerne on wheat grain yields and protein content have been measured for up to 8 years, with increases in yield over the first 4 years of approximately 0.5 t/ha.[101] Benefits from medic leys on soils of low N status depend on winter rainfall for growing the medics; input of N being related to growth of the legume.[101] The benefit to subsequent grain yields can be largely attributable to N input to the system, but reductions in pests and diseases must also play a part, since the benefits are longer lasting than when N fertilizer is applied to wheat crops.

In the semiarid tropics of northern Australia, experimental ley pastures based on the legumes *Stylosanthes hamata* cv. Verano, *Alysicarpus vaginalis*, and *Centrosema pascuorum* contributed between 15 and 85 kg N/ha to a sorghum crop after short and long leys, respectively, compared with the nitrogen uptake following a grass ley.[103]

In this system, the ley provided high-quality feed for cattle, residues for a mulch in a no-till system, as well as N for the following crop.[103] Although the ley system provided the benefits expected from legume leys in southern Australia, the system is not widely adopted due to the requirements for high capital and managerial inputs, coupled with the current poor economic viability of grain cropping in the semiarid tropics.

3. Improvement Objectives

A key improvement objective for grass/legume associations, particularly in the subtropics, is persistence of the legume component at a reasonable level under grazing.[104,105] To fix 100 kg N/ha/yr the legume has to produce about 3 t DM/ha/yr in mixture with grasses.[106] There are many factors that constrain legume production in mixed swards. Some of these factors relate specifically to the legume, e.g., genotype, seed production, disease susceptibility, insect attack, frost tolerance, grazing tolerance, and others that interact with the associate grass, e.g., competition for water, nutrients, light, and relative palatability.[107-109]

Clearly, any improvement in the legume component by way of resistance to diseases or pests, improved seedling vigor, drought tolerance, or the provision of a better *Rhizobium* symbiont, should improve its persistence in mixed swards. Traditionally, the best grass from an evaluation program and the best legume from the legume evaluation program are combined in a pasture mixture. This may not always prove to be the best combination, as they may not be compatible.

Compatibility can be very dependent upon nutrient availability. In mixtures of *Setaria sphacelata* cv. Nandi/*Desmodium intortum*, the supply of potassium (K) was vital to the survival of the *D. intortum*, whereas the *D. intortum* grew well with no additional K in pure culture.[110,111] Grass species that can substitute sodium (Na) for K may associate better with legumes than grasses that cannot use Na effectively.[112]

With increasing P fertilizer levels applied to a *S. hamata* cv. Verano/*Urochloa mosambicensis* cv. Nixon mixture, the grass increased in proportion as the P level increased. At zero P the plots were legume dominant, and at high P they were grass dominant.[78] Some *Stylosanthes* species are strongly mycorrhizal under low P conditions and are able to compete effectively for soil P. Many grasses on the other hand, require moderate levels of soil P to grow well.

Strategic management in addition to nutrient manipulation can also aid persistence of the legume in mixture with grasses. Maintenance of soil seed banks by timely resting of the pastures in autumn to allow seeding and/or the use of heavy grazing in spring, when grass is preferred by stock, to aid reestablishment of seedlings are examples. An understanding of the life cycle of the legumes including plant survival and recruitment and response to management input are vital to the development of appropriate strategies.[108]

D. Successful Uses of Introduced Tropical Pasture Species

Commercial adoption of improved tropical pasture is increasing steadily, particularly in Queensland. Four examples of such development are described illustrating both extensive development and intensive development.

Development of extensive stylo-based pastures has been very successful on the Wrotham Park property at the base of Cape York Peninsula, North Queensland.[113] Owned by the Australian Agricultural Company from 1963, the one million ha property has a MAR of 943 mm (range 408 to 1407 mm), most of which falls in summer (December through March). The infertile yellow earth and yellow duplex soils support a native grass understory (*Sorghum*, *Chrysopogon*, *Schyzachyrium*) to eucalypt woodland. In the unimproved state, steers reach 210 to 220 kg dressed weight at 5 to 6 yr at a carrying capacity of 1 steer to 25 ha. Pasture establishment was initially with Townsville stylo sown into plowed, cleared country. More recently Verano stylo (2 kg/ha) and Seca stylo (1 kg/ha) have been oversown by air with 125 kg of superphosphate/ha onto uncleared country, which has been burned or heavily grazed to control the vigor of native grasses before sowing. Development has occurred in 2,400 ha paddocks capable of carrying 1000 head of cattle throughout the year.

Costs per paddock of $191,454 have been incurred for the development, including fencing and stockwater, or $191.45 per beast area. These pastures carry 1 beast to 2.4 ha and steers achieve 245 to 265 kg dressed weight at 3 to 4 years of age.

Over 37,000 ha of improved pasture have been developed. Currently, this area is used for breeding cattle to supply young stock to another fattening property owned by the company near Townsville developed using the same technology.

At "Kuttabul," near Mackay on the Queensland coast, undeveloped land on sandy coastal lowland soils receiving 1750 mm (792 to 2465) MAR carries one breeder to 10 ha. With development, pastures carry 1 breeder to 1 ha on the 534 ha property with improved weaning percentage and weaning weights.[114] Although development costs for clearing of the woodland, plowing, and sowing are high at $430 to 640/ha, of which $120 is for seed and fertilizer, the outcome is profitable. Again, the key is persistent legumes including Seca, Verano, and Cook stylos and Glenn joint vetch sown in mixtures with Kazungula setaria and *Paspalum plicatutum* cv. Rodd's Bay.

The third example, "Rowanlea," near Calliope, in cooler central Queensland, occupies 7060 ha of hilly country with fertile creek flats alternating with less fertile granite ridges.[115] Rainfall averages 873 mm (430 to 1403) a year and the good native pastures consist of *Heteropogon contortus*, *Themeda triandra*, *Bothriochloa,* and some *Paspalum dilatatum* on the creek flats.

Three intensities of pasture development have occurred. High input pasture development on the better cropping country involves Callide Rhodes grass and Siratro sown onto fully cultivated land. Medium input involves two discings over the native pasture and sowing with Callide Rhodes grass, *Bothriochloa insculpta*, Siratro, Wynn Cassia, and Seca stylo. The low input development is simply broadcasting the legumes Seca, Wynn, and Siratro onto the native pasture. Cost of seed varies from $38 to 58/ha. Pastures are spelled from grazing for 4 months over summer every three years to enable good seeding to improve persistence, particularly of the legumes. These improved pastures carry 1 beast to 2 ha compared with 1:3 ha for native pasture, and steers are heavier than those from native pastures when slaughtered.

The last example, "Minnie Plains," near Blackwater in central Queensland, is a 10,000 ha cattle property on slightly undulating country. The 600 mm annual rainfall is mainly of summer incidence, but with some winter component in most years. Much of the property is cleared Brigalow (*Acacia harpophylla*) scrub. The soils are mainly fertile black cracking clays with ridges of lighter red soil.

Although cropping and feedlot fattening have been tried in the past, together with the use of irrigated oats for grazing, these ventures were not very profitable.

The major management option to improve productivity was the use of the leguminous tree leucaena (*Leucaena leucocephala*) from the mid 1980s. This is established in rows 10 m apart, kept weed free by cultivation and the use of herbicides, and only grazed on a regular basis when well established, about 2 years after sowing. A mixture of grasses, including *Panicum maximum* var *trichoglume*, *Cenchrus ciliaris* cv. *Gayndah*, *Sorghum* sp. cv. Silk, *Setaria incrassata* cv. Inverell, and *Bothriochloa insculpta* cv. Hatch is sown between the leucaena rows in the summer following leucaena planting. This enables the leucaena to become well developed before competition from the associated grasses.

When fully established, the leucaena can carry much higher stocking rates (up to 2 steers/ha) than the native pasture (1 steer/4 ha) and gains per steer of about 1 kg/day can be achieved following the use of the rumen bacterium, *Synergistes jonesii*, to detoxify the mimosine in the leucaena. The aim is to meet Japanese ox specifications of about 650 kg at 2.5 years of age.

Currently about 520 ha of leucaena are grown, but in view of its survival through three consecutive years of drought, when only half the annual rainfall was received, the plan is to increase this to 1,000 ha. In addition to the leucaena, the lighter textured red soil areas are being oversown with *S. scabra* cv. Seca to further improve productivity.

The control of Brigalow regrowth and the use of improved pastures have enabled the cattle herd to be increased from about 1,300 head to 2,300 head in a ten-year period.

V. ADDITIONAL USES OF GRASS OR LEGUME COMPONENTS

A. Silage

Very little silage is made in the tropics and subtropics of Australia. Only 3% of dairy farms in Queensland feed silage as a supplement, and two thirds of these feed maize silage.[116] In general, tropical pasture grasses and legumes are not readily made into silage. Unlike silage from temperate pasture species, silages from tropical species are characterized by acetate fermentation, often with the development of high ammonia-N concentrations.[117] The typical lactate fermentation can be produced reliably only by substantial additions of soluble carbohydrate such as molasses.[117]

With newer techniques of silage making involving wilting, fine chopping to give better compaction, and the use of silage additives, better silage may be possible.[116] The technique of making plastic coated rolls may also favor greater use of silage. However, tropical pasture silages are unlikely to become widely used in Australia. Maize silage is, however, gaining favor among dairy farmers in both northern New South Wales and Queensland. At a cost of 6 to 13 cents/liter of milk produced, it is competitive with alternative feeds.[118]

B. Hay

Lucerne is the major hay crop in the subtropical areas, often grown with irrigation. The advent of lucerne aphids in the 1980s had severe effects on the widely adapted standard cultivar Hunter River, with reductions in areas sown to lucerne. With the development of pest and disease resistant (leaf and root diseases) lucerne cultivars, this trend is now being reversed. Use of large round bales for haymaking facilitates mechanical movement of hay and ease of feeding out in the paddock, and as a result this approach is increasing. In other areas, hay making is a means of conserving excess forage in good seasons and so tends to be rather opportunistic involving a wide range of pasture species and forage crops.

C. Ponded Pastures

These are used to supplement native pastures in the dry season. The aim is to construct shallow dams to impound runoff water in the wet season to extend growth of flooding-tolerant grasses in the dry season. These shallow dams, sited on areas of low relief, can also be filled with water stored in deeper larger dams if necessary. Paragrass (*Brachiaria mutica*) is commonly found along water courses and in low-lying areas along the east coast of Queensland. It is used successfully in these ponded situations. This additional forage has given increased cattle gains compared with native pasture alone.[119] Paragrass is not adapted to water depths exceeding 600 mm and, consequently, the amount of stored water in such situations is limited. Two new grasses, however, are able to tolerate water up to 1.2 m, *Hymenachne amplexicaulis* cv. Olive and *Echinochloa polystachya* cv. Amity. These grasses have extended the ponded pasture technology to drier areas away from the coast with resultant benefits to stock through access to green feed through the dry season.[119] Use of fertilizer nitrogen on these pastures improves their yield and quality, but an effective legume to incorporate in these systems could be economically more attractive.

D. Stand-Over Feed

The predominantly summer rainfall in tropical Australia, the yearlong grazing of the stock, and the relatively minor role overall of conserved forages, highlights the contribution made by stand-over feed. With native pastures, such stand-over feed is inadequate to maintain animals through the dry season, and they inevitably lose weight. On sown grass/legume pastures, weight gain continues longer into the autumn period, and the cattle are generally maintained for most of the dry season on stand-over feed.[46,93,120] Weight change in the dry season is often dependent on legume availability.[121] The stand-over feed retains its nutritive value well under low humidity conditions and in the absence of dew, but serious deterioration can occur if rainfall and low evaporative conditions prevail or under conditions of heavy dewfall. Legumes are more susceptible than grasses as they can support more active fungal growth on the plant tissues, especially the leaves.[122,123] Areas that could suffer from these losses with *Stylosanthes hamata* cv. Verano have been modeled from climatic data and mapped for northern Australia.[124] In central and southern Queensland, frost can have dramatic effects on stand-over feed quality, particularly if this is followed by rain. Some frost tolerance and the ability to grow under cool conditions, e.g., *Paspalum dilatatum*, *Pennisetum clandestinum,* and *Setaria sphacelata* cv. Narok are advantages under these situations. In heavily frosted situations, the accumulation of stand-over feed by the use of N fertilizer can predispose the grass to killing or poor growth in the following spring. With N-fertilized pastures of *Chloris gayana* cv. Samford and *Setaria sphacelata* cv. Nandi, lenient grazing, which resulted in the accumulation of ≈ 4 t DM/ha in autumn, resulted in reduced spring regrowth of 19 to 55% compared with more heavily stocked paddocks after heavy winter frosting.[125]

To avoid the damaging effects of frosting in low lying areas, the pastures should be grazed to prevent accumulation of senescing material or sown with frost-tolerant species. Stand-over feed in the subtropics is best located on ridges not prone to heavy frosting and should contain legumes to improve quality. *Leucaena* under such conditions provides quality feed into autumn. Its height also enables it to escape ground frosts and so retain feed quality.

VI. LATEST DEVELOPMENTS IN THE USE OF BIOTECHNOLOGY FOR FORAGE IMPROVEMENT IN AUSTRALIA

No pasture cultivars released to the industry in Australia have been developed using biotechnology or genetic engineering approaches. However, the prediction is that "genetic engineering will catalyze an era of increased sophistication in both pasture plant breeding and in the management of pastures in the farm system."[126]

The link between plant breeders, biotechnologists, and genetic engineers will be vital to rapid progress in producing new cultivars. The use of DNA sequence tags is already enabling breeders to follow specific genes through the breeding program using DNA hybridization. Such markers are equivalent to following a particular phenotype, but with much more precision.

In Australia, the improvement of feed quality is a specific aim of these new approaches. The focus is currently on legumes, since gene transfer systems for these have been developed and are less difficult than with grasses. In Canberra, the CSIRO is seeking to introduce into subterranean clover and lucerne proteins rich in sulphur amino acids and resistant to rumen degradation to improve wool growth when fed to sheep.[127] In transgenic plants, these foreign proteins, ovalbumin, and sunflower albumin, have yet to be expressed at a high and stable level to confer beneficial effects. However, the construction of genes incorporating flanking DNA sequences may enable high expression of the desired proteins from the current level of 0.3 to 3% of the soluble leaf protein.[127]

In Canberra and Brisbane, the aim is to develop systems that will enable a reduction in the lignin content of pasture plants. Such a reduction should improve nutritive value of the fibrous tropical species.[128] The species of interest are *Stylosanthes humilis* and *Medicago sativa*, and the strategies are to target specific enzymes in the biosynthetic pathway for lignin synthesis with a view to reducing their expression. Ribozyme and antisense technologies are being used to achieve these objectives. Furthermore, specific promoters are being isolated to direct the genes to work specifically in the xylem of the stems.[126,128,129] In the case of *S. humilis*, an increase in stem digestibility of 10 units has been achieved, even though estimates of lignin by standard analytical techniques were not reduced in the transformed plants.[130]

In the future, the ribozyme, ribosome, and antisense technologies will be used to protect pasture plants against pests and diseases, and to improve their tolerance to environmental stresses, including uptake of nutrients, for improved growth on nutrient deficient soils, which are widespread in the tropics.[126]

In addition to the potential for improving forage quality by modification of pasture species, work in Australia has also focused on modification of the rumen microorganisms to achieve better utilization of the fiber in the consumed forage. Of particular interest is the finding that enzymes from the anaerobic rumen fungus *Neocallimastix patriciarum* have high capacities for cellulose degradation. Furthermore, cellobiohydrolase may be rate limiting in certain rumen bacteria. If the enzymes from the rumen fungus could be incorporated in rumen bacteria, then the rate of degradation of plant fiber could be enhanced with consequent improvement in digestibility and feed intake. Genes encoding for high cellulose activity have been isolated from *Neocallimastix patricianum*,[131] and the prospects for improving fiber digestion in the rumen by genetic engineering of rumen bacteria look promising.[132]

A novel approach to overcoming antinutritive factors in forages is the modification of bacteria to degrade plant toxins in the rumen, following on from the successful solution to the leucaena toxicity problem by using DHP-degrading bacteria, which occur naturally in some tropical countries but not in many others.[59,133] Fluoroacetate poisoning in ruminants, which consume the leaves of *Gastrolobium* and *Acacia* species in tropical Australia,[134] may also be solved by using modified rumen bacteria. Bacterial isolates of the genus *Moraxella* from soil beneath the toxic tree species have been found that can metabolize fluoroacetate under aerobic conditions. This ability has been transferred to the rumen bacterium *Butyrivibrio fibrisolvens* OB156 with the expression of the fluoroacetate dehalogenase gene and the subsequent protection of sheep from fluoroacetate poisoning.[135]

Possibilities also exist for improving the feeding value of tannin-containing tropical shrub legumes by the use of tannin-degrading bacteria, which occur naturally in some browsing ruminants,[136,137] or by genetic engineering of rumen bacteria to achieve this end. Using this approach, it may be possible to overcome any associated antinutritional characteristics of plants that have chemical defense mechanisms for preventing pest and disease attacks. This could be a better option than seeking to remove the antinutritional factors in the plant by breeding or other biotechnology approaches, since the plants may then be more susceptible to pests and diseases.

REFERENCES

1. Jennings, J. N. and Mabbutt, J. A., Physiographic outlines and regions, in *Australia, A Geography, Volume One, The Natural Environment*, Jeans, D.N., Ed., Sydney University Press, Sydney, 1986, 80.
2. Isbell, R. F., Soils, in *Rural Research in Northern Australia*, Commonwealth Council for Rural Research and Extension, Australian Government Publishing Service, 1981, 12.
3. Hubble, G. D., Soils, in *Australian Grasslands*, Moore, R. M., Ed., Australian National University Press, Canberra, 1970, 44.
4. Fitzpatrick, E. A. and Nix, H. A., The climatic factor in Australian grassland ecology, in *Australian Grasslands*, Moore, R. M., Ed., Australian National University Press, Canberra, 1970, 3.
5. McCown, R. L. and Williams, J., The water environment and implications for productivity, in *Savanna Ecology and Management, Australian Perspectives and Intercontinental Comparisons*, Werner, Patricia A., Ed., Blackwell Scientific Publications, Oxford, 1991, 169.
6. Shaw, N. H. and Norman, M. J. T., Tropical and subtropical woodlands and grasslands, in *Australian Grasslands*, Moore, R. M., Ed., Australian National University Press, Canberra, 1970, 112.
7. Coaldrake, J. E., The Brigalow, in *Australian Grasslands*, Moore, R. M., Ed., Australian National University Press, Canberra, 1970, 123.
8. Lloyd, P. L. and Burrows, W. H., The importance and economic value of native pastures to Queensland, in *Native Pastures in Queensland — The Resources and Their Management*, Queensland Department of Primary Industries Information Series Q187023, 1988, 1.
9. Norman, M. J. T., *Katherine Research Station 1956-64, A Review of Published Work*, CSIRO Australia, Division of Land Research, Tech. Paper 28, 1966.
10. Mclvor, J. G., Seasonal changes in the growth, dry matter distribution and herbage quality of three native grasses in northern Queensland, *Australian Journal of Experimental Agriculture and Animal Husbandry*, 21, 600, 1981.
11. McLennan, S. R., Hendricksen, R. E., Beale, I. F., Winks, L., Miller, C.P., and Quirk, M. F., Nutritive value of native pastures in Queensland, in *Native Pastures in Queensland — The Resources and Their Management*, Queensland Department of Primary Industries Information Series Q 187023, 1988, 125.
12. Davies, J. G. and Eyles, A. G., Expansion of Australian pastoral production, *Journal of the Australian Institute of Agricultural Science*, 31, 77, 1965.
13. Walker, B. and Weston, E. J., Pasture development in Queensland — A success story, *Tropical Grasslands*, 24, 257, 1990.
14. Harlan, J. R., The scope for collection and improvement of forage plants, in *Genetic Resources of Forage Plants*, Mclvor, J. G. and Bray, R. A., Eds., CSIRO, Australia, 1983, 3.
15. Rees, M. C., Jones, R. M., Brown, A. H. D., and Coote, J. N., *Glycine latifolia* — a potentially useful native legume for clay soils in tropical and subtropical Australia, in *Proceedings of the XVII International Grassland Congress*, 1993, 2134.
16. Eyles, A. G. and Cameron, D. G., *Pasture Research in Northern Australia — Its History, Achievements and Future Emphasis*, Hacker, J. B., Ed., CSIRO, Australia, Division of Tropical Crops and Pastures, Research Report No. 4, 1985, 222.
17. Marshall, D. R. and Brown, A. H. D., Theory of forage plant collection, in *Genetic Resources of Forage Plants*, Mclvor, J. G. and Bray, R. A., Eds., CSIRO Australia, 1983, 136.
18. Reid, R. and Strickland, R. W., Forage plant collection in practice, in *Genetic Resources of Forage Plants*, Mclvor, J. G. and Bray, R. A., Eds., CSIRO, Australia, 1983, 149.
19. Clements, R. J. and Cameron, D. G., Eds., *Collecting and Testing Tropical Forage Plants*, CSIRO, Australia, 1980.
20. Jones, R. J. and Walker, B., Strategies for evaluating forage plants, in *Genetic Resources of Forage Plants*, Mclvor, J. G. and Bray, R. A., Eds., CSIRO, Australia, 1983,185.
21. Edye, L. A. and Grof, B., Selecting cultivars from naturally occurring genotypes: evaluating *Stylosanthes* species, in *Genetic Resources of Forage Plants*, Mclvor, J. G. and Bray, R. A., Eds., CSIRO, Australia, 1983, 217.
22. Burt, R. L., Williams, R. J., and Williams, W. T., Observation, description and classification of plant collections, in *Collecting and Testing Tropical Forage Plants*, Clements, R. J. and Cameron, D. G., Eds., CSIRO, Melbourne, 1980, 40.

23. Williams, R. J., Tropical legumes, in *Genetic Resources of Forage Plants*, Mclvor, J. G. and Bray, R. A., Eds., CSIRO, Australia, 1983, 18.
24. Hacker, J. B., Breeding tropical grasses for ease of vegetative propagation and for improved seed production, in *Proceedings of the XVth International Grassland Congress*, Science Council of Japan and the Japanese Society of Grassland Science, Nishi-nasumo, Japan, 1985, 251.
25. Rains, J. P., Hopkinson, J. M., and Tregea, S. P., Tropical pasture establishment. 10. Satisfying industry's pasture seed requirements, *Tropical Grasslands*, 27, 359, 1993.
26. Eyles, A. G., Forage cultivars released for use in Queensland, *Tropical Grasslands*, 13, 176, 1979.
27. Eyles, A. G., Forage cultivars released for use in Queensland, *Tropical Grasslands*, 23, 115, 1989.
28. Oram, R. N., *Register of Australian Herbage Plant Cultivars*, Third Edition, CSIRO Division of Plant Industry, 1990.
29. Hacker J. B., Priorities and activities of the Australian Tropical Forages Genetic Resource Centre, *Tropical Grasslands*, 31, 243, 1997.
30. Martin, G., Returns from research: meat and livestock, *Agricultural Science*, 4(2), 21, 1991.
31. Healy, T. and I'Ons, J., Returns from CSIRO Research, *Agricultural Science*, 4(2), 26, 1991.
32. Edye, L. A. and Gillard, P., Pasture improvement in semi-arid tropical savannas : a practical example in northern Queensland in *Ecology and Management of the World's Savannas*, Tothill, J. C. and Mott, J. J., Eds., Australian Academy of Science, Canberra, 1985, 303.
33. Cameron, D. F., Miller, C. P., Edye, L. A., and Miles, J. W., Advances in research and development with *Stylosanthes* and other tropical pasture legumes, in *Proceedings of the XVII International Grassland Congress*, 1993, 2109.
34. Stace, H. M. and Edye, L. A., Eds., *The Biology and Agronomy of Stylosanthes*, Academic Press, Sydney, 1984, 636.
35. Lenne, J. M. and Calderon, M. A., Disease and pest problems of *Stylosanthes*, in *The Biology and Agronomy of Stylosanthes*, Stace, H. M. and Edye, L. A., Eds., Academic Press, Sydney, 1984, 279.
36. Williams, R. J., Reid, R., Schultze-Kraft, R., Sousa Costa, N. M., and Thomas, B. D., Natural distribution of *Stylosanthes*, in *The Biology and Agronomy of Stylosanthes*, Stace, H. M. and Edye, L. A., Eds., Academic Press, Sydney, 1984, 73.
37. Date, R. A., *Rhizobium* for *Stylosanthes*, in *The Biology and Agronomy of Stylosanthes*, Stace, H. M. and Edye, L. A., Eds., Academic Press, Sydney, 1984, 243.
38. Edye, L. A. and Hall, T. J., Development of new *Stylosanthes* cultivars for Australia from naturally occurring genotypes, in *Proceedings of the XVII International Grassland Congress*, 1993, 2159.
39. Sutherst, R. W., Jones, R. J., and Schnitzerling, H. J., Tropical legumes of the genus *Stylosanthes* immobilize and kill cattle ticks, *Nature*, London, 295, 320, 1982.
40. Edye, L. A., Hall, T. J., Clem, R. L., Graham, T. W. G., Messer, W. B., and Rebgetz, R. H., Sward evaluation of eleven "*Stylosanthes seabrana*" accessions and *S. scabra* cv. Seca at five subtropical sites, *Tropical Grasslands*, 32, 243, 1998.
41. Hutton, E. M., Siratro. A tropical pasture legume bred from *Phaseolus atropurpureus* D.C., *Australian Journal of Experimental Agriculture and Animal Husbandry*, 2, 117, 1962.
42. Jones, R. J., The effect of rust (*Uromyces appendiculatus*) on the yield and digestibility of *Macroptilium atropurpureum* cv. Siratro, *Tropical Grasslands*, 16, 130, 1982.
43. Jones, R. J., Hall, T. J., and Reid, R., Rust on *Macroptilium atropurpureum*, in *CSIRO Division of Tropical Crops and Pastures Annual Report 1980–1981*, 1981, 113.
44. Bray, R. A., Breeding for rust resistance in *Macroptilium atropurpureum*, in *Proceedings of the XVII International Grassland Congress*, 1993, 2128.
45. Jones, R. J. and Jones, R. M., The ecology of Siratro-based pastures, in *Plant Relations in Pastures*, Wilson, John R., Ed., CSIRO, Australia, 1978, 353.
46. Mannetje, 't., L. and Jones, R. M., Pasture and animal productivity of buffel grass with Siratro, lucerne or nitrogen fertilizer, *Tropical Grasslands*, 24, 269, 1990.
47. Bishop, H. G. and Hilder, T. B., The *Aeschynomene* genus as a source of pasture legumes for tropical and subtropical Australia, in *Proceedings of the XVII International Grassland Congress*, 1993, 2153.
48. Bishop, H. G., Pengelly, B. C., and Ludke, D. H., Classification and description of a collection of the legume genus *Aeschynomene*, *Tropical Grasslands*, 22, 160, 1988.
49. Jones, R. M., McDonald, C. K., and Bunch, G. A., Persistence of roundleaf cassia in grazed pastures — a conceptual model, in *Proceedings of the XVII International Grassland Congress*, 1993, 1902.

50. Partridge, I. J. and Wright, J. W., The value of round-leafed cassia (*Cassia rotundifolia* cv. Wynn) in a native pasture grazed with steers in south-east Queensland, *Tropical Grasslands*, 26, 263, 1992.
51. Burt, R. L., *Desmanthus*: A tropical and subtropical forage legume. Part 1. General review, *Herbage Abstracts*, 63, 401, 1993.
52. Cook, B. G., Graham, T. W. G., Clem, R. L., Hall, T. J., and Quirk, M. F., Evaluation and development of *Desmanthus virgatus* on medium to heavy-textured soils in Queensland, in *Proceedings of the XVII International Grassland Congress*, 1993, 2148.
53. Hegarty, M. P. and Pound, A. W., Indospicine, a hepatotoxic amino acid from *Indigofera spicata*: isolation, structure, and biological studies, *Australian Journal of Biological Science*, 23, 831, 1970.
54. Aylward, J. H., Court, R. D., Haydock, K. P., Strickland, R. W., and Hegarty, M. P., *Indigofera* species with agronomic potential in the tropics: rat toxicity studies, *Australian Journal of Agricultural Research*, 38, 177, 1987.
55. Strickland, R. W., Forage species adaptation in western Queensland, *CSIRO Division of Tropical Crops & Pastures Annual Report 1984–85*, 1985, 23.
56. Cook, B. G. and Loch, D. S., Commercialisation of Arachis pintoi cv. Amarillo in northern Australia, in *Proceedings of the XVII International Grassland Congress*, 1993, 2140.
57. Cook, B. G., Jones, R. M., and Williams, R. J., Regional experience with forage Arachis in Australia, in *Biology and Agronomy of Forage Arachis*, Kerridge, P. C. and Hardy, Bill, Eds., CIAT, Cali, Colombia, 1994, 158.
58. Middleton, C. H., Jones, R. J., Shelton, H. M., Petty, S. R., and Wildin, J. H., *Leucaena* in Northern Australia, in *Leucaena — Opportunities and Limitations*, Shelton, H. M., Piggin, C. M.., and Brewbaker, J. L., Eds., ACIAR Proceedings No. 57, 1995, 214.
59. Jones, R. J., Management of anti-nutritive factors — with special reference to *Leucaena*, in *Forage Tree Legumes in Tropical Agriculture*, Gutteridge, R. C. and Shelton, H. M., Eds., CAB International, Wallingford, U.K., 1994, 216.
60. Bray, R. A., The *Leucaena* psyllid, in *Forage Tree Legumes in Tropical Agriculture*, Gutteridge, R. C. and Shelton, H. M., Eds., CAB International, Wallingford, U.K., 1994, 283.
61. Wildin, J. H., Major beef production gains from commercial rain-fed leucaena-grass pastures in central Queensland, Australia, in *Proceedings of the XVII International Grassland Congress*, 1993, 2070.
62. Bray, R. A., Diversity within tropical tree and shrub legumes, in *Forage Tree Legumes in Tropical Agriculture*, Eds., Gutteridge, R. C. and Shelton, H. M., CAB International, Wallingford, U.K., 1994, 119.
63. Mullen, B. F., Shelton, H. M., Basford, K. E., Castillo, A. C., Bino, B., Victorio, E. E., Acasio, R. N., Tarabu, J., Komolong, M. K., Galgal, K. K., Khoa, L. V., Co, H. X., Wandera, F. P., Ibrahim, T. M., Clem, R. L., Jones, R. J., Middleton, C. H., Bolam, M. J. M., Gabunada, F., Stur, W. W., Horne, P. M., Utachak, K., and Khanh, T. T., Agronomic adaptation to environmental challenges in the genus *Leucaena*, in *Leucaena — Adaptation, Quality and Farming Systems*, Shelton, H. M., Gutteridge, R. C., Mullen, B. F., and Bray, R. A., Eds., ACIAR Proceedings No. 86, Canberra, 1988, 39.
64. Mullen, B. F., Gabunada, F., Shelton, H. M., Stur, W., and Napompeth, B., Psyllid resistance in *Leucaena*, in *Leucaena — Adaptation, Quality and Farming Systems*, Shelton, H. M., Gutteridge, R. C., Mullen, B. F., and Bray, R. A., Eds., ACIAR Proceedings No. 86, Canberra, 1988, 51.
65. Shelton, H. M. and Brewbaker, J. L., *Leucaena leucocephala* — The most widely used forage tree legume, in *Forage Tree Legumes in Tropical Agriculture*, Gutteridge, R. C. and Shelton, H. M., Eds., CAB International, Wallingford, U.K., 1994, 15.
66. Dalzell, S. A., Stewart, J. L., Tolera, A., and McNeill, D. M., Chemical composition of *Leucaena* and implications for forage quality, in *Leucaena — Adaptation, Quality and Farming Systems*, Shelton, H. M., Gutteridge, R. C., Mullen, B. F., and Bray, R. A., Eds., ACIAR Proceedings No. 86, Canberra, 1988, 227.
67. Palmer, B., Macqueen, D. J., and Gutteridge, R. D., *Calliandra calothyrsus* — a multipurpose tree legume for humid locations, in *Forage Tree Legumes in Tropical Agriculture*, Gutteridge, R. C. and Shelton, H. M., Eds., CAB International, Wallingford, U.K., 1994, 65.
68. Jones, R. J. and Palmer, B., A comparison of steer gains on Cunningham leucaena and some alternative psyllid-tolerant shrub legumes, in *Proceedings of the XVIII International Grassland Congress*, 1997, Session 23, 35.
69. Gutteridge, R. C., The perennial *Sesbania* species, in *Forage Tree Legumes in Tropical Agriculture*, Gutteridge, R. C. and Shelton, H. M., Eds., CAB International, Wallingford, U.K., 1994, 49.

70. Skerman, P. J., The Brigalow country and its importance to Queensland, *Journal of the Australian Institute of Agricultural Science*, 19, 167, 1953.
71. Peake, D. C. I., Myers, R. J. K., and Henzell, E. F., Sown pasture production in relation to nitrogen fertilizer and rainfall in southern Queensland, *Tropical Grasslands*, 24, 291, 1990.
72. Myers, R. J. K. and Robbins, G. B., Sustaining productive pastures in the tropics. 5. Maintaining productive sown grass pastures, *Tropical Grasslands*, 25, 104, 1991.
73. Skerman, P. J. and Riveros, F., *Tropical Grasses*, FAO Plant Production and Protection Series, No. 23, FAO, Rome, 1990.
74. Mannetje, 't, L. and Jones, R. M., Eds., *Plant Resources of South-East Asia 4, Forages*, Pudoc Scientific Publishers, Wageningen, 1992.
75. Hacker, J. B., Williams, R. J., and Coote, J. N., Productivity in late winter and spring of four cultivars and 21 accessions of *Cenchrus ciliaris* and *Digitaria eriantha* cv. Premier, *Tropical Grasslands*, 29, 28, 1995.
76. Bisset, W. J., Indian blue grass has special uses, *Queensland Agricultural Journal*, 106, 507, 1980.
77. Jones, R. J. and Kerr, A., *Bothriochloa pertusa* — a useful grazing-tolerant grass for the seasonally dry tropics?, in *Proceedings of the XVII International Grassland Congress*, 1993, 1897.
78. Mclvor, J. G., Phosphorus requirements and responses of tropical pasture species: native and introduced grasses, and introduced legumes, *Australian Journal of Experimental Agriculture and Animal Husbandry*, 24, 370, 1984.
79. Ludlow, M. M., Photosynthesis and dry matter production in C3 and C4 pasture plants, with special emphasis on tropical C3 legumes and C4 grasses, *Australian Journal of Plant Physiology*, 12, 557, 1985.
80. Stobbs, T. H., Seasonal changes in the preference by cattle for *Macroptilium atropurpureum* cv. Siratro, *Tropical Grasslands*, 11, 87, 1977.
81. Gardener, C. J., The dynamics of *Stylosanthes* pastures, in *The Biology and Agronomy of Stylosanthes*, Stace, H. M. and Edye, L. A., Eds., Academic Press, Sydney, 1984, 333.
82. Coates, D. B., Kerridge, P. C., Miller, C. P., and Winter, W. H., Phosphorus and beef production in northern Australia. 7. The effect of phosphorus on the composition, yield and quality of legume-based pasture and their relation to animal production, *Tropical Grasslands*, 24, 209, 1990.
83. Cook, B. G. and Benjamin, A. K., *Vigna parkeri* Baker, in *Plant Resources of South-East Asia 4, Forages*, Mannetje, 't, L. and Jones, R. M., Eds., Pudoc Scientific Publishers, Wageningen, 1992, 232.
84. Mott, J. J., Patch grazing and degradation in native pastures of the tropical savannas in northern Australia, in *Grazing-Lands Research at the Plant-Animal Interface*, Horn, F. P., Hodgson, J., Mott, J. J., and Brougham, R. W., Eds., Winrock International, Morrilton, Arkansas, 1987, 153.
85. Hughes, C. E. and Jones, R. J., Environmental hazards of *Leucaena*, in *Leucaena — Adaptation, Quality and Farming Systems*, Shelton, H. M., Gutteridge, R. C., Mullen, B. F., and Bray, R. A., Eds., ACIAR Proceedings No. 86, Canberra, 1988, 61.
86. Jones, R. J., Mclvor, J. G., Middleton, C. H., Burrows, W. H., Orr, D. M., and Coates, D. B., Stability and productivity of *Stylosanthes* pastures in Australia. 1. Long-term botanical changes and their implications in grazed *Stylosanthes* pastures, *Tropical Grasslands*, 31, 482, 1997.
87. Noble, A.D., Cannon, M., and Muller, D., Evidence of accelerated soil acidification under *Stylosanthes*-dominated pastures, *Australian Journal of Soil Research*, 35, 1309, 1997.
88. Noble, A. D., Thompson, C. H., Jones, R. J., and Jones, R. M., The long-term impact of two pasture production systems on soil acidification in southern Queensland, *Australian Journal of Experimental Agriculture*, 38, 335, 1998.
89. Minson, D. J., *Forage in Ruminant Nutrition*, Academic Press, San Diego, California, 1990.
90. Shaw, N. H., Increased beef production from Townsville lucerne (*Stylosanthes sundaica* Taub.) in the spear grass pastures of central coastal Queensland, *Australian Journal of Experimental Agriculture and Animal Husbandry*, 1, 73, 1961.
91. Bowen, E. J. and Rickert, K. G., Beef production from native pastures sown to fine-stem stylo in the Burnett region of south eastern Queensland, *Australian Journal of Experimental Agriculture and Animal Husbandry*, 19, 140, 1979.
92. Gillard, P., Edye, L. A., and Hall, R. L., Comparison of *Stylosanthes humilis* with *S. hamata* and *S. subsericea* in the Queensland dry tropics: effects on pasture composition and cattle liveweight gain, *Australian Journal of Agricultural Research*, 31, 205, 1980.

93. Gillard, P. and Winter, W. H., Animal production from *Stylosanthes* based pastures in Australia, in *The Biology and Agronomy of Stylosanthes*, Stace, H. M. and Edye, L. A., Eds., Academic Press, Sydney, 1984, 405.
94. Winter, W. H., Removing limitations to cattle production in the semi-arid tropics, in *XVI International Grassland Congress*, the French Grassland Society, Versailles-en-Cedex, France, 1989, 1175.
95. Coates, D. B., Miller, C. P., Hendricksen, R. E., and Jones, R. J., Stability and productivity of *Stylosanthes* pastures in Australia. 2. Animal production from *Stylosanthes* pastures, *Tropical Grasslands*, 31, 494, 1997.
96. Evans, T. R. and Biggs, J., Breeding performance of Hereford cows grazing tropical pastures at Beerwah south-east Queensland, *Tropical Grasslands*, 13, 129, 1979.
97. Holroyd, R. G., Allen, P. J., and O'Rourke, P. K., Effect of pasture type and supplementary feeding on the reproductive performance of cattle in the dry tropics of north Queensland, *Australian Journal of Experimental Agriculture and Animal Husbandry*, 17, 197, 1977.
98. Holroyd, R. G., O'Rourke, P. K., Clarke, M. R., and Loxton, I. D., Influence of pasture type and supplement on fertility and liveweight of cows, and progeny growth rate in the dry tropics of northern Queensland, *Australian Journal of Experimental Agriculture and Animal Husbandry*, 23, 4, 1983.
99. Coates, D. B and Mannetje, 't, L., Productivity of cows and calves on native and improved pasture in subtropical, subcoastal Queensland, *Tropical Grasslands*, 24, 46, 1990.
100. Dalal, R. C., Strong, W. M., Weston, E. J., and Gaffney, J., Sustaining multiple production systems. 2. Soil fertility decline and restoration of cropping lands in subtropical Queensland, *Tropical Grasslands*, 25, 173, 1991.
101. Lloyd, D. L., Smith, K. P., Clarkson, N. M., Weston, E. J., and Johnson, B., Sustaining multiple production systems. 3. Ley pastures in the subtropics, *Tropical Grasslands*, 25, 181, 1991.
102. Puckridge, D. W. and French, R. J., The annual legume pasture in cereal-ley farming systems in southern Australia: a review, *Agriculture, Ecosystems and Environment*, 9, 229, 1983.
103. Jones, R. K., Dalgliesh, N. P., Dimes, J. P., and McCown, R. L., Sustaining multiple production systems. 4. Ley pastures in crop-livestock systems in the semi-arid tropics, *Tropical Grasslands*, 25, 189, 1991.
104. Gramshaw, D., Read, J. W., Collins, W. J., and Carter, E. D., Sown pastures and legume persistence: An Australian overview, in *Persistence of Forage Legumes, Proceedings of a Trilateral Workshop*, Marten, G. C., Matches, A. G., Barnes, R. F., Brougham, R. W., Clements, R. J., and Sheath, G. W., Eds., ASA-CSSA-SSSA, Madison, Wisconsin, 1989, 1.
105. Clements, R. J., Developing persistent pasture legume cultivars for Australia, in *Persistence of Forage Legumes, Proceedings of a Trilateral Workshop*, Marten, G. C., Matches, A. G., Barnes, R. F., Brougham, R. W., Clements, R. J., and Sheath, G. W., Eds., ASA-CSSA-SSSA, Madison, Wisconsin, 1989, 505.
106. Jones, R. J., *Tropical Legumes — Their Growth and Response to Management Variables in a Subtropical Environment*. Ph.D Thesis, University of New England, Armidale, New South Wales, Australia, 1971.
107. Fisher, M. J. and Thornton, P. K., Growth and competition as factors in the persistence of legumes in pastures, in *Persistence of Forage Legumes, Proceedings of a Trilateral Workshop*. Marten, G. C., Matches, A. G., Barnes, R. F., Brougham, R. W., Clements, R. J., and Sheath, G. W., Eds., ASA-CSSA-SSSA, Madison, Wisconsin, 1989, 293.
108. Jones, R. M. and Carter, E. D., Demography of pasture legumes, in *Persistence of Forage Legumes, Proceedings of a Trilateral Workshop*, Marten, G. C., Matches, A. G., Barnes, R. F., Brougham, R. W., Clements, R. J., and Sheath, G. W., Eds., ASA-CSSA-SSSA, Madison, Wisconsin, 1989, 139.
109. Curll, M. L. and Jones, R. M., The plant-animal interface and legume persistence — an Australian perspective, in *Persistence of Forage Legumes, Proceedings of a Trilateral Workshop*, Marten, G. C., Matches A. G., Barnes, R. F., Brougham, R. W., Clements, R. J., and Sheath, G. W., Eds., ASA-CSSA-SSSA, Madison, Wisconsin, 1989, 339.
110. Hall, R. L., The influence of potassium supply on competition between Nandi setaria and Greenleaf desmodium, *Australian Journal of Experimental Agriculture and Animal Husbandry*, 11, 415, 1971.
111. Hall, R. L., Analysis of the nature of interference between plants of different species. II. Nutrient relations in a Nandi setaria and Greenleaf desmodium association with particular reference to potassium, *Australian Journal of Agricultural Research*, 25, 749, 1974.
112. Hall, R. L., The analysis and significance of competitive and non-competitive interference between species, in *Plant Relations in Pastures*, Wilson, John R., Ed., CSIRO, Melbourne, 1978, 163.

113. Arnold, A., Commercial development on "Wrotham Park," via Chillagoe (1963–1988), in *Sown Pastures for the Seasonally Dry Tropics*, Partridge, I. J. and Miller, C. P., Eds., Queensland Department of Primary Industries Conference and Workshop Series QC91002, 1991, 44.
114. McLean, R., Commercial development on "Kuttabul," via Mackay, in *Sown Pastures for the Seasonally Dry Tropics*, Partridge, I. J. and Miller, C. P., Eds., Queensland Department of Primary Industries Conference and Workshop Series QC91002, 1991, 75.
115. Chapman, B., Commercial development on "Rowanlea," Calliope, in *Sown Pastures for the Seasonally Dry Tropics*, Partridge, I. J. and Miller, C. P., Eds., Queensland Department of Primary Industries Conference and Workshop Series QC91002, 1991, 62.
116. Kaiser, A. G., Havilah, E. J., Chopping, G. D., and Walker, R. G., Northern dairy feedbase 2001. 4. Feeding systems during winter and spring, *Tropical Grasslands*, 27, 180, 1993.
117. Catchpoole, V. R. and Henzell, E. F., Silage and silage-making from tropical herbage species, *Herbage Abstracts*, 41, 213, 1993.
118. Cowan, R. G., Kerr, D. V., and Davison, T. M., Maize silage for dairy systems in northern Australia, in *Proceedings of the 1st Australian Maize Conference*, Moama, 1991, 228.
119. Wildin, J. H., Ponded pasture systems for beef production in the seasonally dry zones of northern Australia, in *Proceedings of the XVII International Grassland Congress*, 1993, 527.
120. Winter, W. H., Coates, D. B., Hendricksen, R. E., Kerridge, P. C., McLean, R. W., and Miller, C. P., Phosphorus and beef production in northern Australia. 4. The response of cattle to fertilizer and supplementary phosphorus, *Tropical Grasslands*, 24, 170, 1990.
121. Miller, C. P. and Hendricksen, R. E., Cattle growth and diet responses to legume and phosphorus supply, in *Proceedings of the XVII International Grassland Congress*, 1993, 1984.
122. McCown, R. L. and Wall, B. H., The influence of weather on the quality of tropical legume pasture during the dry season in northern Australia. II. Moulding of standing hay in relation to rain and dew, *Australian Journal of Agricultural Research*, 32, 589, 1981.
123. McCown, R. L., Wall, B. H., and Harrison, P. G., The influence of weather on the quality of tropical legume pasture during the dry season in northern Australia. I. Trends in sward structure and moulding of standing hay at three locations, *Australian Journal of Agricultural Research*, 32, 575, 1981.
124. Wall, B. H. and McCown, R. L., The influence of weather on the quality of tropical legume pasture during the dry season in northern Australia. IV. Geographic variation in risk of spoilage of standing hay, *Australian Journal of Agricultural Research*, 40, 570, 1989.
125. Jones, R. J., Effect of pasture management on grass and animal production following frosting of nitrogen fertilized sub-tropical grass pastures, *Tropical Grasslands*, 22, 57, 1988.
126. Peacock, W. J., Genetic engineering for pastures, in *Proceedings XVII International Grassland Congress*, 1993, 29.
127. McNabb, W. C., Higgins, C., Tabe, L., and Higgins, T. J. V., Transfer of genes encoding proteins with high nutritional value into pasture legumes, in *Proceedings of the XVII International Grassland Congress*, 1993, 1085.
128. McIntyre, C. L., Abrahams, S. L., Bettenay, H. M., Sandeman, R. A., Hayes, C., Sharp, D., Elliot, A., Manners, J. M., and Watson, J. M., Improving pasture digestibility: low lignin forages, in *Proceedings of the XVII International Grassland Congress*, 1993, 1081.
129. McIntyre, C. L., Manners, J. M., Wilson, J. R., Way, H., and Sharp, D., Genetic engineering of pasture legumes and grasses for reduced lignin content and increased digestibility, in *Proceedings of the XVII International Grassland Congress*, 1993, 1100.
130. Rae, A. L., McIntyre, C. L., Jones, R. J., and Manners, J. M., Antisense suppression of the lignin biosynthetic enzyme, o-methyl transferase, improves forage digestibility, in *Proceedings of the 40th Australian Society of Biochemistry and Molecular Biology and the 36th Australian Society of Plant Physiologists Annual Combined Conference*, Canberra, 1996, Sym-21-07.
131. Xue, Gang-Ping, Gobius, K. S., and Orpin, C. G., Isolation of a multi-functional cellulase cDNA (celE) from the rumen fungus *Neocallimastix patriciarum*, in *Proceedings of the XVII International Grassland Congress*, 1993, 1221.
132. Orpin, C. G. and Xue, G., Genetics of fibre degradation in the rumen, particularly in relation to anaerobic fungi, and its modification by recombinant DNA technology, in *Proceedings of the XVII International Grassland Congress*, 1993, 1209.

133. Jones, R. J., *Leucaena* toxicity and the ruminal degradation of mimosine, in *Plant Toxicology, Proceedings of the Australia-USA Poisonous Plants Symposium*, Seawright, A. A., Hegarty, M. P., James, L. F., and Keeler, R. F., Eds., Brisbane, Australia, 1985.
134. Gregg, K. and Sharpe, H., Enhancement of rumen microbial detoxification by gene transfer, in *Physiological Aspects of Digestion and Metabolism in Ruminants, Proceedings of the Seventh International Symposium on Ruminant Physiology*, 1991, 719.
135. Gregg, K., Hamdorf, B. J., Henderson, K. P., Kopecny, J., and Wong, C., Genetically modified rumen bacteria protect sheep from fluoroacetate poisoning, *Applied Environmental Microbiology*, 64, 3496, 1998.
136. Matthew, J., Brooker, J. D., Clarke, K., Lum, D. K., and Miller, S. M., A ruminal bacterium that degrades tannic acid, in *Australian Society for Microbiology, Annual Scientific Meeting*, Gold Coast, Australia, 1991, Poster No. 59.
137. Brooker, J. D., O'Donovan, L. A., Skene, I., Clarke, K., Blackall, L., and Musiera, P., *Streptococcus caprinus* sp. nov., a tannin-resistant ruminal bacterium from feral goats, *Letters in Applied Microbiology*, 18, 313, 1994.
138. Mott, J. J., Williams, J., Andrew, M. H., and Gillison, A. N., Australian savanna ecosystems, in *Ecology and Management of the World's Savannas, Proceedings of the International Savanna Symposium*, Tothill, J. C. and Mott, J. J., Eds., Australian Academy of Science, Canberra, 1985, 56.
139. Walker, B., Baker, J., Becker, M., Brunckhorst, R., Heatley, D., Simms, J., Skerman, D. S., and Walsh, S., Sown pasture priorities for the subtropical and tropical beef industry, *Tropical Grasslands*, 31, 266, 1997.

18 Current Developments from Tropical Forage Research in Africa

Martin B. Adjei and James P. Muir

CONTENTS

0-8493-2318-5/01/$0.00+$.50

I. INTRODUCTION

The continent of Africa covers 30.3 million km^2 of land, approximately one fifth of the world's land area. It is the second largest continent in area and the third largest in human population. Only Asia covers more land and only Asia and Europe have more people. Africa's huge size and broad latitudinal range help to explain not only the variety of natural resource endowments but also the variety of climates, altitudes, and resultant agro-ecological zones. In addition, its historically inaccessible interior, ethnocultural diversity, and human or domesticated animal endemic diseases have all earned Africa acronyms such as the "dark continent" and "green sea of darkness." All these factors have contributed to limiting the scientific exploration of its rich native forage germplasm or the importation of exotic species.

Most animal husbandry systems (including managed wildlife) in Africa are based on range. The rich diversity of natural pasture parallels the varied climates, soils, topography, and agricultural systems. Dependence on range is also a result of harsh climates and endemic animal diseases that preclude forage cultivation. In recognition of this fact, much of this chapter focuses on native pastures. Early in the sixties, most range research in Africa concentrated on assessing botanical composition of African grasslands and identifying the major plant constituents. The extensive vegetation mapping by the Portuguese in Mozambique[1,2] is a good example of this early work. Vegetation surveys, range evaluation, range management studies to determine carrying capacity, the effects of fire on range, and other forms of brush control were carried throughout Africa during colonial and early postcolonial periods. Much valuable information on major grass associations accrued, including vegetation survey maps describing vegetation zones, botanical composition, and their potential use.

Then came the advent of cultivated species selection, introduction, and evaluation, which spanned four decades in sub-Saharan Africa. Evaluation criteria for selected grasses and legumes included dry matter production, fertility requirements, and tolerance to grazing or drought. Examples are Rebelo and Rodrigues[3] in Mozambique, Anderson and Naveh[4] in Tanzania, and Clatworthy[5] in Zimbabwe.

There has been a recent surge in research focused on understanding communal range management in order to more effectively increase sustainability of these systems. Kusekwa et al,[6] for example, described the use of natural foggage (reserved standing hay) reserves by agropastoralists in Tanzania. The effects of deforestation on communal pastures[7,8] is another aspect of this effort.

Cultivated pastures, where utilized or studied in tropical Africa, have played a supporting role to range-based systems (for example, dry season feed banks) or are spin-offs of multiple-use crops and forestry systems (for example, the use of crop residues or fodder trees). The current trend is toward development of feeding systems that utilize all available resources in range-based animal husbandry. These include protein supplements with industrial by products,[9] crop residues,[10] more digestible grasses and pastures,[11] intensive use of fertilizers,[12] persistent legumes,[13-16] and browse legumes high in bypass protein.[17] Large-scale forage seed production is also undertaken by some multinational institutions like the International Livestock Research Institute (formerly ILCA) for distribution to national agricultural research institutions and, ultimately, to farmers.

In this chapter, the geography, soils, climate, and vegetation types of this huge continent are summarized in Section II. The natural grassland associations, dominant grass species, and little-studied legumes are presented in Section III. This is followed by a discussion of the prevalent forage-livestock systems in Section IV and the national and multinational approaches to pasture and forage research in Section V. The chapter is concluded with an exposition on the future outlook of pasture research and its developmental role in Africa.

II. DESCRIPTION OF AFRICA

A. Geography

Geographically, Africa is a vast plateau that straddles the equator from the Mediterranean (37° N) to the Cape of Good Hope (35° S), a distance of 8047 km. Longitudinally, it stretches for 7564 km from Dakar, Senegal in the west to Mogadishu, Somalia in the East. Since the landmass extends almost equidistant from the equator toward either pole, Africa is the most tropical of all continents. Africa is divided into two major land regions by Oliver and Crowder:[18] (1) Low Africa and (2) High Africa. In their classification, Low Africa consists of northern, western, and central Africa (see Figure 18.1). This region lies mainly from 150 to 610 meters above sea level and is broken only by a few mountain ranges and bordered in some areas by a narrow coastal plain. The most striking features of Low Africa are: (1) the Sahara desert — the world's largest desert (9 million km^2) — which covers most of northern Africa; (2) the coastal lowlands, which border most of northern Africa and the bulge of western Africa, which are covered with fertile farmland, forests, sandy beaches, and swamps; (3) the Northern Highlands containing the Atlas Mountains, which form Africa's longest mountain chain from Morocco to Tunisia and contain rich deposits of phosphate rock, iron ore, and manganese; (4) the Western plateau lying south of the Sahara desert, consisting of forests and grasslands and through which rivers like the Niger and Benue flow; (5) the Nile Basin that borders the Nile River and its tributaries in northeastern Africa with fertile farmland and huge swamps; and (6) the Congo Basin in west central Africa including most of the land drained by the Congo River with tropical rain forest cover.

High altitude Africa dominates the eastern and southern portions of the continent where land is usually more than 910 meters above sea level. Prominent subregions within High Africa are: (1) the Great Rift Valley System, which is the world's deepest continental crevice with deep steep-sided valleys, extends from Ethiopia to Mozambique, and includes Africa's largest lakes (Tanganyika, Nyasa, Albert, Turkana, and Victoria) and highest mountains; (2) the Eastern Highlands contain the two highest African peaks (Kilimanjaro at 5895 m and Kenya at 5199 m), the Ethiopian Highlands and grassy plains that provide grazing land to livestock and wildlife (although Mt. Kilimanjaro and Mt. Kenya rise near the equator in eastern Africa, both mountains have glaciers and are covered with snow much of the year. The Rift Valley System cuts through the Eastern Highlands.); (3) the Southern Plateau covers most of the flat or rolling grassland used for crops and pasture in southern Africa and contains the Namib and Kalahari Deserts, swamps, and forests; (4) the Coastal Lowlands, which border the highlands of eastern and southern Africa, and include productive farmland, swamps, and beaches; and (5) Madagascar, which lies 390 km southeast of the mainland in the Indian Ocean and is the world's fourth largest island.

The political geography of Africa consists of 52 independent countries most containing multiple ethnic groups and numerous languages and dialects. These countries vary enormously in size and resources (see Figure 18.2). The largest country, Sudan (2.5 million km^2), is bigger than Alaska and Texas combined. The smallest country, Seychelles (287 km^2), is an island nation with land area half the size of the city of New Orleans. Four regional groupings of sub-Saharan African countries are recognized. These are: **West Africa:** Benin, Burkina Fasso, Chad, Ivory Coast, Gambia, Ghana,

FIGURE 18.1 Africa relief map. (From Khapoya.[19] With permission of Prentice-Hall, Inc.)

Guinea, Guinea Bissau, Liberia, Mali, Mauritania, Niger, Nigeria, Senegal, Sierra Leone, and Togo; **Central Africa:** Angola, Burundi, Cameroon, Central African Republic, Congo, Gabon, Rwanda, and Zaire; **East Africa:** Ethiopia, Kenya, Somalia, Sudan, Tanzania, and Uganda; and **Southern Africa:** Botswana, Lesotho, Madagascar, Malawi, Mauritius, Mozambique, Namibia, Swaziland, South Africa, Zambia, and Zimbabwe. According to the International Livestock Center for Africa[21] (ILCA, 1987) these regions, defined as clusters of countries within a geographic location, have significance for livestock industry in diverse ways: (1) countries within a region share a common pattern of rainfall; (2) they usually are linked by trade and other factors of economy (labor, currency); and (3) some share a common colonial experience, official language, system of education, and regional institutions.

B. Soils

Africa is geologically and topographically the most uniform compared with other continents. In addition to the folds of the Atlas Mountains in the north and the Cape Province ranges in the

FIGURE 18.2 Africa political map. (From Africa Report[20] © 1984. With permission of the African-American Institute.)

extreme south (both outside the tropics), the rest of the continent is one vast rigid block of ancient rock. This crystalline parent rock is chiefly schist and gneiss of Precambrian origin. Approximately two-thirds of the continental outcrop of rock is covered by sediments, chiefly sand deposits and alluvium from the Pleistocene age. The continental plateau is modified by volcanic eruptions, ranging from Precambrian to recent, at the related fault lines of the Rift Valley system.

Because of the lack of folding and constant exposure to high temperature and high rainfall, African soils are generally highly weathered and leached. According to the USDA legend of soil classification,[22] six major soil orders occur on the continent. Alfisols with ulstic soil moisture regimes and plinthite layers are found mainly in the savanna and forest-savanna transition zones of West Africa and the highlands of East and southeastern Africa. These soils are formed in well-drained upland situations with a coarse or medium-textured surface layer and a base saturation of more than 50%.[23] Alfisols constitute about 22% of African soil cover. Ultisols with udic soil moisture regimes and a base saturation of the B horizon below 50% make up only 4% of soils. They are the dominant soils in Liberia, Sierra Leone, Ivory Coast, and eastern Angola. Oxisols belonging to the Orthox suborder are the dominant soils found in southern West Africa and the central part of the African equatorial belt with less than 90 days dry season in the year. The Ustox type of Oxisols occur in east Central Africa, which experiences more than three consecutive dry months.

The Oxisols are strongly weathered acid soils of tropical lowlands and constitute about 22% of African soils. The Oxisols depend mostly on the quality and amount of the organic matter for retention of cations. Without fertilizer, they can support extensive agriculture only under shifting cultivation or with tree crops that protect the soil. If extensive grazing is practiced, pastures are generally covered with poor quality grass. Entisols and Inceptisols are recent soils of alluvial and colluvial origin. They occur in the valley bottoms as moving sand dunes (Psamments) such as around the oases in the Sahara desert or in hydromorphic situations on river flood plains. They constitute the main fertile soils of Africa including the Congo and Nile Basins and the volcanic deposits of the Rift Valley System. Entisols and Inceptisols together constitute 15% of all African soil cover. Large areas of black, cracking clay soils (Vertisols) rich in swelling clay and difficult to work when wet, are found in Sudan, Ethiopia, and on the Accra Plains of Ghana. Overall, however, Vertisols constitute a mere 2% of all African soils. Desert soils (Aridisols) are by far the predominant African soils (33%) and are found on the Sahara, Namib, and Kalahari Deserts.

C. Climate

Insolation, day length, and temperature in Africa are generally favorable for plant growth. Approximately 90% of the African continent lies between the tropics of Cancer and Capricorn and may be classified as tropical. This latitudinal location of Africa astride the equator ensures that most of the continent receives high levels of insolation year round. Solar radiation varies from about 330 cal/cm^2/day in the rain forest areas close to the equator to over 450 cal/cm^2/day inland away from the equator.[24] At the equator, the day and night are of approximately equal duration (12 hours) throughout the year, while at the northern and southern extremities of the continent, day length ranges between 14 hours and 29 minutes in summer and 9 hours and 51 minutes in winter. On an annual basis, two zones of maximum sunshine duration exist: Central Sahara receiving over 4000 hours and the Kalahari over 3600 hours. Because of increasing cloudiness, there is a decrease in insolation southward from the Sahara to the coast of West Africa in the Northern Hemisphere. In the Southern Hemisphere there is an east-west increase with values ranging from 2000 hours in Madagascar to over 3600 hours in Namibia.[18]

Two main factors influence temperature patterns in Africa: solar radiation and elevation. Since the annual variation of radiation is very small, the range of monthly average temperatures is generally also less pronounced in tropical Africa than in temperate latitudes. The range of monthly average temperature is between 3°C and 6°C. The daily range of temperature is much greater, ranging from less than 10°C in many coastal areas to more than 15°C in southwest Libya. With the exception of the highland areas, mean annual temperature, even in the coldest months, is above 18°C. The highest temperatures occur in the Sahara Desert and parts of Somalia where daily July temperatures may soar to 46°C. The Sahara also exhibits the greatest seasonal range of temperatures in Africa. Winter daily temperature in the Sahara averages from 10 to 16°C. Near the equator, temperatures are more constant and may average 24°C year round. The coolest regions in tropical Africa are the highland areas in the east where, in reality, subtropical to temperate conditions may be encountered.

The dominant climatic factor in tropical Africa is the amount and temporal distribution of rainfall. The heaviest rainfall (over 4000 mm) occurs in the area astride the equator, especially from the Niger Delta to the Zaire River Basin and Central Zaire. Along the coast of Sierra Leone and Liberia and along the eastern coast of Madagascar, annual totals exceed 2000 mm. Rainfall decreases in both amount and frequency northward from the equator until about latitude 18°N where the mean annual precipitation is less than 250 mm in the Sahara Desert. In the Southern Hemisphere, there is a north-south decrease in rainfall from the equator to the Tropic of Capricorn; beyond which an east-west pattern of decline prevails. For example, the eastern coast of Madagascar receives more than 2000 mm of rainfall, while the coast of Namibia in the same latitude receives less than 250 mm per year.

Africa is a continent of well-defined wet and dry seasons. This has provided the basis of schemes for classifying the climate. ILCA[21] has modified the classification proposed by Jahnke[25] and divided sub-Saharan Africa into five ecological zones on the basis of plant-growth days (PGD) and/or altitude. The definitions of these zones and the land area they occupy within the different sub-Saharan regional groupings are reproduced in Table 18.1. The dry areas (arid and semiarid zones), which receive 100 to 1000 mm unimodal annual rainfall and 0 to 90 PGD, make up more than 50% of sub-Saharan Africa. These areas lie beyond latitude 9° N and 18° S and are dominant in every region with the exception of Central Africa. Rainfall in the dry unimodal areas occurs between May and September in the Northern Hemisphere and October to November in the Southern Hemisphere. The mean annual rainfall is less than 100 mm in parts of the Sahara and Kalahari deserts with a coefficient of variation that is more than 50%. The subhumid zone (1000 to 1500 mm bimodal rainfall and 180 to 270 PGD) stretches from the equator to approximately latitudes 9° N and 18° S and is well represented in all regions. Approximately 75% each of the highland and humid zones occur in East Africa and Central Africa, respectively.

D. Vegetation

1. Rain Forest

The bands of African vegetation zones (Figure 18.3) mimic the rainfall gradient and lie parallel to one another and to the equator in an east-west course of the coastline. In the humid zone bordering the equator, high rainfall and constant high temperature throughout the year, except in the mountains of East Africa, enable tropical rain forest to flourish. Most of the Zaire River Basin and the southern fringe of West Africa are covered by tropical rain forest. Contrary to popular perception, only 9% of the continent's landmass can be classified as tropical rain forest. Additionally, the interior of this forest is usually not an impenetrable jungle since the shade cast by three strata of trees prevents other plant growth on the forest floor. The highest of these trees may exceed 40 meters. Further description of the range of plant adaptation to this constantly warm humid environment and a uniquely prolific variety of trees and fauna that made the African tropical rain forest one of the richest in the world can be obtained from Oliver and Crowder.[18] Due to increased population pressure and lumbering and shifting cultivation, many parts of the original forest cover have been converted into permanent tree cropping (cocoa, coffee, and coconut). This forest-savanna mosaic is known as "derived savanna" in West-Africa or "miombo" around the lake region of southern Central Africa.

2. Savanna

Away from the equatorial region where rainfall drops below 1500 mm, the climate becomes increasingly seasonal and unable to support evergreen forest. This rainfall gradient results initially in occurrence of an open tropical woodland savanna. With increasing severity of the dry season, tropical grassland savanna occurs where trees are more open and spreading, their bark becoming thick and fire resistant, and thorns and hooks appear on their branches. With further increase in aridity, individual tussocks of grasses are more widely spaced and rarely grow as tall. Eventually, a zone of short annual grasses is found in the subdesert steppe such as the Sahel. The savannas, therefore, occupy a long climatic gradient covering 57% of tropical Africa.[27]

Three general types of savanna are recognized in Africa.[28,29] The "tall grass/low tree" type flanks the rainforest and occurs extensively in the subhumid to humid zones of Africa. This type is referred to as "elephant grass" (*Pennisetum purpureum*) or "guinea grass" (*Panicum maximum*) savanna. The grasses form a dense canopy 2 to 4 m high during the rainy season interspersed with short trees 10 to 15 m high. The main grasses belong to the genera *Pennisetum, Andropogon, Hyparrhenia,* and *Panicum*. Countries that experience this type of savanna include Benin, Cameroon, Central African Republic, Gabon, Gambia, Ghana, Guinea, Ivory Coast, Liberia, Nigeria, Sierra Leone, Togo, Tanzania, Uganda, and Zaire.[30]

TABLE 18.1[a]
Major Vegetation Types and Dominant Grass Genera in the Main Ecoclimatic Zones of Sub-Saharan Africa

Zone	Plant-Growth Days (pgd) and Rainfall	Region	Area (10^6 km^2)	Vegetation Type	Dominant Grass Association	Country
Arid	< 90 pgd[a] < 250 mm	West Africa	3.94	Saharan, Sahel	*Aristida, Panicum turgidium, Cenchrus*	Chad, Mali, Mauritania, Niger
		East Africa	3.02	Saharan, Desert steppe (Horn of Africa)	*Aristida, P. turgidium, Eremopogon foveolatus, Eragrostis, Chrysopogon aucheri*	Sudan, Ethiopia, Somalia
		Southern Africa	0.64	Hunting Veld	*Eragrostis, Cenchrus ciliaris*	Namibia, Botswana, South Africa
Semiarid	90–180 pgd 250–600	West Africa	1.46	Sudanian	*Andropogon*	Burkina Fasso, Chad, Mali, Mauritania, Niger
		Central Africa	1.35	Sudanian and Coastal savanna	*Andropogon, Rhynchelytrum, Setaria, Eragrostis*	Cameroon, Central African Republic, Gabon, Congo, Angola
		East Africa	1.04	Sudanian and subdesert steppe	*Andropogon, Chrysopogon aucheri, Cenchrus, Chloris*	Sudan, Ethiopia, Somalia, Kenya
		Southern Africa	1.09	Dry or "Sweet" Veld	*Eragrostis, Aristida*	Namibia, Botswana, Zimbabwe, Swaziland, South Africa
Subhumid	180–270 pgd 600–1250 mm	West Africa	1.17	Guinean	*Hyparrhenia*	Benin, Burkina Fasso, Chad, Gambia, Ghana, Guinea, Ivory Coast, Mali, Mauritania, Nigeria, Senegal, Togo
		Central Africa	1.54		*Hyparrhenia, Loudetia*	Angola, Cameroon, Central African Republic, Gabon Congo, Zaire
		East Africa			*Hyparrhenia, Sorghum,*	Sudan, Ethiopia, Tanzania
		Southern Africa	1.22	Semisour Veld	*Hyparrhenia*	Mozambique, Malawi, South Africa, Zambia, Zimbabwe
Humid	> 270 pgd > 1250 mm	West Africa	0.73	Guinean Equatorial	*Pennisetum, Panicum, Hyparrhenia*	Ghana, Ivory Coast, Liberia, Nigeria, Sierra Leone
		Central Africa	3.13	Equatorial	*Loudetia, Hyparrhenia*	Cameroon, Congo, Gabon, Zaire,
		East Africa	0.12		*Pennisetum, Themeda*	Uganda, Kenya
		Southern Africa	0.22	Sour Veld	*Themeda*	South Africa, Zimbabwe

[a] Modified after International Livestock Center for Africa ILCA[21] 1987, Skerman and Riveros,[30] 1990 and Rattray,[35] 1960.

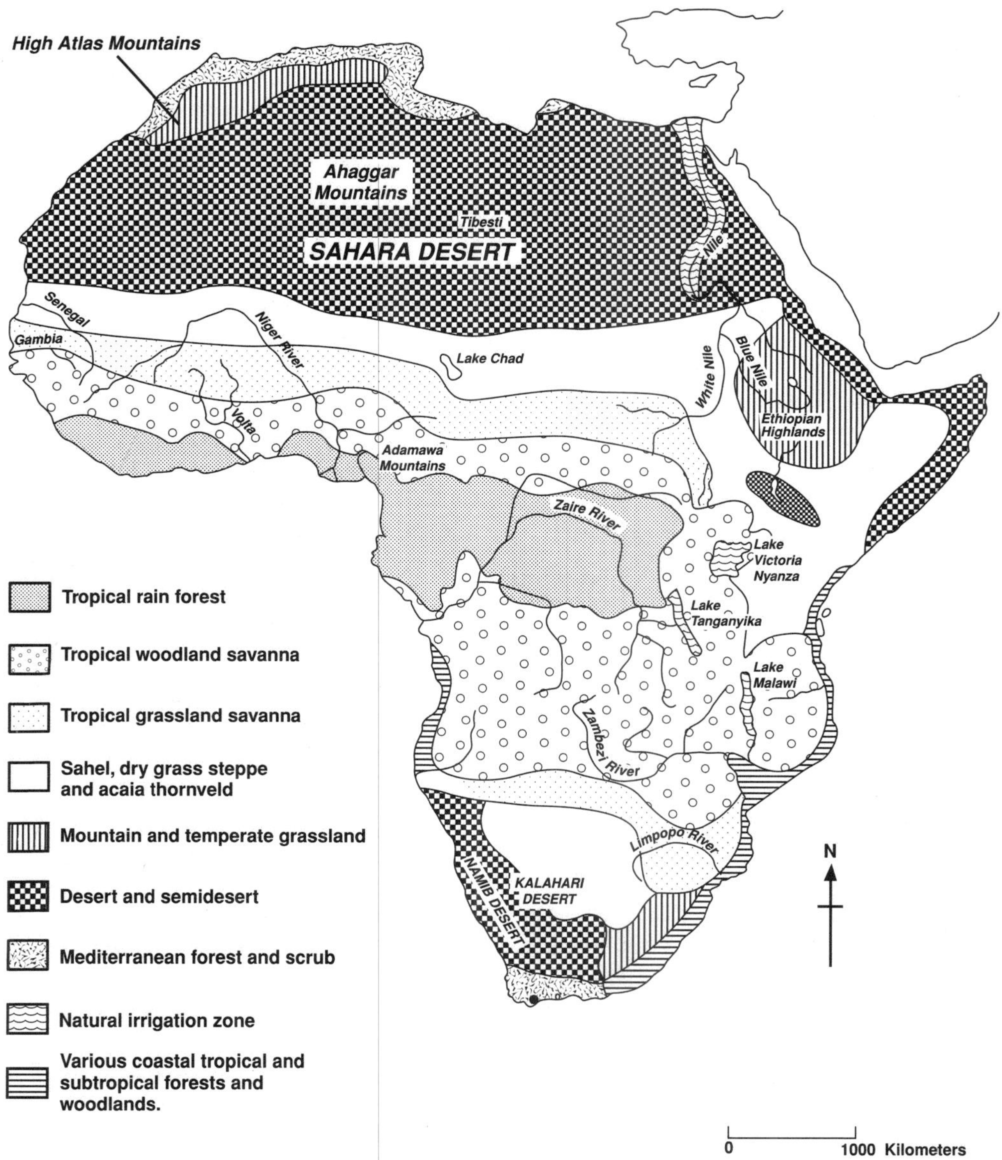

FIGURE 18.3 Africa vegetation map. (From Shillington.[26] With permission of MacMillan Press, Ltd.)

The second savanna type is the "Acacia tall grass" savanna, which occurs in semiarid (600 to 1250 mm rainfall) areas. Tussock (bunch) grasses form a complete cover reaching up to 1 to 2 m high during the rainy season and are sometimes called "Sudan savanna." Dominant grass genera include *Hyparrhenia, Andropogon, Themeda*, and *Sorghum*. Countries bordering the Sahara and Kalahari deserts such as Burkina Fasso, Botswana, Chad, Malawi, Mali, Namibia, Niger, northern Nigeria, Senegal, Sudan, and South Africa are partially covered with this vegetation. The third type is the "Acacia-desert grass" savanna occupying the fringes of the African deserts (250 to 600 mm rainfall). Prominent grass genera here are *Cenchrus, Aristida, Eragrostis,* and *Panicum*. Chad, Mali, Mauritania, Niger, Sudan, and large parts of Botswana, Namibia, and southern portions of Zimbabwe and Mozambique are some of the countries where this vegetation is found.

In summary, tropical Africa is a region of considerable environmental diversity derived primarily from variations in rainfall, but also modified by differences in soils, slope, and altitude. Whereas the vegetation can be broadly simplified into distinct zones based on rainfall, it should be noted that the myriad of microenvironments gives rise to significant local variations in specie composition within the same ecological zones. Superimposed on this picture are biotic factors such as grazing, fire, and activity of man, which keep the vegetation in several parts of the continent in serial stages of succession.

III. THE MAJOR GRASS ASSOCIATIONS OF AFRICA

Africa is the continent with the largest number of endemic grass genera and, hence, the most important center of grass variation in the world. The joint historical evolutionary pressures of fire and heavy ungulate grazing or trampling have produced some of the world's most grazing-tolerant forages. Although only four (*Andropogoneae, Chloridae, Eragrosteae,* and *Panicaeae*) of the 28 tribes of grasses predominate in the African continent, Clayton[31] indicated that 27 out of 45 important tropical grass species have their distribution centered in Africa, seven in the Americas, and one each in the Mediterranean area and Asia. The vigorous growth, response to fertility, and resistance to grazing of these African grass genotypes have made them the most successful forage species in the subtropical and tropical world, especially in the Americas and Australia.[32-34] Genera such as *Panicum, Hyparrhenia, Andropogon, Brachiaria,* and *Digitaria* are sufficiently well adapted and aggressive to have become widely naturalized and pantropical. The major African grass associations are listed and the important species briefly described in this section.

The grass cover of Africa, including a relevant map, was authored by Rattray (1960).[35] Whyte (1968),[29] using Rattray's survey and map, produced a chart of the ecoclimatic gradient in western and equatorial regions of Africa in his book *Grasslands of the Monsoon.* Skerman and Riveros (1990)[30] simplified this chart into a table that contained a list of the prominent grasses. A further modification of the table of Skerman and Riveros using ILCA's[21] concept of plant-growth-days is shown in Table 18.1. Predominant grass associations of the Saharan-Sahelian zones (arid) included *Aristida* species, *Panicum turgidum,* and *Cenchrus* species. Grasses for the Sudanian and Guinean zones (semiarid and subhumid) were chiefly *Andropogon* species and *Hyparrhenia* species, respectively. The Guinean equatorial zone (humid) was dominated by *Pennisetum* species and *Panicum maximum.* Vessey-Fitzgerald,[34] in his article covering the Central African plateau including the headwater catchment area of the Congo River, the interior drainage basins of the Rukwa Valley, the Great Rift Valley and associated highland and depressions, described the headwater valley grassland or "dambos" as containing perennial bunch grasses of *Andropogon* species, *Hyparrhenia* species, and *Loudetia* species. Riverine grasslands, zoned in relation to drainage, contained *Hyparrhenia* species, *Loudetia* species, and *Themeda triandra* in better drained areas and *Acrocerca, Leersia, Oryza, Vossia,* and *Echinochloa* on seasonally flooded wetter areas.

The East African highlands of Ethiopia and central Kenya carry natural grassland of *Pennisetum clandestinum* (Kikuyu grass), while in Zaire and the southern Highlands of Tanzania, the savannas are dominated by *Loudetia* species, *Themeda triandra, Hyperrhenia, Cymbopogon, Setaria, Digitaria,* and *Exotheca.* In the fertile crescent around Lake Victoria in Uganda, *P. purpureum* is indigenous in areas of higher fertility and rainfall. In the arid Horn of Africa the grasses on the Ethiopian Plains are mainly *Eremopogon foreolatus* with species of *Eragrostis, Panicum,* and *Aristida.* On the Red Sea Coast, the main grass is *Panicum turgidium. Chrysopogon aucheri* is the dominant grass in the arid areas of Somalia and Northern Kenya.

The eastern savanna woodland stretching from Tanzania through Kenya to Ethiopia, just east of the extensive Miombo of Central Africa, carries *Themeda triandra* on red latosolic soils, *Hyparrhenia* species on the sandy surface soils, and *Cenchrus* species on the heavy black soils with *Cynodon* species in the Rift Valley.

In Southern Africa, the "sweet" and "sour" velds (range) are recognized. Sweet veld is found on low-rainfall, fertile soils and is made up of summer grasses that retain their palatability and thus provide useful grazing during the dry season. Species *of Panicum,Urochloa, Eragrostis, Themeda, Setaria, Bothriochloa, Brachiaria,* and *Digitaria* abound. Sour veld is dominated by infertile, acid soils with high rainfall where, at maturity, grasses provide poor winter grazing. This veld contains species of *Hyparrhenia, Cymbopogon, Heteropogon, Andropogon,* and *Loudetia.*

A. Major Grasses

Brief traits of the following grasses of ecological importance in native African grasslands are described: *Andropogon, Aristida, Cenchrus, Eragrostis, Hyparrhenia, Loudetia, Panicum, Pennisetum,* and *Themeda.*[30,36]

Andropogon gayanus (gambagrass) consists of four varieties, all being tall, coarse, erect perennial bunch grasses with numerous flowering branches from the upper culms bearing spathate panicles of pale paired racemes. Variety *bisquamulatus* is found in Sudanian savanna woodlands from Senegal to Sudan. It is a coarse and robust (0.8 to 3.6 m high) perennial bunch grass colonizing denuded areas and wasteland. It persists under conditions of drought, flood, seasonal burning, and continuous grazing. Variety *gayanus* is a stout (1.0 to 3.5 m high), glaucous, prop-rooted, and rhizomatous type that is dominant on seasonally flooded marshes throughout the Sudanian savanna. It is suited for erosion control in damp sites. Variety *squamulatus* is also coarse (1.5 to 3.0 m high). It occurs in primary and secondary succession sites, either in loose disturbed sandy soils on road embankments, spoil heaps, old fallows, or in marsh Sudanian savanna flood plains. This variety also occurs in Tanzania, Zimbabwe, Mozambique, and Angola. Variety *tridentatus* is a coarse (2.7 m high) type that is naturally confined to the Nakpanduri area of Ghana.

Aristida adscensionis (common needlegrass) is a straggling, well-branched, annual bunch grass bearing slender, erect culms and rooting at some lower nodes. It grows to 1 m high with a panicle of green to purplish racemes. The genus is widespread in Africa as the first pioneer to inhabit disturbed sites in both low and high rainfall belts. It is the predominant, but a secondary species in most drier parts of Africa, especially on the fringes of the Sahara Desert, where continuous heavy grazing by nomadic herdsmen reduced an already sparse vegetation to a few scattered plants of *Aristida.* It is mainly used as a pioneer species on disturbed soil, waste land, rocky places, and fallows.

Cenchrus ciliaris (African foxtail or buffelgrass) is a dominant species in hotter parts of Africa that experience 300 to 700 mm rainfall, warmer temperatures, and higher evapotranspiration. It is a widespread perennial species fringing the desert from Mauritania to Sudan and commonly associated with *Eragrostis. Cenchrus ciliaris* is hardy and able to grow in semiarid conditions due to a deep rooting system and a free seeding ability. A valuable pasture grass, many varieties have been selected and propagated throughout the tropics and subtropics.

Eragrostis species (teff or love grass) form a large and widespread genus that is prevalent in medium to low rainfall areas of Africa south of 16° S. Prominent species in West Africa are *E. aspera, E. cilianensis, E. ciliaris,* and *E. cylindrifolia,* all of which are annuals. *E. cilianensis, E. caespitosa,* and *E. curvula* originated from East Africa. Grasses such as *Cenchrus* species, *Digitaria* species, and *Urochloa* species are typical under the same climatic conditions as *Eragrostis* species and, at the lower rainfall limits, *Eragrostis* yields to species of *Aristida.*

Hyparrhenia species (thatching grass) are robust erect perennial bunch grasses that grow 2.5 to 3.0 m high. Grasses in this genus cover a large portion of Africa where rainfall is between 700 and 1500 mm. The species *H. filipenndula* and *Hyperthelia dissoluta* are most common at the lower limits of rainfall, but, as rainfall increases, *H. variabilis, H. gazensis, H. bracteata,* and *H. confinis* become more prominent. *H. rufa, H. subplumosa, H. diplandra, H. chrysargyrea,* and *H. cyanescens* are dominant in marshy areas. The genus is widely used for rough pasture throughout the tropics. Growth is rapid in the rainy season leading to a fast decline in nutritive value. Thus, rangeland dominated by *Hyparrhenia* species requires burning every 2 to 3 years to remove ungrazed rank stubble.

Loudetia species (Besemgrass or common russetgrass) are widely distributed and locally abundant on shallow sandy soils overlying impermeable ironstone hardpan or bedrock in the wooded savannas with 750 to 1000 mm of rainfall. Commonly associated with *Hyparrhenia* type of grass cover, *Loudetia* is indicative of shallow, sandy, or stony soil conditions. The most extensive *Loudetia* stands occur on the Kalahari sands experiencing greater than 750 mm annual rainfall.

Panicum maximum (guinea grass) is found throughout the humid tropics (>1000 mm rainfall), especially on the fringes of forests where soils are fertile. It is a very productive and palatable forage, valued in both range and cultivated pastures for domesticated cattle and wildlife. It is especially known as shade tolerant and is commonly associated with savanna tree understories. Due to its palatability, however, it is very susceptible to overgrazing. Another species, *P. coloratum*, grows in medium rainfall (>650 mm) tropics on red and black clay soil, while *P. turgidium* is characteristic of very arid areas throughout the Sahelian region.

Pennisetum purpureum (elephant grass) grows in thick, tall stands on the damp fringes of the forest belt in tropical Africa and near streams and rivers. It is a characteristic grass of savannas derived from clearing of tropical evergreen forests. With adequate soil moisture, it is the most productive tropical grass capable of yielding 50 Mg ha^{-1} with moderate nitrogen fertilization. It produces palatable forage and is commonly used in "cut-and-carry" systems for dry season stall feeding.

Themeda triandra (red oat grass), a ubiquitous species, grows under a wide variety of soil and climatic conditions in natural grasslands of East, Central, and Southern Africa. In East Africa this species constitutes about 16% of the grasslands. *Themeda* grasslands generally provide good grazing when properly maintained by annual burning, but they are easily eliminated by overgrazing in winter and early in the rainy season.

B. Major Legumes

Despite an abundance of native herbaceous legumes in the tropical African range,[37,38] these, in general, have not caught the attention of forage germplasm collectors worldwide. The few cultivated tropical legumes presently used in Africa usually originate outside the continent where historical grazing and environmental pressures have not existed. As a result, introduced herbaceous legumes in tropical Africa have had establishment and persistence problems under grazing. Once plant collectors and African pasture scientists realize that the same pressures that resulted in Africa's now pan-tropical grasses also produced equally promising legumes, tropical African forage legumes will become more widely studied.

The following is a partial list of African tropical legumes, both herbaceous and short woody types, that have been identified as having some forage potential. Skerman et al[39] cataloged many of these, but often failed to specify species as African in origin due to the paucity of literature. This is by no means a complete list since many tropical African legumes have not been recognized in the literature for their forage potential, much less developed for pasture use.

Alysicarpus is one of the better known African genera.[40,41] Species are widely spread throughout higher altitudes and drier climates. Species already studied include *A. vaginalis*, *A. monilifer*,[39] and *A. ovalifolius*.[42]

Cassia/Chamaecrista is another drought-tolerant genus found throughout Africa.[41,43] Although *C. rotundifolia* is the better known species, especially the Australian cultivar Wynn, others exist with as yet unstudied potential.

Crotolaria is widespread throughout the African continent.[41,44,45] Some of these species have multiple uses such as human consumption of *C. brevidens* leaves[46] and cut-and-carry fodder of *C. goodiiformis*[47] while others, such as *C. senegalensis*, are important to range only.[42]

Desmodium barbatum and *D. canum* are recognized as native to Africa, but many other species of the genus can be considered at least naturalized to the African tropics.[40,44,41] This genus generally requires greater rainfall and produces very well in moist soils. Exceptions include *D. velutinum* and *D. scorpiurus*, both of which showed promise as dry season fodder in Nigeria,[48,49] and *D. scorpiurus* in Cape Verde.[50]

Dolichos sericeus has been widely studied in higher rainfall regions of tropical Africa. Due to its climbing growth type, it is able to compete with tall grasses such as *Panicum maximum*.[4] Other species have been identified, but their forage potential is unknown, including *D. biflorus*[51] and *D. kilimandscharicus*.[52]

Indigofera has been identified in nearly all of Africa.[41,44,45] Despite being an annual, *I. hirsuta*, the best known species, has a tendency to become lignified and unpalatable.[53] Perhaps because of this and a propensity for heavy seed production, persistence under heavy grazing of communal lands is good. Numerous other species with greater forage potential, especially for wildlife and goat browse, exist, as in the case of *I. schimperi*[54] and *I. spinosa*.[55]

Lablab purpureus is perhaps the most successful drought-tolerant African forage legume.[39] Although some authors consider it originally a pulse crop thought to be of Indian origin,[56,57] it is now grown widely throughout Africa as a fast-growing green manure and annual forage.[56]

Lotononis bainesii is considered native to Central Africa.[58] It has a narrow area of adaptation with only moderate drought tolerance.

Macrotyloma is mostly an African genus.[37] At least two species, *M. axillare* and *M. uniflorum*, include released cultivars from Australia.[39]

Neonotonia wightii is entirely African, but numerous cultivars have been released in other regions of the tropics. Known variously as perennial soybean or soja perene in anglo or lusophone Africa, respectively, it has wide adaptation to latitude, altitude, soil types, and rainfall amounts.[52]

Rhynchosia species are native to much of Africa. They often predominate in well-watered, fertile soils.[59]

Stylosanthes fruticosa is the only cultivated species in this genus that is African in origin. Due to its susceptibility to anthracnose and overgrazing in other parts of the tropics, it has not been fully studied in Africa. In Africa, however, the fact that it is widespread in range situations[41,45] indicates good adaptation.

Tephrosia is an often neglected genus because of its low biomass production. However, it can still be important to animal nutrition, and species have been identified throughout the continent for their forage potential.[45,50,41]

Teramnus labialis and *T. micans* are among the species of this genus native to Africa.[39] They have shown promise as productive, palatable forage on other continents.

Trifolium species native to Africa exist mostly at higher altitudes and cooler latitudes, both north and south.[39] The species *T. steudneri, T. auaretinianum,* and T. *reuppellianum* are particularly important in Kenya and Ethiopia.[60]

Vigna is a genus with widespread native distribution in tropical Africa. Besides the better known *V. luteola, V. parkeri, V. nuda,* and *V. vexillata*, many other species are native to Africa.[61,52,62] *Vigna unguiculata*, important as a pulse as well as a livestock stover, is also believed to be of African origin.[63]

Zornia glochidiata is the only well-known African species[42] in this genus, although other African species exist.[41]

Many tropical browse legumes have developed in Africa.[39] These are generally low in accessibility and palatability for grazers, but very important to domesticated browsers and the wild ungulates native to Africa.[64] When dealing with savanna revegetation, dry season forage, wildlife feed, or multispecies herds/flocks, these native browse shrubs and short trees become very important as producers of both leaves and edible pods.

Acacia is one of the best known African browse genera for both domestic livestock and wildlife, which ingest its leaves, flowers, and pods.[65-67] Important species include *A. albida, A. brevispica, A. gerrardii, A. giraffae, A. litakunensis, A. macrothyrsa, A. mellifera, A. nigrescens, A. nilotica, A. nubica, A. polycantha, A. seyal, A. sieberiana*, and *A. tortilis.*

Albizia amara, A. adianthifolia, A. stipulata, A. harveyi, and many others in this genus are all native to Africa and are particularly important as dry season leaf and pod feed for both wildlife and domestic flocks.[65,68]

Brachystegia speciformis twigs and leaves are browsed by cattle and goats grazing miombo woodlands.[64,68]

Cajanus cajan, although better known as a pulse with African origins, is also used as a fodder tree.[39]

Dichrostachys cinerea is a very common tree in the savannas of southern Africa, but is found throughout the tropical regions of the continent.[66] It is particularly important for goats and wildlife during the late dry season when its high concentrations of crude protein contribute to browse quality[69] and its edible pods are ingested.[67]

Sesbania sesban and *S. grandiflora* are both widely distributed in stream banks and poorly drained soils of tropical Africa. Cattle and goats avoid the unpalatable browse until the dry season when its remaining green forage is ingested.[70] This genus is particularly popular as a green manure in relay with grain crops.[71] Other *Sesbania* species exist, however, such as *S. leptocarpa* in Senegal.[42]

Tamarindus indica is believed to be of African origin despite being pan-tropical.[72] Its fodder is used as a secondary by-product since the tree is usually grown for human consumption of the pods or as a shade tree.

Other native African browse species include *Colophospermum mopane, Griffonia simplicifolia, Lonchocarpus capassa, Millettia thonningii*, and *Piliostigma thonningii*.[39]

IV. FORAGE-LIVESTOCK SYSTEMS

The animal husbandry systems that have developed to match the diverse climates in Africa are a reflection of the forages on which they depend (see Table 18.2). Both natural grasslands and cultivated pastures are limited by rainfall, soils, ambient temperatures, and cropping systems that dictate animal species utilized and methods for harvesting forages. Both utilizers of these resources as well as researchers who address production bottlenecks will have greater success when they look at pastures as integral parts of animal husbandry systems.

A. Transhumance

Transhumance, or migratory herding in which animals are moved over large distances or altitudes in a response to seasonal forage production, is still common in many areas of drier Africa, mostly the arid zones. Adapting cultivated pastures to such systems in which communal lands are utilized and highly variable rainfall patterns prevail, is a difficult and often unnecessary task. Transhumance is developed specifically to afford herders flexibility in searching out sparse vegetation. Cultivated forages, by definition, require herders to stay in one location long enough for herds to harvest forage as it is produced.

B. Pastoral

Pastoral systems exist where crops too often fail to provide household food security. Ruminants, able to glean sparse range vegetation and store energy and protein from high rainfall years to drought years or seasons, serve as a buffer between herders and an unforgiving, unpredictable climate. These systems work quite well in semiarid and subhumid regions. Here, cultivated forages, especially deep-rooted perennials preserved as forage banks for drier years and dry seasons, can be very useful management tools. Forage cultivation must overcome a lack of cropping tradition, but has the advantage of focusing on the most important component of the production system, the animal.

TABLE 18.2
Potential "End-Users" of Cultivated Forages in Tropical Africa and Their Forage Type Requirements

Species	Class	Forage Characteristic	Forage Types
Guinea Pig		High energy and protein	Vines, legumes
Rabbit	Growing	Low fiber	Vines, legumes
	Reproduction	High protein	Legumes
Pigs		Low fiber, high protein	Roots, stalks, vines
Water buffalo	Growing	High energy and protein	Fertilized grasses
	Lactating	High fiber and quality	Fertilized grasses
Fat-tailed sheep		Low fiber	Grasses, legumes
Meat goats		High protein	Browse legumes
Milk goats		High digestibility	Legumes, grasses
Beef cattle	Growing	High digestibility	Grasses, legumes
	Reproduction	High energy and protein	Young grasses
Dairy cattle	Lactating	High fiber and quality	Fertilized grasses
Wild browsers		Lower canopy high protein	Bushy legumes
		Mid-canopy high protein	Bush/tree legumes
Wild grazers	Selective	High quality	Young grasses
	Bulk	High quantity	Grasses

C. Mixed

Mixed production systems, common in the semiarid to subhumid regions, place varying degrees of importance on the crop and the animal component of agricultural production. In general, as climates become drier, animals gain importance. The animals and the crops often overlap and are interdependent as in the case of feeding crop residues to ruminants and using oxen to cultivate grain fields. Multiple-use forages, agroforestry systems, and improved use of crop residues or components in ruminant nutrition are all aspects of forage production that can become important in a mixed production system.

D. Commercial

Large-scale, privately owned animal production systems, which can focus exclusively on improving animal productivity per area, are perhaps the easiest systems for the introduction of forage cultivation. Any improvement to the land improves the value of the real estate and those improvements can be controlled and utilized in the best interest of the person who invests in the forage cultivation. Justifying the economic return of that investment may be the most difficult aspect of forage research in these systems that can be found in every climatic and vegetation zone. Rangeland is relatively inexpensive in most of tropical Africa so improved management of those resources often provides a greater return than intensified pasture production.

E. Peri-Urban

Peri-urban or suburban homesteads often utilize the majority of the land around the homes for agricultural production to supplement meager wages earned in the urban centers. Animals are often a very important component of such systems and account for a large part of a family's secondary income. With small-scale irrigation often a viable option and horticultural by-products abundant, cultivated forages can often be integrated. Focus on multiple uses such as fuel, grain, fruit, or green manure can make forages very attractive.

F. Tsetse-Infested Regions

Access to a large portion of sub-Sahara's grasslands is limited by the presence of tsetse flies (*Glossina* species). Tsetse flies do not limit the presence of wildlife, but do make the husbandry of domesticated ruminants difficult. The use of insecticides, trypanocides, habitat (brush) removal, domesticated animal species waves (donkeys, goats, then cattle have worked well in Zimbabwe) allows for use of tsetse pastures, but increases cost and inputs. This has implications for pasture research since herbaceous forages are better suited to tsetse exclusion than brush and tree forages.

G. Game Farming

Cultivated pastures are not likely to become a large factor in game parks and private ranches that specialize in hunting or tourism. The "natural" vegetation look is an important part of the package. The learning curve is very steep in these areas when it comes to native range vegetation management, however. Park rangers and safari operators have concentrated on animal ecology and are only now learning that ignorance of the vegetation component can negatively affect the animal component. As these range areas, especially in *Glossina* species infected regions, are degraded, cultivation of native forages may be needed to restore the range and forests. For these areas, local germplasm provides the species of choice since a natural look is essential for client satisfaction.

In game farms (game ranches), where meat production is the main objective, cultivated forages will definitely have a role to play in the future. Currently, African game farmers rarely modify or cultivate their range to increase meat output. The present recommendation is to match game to native vegetation.[73] But, as similar efforts in New Zealand and Europe have shown, cultivated pastures can increase meat production per area using wild animals or mixtures of wild and domesticated herds. Much research needs to be undertaken focusing on wild ungulates as end users. With the wide variety of wildlife in Africa, this will involve juggling a mind-boggling array of small to tall browsers and bulk to highly selective grazers. Since the great advantage of game farming is the wide array of forage niches to be harvested by ungulates, the interaction of herbaceous, brush, and tree canopy forage species needs to be better understood as well.

H. Forestry

Trees as forage have a unique niche in Africa due to the use of this resource by African wildlife. The elephant and giraffe are the most commonly known, but many others, including the gerenuk, kudu, nyala, and impala are also heavily dependent on tree browse. Domesticated herbivores have also developed a dependence on tree forage in more recent history. These can browse lower branches, climb limbs, or be assisted by lopping or "cut-and-carry" systems. Range browse species vary considerably in quality, however, based on the opposite effects of crude protein and tannin concentrations.[74] Cultivated forage trees, in contrast, should be studied in the multiple-use context in which they are grown for green manure, fuel, human consumption, shade, and soil conservation, among others.

IV. PASTURE AND FORAGE RESEARCH APPROACHES

Pasture research in tropical Africa has unique requirements that reflect the equally unique production systems found in the continent. To ignore these factors is to invite failed adoption of new pasture germplasm and technology.

A. Communal Vs. Commercial Pasture Systems

Much of the forage research undertaken during the colonial period was directed at commercial enterprises. In some countries, such as Zimbabwe, South Africa, and Namibia, where large commercial interests still prevail, pasture scientists still focus largely on commercial systems where land is individually owned and managed. This has resulted in a large body of research focused on the European tradition of cultivated paddocks for the semiarid climates that dominate southern Africa.

In contrast, forage research in other regions of Africa since independence and, to an increasing extent in southern Africa, has focused on communal rangelands.[6,75] Since there is little incentive for individuals to cultivate pastures for communal use, most forage research has looked at forage banks, "cut-and-carry" systems, or hay/stover production. These systems provide strategic forage supplementation in such a way that individual efforts benefit individual animals.

B. Vleys

Vleys, or the organic and moisture-rich depressions and valleys found in otherwise dry savannas, have been the focus of recent forage research efforts. These areas are considered key to producing high quality, nutritious forages for hay and dry-season supplementation in a larger range-based landscape. The use of foggage from these vleys, using the correct varieties and fertilization, will continue to be a research focus into the future.[76]

C. Leguminous Forage Banks

A large research effort has focused on establishing high quality, deep-rooted legume forage "banks" that can be utilized to supplement key animal classes during specific times.[77,78] The classic example is the deep-rooted brush or tree legumes that can be browsed daily for short periods at the end of the dry season by lactating females, recently weaned animals, or working draught animals.[17] In many areas, standing hay (foggage) of low quality exists during the dry season, but lacks the necessary nutrient content to prevent animal weight losses.[79] The use of introduced grasses that retain better quality during the dry season may be one approach.[80] The strategic use of legumes during periods when native vegetation is scarce and of low quality, not only adds protein to the diet, but also increases the digestibility of the ingested foggage.[81] Establishment and persistence of these banks are of particular continuing concern.[13,15]

D. Browsers — Goats

As in other continents, goats and other wild or domesticated browsers have been neglected by forage researchers. Only recently has this begun to change as both farmers and researchers realize that these animals have special needs related to their diet preferences,[82] including seed consumption.[83] Rather than clear-cutting brush-invaded paddocks or systematically destroying volunteer forbs and trees in cultivated pastures, research can focus on the use of mixed browser/grazer herds that emulate the native ungulate herds that previously dominated the African landscape. This will eventually evolve into a greater understanding of stocking rates based on a wide divergence of both animal and plant landscapes,[84] and will help control the invasion of woody plants in paddocks used solely by bulk grazers.[85]

E. Wildlife

Most nutrition work with wild ungulates in Africa has focused on native vegetation. Since native African range has co-evolved with its native fauna, an explosion of research in southern Africa aimed at gaining greater understanding of both native flora and fauna is currently taking place.[84] This research has led to a greater understanding of the mutual interdependence of native faunal and floral biodiversity in Africa.[86] Removing any of the components invariably results in deleterious effects on the remaining components.

Unlike North America, where forages are widely cultivated for game consumption, this research has yet to be widely undertaken in Africa. Intensification of game ranching, market demand for higher quality meats, and increase in land value will likely have to occur before interest in modifying the vegetative landscape intensifies. Since native wildlife is so well adapted to the native vegetation with which it co-evolved, cultivated pastures for these systems will likely be useful only where wildlife numbers are artificially high or species composition unbalanced. In these situations, cultivated forages specifically developed for particular wildlife species or mixes have an as yet, unexploited potential in tropical Africa.

F. Local Grass Germplasm

African grass germplasm is widely distributed throughout the world. In both the Australian and American subtropics and tropics, introduced African grasses compete with native germplasm in cultivated pastures. The better known species include *Heteropogon contortus, Chloris gayana, Cenchrus ciliaris, Hemarthria altissima, Pancium coloratum, Panicum maximum, Digitaria decumbens,* and *Pennisetum purpureum.*[30] Most of these have been selected for adaptation to their new environment and are often returned, as commercial cultivars, to the African tropics. In some cases, these commercially available seeds cannot compete with local vegetation, often the same species. Regional research focused on locally adapted forage ecotypes is therefore necessary in order to provide the nascent African seed industries with cultivars for local sale.

G. African Legumes

The collection, evaluation, and cultivar development of African legume germplasm for African use is even less evolved. African rangelands are rich in legumes, herbaceous[48,41] and arboreal,[87] that have evolved in the presence of ungulate herbivory, climatic stresses, local edaphic conditions, pests, pathogens, and vegetative competition. Concerted efforts to evaluate the agronomic characteristics and feed quality of this rich, locally adapted resource may yield species and ecotypes that better answer local forage needs than species and cultivars developed elsewhere in the tropics.[88] Widespread adaptation, long-term persistence under grazing/browsing, compatibility with local soil *Rhizobium* species, and productivity in infertile soils, especially low phosphorus, are all legume research topics to be addressed.

H. Multiple-Use Forages

Forages in tropical Africa are rarely a priority when labor and land are limiting. Animals are usually expected to fend for themselves in a very competitive and often over-stocked communal livestock system. The introduction of multiple-use forages can make cultivation of animal feed more attractive.[47,89] Research may need to focus on balancing two or more production facets, often looking at the forage as a by-product or bonus. These additional uses are as myriad as the farming systems they represent.

1. Fuel

With the steady increase in population, most areas in Africa are already experiencing a shortage in firewood, the primary fuel in most rural and peri-urban areas. Natural forests have been largely cut down near urban markets and even distant forests are in danger of succumbing to charcoal manufacturers. Cultivation of fast-growing forage species, especially legumes, with the goal of using leaves for ruminant feed and wood for fuel may generate income opportunities for rural, peri-urban, and even urban families.[17,90] In addition, this cultivation may alleviate some of the pressure on the few remaining natural African forests.

2. Wood

Construction material for many African fences, houses, and pens depends on trees and shrubs. Forage species that produce poles and posts thick enough to be harvested for building, therefore, become very attractive for producer consumption as well as for sale. Wood density and termite resistance may be as much of a factor as wood yield.[91]

3. Food

Numerous horticultural and grain species have forage potential. This list includes fruit trees and shrubs, plants grown for fiber, tuberous plants whose leaves are palatable to animals, pulse and grain crops with high quality stover, as well as many vegetables whose leaves are collected for human consumption.[92]

4. Shade

Most households in warm climates need to plant trees for shade. Cattle pastures likewise require some shade. Rather than simply establish ornamental trees, researchers need to identify fast-growing shade trees whose leaves and bark can be used for animal feed. Mulberry trees for rabbit feed, leguminous trees for high protein supplements, and even many short-lived fruit trees such as bananas can provide both shade and forage for "cut-and-carry" systems.

5. Soil Conservation

Vegetative cover protects soil from surface wind and water erosion.[93] In addition, root systems will stabilize soils, such as sands, whose structure tends to break down.[94] The use of forages as nurse crops, fallow plantings, shelter belts or filter/contour strips can improve soil conservation and the long-term sustainability of agricultural systems.

6. Green Manure

Green manure crops, especially fast-growing annual legumes, can be partially harvested by animals or "cut-and-carried" as forages.[95] Due to nitrogen sequestration in the added soil organic material, positive effects of such green manure may be delayed. Long-term soil fertility benefits to intensely cultivated lands where prolonged fallow is not an option make the use of green manure attractive,[51] especially for nitrogen demanding crops such as maize.[96,97]

7. Land Tenure

Land tenure is often tenuous in traditional African systems and grasslands are rarely considered individual property. According to Behnke,[98] understanding this in a pastoral setting is the first step to effective range and pasture research. This presents a disincentive for individuals interested in cultivation of forages in tropical Africa, but can be overcome if the community as a whole wishes to invest in improved pasture. Alternatively, planted trees and crops often serve to demarcate temporary land ownership where natural forests or range are communally owned. Planted forages such as grass bands, shrubs, or leguminous trees can therefore serve in crop-field demarcation. Orchards in many societies become inheritable as well, where tenants cannot sell land, but have the right to sell the planted trees on the same land. In silvo-pastoral systems such as coconut or cashew systems, planted trees for forage production can accomplish a de facto transfer ownership and all the stewardship implications that accompany such a change in land tenure. Some traditions do allow for individually owned reserved pastures,[6] while others are changing to encourage individual stewardship. The general conclusion is that local land tenure customs and knowledge of pasture ecology become important to any forage research that hopes to have a sustainable and beneficial impact.[99]

8. Weed Control

The use of some forages as weed suppressants, namely competition for nutrients, sunlight, or moisture, is often associated with crop rotations or tree plantations. More recently, however, the use of specific forages such as *Aeschynomene histrix* in the control of *Striga hermothica*[89] indicates a more specific multiple-use of forages.

9. Crop By-Products

Crop by-products, either stovers, stubbles, or processing by-products are already widely utilized as forages in Africa, mostly to add protein to low-quality roughage diets.[9,100] Research into pulse stovers[10,101] and grain stovers,[102] including urea treatment, are among those already undertaken. Wide genetic variability among and within crop species allows for animal nutrition and agronomic study of yields, conservation, and quality differences among species, varieties, or ecotypes. Although the feed value will not likely be the principal factor in farmer choice of planting material, it could be an important secondary factor.

10. Animal Specific

Forage researchers in Africa need to realize that the bovine is not the only animal consuming forages. The "end-user" of the forage should, in great part, determine which plant species are studied and how these are to fit into the production system (see Table 18.2). For example, forages for rabbits will be a world apart from those destined to feed an eland or a goat.[82,103] Even within animal species, an effort to address specific class or market requirements precludes general recommendations in many cases.

I. Interdisciplinary Research Teams

Forages in Africa are rarely grown as monocrops on farms with single commodity focus. Forage research undertaken by an interdisciplinary team capable of looking at all production and socioeconomic facets of the system or farm family will likely produce more widely applicable results. For example, the study of traditional fodder management systems in Tanzania by Otsyina et al.[104] will more likely result in successful agroforestry interventions by development teams. The use of participatory methods, especially the inclusion of the farmer in teams,[105] is particularly important.

J. In-System Research

A good understanding of the production system for which forages are intended is a prerequisite to initiating a successful research program. This, followed by on-farm evaluation, may break the cycle of plot research that gets international publication but no local adoption.[88] Farmer priorities, animal species and classes, vegetation characteristics, land tenure, and market demands are only a few of the factors that might contribute to forage production bottlenecks that research teams should consider.[106] Tropical Africa has some very unique systems whose forage research needs will not likely be met by results from elsewhere in the tropics. These include:

1. Game Farms, Ranches, or Parks

The wide range in size, forage preferences, and climatic adaptations of African wildlife requires an in-depth knowledge of not only the animals themselves but the native vegetation that has coevolved with those animals. This may involve looking at individual animal preferences[82] as well as carrying capacity of wild species.[84] Once these two factors are put in the climatic and market context, research into forages appropriate for African ungulates will likely have greater success.

2. Communal Lands

Land tenure and traditional management of communal grazing lands are interwoven with changing African society and market pressures. Research into cultivated forages in this dynamic and complex situation goes far beyond simple plant and animal science. Land tenure and resulting economic models may be as important to forage research as plant science.[98]

3. Animal Focused

Although the tendency for forage researchers is to focus on plants, there is a need to study forage production within the animal husbandry system in which they are expected to contribute. If a browse species is expected to sustain goat damage, for example, the effect of the browser on the forages must be assessed within production systems.[69,107] The wide range of animals that utilize forages in tropical Africa requires a careful selection of the equally large variety of cultivated forages available for introduction. That may mean working with animals such as rabbits[103] or pigs[108] as well as the more traditional forage end-users.

VI. OUTLOOK FOR FUTURE PASTURE RESEARCH

Livestock productivity in Africa is still low compared to the rest of the world. Notwithstanding, sub-Saharan Africa has seen a remarkable increase in ruminant livestock population since the early 1960s as a result of better disease control and husbandry practices. This dramatic increase in livestock numbers in association with large increases in human population, without any concomitant changes in the area of available land, has made it urgent to better manage the natural grassland resources. Inadequate nutrient supply is an overriding factor in low animal production on the continent and improved use of native pastures has become pivotal to livestock development.

As laudable as previous research efforts have been, the fundamental barrier to new forage germplasm adoption and management in the region remains unresolved: the land tenure systems. Who has access to the land, who makes management decisions, and who is ultimately responsible for its conservation? Most pasture in sub-Saharan Africa is still under communal ownership and exclusive rights to arable land for forage cultivation is not a part of most traditional systems. Individual grazing rights to specific areas are largely nonexistent, resulting in little incentive for range improvement via reseeding. Future forage research in tropical Africa must take these factors into account if it is to have a lasting effect on animal husbandry systems. In the foreseeable future, Africa will continue to furnish valuable germplasm resources in support of forage improvement throughout the subtropics and tropics. These germplasm resources could contribute much more to Africa under appropriate circumstances.

REFERENCES

1. Balsinhas, A., Aspectos floristicos da vegetacao do concelho de Lourenco Marques, *Informaco Tecnica 6*, IIAM, Maputo, Mozambique, 1969.
2. Myre, M., Esboco sumario da vegetacao do Distrito de Lourenco Marques, *Agronomia Mocambicana*, 2, 149, 1968.
3. Rebelo, D. C. and Rodrigues, A. F., Subsidio para o estudo da producao de algumas leguminosas forrageiras tropicais, *Agronomia Mocambicana*, 8, 121, 1974.
4. Anderson, G. D. and Naveh, Z., Promising pasture plants of Northern Tanzania. V. Overall comparisons of promising plants, *East African Agric. Forestry J.*, 34, 84, 1968.
5. Clatworthy, J. N., Effect of reinforcement of native grazing with silverleaf *desmodium (Desmodium uncinatum)* on dry season performance of beef steers in Zimbabwe, *Tropical Grasslands*, 18, 198, 1984.
6. Kusekwa, M. L., Mwilawa, A. J., and Msechu, J. K. K., Reserving pastures in communal grasslands by agropastoralists of Tanzania, in *Proc. XVIII Int. Grassland Cong.*, Vol. 1, Winnipeg, 1997, 15–13.
7. Smit, G. N., Rethman, N. F. G., Moore, A., Le Roux, A., and Swart, J. S., The influence of intensity of tree thinning on the redistribution of soil water in Southern African mopani veld, in *Proc. XVIII Int. Grassland Cong.*, Vol. 1, Winnipeg, 1997, 7–9.

8. Moyo, C. S. and Campbell, B. M., Grass species composition, yield and quality under and outside tree crowns in a semi-arid rangeland in south-western Zimbabwe, *African J. Range and Forage Science*, 15, 23, 1998.
9. Ketelaars, J. J. M. H., Kaasschieter, G. A., and Kane, M., Effects on intake of supplementing low-quality roughage with protein-rich feeds, in *Proc. XVIII Int. Grassland Cong.*, Vol. 1, Winnipeg, 1997, 2–11.
10. Apori, S. O., Effect of storage on quality of cowpea haulm hay under warm humid conditions, in *Proc. XVIII Int. Grassland Cong.*, Vol. 1, Winnipeg, 1997, 14–11.
11. van Niekerk, W. A. and Rethman, N. F. G., Intake and partial digestibility of Panicum maximum (cv. Gatton) at three stages of maturity during summer, in *Proc. XVIII Int. Grassland Cong.*, Vol. 1 Winnipeg, 1997, 2–5.
12. van Niekerk, W. A. and Rethman, N. F. G., Intake and quality of Panicum maximum (cv. Gatton) receiving different levels of nitrogen during autumn, in *Proc. XVIII Int. Grassland Cong.*, Vol. 1, Winnipeg, 1997, 2–7.
13. Muir, J. P., Establishment of ten forage legumes under three grazing regimes in Southern Mozambique, *African J. Range and Forage Science*, 10, 135, 1993.
14. Njwe, R. M. and Awaounfouet, B., Performance of perennial peanut (Arachis glabrata Benth.) in Camaroon, in *Proc. XVIII Int. Grassland Cong.*, Vol. 1, Winnipeg, 1997, 1–37.
15. Muir, J. P. and Abrao, L., Productivity of 12 *Stylosanthes, Tropical Grasslands*, 33, 40, 1999.
16. Peters, M., Tarawali, S.A., Schultze-Kraft, R., and Musa, A., Legume-legume complementarity for sustainable pasture development in the tropics, in *Proc. XVIII Int. Grassland Cong.*, Vol. 2, Winnipeg, 1997, 22–123.
17. Muir, J. P., Effect of cutting height and frequency on *Leucaena leucocephala* wood and forage production, *African J. Range and Forage Science*, 15, 7, 1999.
18. Oliver, R. and Crowder, M., *The Cambridge Encyclopaedia of Africa*, Press Syndicate of the University of Cambridge, Cambridge, 1981.
19. Khapoya, V. B., *The African Experience: an Introduction*, Prentice-Hall, Inc, Englewood Cliffs, New Jersey, 1994.
20. Africa-America Institute. *Africa Report, 44th Anniversary Issue,* Africa-America Institute, New York, NY, 1984.
21. International Livestock Center for Africa, *ILCA's Strategy and Long-Term Plan*, ILCA, Addis Ababa, Ethiopia, 99, 1987.
22. United States Department of Agriculture, *Manuscript of Soil Map Prepared by Soil Geography Unit*, Soil Conservation Service, USDA, Washington, D.C., 1971.
23. Okigbo, B. N., Agriculture and food production in tropical Africa, in *Proc. of the Seminar on Development Effectiveness of Food Aid in Africa*, New York Agricultural Research Council, New York, 1982, 11.
24. Okigbo, B. N. and Sabiiti, E. N., Sustainability of African farming systems with particular reference to soil fertility, multiple cropping systems and weed ingress in small holder systems in humid tropical Africa, in *Proc. XVII International Grassland Cong.*, Rockhampton, Australia, 1993, 473.
25. Jahnke, H. E., *Livestock Production Systems and Livestock Development in Africa*, Kieler Wissenchaftsvedag, Kiel, Germany, 1982.
26. Shillington, K., *History of Africa*, St. Martin's Press, Inc., New York, 1989.
27. Sanchez, P. A., Comparison of jungle and savanna soil conditions between South America and West Africa, *Agronomic and Economic Research on Tropical Soils Annual Report*, Soils Department, North Carolina State University, Raleigh, 1976, 287.
28. Eyre, S. R., *World Vegetation Types*, Columbia University Press, New York, 1971.
29. Whyte, R. O., *Grasslands of the Monsoon*, Faber and Faber, London, 1968.
30. Skerman, P. J. and Riveros, F., *Tropical Grasses*, FAO, Rome, 1990.
31. Clayton, W. D., Tropical grasses, in *Genetic Resources of Forage Plants*, McIvor, J. G. and Bray, R. A., Eds., CSIRO, East Melbourne, Australia, 1983, 39.
32. Parsons, J. J., Spread of African pasture grasses to the American tropics, *J. Range Management*, 25, 12, 1972.
33. Harlan, J. R., The scope for collection and improvement of forage plants, in *Genetic Resources of Forage Plants*, McIvor, J. G. and Bray, R. A., Eds., CSIRO, East Melbourne, Australia, 1983, 3.

34. Vessey-Fitzgerald, D. F., Central African grasslands, *J. Ecology*, 57, 243, 1963.
35. Rattray, J. M., *The Grass Cover of Africa*, FAO, Rome, 1960.
36. Gibbs Russell, G. E., Watson, L., Koekemoer, M., Smook, L., Barker, L. P., Anderson, H. M., and Dallwitz, M. J., *Grasses of Southern Africa*, National Botanical Institute, Pretoria, 1990.
37. Hutchinson, J. and Dalziel, J. M., *Flora of West Tropical Africa*, Crown Agents for Overseas Governments and Administration, London, 1954.
38. Ross, J. H., *Fabaceae* Part 1 *Mimosoideae*; Part 2 *Caesalpiniodeae*, FSA Vol. 16, National Botanical Institute, Pretoria, 1977.
39. Skerman, P. J., Cameron, D. G., and Riveros, F., *Tropical Forage Legumes*, FAO, Rome, 1988.
40. Schrere, B. D., A synopsis of the tribe *Desmodieae (Fabaceae)* in southern Africa, *Bothalia*, 18, 11, 1988.
41. Kaleja, V., Tarawali, S. A., Peters, M., Onifade, O. S., and Schultze-Kraft, R., Collecting indigenous forage legumes in northern Nigeria, *Plant Genetic Resources Newsletter*, 107, 31, 1996.
42. Buldgen, A., Hellemans, P., and Compere, R., Performance of some species of grasses and legumes characteristic of natural sahelo-sudan pastures in Senegal, *Bulletin des Recherches Agronomiques de Gembloux*, 23, 51, 1988.
43. Lock, J. M., *Cassia* sens. Lat. (*Leguminoseae-Caesalpinoideae*) in Africa, *Kew Bulletin*, 43, 333, 1988.
44. Pamo, E. T. and Yonkeu, S., Reponse du paturage naturel a la fertilisation azotee sur sol basaltique en Adamaoua camerounais, in *Proc XVI Int. Grassland Cong*., Nice, 1989.
45. Muir, J. P., Alage, A., and Maposse, I. C., Species, biomass and crude protein of herbage as affected by the canopies of four dominant trees in a savanna of southern Mozambique, in Mattos, W., Ed., *Proc. XIX Int. Grassland Con*., Sao Paulo, 2001.
46. Uiso, F. C. and Johns, T., Consumption patterns and nutritional contribution of *Crotalaria brevidens* (mitoo) in Tarime District, Tanzania, *Ecology Food and Nutrition*, 35, 59, 1996.
47. Roothaert, R. L. and Paterson, R. T., Recent work on the production and utilization of tree fodder in East Africa, *Animal Feed Science and Technology*, 69, 39, 1997.
48. Asare, E. O., Shehu, Y., and Agishi, E. A., Preliminary studies on indigenous species for dry season grazing in the northern Guinea savanna zone of Nigeria, *Tropical Grasslands*, 18, 148, 1984.
49. Mzamane, N. and Agishi, E. C., Desmodium velutinum (Wild) D.C.: a promising leguminous browse shrub of Nigeria's savannas, *PGRC/E.ILCA Germplasm Newsletter*, 12, 24, 1986.
50. Nogueira, I. and Ormonde, J., Plants collected by Eng. L.A. Grandvaux Barbosa in the Cape Verde archipelago-X. Additions and corrections to parts I to VII, *Garcia de Orta, Botanica*, 6, 163, 1983.
51. Rao, M. R., Niang, A., Kwesiga, F., Duguma, B., Franzel, S., Jama, B., and Buresh, R., Soil fertility replenishment in sub-Saharan Africa, *Agroforestry Today*, 10, 3, 1998.
52. Gubba, A., A note on wild legumes as potential reservoirs of the necrotic (NL3) strain of bean common mosaic virus in Zimbabwe, *Zimbabwe Journal of Agricultural Research*, 32, 183, 1994.
53. Kalmbacher, R. S., Hodges, E. M., and Martin, F. G., Effect of cutting height on yield and quality of Indigofera hirsuta, *Tropical Grasslands*, 14, 14, 1980.
54. Cooper, S. M. and Palmer, T., Observations on the dietary choice of free-ranging juvenile ostriches, *Ostrich* 65, 251, 1994.
55. Oba, G., Stenseth, N. C., and Lusigi, W. J., New perspectives on sustainable grazing management in arid zones of sub-Saharan Africa, *BioScience* 50, 35, 2000.
56. Murphy, A. M. and Colucci, P. E., A tropical forage solution to poor quality ruminant diets: a review of *Lablab purpureus, Livestock for Rural Development*, 11, 1999.
57. Liu, C. J., Genetic diversity and relationships among *Lablab purpureus* genotypes evaluated using RAPD as markers, *Euphytica*, 90, 115, 1996.
58. Norris, D. O., A red strain of Rhizobium from *Lotononis bainesii* Baker, *Australian J. Agric. Res*., 9, 629, 1958.
59. Germishuizen, G., Fabaceae-a new species of *Rhynchosia* from South Africa, *Bothalia* 28, 58, 1998.
60. Shadeyamakundabweni, L. and Njuguna, S. K., Seed production of native hay clovers in the highlands of Eastern Africa, *Tropical Grasslands*, 30, 257, 1996.
61. Rao, S. A., Monyo, E. S., House, L. R., Mengesha, M. H., and Negumbo, E., Collecting germplasm in Namibia, *Plant Genetic Resources Newsletter*, 90, 42, 1992.
62. Padulosi, S. and Ng, N. Q., Collecting wild *Vigna* and *Oryza* in Tanzania, *Plant Genetic Resource Newsletter*, 39, 78, 1989.

63. Ortiz, R., Cowpeas from Nigeria — a silent food revolution, *Outlook on Agriculture*, 27, 125, 1998.
64. van Rensburg, H. J., *Pasture Legumes and Grasses in Zambia*, Government Printer, Lusaka, 1967.
65. van Wyk, P., *Field Guide to the Trees of the Kruger National Park*, Struik Publishers, Pretoria, 1990.
66. Makkar, H. P. S. and Becker, K., Do tannins in leaves of trees and shrubs from Africa and Himalayan regions differ in level and activity?, *Agroforestry Systems*, 40, 59, 1998.
67. Shayo, C. M. and Uden, P., Recovery of seed of four African browse shrubs ingested by cattle, sheep and goats and the effect of ingestion, hot water and acid treatment on the viability of the seeds, *Tropical Grasslands*, 32, 195, 1998.
68. Ngulube, M. R., Hall, J. B., and Maghembe, J. A., Ecology of a miombo fruit tree — *Uapaca kirkiana (Euphorbiaceae), Forest Ecology and Management*, 77, 107, 1995.
69. Muir, J. P. and Massaete, E., Comparative growth of tethered and free-ranged goats supplemented with *Leucaena leucocephala, Small Ruminant Res.*, 23, 103, 1996.
70. Faftine, O., Alage, A., and Muir, J. P., Quality characteristics of seasonally selected pasture species in two range types of Manhica, Mozambique, in *Proc. XIX Int. Grassland Con.*, Mattos, W., Ed., Sao Paulo, Brazil, 2001.
71. Phiri, A. D. K., Kanyama-Phiri, G. Y., and Snapp, S., Maize and sesbania production in relay cropping at three landscape positions in Malawi, *Agroforestry Systems*, 47, 153, 1999.
72. El-Shiddig, K., Ebert, G., and Ludders, P., Tamarind (*Tamarindus indica* L.): a review on a multipurpose tree with promising future in the Sudan, *J. Applied Botany-Angewandt Botanik*, 73, 202, 1999.
73. van Rooyan, N. and Theron, J., Vegetation, veld types and game ranching regions of South Africa, in *Game Ranch Management*, Bothma, J. du P., Ed., J. L. van Schaik, Pretoria, 1989, 18.
74. Aganga, A. A., Tsopito, C. M., and Morake, K. M., Tannin contents of some indigenous browse plants of Botswana, in *Proc. XVIII Int. Grassland Cong.*, Vol. 1, Winnipeg, 1997, 8–3.
75. Peel, M. J. S., Biggs, H., and Zacharias, P. J. K., The evolving use of stocking rate indices currently based on animal number and type in semi-arid heterogeneous landscapes and complex land-use systems, *African J. of Range and Forage Sci.*, 15, 117, 1998.
76. Rethman, N. F. G. and Steenekamp, W. A. J., Nutritive value of guineagrass cultivars in the winter as influenced by nitrogen fertilization, in *Proc. XVIII Int. Grassland Cong.*, Vol. 2, Winnipeg, 1997, 17.
77. Adjei, M. B. and Fianu, F. K., The effect of cutting interval on the yield and nutritive value of some tropical legumes on the coastal grassland of Ghana, *Tropical Grasslands*, 19, 164, 1985.
78. Larbi, A., Adekunle, I. O., Awojide, A., and Akinlade, J., Identifying *Chamaecrista rotundifolia* accessions and *Centrosema* species for bridging seasonal feed gaps in smallholder mixed farms in the West African derived savanna, *Tropical Grasslands*, 33, 91, 1999.
79. Rackling, I. F. and Kruger, G. H. J., The effect of different winter utilization periods of a C_4 subtropical foggage on the growth of weaners, in *Proc. XVIII Int. Grassland Cong.*, Vol. 2, Winnipeg, 1997, 23–11.
80. Muir, J. P. and Abrão, L., Agronomic evaluation of 10 cultivated grasses, *Tropical Grasslands*, 33, 34, 1999.
81. Mureithi, J. G., Njunie, M. N., Muinga, R. W., Ali, R., Thorpe, W., and Mwatate, C. D., Adoption of planted forages by smallholder dairy farmers in coastal lowland Kenya, *Tropical Grasslands*, 32, 221, 1999.
82. Haschick, S. L. and Kerley, G. I. H., Factors influencing forage preference of bushbuck and goats for subtropical thicket plants, *African J. Range Forage Science*, 14, 45, 1997.
83. Aganga, A. A., Tsopito, C. M., Yeboah, S. O., Mokgoko, C., and Manne, S. N., Evaluation of the chemical composition of some locally available Acacia seeds as animal feed in Botswana, in *Proc. XVIII Int. Grassland Cong.*, Vol. 2, Winnipeg, 1997, 23–7.
84. Dekker, B., Calculating stocking rates for game ranches: substitution ratios for use of the Mopani veld, *African J. Range Forage Science*, 14, 62, 1997.
85. Smit, G. N. and Rethman, N. F. G., The effect of long-term grazing treatments on the establishment of woody plants in a southern African savanna, in *Proc. XVIII Int. Grassland Cong.*, Vol. 2, Winnipeg, 1997, 23–7.
86. Du Toit, J. T. and Cumming, D. H. M., Biodiversity in African savannas: functional significance and implications for animal production, in *Proc. XVIII Int. Grassland Cong.*, Vol. 3, Winnipeg, 1997, 233.
87. Otsyina, R. M., Norton, B. W., and Djimde, M., Fodder trees and shrubs in arid and semi-arid livestock production systems, in *Proc. XVIII Int. Grassland Cong.*, Vol. 3, Winnipeg, 1997, 429.
88. Thomas, D. and Sumberg, J. E., A review of the evaluation and use of tropical forage legumes in sub-Saharan Africa, *Agriculture Ecosystems and Environment*, 54, 151, 1995.

89. Muhr, L., Taravali, S. A., Peters, M., Merkel, U., Schultze-Kraft, R., and Berner, D., Multiple uses of tropical forage legumes for sustainable farming in the moist savannas of Africa, in *Proc. XVIII Int. Grassland Cong.*, Vol. 2, Winnipeg, 1997, 19–56.
90. Lindeque, J. P., Rethman, N. F. G., and Botha, C. C., Dry matter allocation and nutritive value of tropical and temperate fodder trees and shrubs in the Gauteng Province of South Africa, in *Proc. XVIII Int. Grassland Cong.*, Vol. 2, Winnipeg, 1997, 23–23.
91. Anane, S. and Twumasi-Ankrah, R., Woodlots: a profitable venture for Ghana's smallholder farmers, *Agroforestry Today*, 10, 19, 1998.
92. Tarawali, S. A., Singh, B. B., Fernandez-Rivera, S., and Peters, M., Optimizing the contribution of cowpea to food and fodder production in crop-livestock systems in West Africa, in *Proc. XVIII Int. Grassland Cong.*, Vol. 2, Winnipeg, 1997, 19–53.
93. Snyman, H. A., The influence of range condition on the hydrological characteristics in a semi-arid rangeland, in *Proc. XVIII Int. Grassland Cong.*, Vol. 2, Winnipeg, 1997, 23–1.
94. Buldgen, A. and Dieng, A., Rehabilitating sandy acid soils in the Sahelo-Sudanese region with Andropogon gayanus Kunth var. Bisquamulatus, a drought resistant savannah grass, in *Proc. XVIII Int. Grassland Cong.*, Vol. 2, Winnipeg, 1997, 16–1.
95. Mohamed Saleem, M. A. and Berhe, K., Forage based fallow management strategies for intensified land use and resource conservation in small-holder farms, in *Proc. XVIII Int. Grassland Cong.*, Vol. 2, Winnipeg, 1997, 23–33.
96. Shehu, Y., Alhassan, W. S., Mensah, G. W. K., Aliyu, A., and Phillips, C. J. C., The effects of green manuring and chemical fertiliser application on maize yield, quality and soil composition, *Tropical Grasslands*, 32, 139, 1997.
97. Jama, B., Buresh, R. J., and Place, F. M., Sesbania tree fallows on phosphorus-deficient sites: maize yield and financial benefit, *Agronomy J.*, 90, 717, 1998.
98. Behnke, R. H., Economic models of pastoral land tenure, *African J. Range Forage Science*, 14, 32, 1997.
99. Hendricks, H. H. and van der Heyden, F., Evaluation of the reliability of local farmer knowledge in Namaqualand, South Africa, *African J. Range Forage Science*, 15, 64, 1998.
100. Naazie, A. and Smith J. W., Modeling feed resources budgets in the moist savannahs of West Africa, in *Proc. XVIII Int. Grassland Cong.*, Vol. 2, Winnipeg, 1997, 19–7.
101. Faftine, O., Muir, J. P., and Massaete, E., Dry season supplementation of goats tethered on range supplemented with cowpea or groundnut residues with or without fresh leucaena leaves, *Animal Feed Science and Technology*, 76, 1, 1998.
102. Khombe, C. T., Dube, I. A., and Nyathi, P., The effects of kraaling and stover supplementation during the dry season on body weights and manure production of Mashona steers in Zimbabwe, *African Livestock Research*, 1, 18, 1992.
103. Muir, J. P. and Massaete, E., Reproductive performance of does fed Leucaena leucocephala in a wheat bran diet, *World Rabbit Science* 3, 91, 1995.
104. Otsyina, R. M., Essai, I., and Asenga, D., Traditional grassland and fodder management systems in Tanzania and potential for improvement, in *Proc. XVIII Int. Grassland Cong.*, Vol. 2, Winnipeg, 1997, 28–3.
105. Waters-Bayer, A. and Bayer, W., Participatory planning, monitoring and evaluation of grassland management in West Africa, in *Proc. XVIII Int. Grassland Cong.*, Vol. 2, Winnipeg, 1997, 18–17.
106. Lecomte, P., Hellemans, P., and Buldgen, A., Integrated management of natural forages resources in a tropical humid savannah of West Africa, in *Proc. XVIII Int. Grassland Cong.*, Vol. 2, Winnipeg, 1997, 18–7.
107. Muir, J. P., Jose, A.B., and Massaete, E. S., *Leucaena leucocephala* bark damage by goats and its effect on subsequent plant development, *Leucaena Research Reports*, NFTA, 12, 70, 1991.
108. Muir, J. P., Massaete, E. S., and Tsombe, H. N., *Leucaena leucocephala* and *Brassica napus* in a wheat bran diet for pigs, *Livestock Res. for Rural Development*, 3, 49, 1992.

19 Contrasts in Current Developments with Tropical Forage Research in Asia and the Americas

W.D. Pitman

CONTENTS

I. INTRODUCTION

Current and historic circumstances influencing developments in tropical forage research have produced distinct and intriguing contrasts between such programs in Asia and those in the Americas. Forage research and pasture development in tropical America expanded through most of the century just completed. Expansive natural grasslands and savannas of generally low productivity across much of South America were recognized as opportunities ready for development. In contrast, tropical grasslands in Asia were largely derived through shifting cultivation of forests and woodlands. Widespread dominance of these derived grasslands by the coarse, low-quality *Imperata cylindrica* in southeast Asia has contributed to their general image as wastelands. Agricultural research emphasis has been largely focused on production of food crops to support the large and expanding human population throughout Asia. As the twenty-first century begins, there is an overwhelming decrease in global support for pasture research in tropical America. At the same time, there is a new enthusiasm and expansion associated with tropical forage research in Asia.

II. BACKGROUND FOR FORAGE AND PASTURE DEVELOPMENT

Highly infertile soils of low productivity characterize the extensive natural grasslands and savannas of tropical America. The native grasses of these areas are also generally not productive pasture plants. Due largely to this lack of productivity of the grasslands, populations of both native Americans and European settlers were primarily concentrated in forests and forest margins. Because of the relatively low human population at the time of European settlement, woodland cleared for cultivation was allowed to progress naturally to weeds, grass, shrubs, and ultimately trees following

0-8493-2318-5/01/$0.00+$.50

only one or two crops.[1] Along with the natural grasslands of minimal productivity, weedy clearings within the forests and woodlands provided an expanding resource for development. Thus, grasses with superior establishment ability, tolerance of grazing, and increased nutritive value could readily enhance the value of the extensive land base for livestock production. The initial introduction of such grasses to fill this apparent void was the product of neither perceptive planning nor extensive research. The largely unintentional introduction initially of guinea grass (*Panicum maximum*) as early as the 1700s followed closely by paragrass (*Brachiaria mutica*), molassesgrass (*Melinis minutiflora*), and jaraguágrass (*Hyparrhenia rufa*) stimulated the expansion of a latent livestock industry.[2] Both the livestock and a culture based on livestock production had already been introduced to the American tropics from Europe.

The tremendous increases in pasture productivity with introduction of these invasive grasses led to their rapid adoption and further distribution by both government officials and livestock producers.[2] Considerable practical knowledge developed from the use of these and subsequently introduced forage plants in tropical America through the 1800s and early 1900s. However, experimental-based assessment of tropical forages and their use were limited prior to the mid 1900s as illustrated by Rocha et al.[3] and Crowder et al.[4] Throughout tropical America, agricultural colleges and government ministries of agriculture have initiated and expanded pasture and forage research programs since the mid 1900s. Internationally funded pasture research in the region culminated in the Centro Internacional de Agricultura Tropical (CIAT), Tropical Pastures Program, Cali, Colombia, which made tremendous contributions in forage germplasm collection, evaluation, and development from the 1970s into the 1990s.

In tropical Asia, as Nitis[5] has indicated, the extensive use of available land for field crops, plantation crops, and forestry gives the impression that not much remains for use as grassland. However, ruminant livestock have been widely produced throughout tropical Asia as sources of milk, meat, and power. In some areas, forage for livestock is primarily provided by crop residues, crop by-products, roadside vegetation, and weeds on field borders. Herbaceous plants under plantation tree crops, especially coconuts, can be major sources of forage in the humid tropics of Asia.[5] Such forage has ranged from naturally occurring, weedy brush, where livestock were used for weed control, to closely managed plantings of forage or cover crops, which were either grazed or hand harvested.[6] Only after their adoption as plantation cover crops were some tropical legumes recognized as potential forage plants. The potential for integrated plantation crop and livestock production provided the impetus for some of the initial tropical forage research in Asia,[5,7] generating early interest in the 1960s.

Extensive, open grasslands occur on low fertility, erosive hills in Malaysia, Indonesia, and the Philippines.[7] These derived grasslands are maintained by frequent fire, grazing, and harvest of woody plants. They provide a resource with an appearance of potential for development, however, both past degradation and potential for rapid continuing erosion are hazards requiring caution.

Increasing demand for animal products and increasing numbers of ruminant livestock are currently supporting the incorporation of forage and browse plants into crop production systems in Southeast Asia.[8] This is a new and somewhat revolutionary development within cropping systems in most of the region. Even small farm enterprises are expanding livestock components from the historic opportunistic harvest of residues and waste to production systems based on a combination of planted forage and other available resources.[8] The current rapid expansion of small-scale forage plantings in tropical Asia has also resulted in a cash crop potential of forage seed production even from extremely small plantings.[9,10] The combined demand for increased crop productivity, increased animal protein, and stabilization of both soil erosion and soil fertility has led to development of specific strategies of integrated farm management including forages. One approach to include trees, shrubs, and grasses on marginal land around field crops and plantations has been called "three-strata-forage" production.[5] Green foliage growth from all three strata is used to feed livestock, while the woody material is in demand as firewood. In humid areas with essentially no suitable cropland, an approach called "Sloping Agricultural Land Technology" (SALT) was developed as a form of alley cropping to allow subsistence farming to succeed on highly erodible soils.[11] From initial efforts led by Harold Watson in the Philippines, this system of contour hedgerows of woody legumes and permanent crops such as coffee and citrus was developed to provide soil stabilization. Prunings from the woody legumes provide both forage for ruminant livestock and mulch for additional erosion control. Once established, the hedgerows provide sufficient stabilization for narrow contour plantings of food crops in integrated, small-scale plantation, forage/livestock, and food crop production systems on otherwise unsuitable sites.

In the more arid region of the Asian tropics, livestock numbers have been extremely high for centuries. Although the need for greater forage production has long been recognized, the continuing urgent need for increased food production overshadows forage needs. Nonetheless, forage research and associated development approaches, especially in India, have provided technical expertise and viable models for enhanced forage and grassland production.[1,12,13] Extreme population pressures, resulting land degradation, and periodic regional drought greatly complicate the process of forage and grassland improvement in the dry portion of tropical Asia.

III. AVAILABLE TECHNOLOGY BASE AND RECENT RESEARCH

In tropical America, forage germplasm resources and information for their use increased rapidly in the 1980s. Germplasm collection programs provided tremendous numbers of accessions of potentially useful species. Seed increase, preliminary assessment, and subsequent selection of potentially useful genotypes proceeded through organized efforts throughout tropical America with leadership and coordination provided by the CIAT Tropical Pastures Program. Tropical legumes adapted to the infertile acid soils were especially meaningful contributions of the extensive research effort. These legumes included accessions of *Desmodium ovalifolium, Pueraria phaseoloides, Centrosema acutifolium, Centrosema macrocarpum, Arachis pintoi, Stylosanthese guianensis, Stylosanthes capitata,* and *Centrosema brasilianum.* Legumes were included in plantings with the already extensively used grass species and with the newly selected accessions of grasses such as *Brachiaria dictyoneura, Brachiaria brizantha, Brachiaria decumbens,* and *Andropogon gayanus.*[14] Programs to validate the practicality of this new tropical pasture technology and contribute to its adoption and use were initiated.[15,16] Throughout tropical America the above-mentioned forages selected for infertile soils, along with grasses which require increased inputs of fertilizer and intensive management such as *Cynodon* spp., *Digitaria* spp., *Hemarthria altissima*, and *Pennisetum purpureum*, continue to be assessed for specific production systems as illustrated by recent publications.[17-24]

Tropical pasture legume technology has been less successful on a commercial scale and thus less widely adopted than the introduced grasses in tropical America. This consequence is despite the American origin of most available tropical pasture legume cultivars. Inherent characteristics of the legumes contribute to their elusiveness in tropical pastures, even in their area of origin. As observed by Crowder et al.,[4] populations of legumes occurred "spontaneously but erratically" in some regions but "generally made no significant contribution to sward production." The almost ephemeral nature of some tropical legumes continues to provide challenges to their successful use in pastures. Recent publications attest to the continuing interest and recognized potential of herbaceous legumes for enhancement of livestock and pasture production throughout the American tropics.[25-32]

As with herbaceous legumes, woody legumes of both shrub and tree types are naturally distributed through most of tropical and subtropical America. A few of these woody legumes are recognized as useful forage plants. *Leucaena leucocephala* has been the most widely evaluated and used of the woody legumes.[33,34] In naturally forested areas, such as its area of origin in Mexico and Central America, *Gliricidia sepium* has a rather long-established, distinctive place on livestock farms for use as living fence posts that also provide high-protein forage from lopped branches.[11,35] Secondary metabolic compounds act as antiquality components to livestock consuming forage of many woody legume species. Such nutritional limitations along with limited and highly specific pest problems, establishment difficulties, and a generally inadequate seed supply have contributed to the lack of widespread use of woody legumes in tropical America. Nevertheless, recognition of the great potential of shrub and tree legumes in the region continues, and some aspects of the limitations are being assessed.[36,37] Also, specialized roles for their effective use continue to be assessed.[37,38]

Some very distinct contrasts in locally available forage technology are evident between the tropics of America and those of Asia. In Asia, forages are generally integrated into farm systems where crops are the priority component. In tropical Asia, land availability is limited and population pressures have been long standing; thus, all available land must be effectively used within the constraints of available technology and labor. In tropical America, a technology associated with extensive grasslands stocked at low animal density producing low levels per animal and per unit of land at relatively low cost is a cultural aspect of the cattle industry. The African grasses, which were inadvertently introduced to tropical America and extensively naturalized with rather minimal assistance in some instances, failed to extensively colonize the Asian tropics. The only low-input forage production throughout much of tropical Asia has been that from waste areas such as field borders and roadsides. Thus, the concept of intensive forage production is generally a more natural and readily adopted component of the technology base in Asia. This intensive approach to forage enterprises is illustrated by the previously mentioned small-scale seed production of tropical forages.[10,39]

The restricted use of herbaceous legumes, largely as green manure and cover crops, and planting of tropical grasses on very small areas in Asia have provided little incentive for development of mechanized pasture establishment or management technology. Thus, the typical forage and pasture production technology is highly labor intensive. The history of tethering individual animals for access to grazing, herding of grazing livestock in unfenced cropland areas, and "cut-and-carry" harvest of roadside vegetation provide a culture amenable to intensive forage production, harvest, and feeding. In addition, woody vegetation has provided a traditional component of the material cut for livestock feed. Thus, woody legume forage systems with high labor inputs are not only more readily accepted in tropical Asia than in other regions, but such approaches are often preferred over grass and herbaceous legume alternatives. The technology required for effective integrated systems that include woody legumes has both an extensive history of use in tropical Asia, as illustrated by *Calliandra calothyrsus* in Indonesia,[40] and a continuing research base. Diversity of woody legume species, production systems, and environments being addressed through recent research efforts in the region is substantial as shown by current literature.[41-45]

Since the 1980s, herbaceous legumes, especially species of *Stylosanthes*, have been more widely evaluated for forage use in tropical Asia.[9,46-52] Recent research has also addressed the use of various tropical grasses[9,52-56] with particular interest in *Pennisetum purpureum* and species of *Brachiaria*. Recent international forage evaluations in Southeast Asia largely funded by Australia and coordinated by the Commonwealth Scientific and Industrial Research Organization of Australia and CIAT are reminiscent of the CIAT Tropical Pastures Program of the past two decades. Research, training, and technology transfer have targeted those farming small areas. The coordination and program support provided by Australia and CIAT have enabled cooperating universities and government agencies from a number of countries to address otherwise overwhelming aspects of forage development. Thus, much new tropical forage technology has recently developed and is currently developing in tropical Asia. Many of the most recent developments have yet to be adequately validated and documented. The process of rapid generation of new information and recognition of the potential represented has provided substantial impetus for program participation and technology adoption. Program continuity is a critical aspect of maintaining progress in such novel developments. Aspects of pasture and forage technology demonstration and adoption have, themselves, been subjects of recent assessments in tropical Asia.[57,58]

IV. FACTORS NOW AFFECTING FORAGE RESEARCH AND DEVELOPMENT

In relation to tropical Asia, the pervasive global perception in recent years has been that adapted tropical forage plants have great potential for restoration of degraded lands, protection of soil from erosion, increased feed supply for undernourished livestock, and general enhancement of human welfare and environmental stability. Increased standards of living throughout Asia in recent years have created substantial demand for animal protein in human diets. The resulting opportunity for profit from livestock production and continuing widespread soil erosion in some cropping systems provide tremendous economic and environmental potential for appropriate integration of forage plants into cropping systems. Despite the general increase in affluence, extreme poverty, inadequate nutrition, and intolerably adverse living conditions still persist in some locations. Integration of small-scale, forage-based livestock enterprises into subsistence farming systems could increase food production, enhance diet quality, stabilize erosive soils, contribute to soil fertility, and provide cash income.

The Australian Center for International Agricultural Research has identified several specific priority areas in tropical Asia that represent distinct challenges and tremendous opportunities for forage research.[59] The subtropical, infertile, red soil grassland of southern China is both an erosion hazard and an opportunity for greatly increased production. The *Imperata cylindrica* grasslands of Indonesia and the Philippines provide a somewhat similar situation with the increased hazard of reinfestation by *Imperata cylindrica*, if introduced forage plants are not appropriately managed. Extensive areas of the drier Asian tropics and subtropics have been abandoned due to salt increase from irrigated crop production. Salt-tolerant forage plants could restore these lands to production and partially meet the tremendous regionwide livestock feed deficit. Plantations producing oil palm, rubber, and coconuts continue to provide opportunity for incorporation of forage crops to increase livestock production, contribute to weed control, and reduce soil erosion. Technology and local availability of resources could increase the use of shrub legumes in alley farming systems to reduce erosion, improve soil fertility, and enhance food production on erosive, marginal soils. Appropriate plant varieties and locally acceptable pasture management could contribute to restoration of degraded grazing lands in tropical and subtropical India, where food deficits and an expanding dairy industry demand more efficient use of land resources. Enhanced technology for productive, intensively managed "cut-and-carry" forage in India and southern China could contribute to subsistence farming systems.

Bhat and Bansil[60] suggested that feeding the millions of an increasing human population in Asia and safeguarding food security depend upon intensification of agriculture and breakthroughs in food and feed biotechnology. They further indicated that there is increasing demand for livestock products and decreasing land area available for agriculture. Traditional dietary habits across Asia are extremely diverse, but rice has generally been the common staple food.[60] Recent changes in dietary patterns throughout the region involve increased consumption of livestock products and decreased rice consumption per person in most countries.[60]

Livestock numbers throughout Asia typically exceed the feed and forage resources for efficient production. In India, 15% of the world's livestock are supported, though only marginally, on 2% of the world's land area.[49] Hazra[49] noted the potential of integration of food and forage crops, planting of highly productive tropical grasses such as *Pennisetum* and *Panicum maximum*, and agroforestry based on shrub legumes in India. Such forage technology could salvage substantial unproductive overgrazed grassland and wasteland resulting from previous cropping of marginal lands.[48] In southern China, recent research has demonstrated substantial potential for improvement of subtropical and tropical grasslands using improved tropical pasture technology.[61] Further research and development efforts have potential to provide a sustainable agriculture to contribute to food needs of a massive population. In southeast Asia, pressure of expanding population drives a demand for more cropland. Moog[7] has recommended forage research in this region directed at production efficiency and environmental protection.

Again, the global perception of tropical forages as a means of improving human welfare and enhancing environmental stability in tropical Asia contrasts with the frequently stated perception of the role of pastures and forages in tropical America. Interrupting a twenty-year period of remarkable progress in pasture research and development in tropical America, the twentieth century ended with pastures and forages widely viewed as environmental problems rather than solutions for the American tropics. At essentially the same time, a sufficient information base, the necessary plant germplasm collections, adequate research resources, a technically sound regionwide forage germplasm assessment strategy[62] and a highly emotional opposition[63] to pasture research and development in the American tropics emerged. The emotional opposition initially targeted the methods and consequences of some extensive pasture planting and ranching interests in tropical forests. Cattle ranching with associated conversion of forest land to introduced pasture grasses was perceived as the predominant development activity invading and displacing the forests of tropical America. The indigenous peoples and early settlers and their access to and use of lands in tropical forests were sometimes drastically affected. Accusations were made of brutal treatment of these people in the process of establishing large ranching operations.[63] Activities supporting or associated with the success of the ranching industry in tropical America were considered a part of the problem. Demands were made for accountability by governments, institutions, and individuals with any perceived role in contributing to the reported unacceptable activities.[63] Pasture and forage research were clearly perceived as contributing to the development and success of the ranching industry.

Highly provocative accusations of flawed technology were directed toward the research and development of tropical forage science itself, rather than simply recognizing the misuse of the technology. A number of additional deficiencies were quickly added to the unacceptable displacement of tropical forests, including direct destruction of vital ecosystems (a major cause of species extinctions), alteration of regional climate, declining rainfall, flooding, drought, global warming, dangerously increased levels of atmospheric carbon dioxide, depletion of soil nutrients, increased soil erosion, environmental contamination from pesticides, siltation of rivers, diminished hydroelectric capacities, and subsequent accumulative effects resulting in melting of polar ice caps, rising sea levels, inundated coastal cities, impaired marine ecosystems, deteriorated fishing industries worldwide, and desertification of major grain producing areas of the world.[63]

Certainly, the excessive and inappropriate use of numerous technologies contribute to many of the above listed adverse environmental consequences of human activities. Tropical forage and pasture technology can be, and has been, a contributing factor to undesired environmental alterations. Both inappropriate use of otherwise acceptable technology and premature use of developing technology, not yet sufficiently refined and validated, are persisting problems with tropical forages. Overwhelming need for the potential benefits in some situations and overanxious pursuit of the potential profits in other circumstances encourage undue haste. The extremely slow pace of thorough research and the typical enthusiasm of researchers for their developing projects contribute to the premature application of initially promising, but inadequately assessed, technology.

Toledo and Formoso[64] stated that inadequate production stability of pastures and pasture degradation are "common phenomena throughout the different ecosystems" of tropical America. Paradoxically, they further stated that pastures have potential for "nutrient cycling and soil improvement" and that they "can be highly sustainable." Well-adapted pasture species may provide sustainability through resilience mechanisms such as seed reserves, even though pasture stability may not be high under periodic drought or mismanagement.[64] Some of the widely reported pasture degradation may be the temporary effects of unusual and extreme circumstances causing temporary reduction of pasture stands. More extensive and well documented pasture degradation has been the result of planting unadapted pasture species such as the early Australian legume cultivars and high-fertility requiring grasses, especially *Panicum maximum*, on highly infertile soils.[64] Inadequate pasture management further complicates discernment of actual lack of sustainability of introduced pasture plants. Seed supplies of both adapted and unadapted forage varieties typically become available in new areas before sufficient technology can be provided for optimal management.[65] In more fertile forest areas, erosion and nutrient depletion often occur before pastures are planted. Access to information, financial resources, and other production inputs are typically lacking in frontier areas where novel pasture species are widely planted.[64] Political, economic, and social instability contribute to insufficient management for optimal pasture sustainability. In some situations, increasing animal numbers is a traditional means of risk management, with resulting overgrazing expected to be only temporary and of little consequence.[65]

Following extensive planting of newly introduced, well-adapted plant species in an area, an initial period of unimpeded growth is often followed by development of plant pests. Planting large areas to a single plant genotype, as has been done with superior tropical forages such as *Brachiaria decumbens* in tropical America, predisposes the development to a catastrophic disease or insect invasion. Susceptibility of *Brachiaria decumbens* to spittlebug (*Aeneolamia, Deois,* and *Zulia* species) resulted in just such disaster. Additional devastating examples with tropical forages are susceptibility of several widely planted *Stylosanthes* varieties to the plant disease anthracnose caused by *Colletotrichum gloeosporioides* and susceptibility of *Leucaena leucocephala* to the psyllid insect (*Heteropsylla cubana*). Pastures planted in these forages have certainly degraded in instances of severe pest damage. Such susceptibility to pests is not unique to tropical forages. In fact, the most important food and forage crops throughout the world are often successful because of continuing research programs to provide new germplasm with resistance to continually emerging novel pest genotypes. Research to overcome pest problems is an accepted component of the required technology for successful agriculture. Because of the early stage of development with its lack of historic occurrence of pest invasions and disease epidemics, these examples have sometimes been naively unexpected and depicted as verification of the unacceptability of tropical pasture technology. Such logic would also dismiss the acceptability of all the major grain crops that feed the world's billions of people. In contrast to the numerous research efforts in various countries around the globe continually developing solutions for food-crop pest problems, limited efforts are in progress to address such problems of tropical forages. Progress has, however, been reported with development of potential solutions for spittlebug susceptibility of *Brachiaria*,[66,67] anthracnose susceptibility of *Stylosanthes*,[68-70] and psyllid susceptibility of *Leucaena*.[71]

In recent years, the savannas of tropical America have generally been perceived as providing the greatest potential for agricultural development in the region.[72] This is despite the longer history of exploitation of the more fertile forests and woodlands. The savannas provide less fragile ecosystems where appropriate pasture development can often provide substantial increases in plant and animal production[64,72] and even substantially increased soil organic matter and carbon sequestration[64,73] compared to the native vegetation. Some tropical grass-legume pastures have been reported to accumulate more carbon than pastures of the grass alone and even amounts comparable to those of tropical forest.[74] Still, pasture degradation, lack of sustainability, and environmental liabilities remain widespread perceptions of these tropical pastures. While the conclusion that "pasture development in the American tropics is a large scale failure" is not without evidence, such pasture development is actually a technology in the formation process. Small-scale initial growth could have supported this process with less dramatic consequences and publicity from each bump along the way. Both desperate circumstances of local people and potential for exorbitant returns to investors and governments contributed to overzealous use of an incomplete technology.

One common result, widely proclaimed as environmental degradation, is primarily a tremendous decline in productivity and profitability of inadequately adapted plant genotypes. Nutrients released from breakdown of the original savanna vegetation support the increased productivity of nutrient-demanding introduced grasses for a short time. Even a slow rate of nutrient loss and nutrient redistribution with localized concentration by grazing livestock gradually reduces pasture productivity. Failure to initially recognize this reduced productivity leads to overgrazing and gradual depletion of the grazing resource. While the overgrazed pasture can be subject to soil degradation through compaction, erosion, loss of water-holding capacity, and excess runoff in the short term,[72] plant successional processes typically result in a stable plant community of reduced forage value.

As demonstrated by the dramatic production increases in the initial years of many pasture plantings, appropriate grazing pressure and adequate soil fertility can contribute to success of many tropical pasture systems. Unfortunately, recognition of the need for stocking rate adjustments and flexibility to allow such adjustments are often not simple and obvious measures.[65] Likewise, availability and economics of pasture fertilizer are not assured, even when responses are dramatic. An alternative to fertilization, especially in relation to soil phosphorus, has been to identify grass germplasm tolerant of low-fertility soils. A further development has been to combine an adapted nitrogen-fixing legume with such grasses to maintain productivity on infertile soils. An additional hazard of this strategy is the potential long-term acidification of soil with the added biologically fixed nitrogen. The various agronomic developments appear to be repeatedly limited by managerial and economic constraints.

Several options to support sustainable development of forage-based agriculture in this vast region of tropical America, where such increased productivity can benefit impoverished people, have been proposed. Farming systems, where pastures are integrated within other enterprises, provide potential.[75] Residual fertilizer, from crops that can be economically fertilized, may allow long-term rotation of pastures and crops.[72] Screening and selection of pasture plants with greater adaptation to low fertility, soil acidity, heavy grazing, moisture stress, plant pests, etc. can continue with progress likely from existing germplasm resources.[64] Assessment of locally viable pasture management alternatives, such as strategies or systems of stocking to benefit both pastures and livestock, has sufficient basis for continuation.[64] Economic and policy analyses are needed to maintain awareness of any increased viability of the various recognized technological means of increasing sustainability and production of existing pastures. The dynamic world market will not likely maintain economic constraints on periodic amendment of tropical pasture soils for an extended time. Infrastructure limitations are gradually being removed in some areas. While current approaches are not without substantial constraints, potential technical, economic, and political developments could readily enhance the value of existing options and availability of additional superior options for pasture management in savannas of tropical America.

As Macedo[72] recently noted, the more fertile tropical forests present greater limitations to pasture development than do the savannas. They are typically more fragile ecosystems, where disturbance can result in rapid nutrient depletion, soil erosion, and drastic biodiversity devastation. Inappropriate and extensive pasture development in the tropical forests for short-term benefits provided the incentive for the negative public attitude toward essentially all pastures in the American tropics. While destruction of an invaluable resource for short-term gain is highly inadvisable, total dismissal of a valid and useful technology because of its misuse is also inadvisable. Along with the extensive lists of devastating environmental consequences, opponents of tropical pasture technology have often acknowledged an appropriate use of these lands for production of locally needed agricultural and forestry products.[63] Just such use of similar tropical forests under similar circumstances of rapidly increasing population pressures led to the current deteriorated state where tropical forage plants provide potential to stabilize soil, contribute to nutrient cycling, feed undernourished livestock, and enhance diet quality of human populations throughout Asia.

Tropical forage plants provide tremendous potential for use in developing sustainable, integrated agricultural production systems in the tropics of Asia, America, and elsewhere. Use of appropriate pasture technology can contribute to productivity and environmental stability as population pressures increase in the fragile tropical forests of South and Central America. The current agricultural approach is quite primitive with substantial loss of topsoil, nutrients, ecosystem function, and other values as tropical forests are converted to agricultural fields. Rather than wait to provide international support in remedial efforts to stop environmental deterioration and feed starving people, in the American tropics opportunity still exists to develop sustainable approaches. Local food production and environmental integrity can be achieved through integration of crop, forage, forestry, and other land uses as are now being developed for restoration of degraded forest lands in Asia. Rather than relegate the recent substantial advancements in forage and pasture science in tropical America to the status of tools of exploitation, they must be recognized and further enhanced for use in productive, sustainable environmental systems.

In Costa Rica, extensive conversion of humid tropical forest to pasture has already occurred. The pasture degradation generally recognized is widespread, but at the local level provides primarily an economic rather than an environmental concern.[76,77] Weed invasion due to natural successional processes is the primary concern.[77] As economic conditions justify, replacing the existing grass pastures with grass-legume pastures has recently been recommended as a viable strategy.[76] From an ecological standpoint, remnant trees in Costa Rican pastures have been found to contribute to biodiversity conservation and habitat preservation.[78] Unfortunately, decreases in such pasture biodiversity are anticipated.[78] Integrated systems and means of providing appropriate economic value to such systems[77] could contribute to biodiversity of existing pastures and allow a more sustainable development of currently endangered tropical forests. Also in Costa Rica, tropical forest cleared for one or two years of cultivation followed by two or three years in pasture has rapidly accumulated soil organic matter to initial levels with secondary forest development.[79] Although secondary forests may not provide suitable replacements for all functions of primary forests, they must be recognized as additional resources to be integrated into crop, forage-livestock, forestry systems of environmental management. As demonstrated by experiences in Costa Rica, typically social, economic, and political constraints to the use of available pasture technology rather than the technology itself, lead to undesired consequences of such land use. Continued research, including pasture and forage research in tropical America, can contribute to the only acceptable solution of productive, sustainable systems.

V. CONCLUSIONS

A major aspect of contrasts in current environmental conditions and the resulting role of forage plants in Asia and the Americas is their differing stages of development. Population pressures through centuries in some areas of Asia have led to such advanced stages of development that the natural ecosystems no longer provide ecosystem functions and substitutes must be provided to stabilize soils and contribute products for human welfare. History demonstrates that the great forests and grasslands of the world in both temperate and tropical regions have succumbed to expanding population pressures. While deliberate negligent destruction of the forests of tropical America cannot be ethically advocated, the historic process of human population expansion in such regions must be recognized. Current technology and the probable progress of continuing investigation hold promise for approaches to development that are much more sustainable than those of the past. Integrated systems of crop, forage, and forest enterprises hold promise for restoring ecosystem functions in degraded Asian environments and for maintaining such ecosystem functions despite current human encroachment in the American tropics.

Crop production is not a valid alternative for forages and pastures, but a component of sustainable integrated systems of agriculture. Throughout history, forages have been produced largely on sites not suitable for optimal crop production. Thus, their roles are complementary not substitutionary. Human welfare and immediate local demands will likely continue to take precedence over long-term global environmental concerns despite increasing awareness of the latter. Technological means of providing for these local needs and concurrently minimizing adverse effects, which have historically resulted from advancement of dense human populations into new areas, are realistic approaches. While there is a general enthusiastic acceptance of the role of forages and the promise of forage research for remediation of extensive tropical and subtropical lands in Asia, the real urgency may be in tropical America.

An urgent aspect of the need to embrace the valid role of forages in tropical America involves offsetting the widespread localized acceptance of inappropriate use, inadequate management, and accepted failures of such misuse. Another aspect of the immediate need for continued international input into forage development in tropical America is the current declining coordination and support of local efforts. Fragmentation, unnecessary duplication, and isolation of widely dispersed independent research contribute to inefficient use of limited resources. Availability of potentially adapted germplasm, seed increase for distribution of such germplasm, and information exchange across national borders are examples of components of the recent CIAT Tropical Pastures Program that will likely be lost to varying extents throughout the region. The existing cadre of forage scientists and technicians can quickly become unavailable for rapid revitalization of the recent regionwide research effort. The need for forage-based technology is increasing with continuing development of fragile ecosystems. Opportunities for progress toward viable solutions to potential degradation and existing degraded ecosystems worldwide involve use of the extensive forage germplasm resources. An impressive base for further progress is now in place. Failure to provide continuity in global tropical forage research and development efforts will contribute to environmental degradation and human misery in impoverished localities around the world.

REFERENCES

1. Semple, A. T., *Grassland Improvement*, Leonard Hill Books, London, 1970.
2. Parsons, J. J., Spread of African pasture grasses to the American tropics, *Journal of Range Management*, 25, 12, 1972.
3. Rocha, G. L. da, Martinelli, D., Silveira Corrêa, H. da, and Cintra, B., Seasonal productivity of molasses grass in Brazil, in *Proc. of the Eighth International Grassland Congress*, The British Grassland Society, Berkshire, England, 1960, 378.
4. Crowder, L. V., Chaverra, H., and Lotera, J., Productive improved grasses in Colombia, in *Proc. XI International Grassland Congress*, University of Queensland Press, St. Lucia, Queensland, Australia, 1970, 147.
5. Nitis, I. M., Present state of grassland production and utilization and future perspectives for grassland farming in humid tropical Asia, in *Proc. XV International Grassland Congress*, the Japanese Society of Grassland Science, Nishi-nasuno, Tochigi-ken, Japan, 1985, 39.
6. Reynolds, S. B., *Pastures and Cattle under Coconuts*, FAO Plant Production and Protection Paper 91, Food and Agriculture Organization of the United Nations, Rome, 1988.
7. Moog, F. A., Sustainability of forage-livestock production in smallholder and plantation systems of Asia, in *Grasslands for Our World*, Baker, M. J., Ed., SIR Publishing, Wellington, New Zealand, 1993, 151.
8. Blair, G. J. and Kerridge, P., Nutrient management in tropical forage systems — what should be, and what is practised, in *Proc. XVIII International Grassland Congress*, Vol. 3, International Grassland Congress, Calgary, Alberta, Canada, 1997, 207.
9. Phaikaew, C., Guodao, L., Abdullah, A., Tuhulele, M., Magboo, E., Bouahom, B., and Stür, W., Tropical forage seed production in Southeast Asia: current status and prospects, in *Proc. XVIII International Grassland Congress*, Vol. 2, International Grassland Congress, Calgary, Alberta, Canada, 1997, 7 (Session 28).
10. Hare, M. D. and Phaikaew, C., Forage seed production in Northeast Thailand, in *Forage Seed Production 2. Tropical and Subtropical Species*, Loch, D. S. and Ferguson, J. E., Eds., CABI Publishing, Wallingford, Oxon, U.K., 1999, 435.
11. Paterson, R. T., Daniel, J., Korwar, G. R., Pathak, P. S., Liyanage, M. De S., Ejercito, M. G., Lesseps, R., Novaho, S., Kato, T., Gill, A. S., Roshetko, J. M., and Perera, A. N. F., Fodder production systems, in *Nitrogen Fixing Trees for Fodder Production: A Field Manual*, Forest, Farm, and Community Tree Network, c/o Winrock International, Morrilton, Arkansas, 1996, 25.
12. Roy, R. D. and Roy, M. M., Communal grazing lands and their importance in India and some other Asian countries, in *Proc. XVIII International Grassland Congress*, Vol. 3, International Grassland Congress, Calgary, Alberta, Canada, 1997, 333.
13. Ghosh, A. N., Protection and regeneration of common property grazing resources: socio-economic issues — the Indian experience, in *Proc. XVIII International Grassland Congress*, Vol. 3, International Grassland Congress, Calgary, Alberta, Canada, 1997, 553.
14. CIAT, Tropical pastures program, in *CIAT Report 1989*, Centro Internacional de Agricultura Tropical, Cali, Colombia, 1989, 53.

15. Toledo, J. M., Pasture development for cattle production in the major ecosystems of the tropical American lowlands, in *Proc. XV International Grassland Congress*, the Japanese Society of Grassland Science, Nish-nasuno, Tochigi-ken, Japan, 1985, 74.
16. Vera, R. R. and Sere, C., Evaluation of tropical pasture species with a farming system perspective, in *Proc. XV International Grassland Congress*, the Japanese Society of Grassland Science, Nish-nasuno, Tochigi-ken, Japan, 1985, 1187.
17. Andrade, J. B., Justo, C. L., Coutinho, J. L. V., Peres, R. M., Ferrari Junior, E., Paulino, V. T., Henrique, W., Werner, J. C., and Mattos, H. B., Effect of nitrogen and potash fertilizations on dry matter yield and mineral composition of Pangolagrass, in *Proc. XVIII International Grassland Congress*, Vol. 1, International Grassland Congress, Calgary, Alberta, Canada, 1997, 11 (Session 10).
18. Andrade, I. F. and Oliveira, M. A., Productivity of grasses in a pasture in the Cerrado area of the "Triangulo Mineiro," in *Proc. XVIII International Grassland Congress*, Vol. 2, International Grassland Congress, Calgary, Alberta, Canada, 1997, 133 (Session 22).
19. Euclides, V. P. B., Macedo, M. C. M., and Oliveira, M. P., Beef cattle production on renovated grass pastures in the savannas of Brazil, in *Proc. XVIII International Grassland Congress*, Vol. 2, International Grassland Congress, Calgary, Alberta, Canada, 1997, 109 (Session 29).
20. Frasinelli, C. A., Petruzzi, H. J., Veneciano, J. H., Ferri, C. M., Jouve, V. V., Stritzler, N. P., and Terenti, O. A., Dry matter production and nutritive value of forage of *Digitaria eriantha* cv. Irene in two locations of the central semi-arid region of Argentina, in *Proc. XVIII International Grassland Congress*, Vol. 2, International Grassland Congress, Calgary, Alberta, Canada, 1997, 101 (Session 22).
21. Medeiros, R. B., Souza, J. M., Zambra, J. E. G., and Pillar, V. D. P., Evaluation of tropical and subtropical forage grasses in the northwest region of Rio Grande do Sul, Brazil, in *Proc. XVIII International Grassland Congress*, Vol. 1, International Grassland Congress, Calgary, Alberta, Canada, 1997, 19 (Session 1).
22. Perez-Perez, J., Hernandez-Velez, J. O., Herrera-Haro, J. G., and Barcena-Gama, R., Live-weight gain of steers grazing African star grass at four herbage allowances, in *Proc. XVIII International Grassland Congress*, Vol. 2, International Grassland Congress, Calgary, Alberta, Canada, 1997, 103 (Session 29).
23. Juarez Lagunes, F. I., Fox, D. G., Blake, R. W., and Pell, A. N., Evaluation of tropical grasses for milk production by dual-purpose cows in tropical Mexico, *Journal of Dairy Science*, 82, 2136, 1999.
24. Faria, J. R., Gonzalez, B., Faria-Marmol, J., and Morillo, D. E., Effect of nitrogen and phosphorus fertilizers on some components of nutritive value of dwarf elephantgrass, *Communications in Soil Science and Plant Analysis*, 30, 2259, 1999.
25. Faria-Marmol, J., Morillo, D. E., Gonzalez, R., and Chirinos, Z., Effect of three cutting frequencies on yield, crude protein and mineral contents of *Centrosema macrocarpum*, in *Proc. XVIII International Grassland Congress*, Vol. 2, International Grassland Congress, Calgary, Alberta, Canada, 1997, 63 (Session 22).
26. Barcellos, A. O., Pizarro, E. A., and Costa, N. L., Agronomic evaluation of novel germplasm under grazing: Arachis pintoi BRA-031143 and Paspalum atratum BRA-009610, in *Proc. XVIII International Grassland Congress*, Vol. 2, International Grassland Congress, Calgary, Alberta, Canada, 1997, 47 (Session 22).
27. Kerridge, P. C. and Franco, L. H., The use of legumes for improvement of fallow land in smallholder farming systems, in *Proc. XVIII International Grassland Congress*, Vol. 2, International Grassland Congress, Calgary, Alberta, Canada, 1997, 25 (Session 16).
28. Lascano, C. E., Ruiz, G. A., Velasquez, J., and Rozo, J., Validation of *Arachis pintoi* as a forage legume in commercial dual purpose cattle farms in forest margins of Colombia, in *Proc. XVIII International Grassland Congress*, Vol. 2, International Grassland Congress, Calgary, Alberta, Canada, 1997, 31 (Session 24).
29. Maass, B. L., Franco, L. H., Ramirez, G., Lascano, C. E., and Velasquez, J. E., Codariocalyx gyroides — a new forage option for the humid tropics, in *Proc. XVIII International Grassland Congress*, Vol. 1, International Grassland Congress, Calgary, Alberta, Canada, 1997, 53 (Session 1).
30. Grof, B., Fernandes, C. D., Santos, A. V. dos, and Almeida, C. B., Selection of Stylosanthes guianensis for the Cerrados of Brazil, in *Proc. XVIII International Grassland Congress*, Vol. 1, International Grassland Congress, Calgary, Alberta, Canada, 1997, 33 (Session 4).

31. Moulaert-Quiros, A., Mueller, J. P., Villarreal, M., Piedra, R., and Villalobos, L., A silvopastoral system in the North Atlantic Zone of Costa Rica: combining indigenous timber species with dairy pasture swards, in *Proc. XVIII International Grassland Congress*, Vol. 1, International Grassland Congress, Calgary, Alberta, Canada, 1997, 17 (Session 6).
32. Ibrahim, M. A. and 't Mannetje, L., Compatibility, persistence and productivity of grass-legume mixtures in the humid tropics of Costa Rica. 1. dry matter yield, nitrogen yield and botanical composition, *Tropical Grasslands*, 32, 96, 1998.
33. Humphreys, L. R., *Tropical Pasture Utilization*, Cambridge University Press, Cambridge, U.K., 1991.
34. Paterson, R. T., Philip, P., and Maynard, P., *A Guide to Improved Pastures for the Drier Areas of the Eastern Caribbean*, Caribbean Agricultural Research and Development Institute, St. John's, Antigua, 1986.
35. Glover, N., Gliricidia — its names tell its story, in *Nitrogen Fixing Trees for Fodder Production: A Field Manual*, Forest, Farm, and Community Tree Network, c/o Winrock International, Morrilton, Arkansas, 1996, 110.
36. Veasey, E. A., Ghishi, O. M. A. A., Valarini, M. J., Otsuk, I. P., Cardelli, M. A., Sanchez, M. J. F., and Beisman, D. A., Early growth and native nodulation of leguminous shrub and tree species in Brazil, *Tropical Grasslands*, 31, 40, 1997.
37. CIAT, *Tropical Grasses and Legumes: Optimizing Genetic Diversity for Multipurpose Use*, Annual Report 1998, Centro Internacional de Agricultura Tropical, Cali, Colombia, 1998.
38. Clavero, T. and Razz, R., Leucaena leucocephala as a supplement for goats on tropical pastures, in *Proc. XVIII International Grassland Congress*, Vol. 2, International Grassland Congress, Calgary, Alberta, Canada, 1997, 9 (Session 23).
39. Kowithayakorn, L. and Phaikaew, C., Harvesting and processing techniques of tropical grass and legume seeds for small farm holders, in *Proc. XVII International Grassland Congress*, Hamilton, New Zealand, 1993, 1809.
40. National Research Council, *Calliandria: a Versatile Small Tree for the Humid Tropics*, National Academy Press, Washington, D.C., 1983.
41. Djogo, A. P. Y., The use of forage tree legumes in Timor, Indonesia, in *Forage Tree Legumes in Tropical Agriculture*, Gutteridge, R. C. and Shelton, H. M., Eds., CAB International, Wallingford, Oxon, U.K., 1994, 374.
42. Sukanten, I. W., Nitis, I. M., Lana, K., Suarna, M., and Uchida, S., Growth and fodder yield of the *Gliricidia sepium* provenances in alley cropping system in dryland farming area in Bali, Indonesia, *Asian-Australasian Journal of Animal Science*, 8, 195, 1995.
43. Pariyar, D. and Grinten, P. van der, New approaches for rehabilitating and developing fodder production from acid mountain soils in Nepal to alleviate poverty and restore the environment, in *Proc. XVIII International Grassland Congress*, Vol. 2, International Grassland Congress, Calgary, Alberta, Canada, 1997, 11 (Session 16).
44. Tudsri, S., Prasanpanich, S., and Swasdiphanich, S., Milk production from ruzi grass alone, ruzi mixed with leucaena and ruzi supplemented with dolichos lablab, in *Proc. XVIII International Grassland Congress*, Vol. 2, International Grassland Congress, Calgary, Alberta, Canada, 1997, 5 (Session 29).
45. Hazra, C. R., Bio-diversity conservation and natural resource regeneration through watershed development at Jhansi, in *Proc. XVIII International Grassland Congress*, Vol. 2, International Grassland Congress, Calgary, Alberta, Canada, 1997, 23 (Session 16).
46. Tudsri, S., Watkin, B. R., Chantkam, S., Chu, A. C. P., and Forde, B. J., Effect of first year grazing management on *Stylosanthes hamata* cv. Verano production at Muak Lek, Saraburi, Thailand, *Tropical Grasslands*, 23, 35, 1989.
47. Guodao, L. and Kerridge, P. C., Selection and utilization of Stylosanthes guianensis for green cover and feed meal production in China, in *Proc. XVIII International Grassland Congress*, Vol. 2, International Grassland Congress, Calgary, Alberta, Canada, 1997, 49 (Session 19).
48. Ramesh, C. R., Mal, B., Hazra, C. R., Sukanya, D. H., Ramamurthy, V., and Chakraborty, S., Status of *Stylosanthes* development in other countries. III. *Stylosanthes* development and utilization in India, *Tropical Grasslands*, 31, 467, 1997.
49. Hazra, C. R., Status of *Stylosanthes* development in other countries. IV. *Stylosanthes* as a component of the forage research network in India, *Tropical Grasslands*, 31, 476, 1997.

50. Shankar, V., Gupta, J. N., and Upadhyay, V. S., Growth of heifers and changes in vegetation in a natural grassland oversown with Caribbean stylo, in *Proc. XVIII International Grassland Congress*, Vol. 2, International Grassland Congress, Calgary, Alberta, Canada, 1997, 113 (Session 22).
51. Guodao, L., Phaikaew, C., and Stur, W. W., Status of *Stylosanthes* development in other countries. II. *Stylosanthes* development and utilization in China and south-east Asia, *Tropical Grasslands*, 31, 460, 1997.
52. Ibrahim, M., Lanting, E., Khemsawat, C., Wong, C. C., Guodao, L., Phimphachanhvongsod, V., Binh, L. H., and Horne, P. M., Forage grasses and legumes with broad adaptation for southeast Asia, in *Proc. XVIII International Grassland Congress*, Vol. 1, International Grassland Congress, Calgary, Alberta, Canada, 1997, 51 (Session 1).
53. Jorgensen, S. T., Tudsri, S., Pookpakdi, A., and Stolen, O., Effects of cutting height on yield and quality of five napier cultivars, in *Proc. XVIII International Grassland Congress*, Vol. 2, International Grassland Congress, Calgary, Alberta, Canada, 1997, 5 (Session 22).
54. Ishii, Y., Ito, K., and Shinohara, S., Relation between growth in summer and the overwintering ability in hybrid *Pennisetum* and napiergrass genotypes, in *Proc. XVIII International Grassland Congress*, Vol. 2, International Grassland Congress, Calgary, Alberta, Canada, 1997, 19 (Session 22).
55. Cai, Q. S., Ishii, Y., and Ito, K., Seasonal changes in photosynthetic rate of 6 Panicum genotypes, in *Proc. XVIII International Grassland Congress*, Vol. 2, International Grassland Congress, Calgary, Alberta, Canada, 1997, 21 (Session 22).
56. Cheng, Y. K. and Chen, C. S., Breeding to improve yield and quality in napiergrass, in *Proc. XVIII International Grassland Congress*, Vol. 2, International Grassland Congress, Calgary, Alberta, Canada, 1997, 11 (Session 17).
57. Gabunada, F. G., Stur, W. W., and Horne, P. M., Development of forage components through farmer participatory research, in *Proc. XVIII International Grassland Congress*, Vol. 2, International Grassland Congress, Calgary, Alberta, Canada, 1997, 33 (Session 24).
58. Singh, R. P. and Mal, B., Strategies for forage research and transfer in India, in *Proc. XVIII International Grassland Congress*, Vol. 2, International Grassland Congress, Calgary, Alberta, Canada, 1997, 9 (Session 24).
59. Clements, R. J., Willett, I. R., Davis, J. S., and Fischer, R. A., ACIAR's priorities for forage research in some developing countries, *Tropical Grasslands*, 31, 285, 1997.
60. Bhat, P. N. and Bansil, P. C., Grains and roughage production and its utilization in Asian- Australasian region — review, *Asian-Australasian Journal of Animal Science*, 12, 481, 1999.
61. Jizhou, R. and Wenlan, J., Sustainable grassland production systems in southern China: an example from Guizhou Province, in *Proc. XVIII International Grassland Congress*, Vol. 3, International Grassland Congress, Calgary, Alberta, Canada, 1997, 499.
62. Toledo, J. M., Pasture development for cattle production in the major ecosystems of the tropical American lowlands, in *Proc. XV International Grassland Congress*, the Japanese Society of Grassland Science, Nish-nasuno, Tochigi-ken, Japan, 1985, 74.
63. Shane, D. R., *Hoofprints on the Forest, Cattle Ranching and the Destruction of Latin America's Tropical Forest*, Institute for the Study of Human Issues, Philadelphia, Pennsylvania, 1986.
64. Toledo, J. M. and Formoso, D., Sustainability of sown pastures in the tropics and subtropics, in *Grasslands for Our World*, Baker, M. J., Ed., SIR Publishing, Wellington, New Zealand, 1993, 710.
65. Costa, F. P. and Rehman, T., Exploring the link between farmer's objectives and the phenomenon of pasture degradation in the beef production systems of Central Brazil, *Agricultural Systems*, 61, 135, 1999.
66. Miles, J. W. and Valle, C. B. do, Advances in breeding apomictic *Brachiaria* in tropical America, in *Proc. XVIII International Grassland Congress*, Vol. 1, International Grassland Congress, Calgary, Alberta, Canada, 1997, 63 (Session 4).
67. Valerio, J. R. and Souza, A. P. de, Screening tropical forage grasses for resistance to the spittlebugs (Homoptera: Cercopidae), in *Proc. XVIII International Grassland Congress*, Vol. 1, International Grassland Congress, Calgary, Alberta, Canada, 1997, 15 (Session 13).
68. Chakraborty, S., Perrott, R., Charchar, M. J. d'A, Fernandes, C. D., and Kelemu, S., Variation in the *Stylosanthes* anthracnose pathogen: implications for Australian cultivars, in *Proc. XVIII International Grassland Congress*, Vol. 1, International Grassland Congress, Calgary, Alberta, Canada, 1997, 1 (Session 13).

69. Grof, B., Fernandes, C. D., Almeida, C. B., and Santos, A. V. dos, Development of a multicross cultivar of *Stylosanthes* spp., in *Proc. XVIII International Grassland Congress*, Vol. 1, International Grassland Congress, Calgary, Alberta, Canada, 1997, 31 (Session 4).
70. Miles, J. W. and Grof, B., Recent advances in studies of anthracnose of *Stylosanthes*. III. *Stylosanthes* breeding approaches in South America, *Tropical Grasslands*, 31, 430, 1997.
71. Mullen, B. F., Shelton, H. M., Gabunada, F., and Stur, W. W., Psyllid resistance in the *Leucaena* genus, in *Proc. XVIII International Grassland Congress*, Vol. 1, International Grassland Congress, Calgary, Alberta, Canada, 1997, 23 (Session 6).
72. Macedo, M. C., Sustainability of pasture production in the savannas of tropical America, in *Proc. XVIII International Grassland Congress*, Vol. 3, International Grassland Congress, Calgary, Alberta, Canada, 1997, 391.
73. Fisher, M. J., Rao, I. M., Ayarza, M. A., Lascano, C. E., Sanz, J. I., Thomas, R. J., and Vera, R. R., Carbon storage by introduced deep-rooted grasses in the South American savannas, *Nature*, 371, 236, 1994.
74. 't Mannetje, L., Potential and prospects of legume-based pastures in the tropics, *Tropical Grasslands*, 31, 81, 1997.
75. CIAT, CIAT and sustainable agricultural production, *CIAT International*, 8 (No. 1), 3, 1989.
76. Bouman, B. A. M., Nieuwenhuyse, A., and Ibrahim, M., Pasture degradation and restoration by legumes in humid tropical Costa Rica, *Tropical Grasslands*, 33, 98, 1999.
77. Japsen, H. G. P., Ibrahim, M. A., Nieuwenhuyse, A., 't Mannetje, L., Joenje, M., and Abarca, S., The economics of improved pasture and silvipastoral technologies in the Atlantic Zone of Costa Rica, *Tropical Grasslands*, 31, 588, 1997.
78. Harvey, C. A. and Haber, W. A., Remnant trees and the conservation of biodiversity in Costa Rican pastures, *Agroforestry Systems*, 44, 37, 1999.
79. Guggenberger, G. and Zech, W., Soil organic matter composition under primary forest, pasture, and secondary forest succession, Region Huetar Norte, Costa Rica, *Forest Ecology and Management*, 124, 93, 1999.

Index

A

B

C

D

E

F

G

H

I

J

M

P

Q

R

S

T

U

V

W

X

Y

Z